Handbook of Petroleum
(Science and Technology)

Handbook of Petroleum (Science and Technology)

Edited by Oliver Haghi

SYRAWOOD
PUBLISHING HOUSE

New York

Published by Syrawood Publishing House,
750 Third Avenue, 9th Floor,
New York, NY 10017, USA
www.syrawoodpublishinghouse.com

Handbook of Petroleum (Science and Technology)
Edited by Oliver Haghi

International Standard Book Number: 978-1-68286-688-7 (Hardback)

Cataloging-in-Publication Data

Handbook of petroleum : science and technology / edited by Oliver Haghi.
    p. cm.
Includes bibliographical references and index.
ISBN 978-1-68286-688-7
1. Petroleum. 2. Petroleum engineering. 3. Petroleum--Prospecting. I. Haghi, Oliver.
TN870 .H36 2019
665.5--dc23

TABLE OF CONTENTS

Preface VII

Chapter 1 Modeling, analysis, and screening of cyclic pressure pulsing with nitrogen in
hydraulically fractured wells 1
Emre Artun, Ali Aghazadeh Khoei, Kutay Köse

Chapter 2 Clastic compaction unit classification based on clay content and integrated
compaction recovery using well and seismic data 19
Zhong Hong, Ming-Jun Su, Hua-Qing Liu and Gai Gao

Chapter 3 The potential of domestic production and imports of oil and gas in China: an
energy return on investment perspective 32
Zhao-Yang Kong, Xiu-Cheng Dong, Qian Shao, Xin Wan, Da-Lin Tang
and Gui-Xian Liu

Chapter 4 Interactions between the fluid and an isolation tool in a pipe: laboratory
experiments and numerical simulation 49
Hong Zhao, Yi-Xin Zhao and Zhi-Hui Ye

Chapter 5 Formation mechanisms and sequence response of authigenic grain-coating
chlorite: evidence from the Upper Triassic Xujiahe Formation in the southern
Sichuan Basin, China 63
Yu Yu, Liang-Biao Lin and Jian Gao

Chapter 6 Sedimentary characteristics and processes of the Paleogene Dainan Formation
in the Gaoyou Depression, North Jiangsu Basin, eastern China 75
Xia Zhang, Chun-Ming Lin, Yong Yin, Ni Zhang, Jian Zhou and Yu-Rui Liu

Chapter 7 Regularized least-squares migration of simultaneous-source seismic data with
adaptive singular spectrum analysis 92
Chuang Li, Jian-Ping Huang, Zhen-Chun Li and Rong-Rong Wang

Chapter 8 The remarkable effect of organic salts on 1,3,5-trioxane synthesis 106
Liu-Yi Yin, Yu-Feng Hu and Hai-Yan Wang

Chapter 9 Effects of ultrasonic waves on carbon dioxide solubility in brine at different
pressures and temperatures 112
Hossein Hamidi, Erfan Mohammadian, Amin Sharifi Haddad, Roozbeh Rafati,
Amin Azdarpour, Panteha Ghahri, Adi Putra Pradana, Bastian Andoni
and Chingis Akhmetov

Chapter 10 Rheology of rock salt for salt tectonics modeling 120
Shi-Yuan Li and Janos L. Urai

Chapter 11 Origin of dolomite in the Middle Ordovician peritidal platform carbonates in
the northern Ordos Basin, western China 133
Xiao-Liang Bai, Shao-Nan Zhang, Qing-Yu Huang, Xiao-Qi Din and Si-Yang Zhang

Chapter 12   **Quantitative characterization of polyacrylamide–shale interaction under various saline conditions**...149
Samyukta Koteeswaran, Jack C. Pashin, Josh D. Ramsey and Peter E. Clark

Chapter 13   **Influence of friction on buckling of a drill string in the circular channel of a bore hole**...160
Valery Gulyayev and Natalya Shlyun

Chapter 14   **Effects of pH on rheological characteristics and stability of petroleum coke water slurry**..174
Fu-Yan Gao and Eric-J. Hu

Chapter 15   **Formation of fine crystalline dolomites in lacustrine carbonates of the Eocene Sikou Depression, Bohai Bay Basin, East China**.....................................180
Yong-Qiang Yang, Long-Wei Qiu, Jay Gregg, Zheng Shi and Kuan-Hong Yu

Chapter 16   **FCC riser quick separation system**..195
Zhi Li and Chun-Xi Lu

Chapter 17   **Experiments on acoustic measurement of fractured rocks and application of acoustic logging data to evaluation of fractures**.......................................201
Bao-Zhi Pan, Ming-Xin Yuan, Chun-Hui Fang, Wen-Bin Liu, Yu-Hang Guo and Li-Hua Zhang

Chapter 18   **Comparative study of HFACS and the 24Model accident causation models**...............210
Gui Fu, Jia-Lin Cao, Lin Zhou and Yuan-Chi Xiang

Chapter 19   **Similarity measure of sedimentary successions and its application in inverse stratigraphic modeling**...219
Taizhong Duan

Permissions

List of Contributors

Index

PREFACE

The main aim of this book is to educate learners and enhance their research focus by presenting diverse topics covering this vast field. This is an advanced book which compiles significant studies by distinguished experts in the area of analysis. This book addresses successive solutions to the challenges arising in the area of application, along with it; the book provides scope for future developments.

Petroleum science deals with the study and production of crude oil as well as its refinement to a usable form. It is an interdisciplinary field that combines the principles of engineering, geology and mineralogy to facilitate the exploration, analysis, drilling and refining of petroleum. A lot of technological advancements in reservoir simulation, formation evaluation, well engineering, etc. have revolutionized the field of petroleum science. The ecological impact of petroleum has become a concern in the past decades and research is being undertaken to develop and adopt practices to mitigate the effects. This book elucidates the most relevant concepts and innovations that have occurred in this field in recent years. While understanding the long-term perspectives of the topics, the book makes an effort in highlighting their impact as a modern tool for the growth of the discipline. It will be beneficial to engineers, geologists and petroleum engineers as well as researchers and students involved in this field of study.

It was a great honour to edit this book, though there were challenges, as it involved a lot of communication and networking between me and the editorial team. However, the end result was this all-inclusive book covering diverse themes in the field.

Finally, it is important to acknowledge the efforts of the contributors for their excellent chapters, through which a wide variety of issues have been addressed. I would also like to thank my colleagues for their valuable feedback during the making of this book.

Editor

Modeling, analysis, and screening of cyclic pressure pulsing with nitrogen in hydraulically fractured wells

Emre Artun[1] · Ali Aghazadeh Khoei[1,2] · Kutay Köse[1]

Abstract Cyclic pressure pulsing with nitrogen is studied for hydraulically fractured wells in depleted reservoirs. A compositional simulation model is constructed to represent the hydraulic fractures through local-grid refinement. The process is analyzed from both operational and reservoir/hydraulic-fracture perspectives. Key sensitivity parameters for the operational component are chosen as the injection rate, lengths of injection and soaking periods and the economic rate limit to shut-in the well. For the reservoir/hydraulic fracturing components, reservoir permeability, hydraulic fracture permeability, effective thickness and half-length are used. These parameters are varied at five levels. A full-factorial experimental design is utilized to run 1250 cases. The study shows that within the ranges studied, the gas-injection process is applied successfully for a 20-year project period with net present values based on the incremental recoveries greater than zero. It is observed that the cycle rate limit, injection and soaking periods must be optimized to maximize the efficiency. The simulation results are used to develop a neural network based proxy model that can be used as a screening tool for the process. The proxy model is validated with blind-cases with a correlation coefficient of 0.96.

Keywords Cyclic pressure pulsing · Nitrogen injection · Hydraulically-fractured wells · Experimental design · Artificial neural networks

1 Introduction

In low-permeability reservoirs, which are dissected by a network of interconnected fractures, solution channels, and vugs, water and gas flooding have been found to be ineffective secondary recovery methods (Raza 1971). The injected fluid tends to the channel through the high-conductivity network and bypass the low-permeability, oil-bearing matrix. In this type of reservoirs, cyclic pressure pulsing using different types of gases as an alternative method to improve recovery has been found to be effective. Injected gas can penetrate and diffuse through the low-permeability matrix with the help of the large contact area, which is created by fractures. High-permeability fractures allow easy delivery of the injected gas and production of oil. Well-to-well connectivity is not required as it is a single-well process. The process is characterized by three stages, which are also illustrated in Fig. 1:

(1) *Injection period* Gas is injected into the reservoir.
(2) *Soaking period* Gas diffuses from fractures into the matrix.
(3) *Production period* The well is put on production. At the beginning of production, gas may be produced at high rates; however, as time passes by, it will decrease. Production may continue until the economic limit is reached, and if necessary, another cycle can be initiated.

Since the 1960s a number of studies have been published on cyclic-pressure pulsing. Initial applications were

✉ Emre Artun
artun@metu.edu

[1] Petroleum and Natural Gas Engineering Program, Middle East Technical University, Northern Cyprus Campus, Kalkanli, Guzelyurt, 99738 Mersin 10, Turkey

[2] Present Address: University of Tulsa, Tulsa, Oklahoma, USA

Edited by Yan-Hua Sun

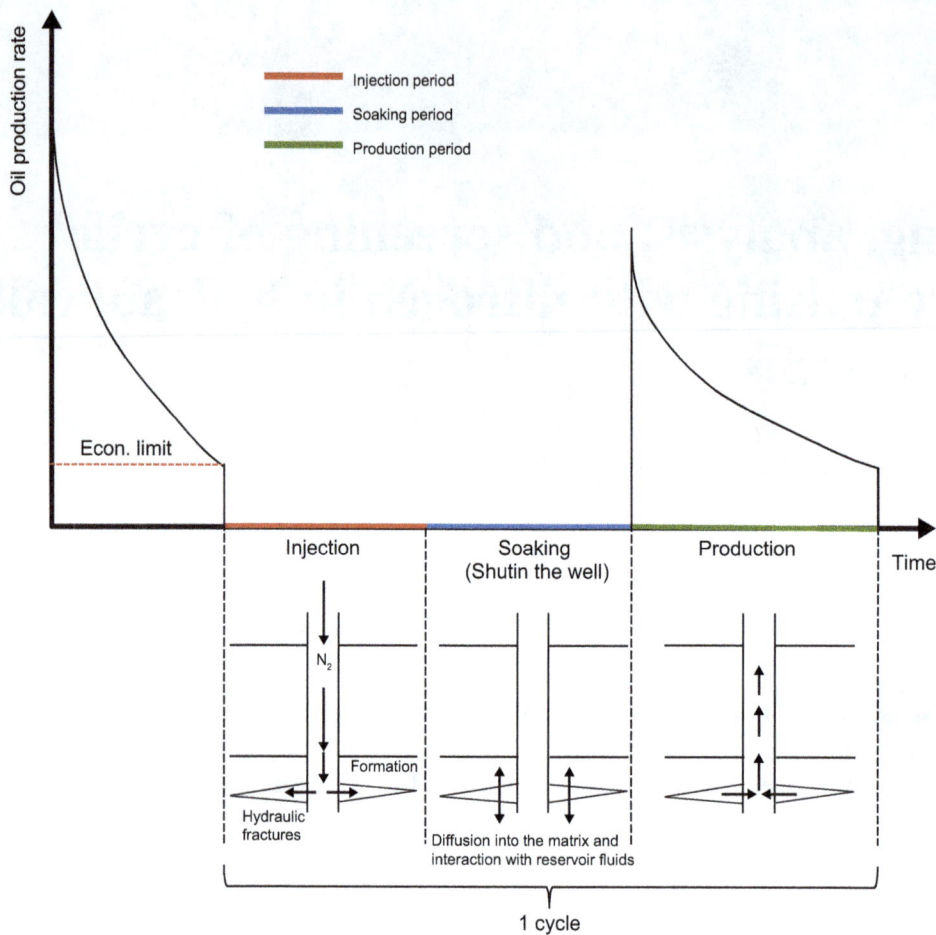

Fig. 1 Overview of the cyclic-pressure-pulsing process and its resulting impact on the produced oil-flow rate

with water as an improved way of waterflooding (Owens and Archer 1966; Felsenthal and Ferrell 1967; Raza 1971). Shelton and Morris (1973) used rich hydrocarbon gases instead of steam to increase the reservoir energy (as a short-term benefit) and reduce oil viscosity (as a long-term benefit). Moreover, they found that soaking mainly affects the peak production rate after injection. Later, cyclic injection of carbon dioxide was utilized for heavy oil in California (Sankur and Emanuel 1983), Arkansas (Khatib et al. 1981), and Turkey (Bardon et al. 1986), and for light and medium types of oil in Kentucky (Bardon et al. 1994), Texas (Haskin and Alston 1989), and Louisiana (Monger and Coma 1988). In the 1990s and 2000s, nitrogen and mixtures with nitrogen were proposed and successfully applied (Shayegi et al. 1996; Miller and Gaudin 2000). Artun et al. (2010, 2011a, b, 2012) performed detailed parametric studies of the process by analyzing a large set of reservoir simulation runs and developed proxy models to be used for screening and optimization of cyclic pressure pulsing with nitrogen and carbon dioxide in naturally fractured reservoirs. These studies showed that cyclic

pressure pulsing can be an effective enhanced oil recovery method in naturally fractured reservoirs. Nitrogen has several advantages over carbon dioxide and other types of gases because of being inert, non-corrosive, environmentally friendly and cost effective (Miller and Gaudin 2000). In fractured systems, the primary mechanism that contributes to the displacement of oil is the gas diffusion through the surface of the fracture network. While naturally fractured reservoirs provide an extensive surface area for diffusion, hydraulic fractures can help to achieve a similar mechanism to some extent. In recent studies, it was shown that the cyclic pressure pulsing with water, nitrogen and carbon dioxide (Gamadi et al. 2014; Sheng and Chen 2014; Sheng 2015) can be an effective method to improve recovery in shale oil and liquid-rich shale reservoirs which benefit from an extensive network of hydraulic and natural fractures.

There are many hydraulically fractured wells in the world, and some of these wells are producing depleted reservoirs with very low production rates. Oil fields of the Appalachian Basin in the North-East USA can be given as examples with majority of the wells being hydraulically fractured and reservoir pressures at depleted levels. This makes most of the wells in the region classified as stripper wells, producing at marginal oil rates (typically less than 10 barrels per day). Such reservoir conditions require enhanced oil recovery methods to increase the production rate while low profit margins make it difficult to justify conventional flooding-type enhanced oil recovery methods. In this study, we propose that the cyclic pressure pulsing with nitrogen as a promising enhanced oil recovery method that can benefit from the existing hydraulic fractures in the well even though there is not an existing natural-fracture network. In addition, low-cost requirements of nitrogen generated from a membrane unit would help the process to be attractive from a feasibility point of view. The primary objectives of the study are understanding the following: (1) Applicability of cyclic pressure pulsing with nitrogen in hydraulically fractured wells in the Appalachian Basin-like reservoirs and others; and (2) Impact of various reservoir and operational parameters on the process efficiency.

To achieve the objectives, the workflow shown in Fig. 2 is followed which starts with a representative reservoir model. The investigation is accomplished using a compositional numerical reservoir model, which is characterized with representative properties of Appalachian Basin sandstones and Mid-Continent crude oil composition. Then, through a systematic experimental-design procedure, certain reservoir and operational parameters are varied. By collecting results and analyzing a critical performance indicator from the simulation outputs, impacts of those parameters are analyzed. The final step is to construct a proxy model that can be used for screening purposes to assess the applicability in cases which were not necessarily studied using the numerical model.

2 Methodology

2.1 Reservoir simulation model

A single-well, compositional, single porosity reservoir with a Cartesian gridblock system is constructed using a commercial simulator (CMG 2013). To represent the component-mass flow from fractures into the matrix caused by compositional gradients, the molecular diffusion option for nitrogen (CMG 2013), which is a critical factor during the soaking period of cyclic pressure pulsing process, is

activated. Sigmund correlation for molecular diffusion (Sigmund 1976) is used with a diffusion coefficient for nitrogen of 0.001 cm^2/s, which is taken from the literature (Silva and Belery 1989) and validated with the Chapman-Enskog binary-diffusion theory (Marrero and Mason 1972).

The square-shaped model consists of 961 gridblocks and it has only one layer (31 blocks in the x and y directions, 1 block in the z direction). There is a production/injection well at the center of the model. For the production well, the minimum bottom hole pressure is specified as 14.7 psia since the study focuses on fully or nearly depleted reservoirs with an average reservoir pressure of 50 psia. Injection constraints are defined as design parameters of the cyclic pressure pulsing process and varied during the study as explained in Sect. 2.2. The reservoir model is mainly characterized with properties from the Appalachian Basin sandstones (Duda et al. 1967; Boswell et al. 1993) that are shown in Table 1. Parameters that define the reservoir volume and initial conditions are kept constant, since the primary objective is to study the effects of other parameters that affect the flow dynamics such as the reservoir permeability and the hydraulic fracture characteristics.

The oil composition used is the Mid-Continent crude oil of 36°API gravity (Abboud 2005) and it is shown in Table 2. Figure 3 shows the phase envelope of the oil mixture around the wellbore after injecting nitrogen into the Mid-Continent crude oil (Farias and Watson 2007). During the operating range of this study, (70 °F and 50–500 psia) the reservoir hydrocarbon-mixture is 100 % in a liquid phase. Therefore, there is not any free gas other than the injected gas. Relative permeability curves for oil–water and gas–oil systems are shown in Fig. 4, which are taken from the commercial simulator used (CMG 2013),

Fig. 2 The workflow followed to achieve the objectives of this study

Reservoir modeling → Experimental design → Performance analysis → Screening

Table 1 Reservoir characteristics of the Appalachian Basin sandstones and its single-well homogeneous reservoir model

Property	Value
Porosity	0.1
Thickness, ft	50
Initial pressure, psia	50
Drainage area, acres	220
Water saturation	0.5
Gas saturation	0
Oil saturation	0.5
Original oil in place, MMSTB	4.14
Reservoir permeability[a], mD	1
Fracture permeability[a], mD	5,000
Fracture half-length[a], ft	550
Fracture width[a], ft	0.1

[a] Parameters those were changed during different parts of the study

with the end-point saturations being consistent with Appalachian Basin characteristics (Boswell et al. 1993). Here k_{rw} and k_{row} are the relative permeability to water and oil, respectively, for the water–oil system; k_{rg} and k_{rog} are the relative permeability to gas and oil, respectively, for the

Table 2 Mid-Continent crude oil composition with an API gravity of 36° (after Abboud 2005)

Component	N_2	C_1	C_2	C_3	i-C_4	i-C_5	C_6	C_{7+}
Molar fraction	0.1	0.2	1.1	5.5	9.6	14.9	5.8	62.8

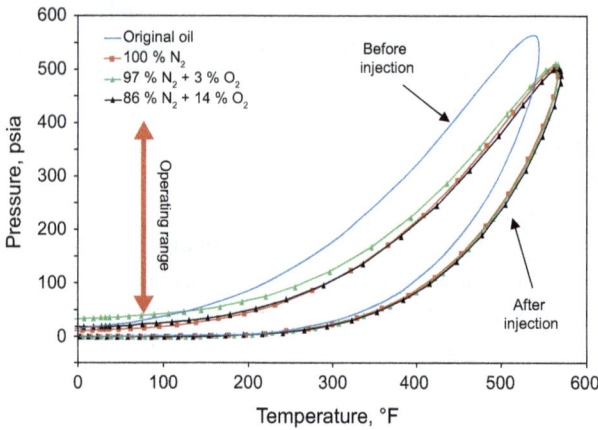

Fig. 3 Phase envelope of the oil mixture around the wellbore after injecting N_2 into the Mid-Continent crude oil defined in the model (Farias and Watson 2007)

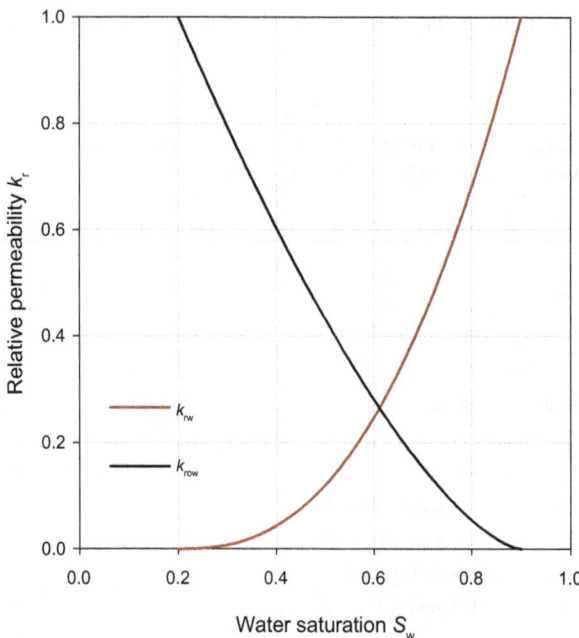

gas–oil system. The production/injection well is hydraulically fractured, and the hydraulic fracture is represented as a high-permeability streak with a local-grid refinement to be able to capture the flow dynamics at the matrix-fracture interface (Fig. 5). A representative gas-saturation distribution during injection, after injection and after soaking around the hydraulic fracture is also seen in Fig. 5.

2.2 Experimental design

To design the simulation cases to run, an experimental design procedure is utilized which consists of selecting the parameters to be varied and selecting their ranges and levels. As a result, the parameters shown in Table 3 are selected which are divided into two groups as operational parameters and reservoir properties. Ranges of these parameters are selected with the objective of having reasonable uncertainties in each parameter. Hydraulic fracture parameters define the overall effectiveness of the fracture-system to represent possible natural fractures around the wellbore that may contribute to the overall surface area for gas diffusion. It was decided to have five levels of variation in each variable. Normally, if the number of factors becomes moderately large, the number of runs may become unmanageable especially with the full-factorial design (Kelton and Barton 2003). However, in this case, a full-factorial design was found to be achievable considering the CPU time of a single run, which is a function of the size and complexity of the reservoir model. Considering four variables and five levels, the total number of runs using a

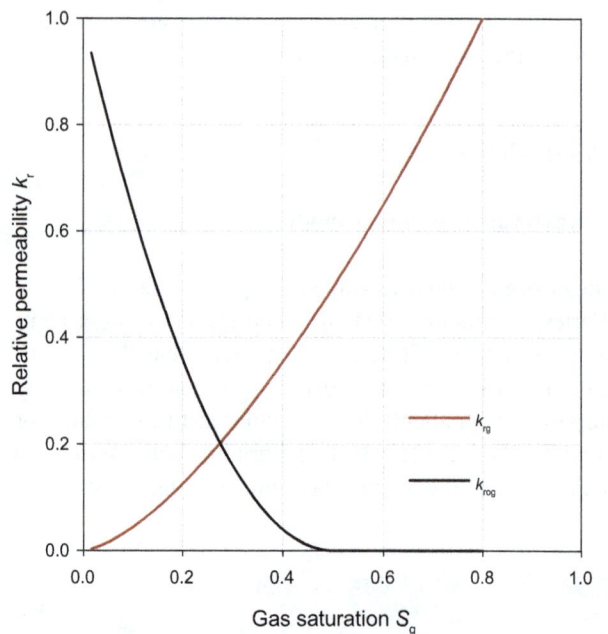

Fig. 4 Relative permeability curves used in the model

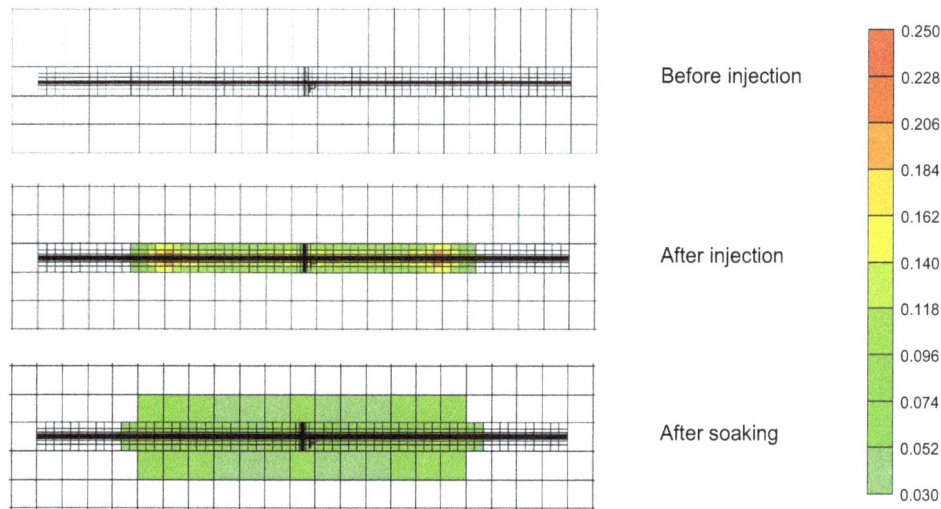

Fig. 5 Gas saturation during injection, after injection, and after soaking

Table 3 Parameters used with their ranges and levels as utilized in the experimental design procedure

Levels	Operational parameters					Reservoir/hydraulic fracture parameters				
	1	2	3	4	5	1	2	3	4	5
Injection rate, MCF/day	40	80	120	160	200	120				
Injection period, day	10	25	50	75	100	50				
Soaking period, day	10	25	50	75	100	50				
Cycle rate limit, STB/day	1.0	2.0	3.0	4.0	5.0	3.0				
Reservoir permeability, mD	1					0.1	1	10	50	100
Fracture permeability, mD	5000					1000	2500	5000	7500	10000
Fracture half length, ft	550					150	350	550	750	1050
Fracture width, ft	0.1					0.01	0.05	0.1	0.25	0.5

full-factorial design would be equal to 625 (5^4). Therefore, for both components of the study (operational and reservoir), the total number of simulation runs required is 1250. As can be seen in Table 3, when the operational parameters are studied, reservoir and hydraulic fracture parameters are kept constant, and vice versa.

2.3 Performance assessment

Using the simulation model, the process performance is analyzed. The incremental oil production and the injected volume of gas are both incorporated into the assessment. The incremental production represents the additional oil produced on top of the base cumulative production that would have been achieved without the injection process (Prats 1982). The incremental oil produced (N_{pin}) during year n in STB is calculated as:

$$N_{pin} = N_{pcn} - N_{pbn} \tag{1}$$

where N_{pcn} is the cumulative recovery during year n when cyclic injection is utilized (subscript c stands for cyclic injection), STB; N_{pbn} is the cumulative recovery during

year n when cyclic injection is not utilized (subscript b stands for base), STB. To represent the overall process performance, the discounted cyclic nitrogen injection efficiency is calculated by incorporating:

- Income generated from cumulative values of incremental oil produced,
- Costs due to nitrogen generation and injection,
- Time value of money through a discounting factor, i.

The present value of the incremental oil produced for 20 years of project time can be calculated from:

$$N_{pi0} = \sum_{n=1}^{20} \frac{N_{pin}}{(1+i)^n} \tag{2}$$

The present value of the cumulative volume of nitrogen injected can be calculated from:

$$G_{i0} = \sum_{n=1}^{20} \frac{G_{in}}{(1+i)^n} \tag{3}$$

where G_i is the cumulative volume of nitrogen injected during year n; i is the interest rate (taken as 10 %, yearly, in this study), and n is the number of years (ranging between 1 and 20, for 20 years of project time). The performance indicator, discounted cyclic nitrogen-injection efficiency (in STB/MCF), is defined as:

$$E_c = \frac{N_{pi0}}{G_{i0}} \qquad (4)$$

which states the incremental volume of oil produced per MCF of gas injected. The economic efficiency can be calculated from:

$$E_{ce} = E_c \times \frac{P_o}{P_{N_2}} \qquad (5)$$

where P_o is the oil price; P_{N_2} is the nitrogen price.

Assuming all other economic parameters are constant, this ratio can be used to identify if the project is feasible or not. Nitrogen generation cost is taken as \$1/MCF of nitrogen for generating nitrogen from a polymeric membrane unit (Miller and Gaudin 2000; Artun et al. 2011a). Because the values are discounted, the numerator is representative of the time-zero value of additional income generated from the incremental oil production, and the denominator is the time-zero value of the cost associated with nitrogen generation and injection:

If $E_{ce} > 1$ then the net present value $NPV > 0$ $\qquad (6)$

If $E_{ce} < 1$ then $NPV < 0$ $\qquad (7)$

This efficiency parameter is not a substitute for a detailed economic analysis, but an indication and a quick estimation of whether the nitrogen generation/injection cost would be justified by the incremental oil produced. Therefore, it only includes the parameters that change from one case to another (i.e., how much money is spent on generating and injecting nitrogen, how much additional money is earned due to additional oil production). Since the same well is used for both injection and production, it is assumed that other operational costs, labor and expenses are not going to change during injection and production.

2.4 Development of a screening tool

This part of the study is aimed to develop a screening tool, to assess the performance of the cyclic nitrogen injection in a hydraulically fractured well in a computationally efficient manner. Intelligent systems have been applied to many different types of optimization problems in the petroleum industry. Most of these problems presented in the literature are based on the development of artificial neural network (ANN) based proxy models that can accurately mimic reservoir models within a reasonable amount of accuracy and computational efficiency. Artificial neural networks (ANN) are very powerful in extracting non-linear and complex relationships between input and output patterns. Several areas of application included reservoir characterization (Artun and Mohaghegh 2011; Raeesi et al. 2012; Alizadeh et al. 2012; Artun 2016), candidate well selection for hydraulic fracturing treatments (Mohaghegh et al. 1996), field development (Centilmen et al. 1999; Doraisamy et al. 2000; Mohaghegh et al. 1996), well-placement and trajectory optimization (Johnson and Rogers 2011, Guyaguler 2002; Yeten et al. 2003), scheduling of cyclic steam injection processes (Patel et al. 2005), screening and optimization of secondary/enhanced oil recovery (Ayala and Ertekin 2005; Artun et al. 2010, 2011b, 2012; Parada and Ertekin 2012; Amirian et al. 2013), history matching (Cullick et al. 2006; Silva et al. 2007; Zhao et al. 2015), underground-gas-storage management (Zangl et al. 2006), reservoir monitoring and management (Zhao et al. 2015; Mohaghegh et al. 2014), and modeling of shale-gas reservoirs (Kalantari-Dhaghi et al. 2015; Esmaili and Mohaghegh 2015).

In this study, the backpropagation algorithm is used to train the neural network. The backpropagation algorithm is a gradient-descent method that minimizes the error during the training process. A given set of inputs is mapped into a set of given outputs which classifies this training algorithm as a supervised training algorithm (Fausett 1994). The training process includes 3 stages:

(1)   Feed-forward of the input training pattern,
(2)   Calculation and back-propagation of the error,
(3)   Adjustment of weights.

In Fig. 6, a fully-connected neural network with one hidden layer is shown. Number of input, hidden, and output neurons are n, p, and m, respectively. The number of inputs and outputs depend on the problem studied and the objective of the developed neural network, which both require knowledge of the subject-matter. The number of hidden neurons and hidden layers are determined as a part of the design process carried out for the neural network. Although there is not a straight-forward recipe for determining the number of hidden layers and neurons, they typically depend on the complexity of the problem as defined by the number of parameters and training patterns involved. There are a number of rules of thumb presented in the literature to determine the number of hidden neurons, and one of them is the following (Neuroshell 1998):

$$p = \frac{n + m}{2} + \sqrt{N_{TP}} \qquad (8)$$

where N_{TP} is the number of training patterns. It should be noted that this equation was developed mostly based on experience and should not be assumed to provide the correct number of hidden neurons for a given problem. However, it can be used as a good start for the optimization

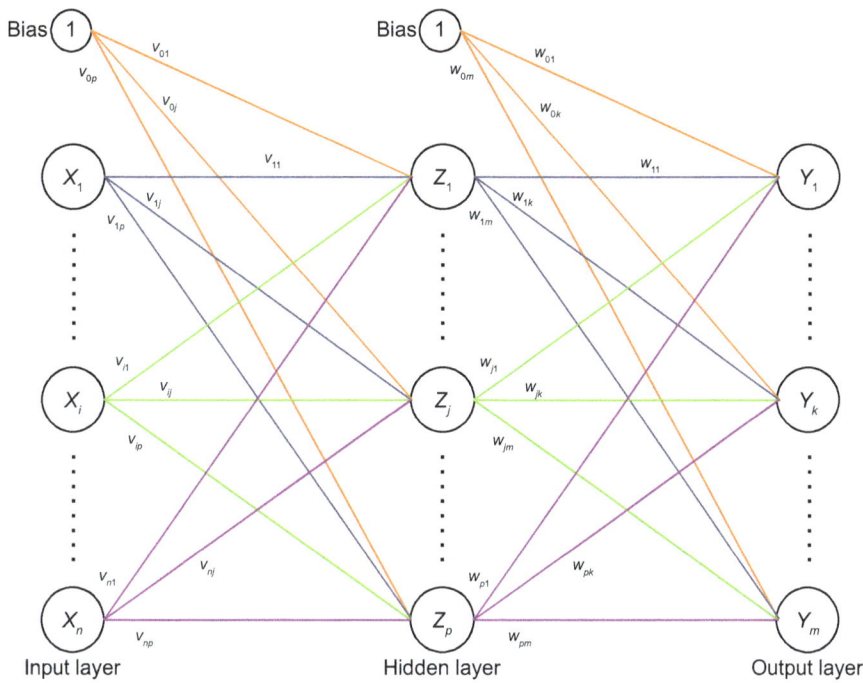

Fig. 6 Architecture of a multilayer network (after Artun 2016)

process. A step-by-step explanation of the backpropagation algorithm is shown below and more details about the terms involved can be found in any book on artificial neural networks such as (Fausett 1994):

squared error is calculated and iterations continue until the stopping criteria is satisfied. Most common stopping criteria include achieving the minimum mean-squared error or maximum number of epochs. After training is completed,

```
initialize weights (small random values)
while stopping condition is false do
        for each training pair do
                Feedforward stage:
                - each input unit receives signal, and broadcasts it to all units in the layer above (hidden
                units)
                - each hidden unit sums its weighted input signals, applies its activation function to
                calculate its output, and sends this signal to all units in the layer above (output layer)
                - each output unit sums its weighted input signals, and applies its activation function
                to calculate its output
                Error backpropagation stage:
                for both hidden units and output units do
                        compute: error, and error information term
                        compute: weight correction term
                end for
                Weight adjustment stage:
                update weights and biases
        end for
        test for stopping condition
end while
```

The inner iterative loop shown in this algorithm is repeated for all training cases included in the training set. When all training cases are processed once, one epoch is completed. After each training iteration, the average mean-

the weights on connection links achieve their optimum states. If the training performance is satisfactory, and if the model is validated with realistic, representative cases, then the trained network can be used as a predictive model. In this study, mapping input–output relationships is achieved with the inputs of operational and reservoir/hydraulic fracture characteristics, and the output of the cyclic nitrogen injection efficiency.

After running all cases and collecting corresponding performance indicators using the numerical model, a knowledge base is obtained. This knowledge base is then fed into an ANN, which has characteristics that have been pre-determined, for training. Once being trained and validated it is expected that the ANN-based proxy model can provide responses within comparable accuracy to a numerical model. Such kind of a model can be used as a screening tool for the cyclic pressure pulsing process with nitrogen, for cases that have not been necessarily run using the numerical model. This workflow is summarized in Fig. 7. In this study, the input parameters are the operational and reservoir/hydraulic fracture parameters as shown in Table 3, and the critical performance

indicator is the discounted cyclic-nitrogen-injection efficiency, E_c. defined in Eq. (4).

3 Results and discussion

3.1 Analysis of operational parameters

In this part of the study, the reservoir parameters are kept constant as shown in Table 3. Matrix permeability of 1 mD is representative of sandstone reservoirs of the Appalachian Basin. Hydraulic fracture properties are the mid-levels of the levels used in the 2nd part of the study.

3.1.1 An overview of results

Minimum and maximum efficiencies obtained among all cases are shown in Table 4. It is observed that in all cases

Injection rate, MCF/d	Injection period, d	Soaking period, d	Production period, months
50	100	50	12
100	25	100	6
75	75	25	3
150	50	75	9
⋮	⋮	⋮	⋮
100	25	25	12

1. Construct a data-base of the parameters to be studied within the ranges of the study

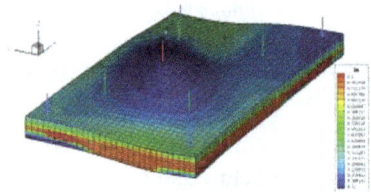

2. Construct a representative numerical reservoir model for the problem studied, and run the model for each case defined in Step 1.

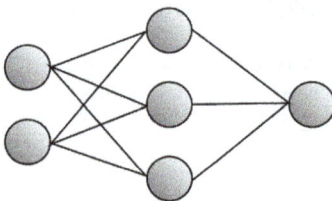

4. Design an artificial neural network and train the neural network by feeding the knowledge base as the training data.

Design scenario	Cumulative oil produced, STB	Peak oil rate, STB/d
1	10000	8
2	12000	15
3	7500	12
4	9000	10
⋮	⋮	⋮
1000	18000	9

3. Collect the corresponding performance indicators in a knowledge base.

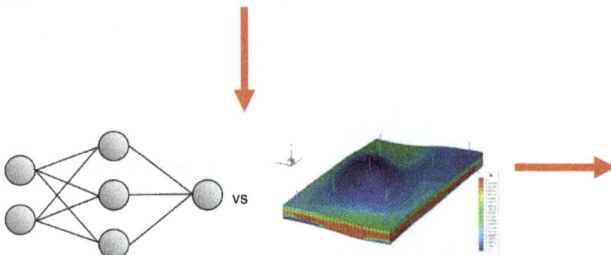

5. Validate the generalization capabilities of the trained neural network by testing with representative blind cases.

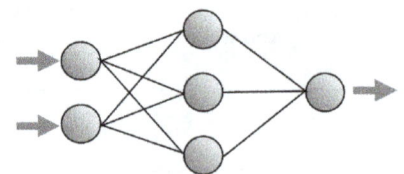

6. Use the trained neural network as a ANN-based proxy model for screening purposes.

Fig. 7 Workflow for constructing an ANN-based proxy model that can be used as a screening tool (after Artun et al. 2012)

Table 4 Minimum and maximum discounted efficiencies obtained when the operational parameters are varied

	Injection rate, MCF/day	Injection period, day	Soaking period, day	Cycle rate limit, STB/day	Efficiency, STB/MCF
Min.	200	100	50	5.0	0.9
Max.	40	10	75	5.0	14.4

the efficiency is greater than 0. This means that discounted oil production with nitrogen injection is always greater than cases without injection, which resulted in an incremental production of greater than zero. Since the cost of nitrogen generation is $1/MCF, the minimum efficiency of 0.9 STB/MCF indicates that the economic efficiency would be greater than 1 as long as the oil price is greater than $1.1/STB. Therefore, for almost any realistic oil price scenario, the 20-year cyclic nitrogen injection project would be feasible within the operational ranges studied.

3.1.2 Analysis of top 100 cases

Cases with the highest 100 efficiencies are analyzed to develop an understanding of ranges of variables that are favorable. Histograms that show the number of occurrences for all variables are plotted. Histograms for injection rate and period for top 100 discounted cyclic nitrogen injection efficiencies are shown in Fig. 8a, b. Results show that most of the top cases (87 %) have a nitrogen injection rate less than or equal to 120 MCF/day and 93 % of cases have an injection period of less than or equal to 25 days. The injected volume of nitrogen can be calculated by multiplying injection rate and time. Figure 8c shows the histogram of injection volume for top 100 discounted efficiencies. It is seen that 99 % of the cases have injection volumes less than 2000 MCF per cycle. These results indicate that nitrogen injection should be kept at the lower ranges that are studied. This may be due to the blockage of flow paths into the hydraulic fractures when higher volumes are injected and relative permeability effects. On the contrary, in the case of naturally fractured reservoirs, there is an interconnected network of fractures that enables easier flow of both gas and oil and higher volumes of gas contributes to higher oil recovery (Artun et al. 2011a). Longer injection periods may affect the process negatively because of dissipation of pressure with time. Therefore, these results show that it is critical to optimize the injection rate, period and volume. Figure 8d shows the histogram of soaking period for the top 100 cases. It is observed that for 72 % of the cases, the soaking period is greater than or equal to 50 days. Therefore, longer soaking periods favor the efficiency of the process. It is known from earlier studies that the typical soaking period needed for naturally fractured reservoirs is around 2–4 weeks. Therefore, a much longer time is necessary for a hydraulically fractured

well than a naturally fractured system. This is due to the smaller contact area for the gas diffusion in a hydraulically fractured well, when there is not a naturally fractured system that has an extensive contact area for the injected gas. Figure 8e shows the histogram of cycle rate limit for the top 100 cases. It is seen that 96 % of the cases realized with 2 STB/day of economic limit or more to stop the production and start the injection. The case that there are only 4 cases with 1 STB/day highlights the necessity of existing reservoir energy for the process to be successful. While there is not a clear indication of the optimum rate to stop the production between 2 and 5 STB/day, the potential risk of losing production time should be noted for higher rates. Amount of production lost by stopping early may not be compensated by the additional production due to injection. Figure 8f shows the total production shut-in time (injection and soaking periods). In this histogram, the maximum number of occurrences is when the time is between 50 and 100 days and less optimum results when the time is less than 50 days and greater than 100 days. This shows that there must be sufficient time of injection and soaking to maximize the process efficiency. However, when the time is too long, the process is clearly affected negatively with no cases above 150 days of shut-in time. This is due to dissipation of the pressure increase by injection at longer periods of time. This highlights another important parameter, the total duration of shut-in, for optimization.

3.1.3 Analysis of all cases

For further analysis, all cases are analyzed by taking the arithmetic average of all levels for each parameter and generating a 2-dimensional table of the averaged values. Figure 9 shows these values with respect to cycle rate limit and injection volume. The results indicate a lower range of injection volumes per cycle is more favorable for the efficiency of the process. This is probably due to the fact that lost production time is not compensated by the incremental oil produced. The low-permeability nature of the reservoir system does not allow gas to be transported into further portions of the reservoir, and therefore fails to displace more volume of oil from the matrix system. When we analyze the cycle rate limit, it is observed that rates higher than or equal to 2 STB/day are favorable. This indicates that existing reservoir energy in the system is

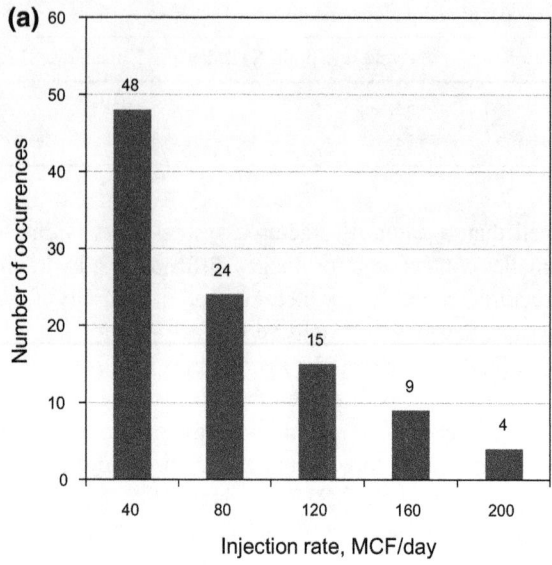

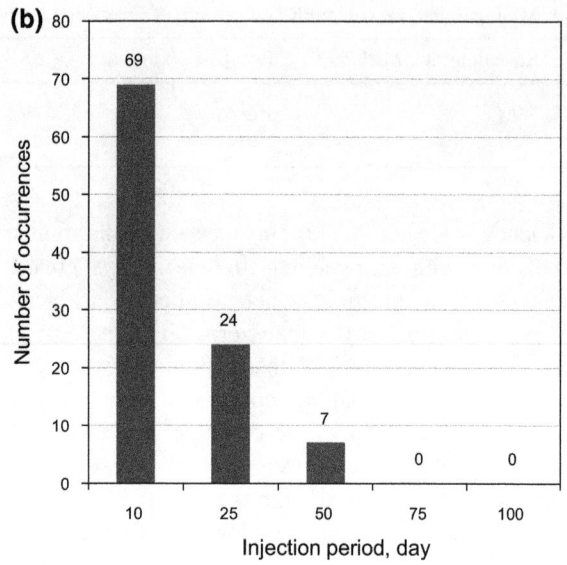

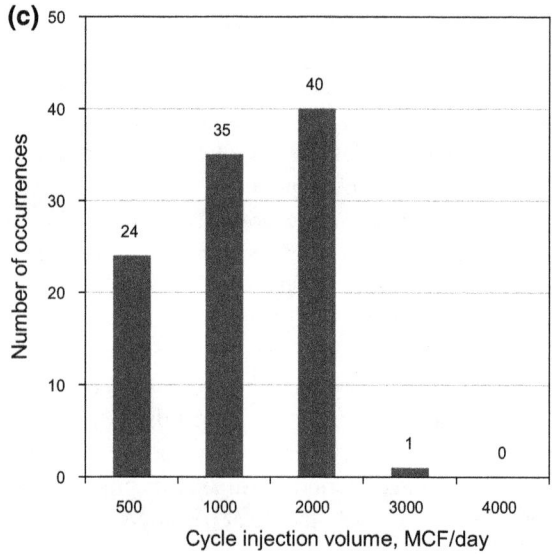

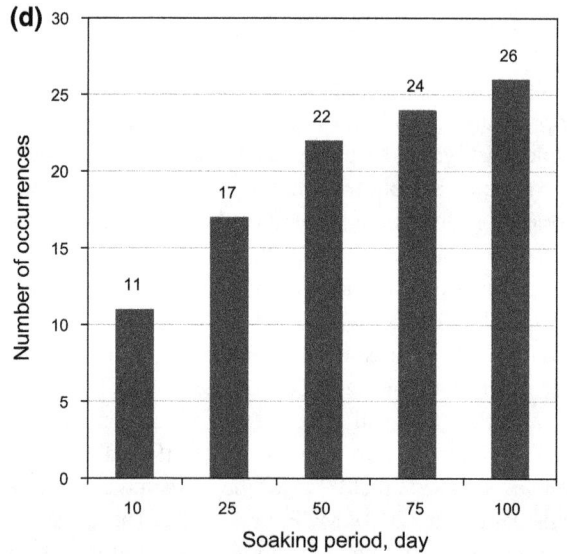

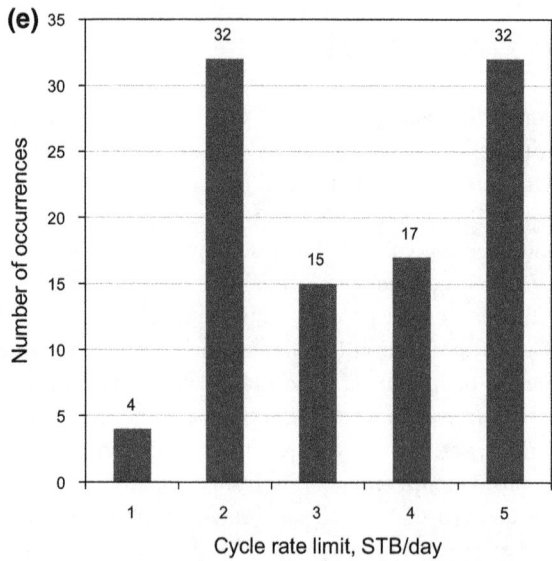

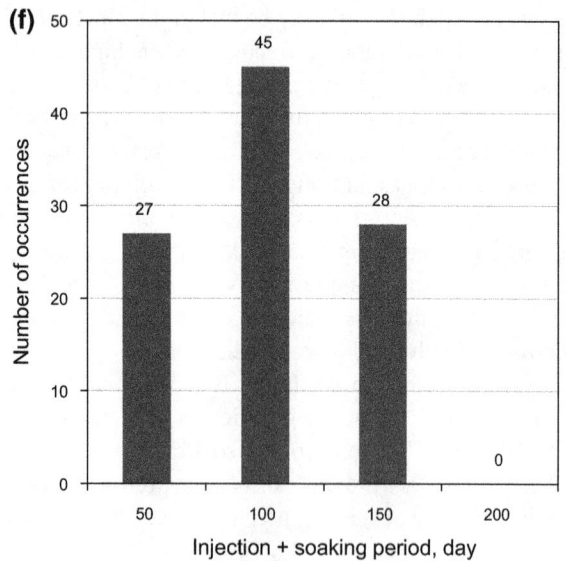

◄**Fig. 8** Histogram of operational parameters for top 100 runs in terms of the process efficiency

critical for the efficiency. Therefore, while the existing energy is critical, shutting in the well for long times when the well is still producing at reasonable rates is not a good practice. This is seen with the wide-range of the low-efficiency area when the limit is 5 STB. However, it should be noted that the best performers are with minimum injection volume and maximum cycle rate limit (400 MCF, and 5 STB/day). In Fig. 10, the efficiency values are mapped with respect to soaking and injection period lengths. The same observations with the top 100 cases also hold in this case. A soaking period is required and longer soaking periods improve the efficiency almost up to 28 % for 10 day of injection period. When the injection period is 100 days, since most of soaking already realized during the injection period itself, the improvement is very small (only 12 %).

3.2 Analysis of reservoir/hydraulic fracture parameters

In this part of the study, the operational parameters are kept constant and reservoir and hydraulic fracture properties are

varied. The injection rate is 120 MCF/day, injection and soaking periods are 50 days, and the economic rate limit for oil production is 3 STB/day.

3.2.1 An overview of results

Minimum and maximum efficiencies obtained among all cases are shown in Table 5. It is observed that in all cases the efficiency is greater than 0. This means that discounted oil production with nitrogen injection is always greater than cases without gas injection. Since the cost of nitrogen generation/injection is $1/MCF, the minimum efficiency of 0.4 STB/MCF indicates that economic efficiency would be greater than one as long as the oil price is greater than $2.50/STB. Therefore, for almost any realistic oil price scenario, the 20-year cyclic nitrogen injection project will generate a net present value greater than zero within the operational ranges studied.

3.2.2 Analysis of top 100 cases

Cases with highest 100 efficiencies are analyzed to develop an understanding of ranges of variables that are favorable. Histograms that show the number of occurrences for all variables are plotted. Histograms for matrix permeability, fracture permeability, fracture width, and half-length are shown in Fig. 11. These results indicate that matrix

	E_c	Cycle rate limit, STB/day					
		1	2	3	4	5	Average
Cycle injection volume, MCF/day	400	3.8	5.2	4.9	8.4	11.9	6.8
	800	3.3	4.5	4.3	5.0	7.9	5.0
	1000	3.1	5.0	3.7	6.3	7.7	5.2
	1200	2.9	4.2	4.2	3.9	5.9	4.2
	1600	2.8	3.9	3.4	3.0	5.0	3.6
	2000	2.7	3.8	3.1	2.7	3.9	3.2
	3000	2.8	3.5	3.0	2.5	2.4	2.8
	4000	2.6	3.0	2.6	2.2	2.1	2.5
	5000	2.4	2.7	2.6	2.2	1.8	2.3
	6000	2.1	2.3	2.2	1.9	1.7	2.1
	8000	1.7	2.1	1.9	1.6	1.4	1.7
	9000	1.7	2.1	1.8	1.5	1.3	1.7
	10000	1.6	2.0	1.7	1.5	1.2	1.6
	12000	1.4	1.7	1.5	1.4	1.2	1.4
	15000	1.3	1.5	1.4	1.3	1.1	1.3
	16000	1.2	1.5	1.4	1.2	1.0	1.3
	20000	1.1	1.2	1.2	1.1	1.0	1.1
	Average	2.3	2.9	2.6	2.6	3.1	

Fig. 9 Average values of efficiencies of all cases with respect to the cycle rate limit and injection volume

E_c	Injection period, day					
	10	25	50	75	100	Average
Soaking period, day 10	3.7	2.8	2.0	1.8	1.5	2.4
25	4.3	3.1	2.2	1.9	1.6	2.6
50	4.9	3.3	2.3	1.9	1.6	2.8
75	4.9	3.5	2.3	1.9	1.6	2.8
100	5.1	3.4	2.4	1.9	1.7	2.9
Average	4.6	3.2	2.2	1.9	1.6	

28 % higher efficiency with soaking for 100 days than soaking for 10 days

Dissipation of pressure with long injection time (too late for soaking)

Fig. 10 Average values of efficiencies of all cases with respect to soaking and injection periods

Table 5 Minimum and maximum discounted efficiencies obtained when reservoir/hydraulic-fracture parameters are varied

	Matrix permeability, mD	Fracture permeability, mD	Fracture width, ft	Fracture half-length, ft	Efficiency, STB/MCF
Min.	0.1	7500	0.05	150	0.4
Max.	10	10,000	0.5	750	3.2

permeability is very critical such that range of permeability between 10 and 100 mD constitutes 98 % of the cases. While higher fracture permeabilities are favorable, their impact is not as large as the matrix permeability. This is probably due to the permeability difference between the fracture and the matrix and its contribution to the diffusion process. The reason that there is not an observable difference between 10, 50 and 100 mD, is due to the high-base recovery of high permeability reservoirs. Since the base recoveries are high, the amount of incremental recovery that could be achieved with injection is also reduced. Effective fracture width values greater than 0.01 ft (between 0.05 and 0.5 ft) are favorable, while 80 % of cases are 0.25 ft or more. For the fracture half-length, we would expect that longer fractures would provide better efficiency because of the greater surface area for gas diffusion. The results indicate that 85 % of the cases are with half-lengths of 550 or 1050 ft.

3.2.3 Analysis of all cases

Figure 12 shows average efficiency values with respect to matrix and fracture permeabilities. A similar observation with the top 100 cases is the fact that matrix permeabilities of 10 mD and higher result in higher efficiencies. It is also observed that the efficiency is not a strong function of fracture permeability but an effective fracture permeability of 10,000 mD can increase the efficiency by 25 % when compared with 1000 mD. This indicates that as long as

there is a fracture, the diffusion process helps to displace the oil in the matrix. For higher matrix permeabilities of 50 and 100 mD, the base recovery is high and the efficiency does not change significantly from 1 or 10 mD reservoirs. A tight-reservoir permeability of 0.1 mD is also less effective than that of a permeability of 1–10 mD. This indicates that the Appalachian Basin sandstones would be good candidates for cyclic nitrogen injection. Figure 13 is a similar plot with the effective fracture width and fracture half length. As long as the fracture width is greater than 0.01 ft, and fracture half-length is greater than 150 ft, the process efficiency appears to be more favorable. This is due to the increased surface area for diffusion with an effective hydraulic fracture. Results indicated that with the same operational parameters, hydraulic fracture effectiveness can double the efficiency.

3.3 Screening tool

An ANN with an architecture shown in Fig. 14 is constructed. In this neural network, there are 12 input parameters, with six of them represent operational parameters, and remaining six parameters represent reservoir and hydraulic fracture parameters. In addition to the base parameters determined in the previous stages of the study, 4 additional parameters are added which are functions of the base parameters to help represent the problem in an improved way. These include:

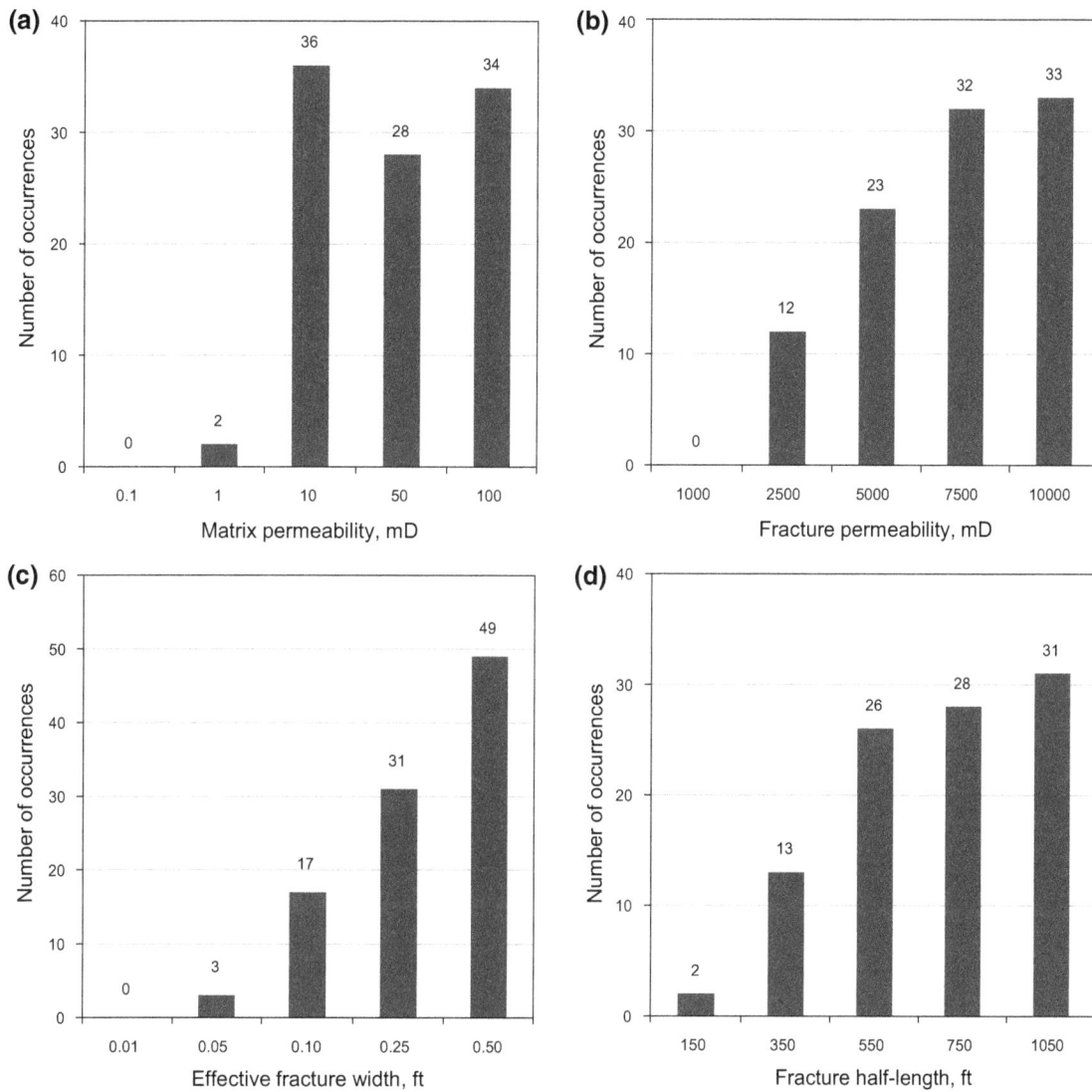

Fig. 11 Histogram of reservoir parameters for top 100 runs in terms of the process efficiency

(1) $t_i + t_s$: summation of injection and soaking periods to account for the total amount of time in each cycle during which the well is not on production,

(2) G_i: multiplication of injection period with injection rate to account for the volume of gas injected in each cycle,

(3) A_f: multiplication of fracture length with fracture width to account for the effective size of the fracture,

(4) k_f/k_m: ratio of fracture permeability to the matrix permeability to account for the contrast between matrix and fracture permeabilities.

The only output is the discounted cyclic-injection efficiency, E_c. Considering the size of the problem (1250 patterns, 12 inputs and 1 output), a single-hidden-layer network with 50 neurons is constructed (Fig. 14). A Levenberg–Marquardt backpropagation algorithm is used for training (MATLAB 2013a), with a hyperbolic tangent sigmoid transfer function between input-hidden layers, and a linear transfer function between hidden and output layers. Among 1250 patterns 70 % of the dataset is used for training (874 cases), 15 % (188 cases) is used for validation during training to prevent over-training problems, and remaining 15 % (188 cases) used for blind-testing purposes to test the generalization capabilities of the trained neural network. The full set of characteristics of the neural network is shown in Table 6.

The training is terminated after 200 validation checks without improvement after 213 epochs. The training error was 0.5 %. The trained neural network is applied to the whole data set, and to the training, validation, and testing sets separately. Figure 15 shows the cross-plots of results obtained from the numerical model and the ANN-proxy model. These comparisons indicate the proxy model has high-accuracy prediction capability, with a correlation

Too tight   High base recovery

E_c	Matrix permeability, mD					
	0.1	1	10	50	100	Average
1000	0.6	1.6	1.9	1.9	1.9	1.6
2500	0.7	1.8	2.2	2.1	2.1	1.8
5000	0.8	1.9	2.3	2.3	2.4	1.9
7500	0.8	2.0	2.4	2.4	2.4	2.0
10000	0.8	2.0	2.5	2.4	2.5	2.0
Average	0.7	1.8	2.2	2.2	2.3	

(Fracture permeability, mD — row label)

Fig. 12 Average values of efficiencies of all cases with respect to matrix permeability and fracture permeability

E_c	Effective fracture width, ft					
	0.01	0.05	0.1	0.25	0.5	Average
150	1.2	1.4	1.4	1.6	1.7	1.5
350	1.4	1.7	1.8	1.9	2.1	1.8
550	1.5	1.9	2.1	2.2	2.3	2.0
750	1.5	1.9	2.1	2.2	2.4	2.0
1050	1.5	1.9	2.1	2.2	2.3	2.0
Average	1.4	1.8	1.9	2.0	2.2	

(Fracture half-length, ft — row label)

Hydraulic fracture effectiveness can double the
process efficiency (from 1.2 to 2.4 STB/MCF)

Fig. 13 Average values of efficiencies of all cases with respect to effective fracture width and fracture half-length

coefficient of 0.96 for the testing set, which is the set that was not shown during training. In Fig. 16, the histogram of the calculated errors is shown. For the efficiency range of 0–14.4, the fact that the great majority of the cases have an error less than 0.5 STB/MCF also highlights the high predictive capability of the trained neural network. Therefore, this model can be used for screening of the cyclic nitrogen injection process for a well with hydraulic fractures, when the quantities of specified operational and reservoir/hydraulic fracture parameters are provided. This helps the practicing reservoir engineer or manager to evaluate a large number of different scenarios and obtain expected process efficiency in a quick and practical manner.

4 Conclusions

The purpose of this study was to develop a better understanding of how operational and reservoir/hydraulic-fracture parameters affect the performance of the cyclic

nitrogen injection in hydraulically fractured wells. This is achieved by building and running a numerical reservoir simulation model. The reservoir fluid is characterized with a Mid-Continent crude oil that can be considered as volatile. By extending the range of certain reservoir properties, a generalized analysis is also carried out. Experimental design methodology is followed to analyze the outcomes of the wide ranges of the properties studied. A screening tool is developed by training a neural network with the knowledge base obtained with the simulation runs. The principal conclusions drawn from this study can be summarized as the following:

(1)   Within the ranges studied, considering a cost of $1/ MCF for nitrogen generation, cyclic injection of nitrogen is a feasible enhanced oil recovery method in hydraulically fractured wells, especially in the Appalachian Basin.

(2)   The economic rate limit for stopping the production and starting the injection must be optimized. The

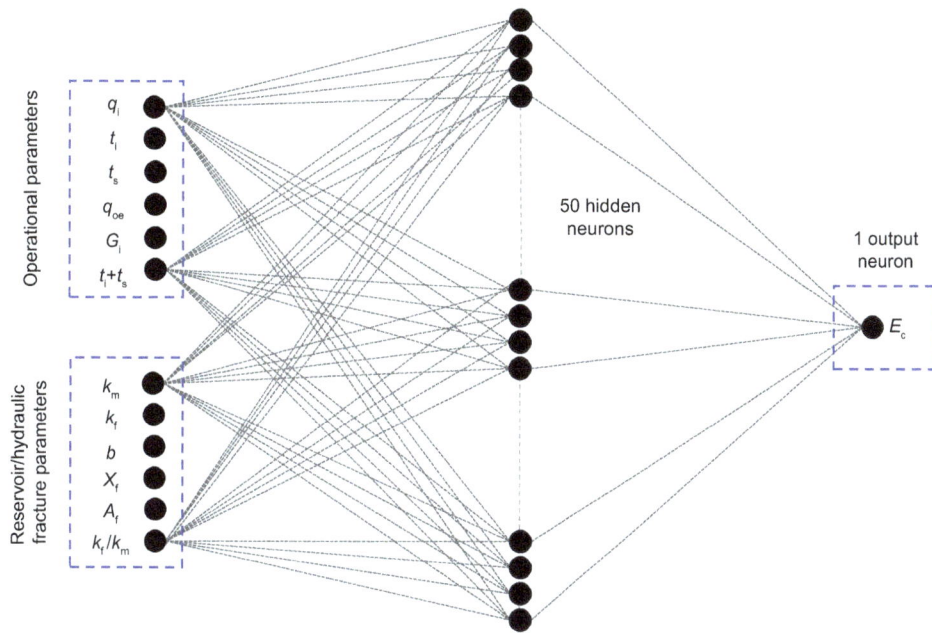

Fig. 14 Architecture of the ANN developed. q_i is the injection rate, MCF/day; t_i is the length of the injection period length, d; t_s is the length of the soaking period length, d; q_{oe} is the economic rate limit for each cycle, STB/day; G_i is the volume of gas injected in each cycle, MCF; k_m is the matrix permeability, mD; k_f is the fracture permeability, mD; b is the effective fracture width, ft; X_f is the effective fracture half-length, ft; A_f is the effective areal size of the fracture, ft^2

Table 6 Characteristics of the constructed artificial neural network

Property	Value
No. of input neurons	12
No. of hidden layers and neurons	1 hidden layer with 50 neurons
No. of output neurons	1
Training algorithm	Levenberg–Marquardt backpropagation
Transfer function between input and hidden layers	Hyperbolic tangent sigmoid
Transfer function between hidden and output layers	Linear
Training set, %	70
Validation set, %	15
Testing set, %	15
Max. No. of epochs	10,000
Min. error, %	0.0001
Max. No. of iterations without improvement	200

amount of existing energy in the reservoir is important for successful application. When the production is stopped at higher rates, the injection and soaking time should be kept to a minimum so not to lose production time that cannot be recovered with the help of injection.

(3) A soaking period is necessary to allow for gas diffusion, but together with the injection period, the optimum time must be determined as long shut-in periods (longer than 100 d) cause dissipation of pressure after injection.

(4) Matrix permeability is very critical such that permeabilities of 10 mD and higher result in more favorable results. The fracture permeability affects the process less than the matrix permeability, and an effective fracture permeability can increase the efficiency by 25 %.

(5) Longer (>150 ft) and wider (>0.01 ft) fractures provide better efficiency because of the greater surface area for gas diffusion. With the same operating conditions, hydraulic-fracture effectiveness can double the efficiency.

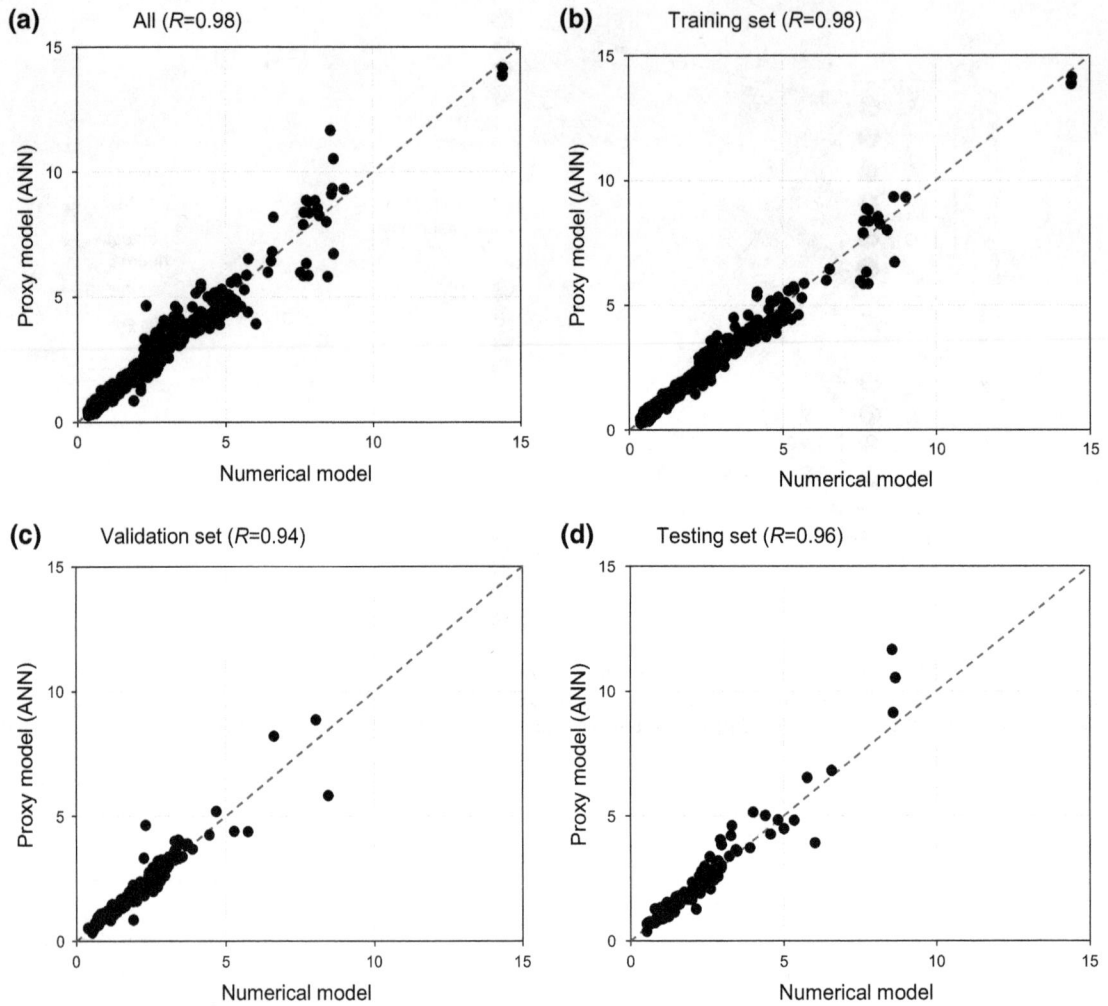

Fig. 15 Accuracy of ANN predictions for training, validation, and testing sets

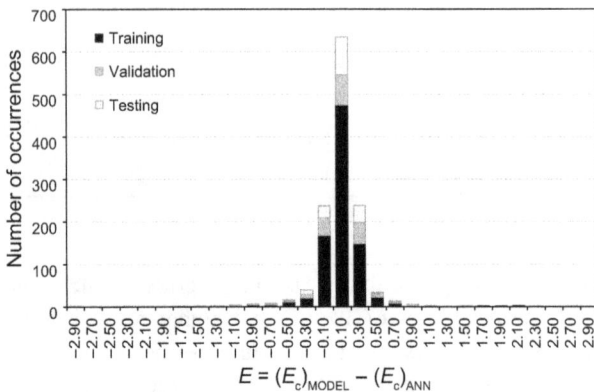

Fig. 16 Histogram of calculated errors of training, validation, and testing sets

(6) An ANN-based proxy model is successfully trained, and the resulting model can be used as a quick screening tool to estimate the discounted process efficiency, once corresponding operational and reservoir/hydraulic-fracture parameters are provided. The model was able to estimate the efficiency of 188 cases that it had not seen before with a correlation coefficient of 0.96.

References

Abboud A. A study of cyclic injection of nitrogen on Mid-Continent crude oil: an investigation of the vaporization process in low pressured shallow reservoirs. M.Sc. thesis, The Pennsylvania State University, University Park, Pennsylvania; 2005.

Alizadeh B, Najjari S, Ali K. Artificial neural network modelling and cluster analysis for organic facies and burial history estimation using well log data: a case study of the South Pars Gas Field, Persian Gulf, Iran. Comput Geosci. 2012;45:261–9. doi:10.1016/j.cageo.2011.11.024.

Amirian E, Leung J, Zanon S, et al. Data-driven modeling approach for recovery performance prediction in SAGD operations. In: SPE heavy oil conference, Calgary, 11–13 June; 2013. doi:10.2118/165557-MS.

Artun E, Ertekin T, Watson R, et al. Development and testing of proxy models for screening cyclic pressure pulsing process in a depleted, naturally fractured reservoir. J Pet Sci Eng. 2010;73(1):73–5. doi:10.1016/j.petrol.2010.05.009.

Artun E, Ertekin T, Watson R, et al. Performance and economic evaluation of cyclic pressure pulsing in naturally fractured reservoirs. J Can Pet Technol. 2011a;50(9–10):24–36. doi:10.2118/129599-PA.

Artun E, Ertekin T, Watson R, et al. Development of universal proxy models for screening and optimization of cyclic pressure pulsing in naturally fractured reservoirs. J Nat Gas Sci Eng. 2011b;3(6):667–86. doi:10.1016/j.jngse.2011.07.016.

Artun E, Mohaghegh S. Intelligent seismic inversion workflow for high-resolution reservoir characterization. Comput Geosci. 2011;37(2):143–57. doi:10.1016/j.cageo.2010.05.007.

Artun E, Ertekin T, Watson R, et al. Designing cyclic pressure pulsing in naturally fractured reservoirs using an inverse looking recurrent neural network. Comput Geosci. 2012;2012(38):68–79. doi:10.1016/j.cageo.2011.05.006.

Artun E. Characterizing interwell connectivity in waterflooded reservoirs using data-driven and reduced-physics models: a comparative study. Neural Comput Appl. 2016. doi:10.1007/s00521-015-2152-0.

Ayala L, Ertekin, T. Analysis of gas-cycling performance in gas/condensate reservoirs using neuro-simulation. In: SPE annual technical conference and exhibition, Dallas, 9–12 October; 2005. doi:10.2118/95655-MS.

Bardon C, Karaoguz D, Tholance M. Well stimulation by CO_2 in the heavy oil field of Camurlu in Turkey. In: SPE enhanced oil recovery symposium, Tulsa, 20–23 April; 1986. doi:10.2118/14943-MS.

Bardon C, Corlay P, Longeron D, et al. CO_2 huff 'n' puff revives shallow light-oil-depleted reservoirs. SPE Reserv Eng. 1994;9(2):92–100. doi:10.2118/22650-PA.

Boswell R, Pool S, Pratt S, et al. Appalachian Basin low-permeability sandstone reservoir characterizations. Final Contractor's Report to the U.S. Department of Energy. Contract No. DE-AC21-90MC26328, Report No. 94CC-R91-003; 1993. doi:10.2172/1030788.

Centilmen A, Ertekin T, Grader A. Applications of neural networks in multiwell field development. In: SPE annual technical conference and exhibition, Houston, 3–6 October; 1999. doi:10.2118/56433-MS.

CMG: Computer Modeling Group Advanced Compositional Reservoir Simulator (GEM) User's Guide. Version 2013. Calgary, Alberta, Canada; 2013.

Cullick A, Johnson D, Shi G. Improved and more rapid history matching with a nonlinear proxy and global optimization. In: SPE annual technical conference and exhibition, San Antonio, 24–27 September; 2006. doi:10.2118/101933-MS.

Doraisamy H, Ertekin T, Grader A. Field development studies by neuro-simulation: an effective coupling of soft and hard computing protocols. Comput Geosci. 2000;26(8):963–73. doi:10.1016/S0098-3004(00)00032-7.

Duda J, Overbey W, Johnson H. Predicted oil recovery by waterflood and gas drive, Bradford Third and Sartwell Sands, Sartwell Oilfield, McKean County, PA. U.S. Department of The Interior, Bureau of Mines, Report of Investigations No. 6943; 1967.

Esmaili S, Mohaghegh S. Full field reservoir modeling of shale assets using advanced data-driven analytics. Geosci Front. 2015;7(1):11–20. doi:10.1016/j.gsf.2014.12.006.

Farias M, Watson R. Interaction of nitrogen/CO_2 mixtures with crude oil. Report to the U.S. Department of Energy. Contract No. DE-FC26-04NT42098; 2007.

Fausett L. Fundamentals of neural networks: architectures, algo-

rithms, and applications. Englewood Cliffs: Prentice-Hall; 1994.

Felsenthal M, Ferrell H. Pressure pulsing: an improved method of waterflooding fractured reservoirs. In: SPE Permian Basin oil recovery conference, Midland, Texas, 8–9 May; 1967. doi:10.2118/1788-MS.

Gamadi T, Sheng J, Soliman M, et al. An experimental study of cyclic CO_2 injection to improve shale oil recovery. In: SPE improved oil recovery symposium, 12–16 April, Tulsa, Oklahoma; 2014. doi:10.2118/169142-MS.

Guyaguler B. Optimization of well placement and assessment of uncertainty. Ph.D. dissertation, Stanford University, Stanford, California; 2002.

Haskin H, Alston R. An evaluation of CO_2 huff 'n' puff tests in Texas. SPE Reserv Eng. 1989;41(2):177–84. doi:10.2118/15502-PA.

Johnson V, Rogers L. Applying soft computing methods to improve the computational tractability of a subsurface simulation-optimization problem. J Pet Sci Eng. 2011;29(3–4):153–75. doi:10.1016/S0920-4105(01)00087-0.

Kalantari-Dhaghi A, Mohaghegh S, Esmaili S. Data-driven proxy at hydraulic fracture cluster level: a technique for efficient CO_2-enhanced gas recovery and storage assessment in shale reservoir. J Nat Gas Sci Eng. 2015;27(2):515–30. doi:10.1016/j.jngse.2015.06.039.

Khatib A, Earlougher R, Kantar K. CO_2 injection as an immiscible application for enhanced recovery in heavy oil reservoirs. In: SPE California regional meeting, Bakersfield, 25–27 March; 1981. doi:10.2118/9928-MS.

Kelton W, Barton R. Experimental design for simulation. In: Winter simulation conference, New Orleans, 7–10 December; 2003.

Marrero T, Mason E. Gaseous diffusion coefficients. J Phys Chem Ref Data. 1972;1(1):3–118. doi:10.1063/1.3253094.

MATLAB: Neural network toolbox; Version 2013a, Mathworks, Inc., Natick, Massachusetts; 2013.

Miller B, Gaudin R. Nitrogen huff and puff process breathes new life into old field. World Oil Mag. 2000;221(9):7–8.

Mohaghegh S, Balan B, McVey D, et al. A hybrid neuro-genetic approach to hydraulic fracture treatment design and optimization. In: SPE annual technical conference and exhibition, Denver, 6–9 October; 1996. doi:10.2118/36602-MS.

Mohaghegh S, Al-Mehairi Y, Gaskari R, et al. Data-driven reservoir management of a giant mature oilfield in the Middle East. In: SPE annual technical conference and exhibition, 27–29 October, Amsterdam; 2014. doi:10.2118/170660-MS.

Monger TG, Coma JM. A laboratory and field evaluation of the CO_2 huff 'n' puff process for light-oil recovery. SPE Reserv Eng. 1988;3(4):1168–76. doi:10.2118/15501-PA.

Neuroshell: Neuroshell 2 tutorial. Ward Systems, Inc., Frederick, Maryland; 1998.

Owens W, Archer D. Waterflood pressure pulsing for fractured reservoirs. J Pet Technol. 1966;18(6):745–52. doi:10.2118/1123-PA.

Parada C, Ertekin T. A new screening tool for improved oil recovery methods using artificial neural networks. In: SPE western regional meeting, Bakersfield, 19–23 March; 2012. doi:10.2118/153321-MS.

Patel A, Davis D, Guthrie C, et al. Optimizing cyclic steam oil production with genetic algorithms. In: SPE western regional meeting, Irvine, 30 March–1 April; 2005. doi:10.2118/93906-MS.

Prats M. Thermal recovery. New York: SPE of AIME; 1982.

Raeesi M, Moradzadeh A, Ardejani F, et al. Classification and identification of hydrocarbon reservoir lithofacies and their heterogeneity using seismic attributes, logs data and artificial neural networks. J Pet Sci Eng. 2012;82–83:151–65. doi:10.1016/j.petrol.2012.01.012.

Raza SH. Water and gas cyclic pulsing method for improved oil recovery. J Pet Technol. 1971;23(12):1467–74. doi:10.2118/3005-PA.

Sankur V, Emanuel A. A laboratory study of heavy oil recovery with CO_2 injection. In: SPE California regional meeting, Ventura, 23–25 March; 1983. doi:10.2118/11692-MS.

Shayegi S, Jin Z, Schenewerk P, et al. Improved cyclic stimulation using gas mixtures. In: SPE annual technical conference and exhibition, Denver, 6–9 October; 1996. doi:10.2118/36687-MS.

Shelton J, Morris E. Cyclic injection of rich gas into producing wells to increase rates from viscous-oil reservoirs. J Pet Technol. 1973;25(8):890–6. doi:10.2118/4375-PA.

Sheng J, Chen K. Evaluation of the EOR potential of gas and water injection in shale oil reservoirs. J Unconv Oil Gas Resour. 2014;5:1–9. doi:10.1016/j.juogr.2013.12.001.

Sheng J. Enhanced oil recovery in shale reservoirs by gas injection. J Nat Gas Sci Eng. 2015;22:252–9. doi:10.1016/j.jngse.2014.12.002.

Sigmund PM. Prediction of molecular diffusion at reservoir conditions. Part 1—measurement and prediction of binary dense gas diffusion coefficients. J Can Pet Technol. 1976;15(2):48–57. doi:10.2118/76-02-05.

Silva FV, Belery P. Molecular diffusion in naturally fractured reservoirs: a decisive recovery mechanism. In: SPE annual technical conference and exhibition, San Antonio, 8–11 October; 1989. doi:10.2118/SPE-19672-MS.

Silva P, Clio M, Schiozer D. Use of neuro-simulation techniques as proxies to reservoir simulator: application in production history matching. J Pet Sci Eng. 2007;57(3–4):273–80. doi:10.1016/j.petrol.2006.10.012.

Yeten B, Durlofsky L, Aziz K. Optimization of nonconventional well type, location, and trajectory. SPE J. 2003;8(3):200–10. doi:10.2118/86880-PA.

Zangl G, Giovannoli M, Stundner M. Application of artificial intelligence in gas storage management. In: SPE Europec/EAGE annual conference and exhibition, 12–15 June; 2006. doi:10.2118/100133-MS.

Zhao H, Kang Z, Zhang X, et al. INSIM: a data-driven model for history matching and prediction for waterflooding monitoring and management with a field application. In: SPE reservoir simulation symposium, 23–25 February, Houston; 2015. doi:10.2118/173213-MS.

Clastic compaction unit classification based on clay content and integrated compaction recovery using well and seismic data

Zhong Hong[1] · Ming-Jun Su[1] · Hua-Qing Liu[1] · Gai Gao[2]

Abstract Compaction correction is a key part of paleo-geomorphic recovery methods. Yet, the influence of lithology on the porosity evolution is not usually taken into account. Present methods merely classify the lithologies as sandstone and mudstone to undertake separate porosity-depth compaction modeling. However, using just two lithologies is an oversimplification that cannot represent the compaction history. In such schemes, the precision of the compaction recovery is inadequate. To improve the precision of compaction recovery, a depth compaction model has been proposed that involves both porosity and clay content. A clastic lithological compaction unit classification method, based on clay content, has been designed to identify lithological boundaries and establish sets of compaction units. Also, on the basis of the clastic compaction unit classification, two methods of compaction recovery that integrate well and seismic data are employed to extrapolate well-based compaction information outward along seismic lines and recover the paleo-topography of the clastic strata in the region. The examples presented here show that a better understanding of paleo-geomorphology can be gained by applying the proposed compaction recovery technology.

Keywords Compaction recovery · Porosity-clay content-depth compaction model · Classification of lithological compaction unit · Well and seismic data integrated compaction recovery technology

1 Introduction

Paleo-geomorphology controls not only the spatial distribution of a depositional system, but also to some extent determines the source, reservoir, and seal units. The effect of mechanical compaction is one of several factors influencing hydrocarbon formation and evolution, with other factors including erosion effects, tectonic activity, diagenesis, and abnormal pressure. As the compaction process is extremely complex, it has garnered wide interest. A considerable number of compaction correction models have been established. Athy (1930) proposed an empirical exponential model of porosity and depth based on the condition of normal pressure, which served as an extensive reference for numerous basins around the world. Falvey and Middleton (1981) held that the evolution of porosity cannot be characterized by a classical exponential model at shallow depths. As a result, they proposed a compaction function based on the relationship between porosity and upper-layer load. In deep strata, porosity does not necessarily change as depth increases. Taking this phenomenon into account, Athy's exponential model was improved (Li and Kong 2001). Li et al. (2000) proposed a compaction correction method based on the principle that the grain volume and mass of formation should remain constant, which has resulted in deep debate among Chinese investigators (e.g., Qi and Yang 2001; Li 2001). Schon (2004) proposed a logarithmic function of compaction between porosity and depth.

✉ Zhong Hong
  hongzhong_go@petrochina.com.cn

[1] PetroChina Research Institute of Petroleum Exploration and Development (RIPED)-Northwest, Lanzhou 730020, Gansu, China

[2] Research Institute of Exploration and Development, PetroChina Changqing Oilfield Company, Xi'an 710018, Shaanxi, China

Edited by Jie Hao

The influence of lithology on porosity rarely has been considered in previous compaction correction studies. Actually, the compaction characteristics of clastic rocks are considerably different as the lithology changes. Consequently, among the present studies, dividing the clastic sedimentary units into just two lithologies, mudstone and sandstone, is too much of a simplification. Such a method is not able to represent and generalize the compaction history of mixed grain-size lithologies such as muddy siltstone and silty mudstone, in addition to pure mudstone and pure sandstone. A clastic compaction study considering only the influence of lithology toward the porosity was undertaken by Ramm and Bjørlykke (1994), who found that according to the statistical relationship between mineral content and porosity, porosity had a higher correlation with the clay-content index. Subsequently, a compaction model involving porosity, depth, and clay-content index has been developed.

Based on former research, a depth compaction model has been proposed that takes porosity and clay content into account to be reliably assessed in terms of rock-physics theory and laboratory experiments. To improve the precision of compaction correction studies, a clastic lithological compaction unit classification method based on clay content has been designed to identify lithological strata and establish compaction units (Fig. 1). Also, to assess the method for industrial applications, two integrated methods are proposed that use both well and seismic data to evaluate 3-D compaction characteristics in clastic strata.

2 Theoretical basis of clastic compaction unit classification based on clay content

Usually, qualitative methods are used to classify clastic compaction units. This is done in terms of well log interpretations that classify lithologies into categories such as pure sandstone, pure mudstone, muddy siltstone, and silty mudstone. In this way, porosity-depth evolution curves can be obtained for various lithologies. However, this method is time-consuming, which makes its wide implementation difficult in industrial practice. In light of this, an automatic optimization classification method based on clay content is proposed to identify various lithological compaction units. The feasibility of the method and its validation are discussed in this paper in terms of rock-physics theory and rock-physics laboratory tests.

2.1 Ideal binary clastic mixture model

The ideal binary clastic mixture model characterizes the inherent correlation, as well as the quantitative functional relationship, between clay content and porosity (Thomas and Stieber 1975, 1977); explicit studies have also been

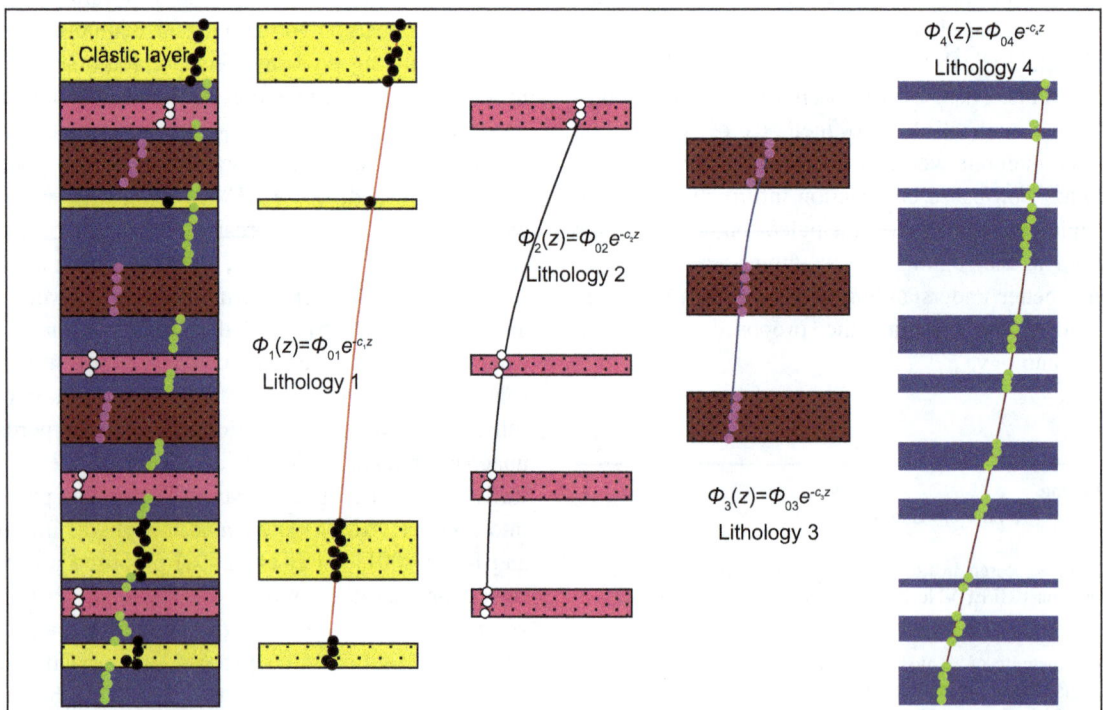

Fig. 1 Schematic map of clastic lithological compaction unit classification

Clastic compaction unit classification based on clay content and integrated compaction recovery...

21

conducted by several others (e.g., Pedersen and Norda 1999; Mavko 2009; Dvorkin et al. 2014).

Figure 2 shows an ideal binary mixture model. ϕ_{ss} is the critical porosity of pure sandstone, ϕ_{sh} is the critical porosity of pure shale, and C is clay content. When the clay particles fill the pore space between grains in a sandstone, some simple equations can be obtained.

Specifically, for $0 \leq C \leq \phi_{ss}$, $\phi = \phi_{ss} - (1 - \phi_{sh})C$, and for $\phi_{ss} \leq C \leq 1$, $\phi = \phi_{sh}C$.

When clay particles serve as part of the framework in sandstone, $\phi = \phi_{ss} + \phi_{sh}C$.

This rock-physics model assumes that the sand is clean and homogeneous with a constant porosity and that any change in porosity owing to cementation or sorting is ignored. Practically, even though the formation and evolution of porosity are extremely complex, the ideal binary mixture model is able to simply describe the relationship between porosity and clay content for different clastic lithologies.

2.2 Rock-physics laboratory relationship between clastic porosity and clay content

Yin's rock-physics experiments presented a qualitative relationship between porosity and clay content under different pressures (Yin 1992). Figure 3a shows that with different pressures (i.e., different depths), the relationship between porosity and clay content presents a V-shape with an obvious inflection point. This is similar to the ideal binary mixture model. Also, as pressure (depth) increases, the V-shape exhibited by the porosity and clay content is preserved. The experiment shows that although porosity changes, due to mechanical compaction as d epth increases, the porosity and clay content still maintain a regular relationship. The experiment also shows that the higher the

clay content is, the more distinct will be the change in porosity as pressure increases; additionally, the mechanical compaction effect will be more intense. This conforms to the practical compaction characterization of clastic rocks.

Porosity data for specific clay contents under different pressures (Fig. 3a) are selected to map the cross-plot between pressure and porosity. Figure 3b shows the evolution of porosity versus pressure for pure sandstone, pure mudstone, and a mixed lithology with 50% clay content. In this way, the rock-physics experiments illustrate that lithologically independent compaction histories can be identified based on clay content.

As a whole, clay content does not change and has no trend in depth. Based on clay content, it is possible to link compaction relationships for the same lithology at different depths. Also, in terms of not only intrinsic features but also dynamic evolution, the porosity and clay content of clastic rocks have an inherent connection between each other. Consequently, it is valid and feasible to classify clastic compaction units based on clay content.

3 Classification method for clastic compaction units based on clay content

3.1 Calculation of initial porosity

The initial porosity is defined as the porosity of sediments at the earth's surface directly after deposition. In the exponential function model of porosity and depth, the initial porosity has a significant influence on the compaction recovery. When initial porosity changes by 10%, the resulting estimated thickness after compaction recovery may differ by more than 10% (Yang and Qi 2003). The formation of initial porosity is extraordinarily complex, and it varies according to depositional background. The initial porosity is affected not only by grain size, sorting, sphericity, roundness, and filling of sediment, but it also involves grain assemblage and consolidation (He et al. 2002).

The initial porosity can be obtained through direct experimental measurements (Atkins and McBride 1992; Beard and Weyl 1973; Pryor 1973) or numerical simulations (Zaimy and Rasaei 2013; Fawad et al. 2011; Alberts and Weltje 2001; Harbaugh et al. 1999; Syvitski and Bahr 2001); however, this is a time-consuming process. In industrial practice, He et al. (2002) proposed a method in which the initial porosity is assigned to fall within a reasonable range of values. Then, on the basis of an assigned porosity within the range, residuals of fitted porosities and original porosities are calculated. An initial porosity is subsequently selected from the range as the one with the smallest residual. Yang and Qi (2003) proposed a method

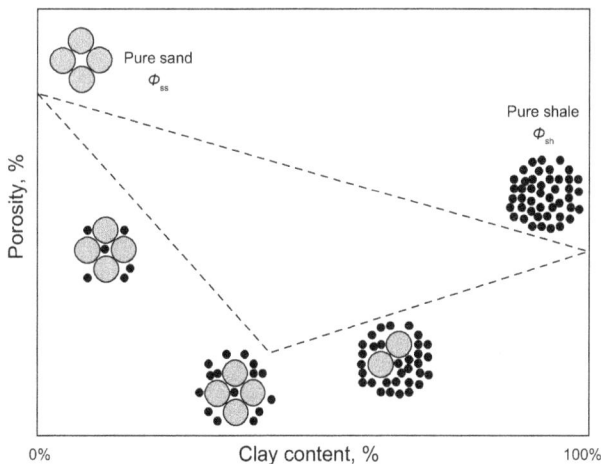

Fig. 2 Ideal binary mixture model (revised according to references of Mavko 2009; Dvorkin et al. 2014)

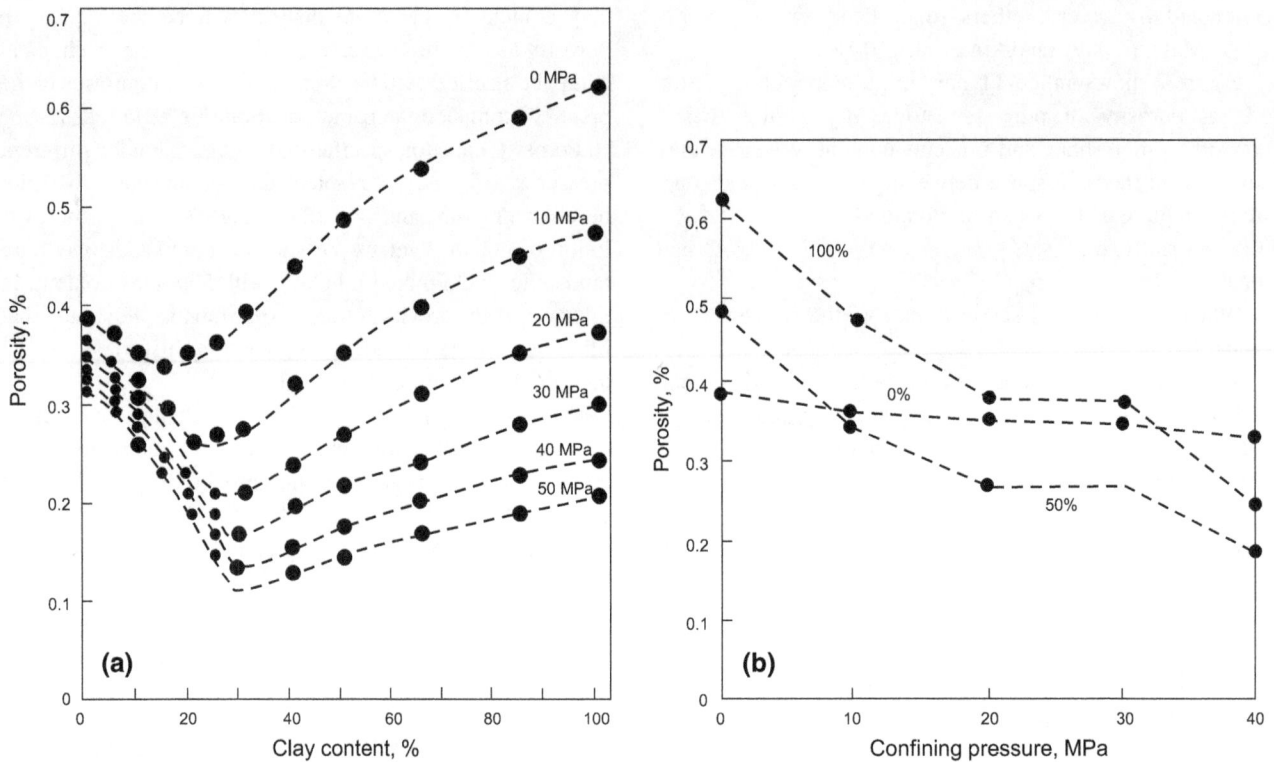

Fig. 3 Rock-physics experiment result maps for the relationship between clay content and porosity (Fig. 3a: relationship map for clay content and porosity under different pressures; Fig. 3b: relationship map for clay content and pressure with different lithologies, Yin 1992)

for directly assigning the initial porosity by either analyzing the grain-size and sedimentary characteristics, or by direct measurements in the laboratory.

Being influenced by the complications of porosity evolution and the reliability of porosity data, the initial porosity fitted by the practical porosity (the porosity value of practical data) always exceeds the critical porosity (it is defined as a porosity that separates mechanical and acoustic behaviors into two distinct domains) of pure shale or sand. Therefore, it is necessary to constrain the initial fitted porosity. Based on a choice of appropriate porosity data, this paper proposes a method for obtaining the initial porosity constrained by the ideal binary clastic mixture model.

First, primary porosity data ought to be selected as reasonably as possible by excluding porosity influenced by tectonic activity, diagenesis effects, and abnormal pressures. Also, because it is calculated by an ideal binary mixture model, the theoretical initial porosity can constrain the fitted initial porosity. For example, when clay particles fill the pore space in a sandstone, the theoretical porosity can be calculated. Specifically, for $0 \leq C \leq \phi_{ss}$, $\phi = \phi_{ss} - (1 - \phi_{sh})C$, whereas for $\phi_{ss} \leq C \leq 1$, $\phi = \phi_{sh}C$. The fitted porosity may be regarded as unreasonable when it exceeds or is less than an appropriate value (e.g., 20%) of

the theoretical porosity of the ideal binary mixture model. As a result, the theoretical porosity that regard the fitted data can be considered as the initial porosity. Besides, when the fitted initial porosity is within an appropriate range, the fitted initial porosity ought to be used.

3.2 Methods

Based on clay content, a method for identifying lithological compaction units was designed to integrate the proposed method above for calculating initial porosity (see flow chart in Fig. 4). The specific procedures followed are outlined below.

Step 1 Data permutation

The data (depth, clay content, porosity) are permuted into units (X(1), X(2), X(3), X(4), ...) according to the clay content from minimum to maximum with a selected window length (e.g., 5%–15%, 15%–25%, 25%–35%, 35%–45%, 45%–55%, ... with a window length of 5%).

Step 2 Data combination

Combine the data units into groups [e.g., X(1), (X(1), X(2)), (X(1), X(2), X(3)), (X(1), X(2), X(3), X(4))] (i.e., 5%–15%, 5%–25%, 5%–35%, 5%–45%, 5%–55%, ...).

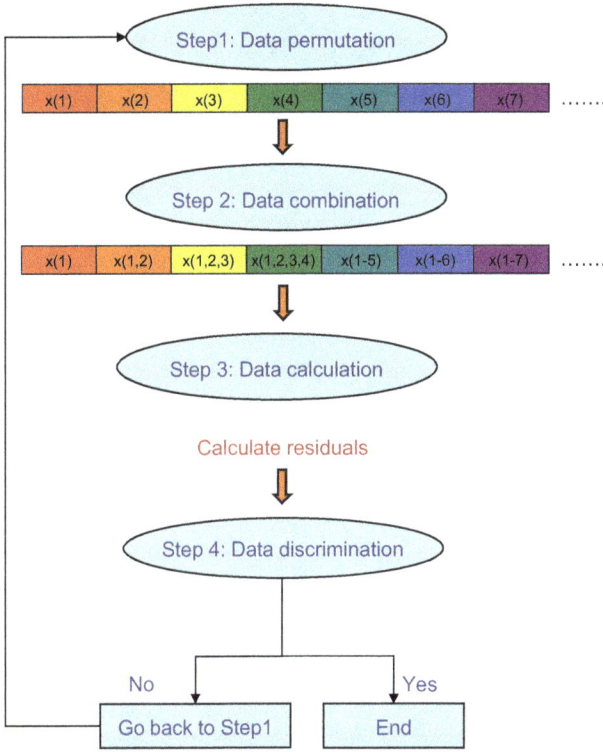

Fig. 4 Flowchart showing automatic classification of clastic compaction units based on clay content

Step 3 Calculation

Fit the original data from different groups according to the exponential model of porosity and depth ($\phi(z) = \phi_0 e^{-cz}$) to obtain the initial porosity (ϕ_0) and compaction coefficient (c) for each group. Also, the average residual (Rm) of each group ought to be returned. Rm is the average residual between practical porosity data and fitted porosity data.

$$Rm = \sum_{i=1}^{n} \left(\varphi_{\text{real}(i)} - \varphi_{\text{fitting}(i)} \right)^2 / n \qquad (1)$$

In the equation, $\varphi_{\text{real}(i)}$, $\varphi_{\text{fitting}(i)}$ are practical porosity and fitted porosity with an index of i. n is the total number of porosity data.

Step 4 Data discrimination

Discriminate the separation group with the minimum average residual among all the groups. Return to Step 2 and calculate the remaining data after the separation group.

For example, if group [X(1), X(2), X(3)] has the minimum average residual, then calculate from X(4) based on step 2, i.e., (X(4)), (X(4), X(5)), (X(4), X(5), X(6)), (X(4), X(5), X(6), X(7)), ….

Step 5 Calculate to the last group unit

The proposed method regards the size of the average residual as a criterion to judge the precision of the

compaction correction for the study. To some extent, the minimum average residual corresponds to the highest precision of the compaction recovery result that we can obtain. Also, the size of the average residual is a criterion that can classify clastic lithological compaction units based on clay content. The method followed in step 4 shows that the separation group, along with the minimum average residual, can be considered as a compaction unit. On the whole, the average residual derived by applying the clastic lithological compaction unit classification technology is less than or equal to that obtained when no clastic lithological compaction units are classified. This means that the precision of the compaction correction based on the proposed methods can be considerably improved.

3.3 Case study

Programming for the proposed method of clastic compaction unit classification based on clay content has been conducted successfully. A large amount of well log data, including neutron porosity and clay-content measurements from the Qikou Sag, was employed to test the feasibility of the proposed method. The well log data were derived from the clastic layers of the Minghuazhen, Guantao, Dongying, and Shahejie Formations. To illustrate the success of this process, consider a typical well. First, porosity data needed to be selected. Abnormal porosity data or porosities that obviously deviate from the normal trend, probably within the abnormal pressure section, were excluded. Then, the selected data containing porosity, clay content, and depth were assessed by the program.

Figure 5 and Table 1 show the results. The blue points are input porosity data with various clay contents, whereas the red points are fitted porosity data. Before applying the proposed method, the average residual between the original porosity and fitted porosity was about 47.3. The automatic compaction unit classification method divided the lithologies into three categories. The average residual of each category is less than 47.3. The practical application illustrates the feasibility of the proposed method and the need to lithologically classify the various compaction units.

4 Integrated well- and seismic-data compaction recovery method

To propagate well-based compaction information into the data assessment, two compaction recovery methods that integrate well and seismic data are proposed. One is based on 3-D clay-content data, and the other one is a 3-D interpolation correction based on compaction degree parameters.

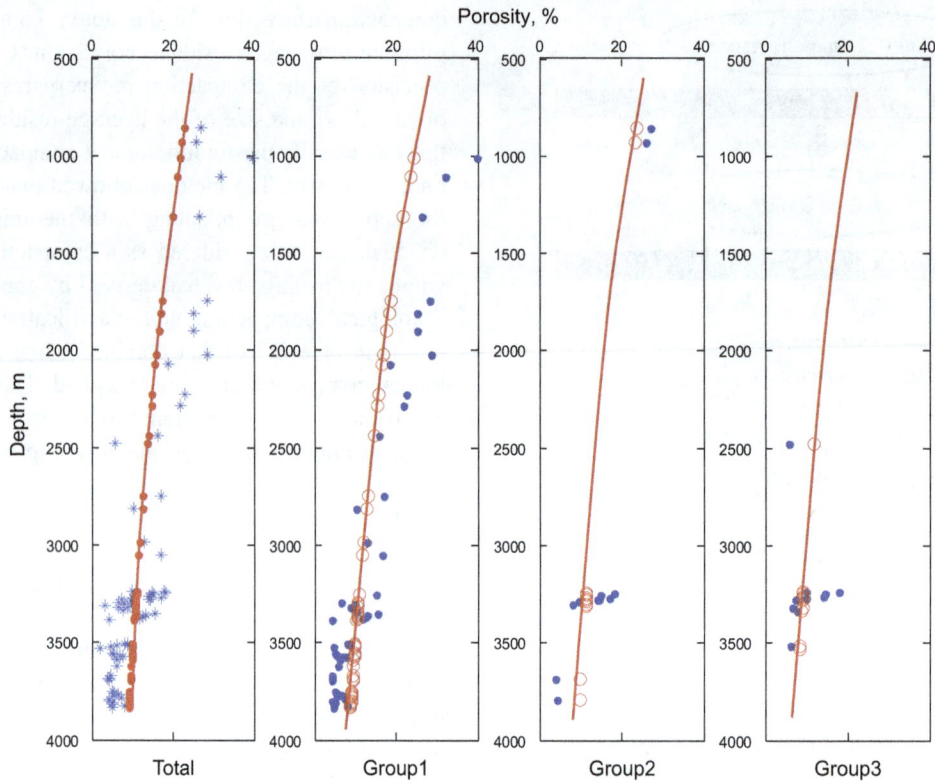

Fig. 5 Automatic classification of clastic lithological compaction units based on the clay content of a single well

Table 1 Automatic classification of clastic lithological compaction units

	Lower bound of clay content, %	Upper bound of clay content, %	Initial porosity, %	Compaction coefficient	Fitted residual
Original	3	48	30	0.00032	47.3
Group 1	3	19.9	35.9	0.00036	25.8
Group 2	19.9	25.5	31	0.00030	22.3
Group 3	25.5	48	27.4	0.00038	26.3

4.1 Compaction recovery method based on 3-D clay-content data

Lithological compaction units identified in well data were determined by assessing their respective clay contents. Then, 3-D clay-content seismic inversion data were incorporated to regionally propagate the 1-D compaction information from the wells in plan view. The detailed calculation procedures are as follows (Fig. 6):

Step 1 Acoustic impedance inversion

First, a seismic acoustic impedance inversion should be conducted.

Step 2 Clay-content inversion based on a probabilistic neural network

This method is based on the statistical relationship between acoustic impedance from seismic data and clay content from wells. Usually, the acoustic impedance and the sand content correlate positively. Therefore, a probabilistic neural network (PNN) method is applied to acquire a nonlinear statistical relationship between the acoustic impedance of a borehole side seismic trace and the sand content derived from a well log curve. A 3-D sand-content inversion cube then can be obtained by propagating the statistical relationship to the whole acoustic impedance inversion cube. A clay-content inversion cube then can be conveniently produced.

Step 3 Lithological compaction units determined for multiple wells

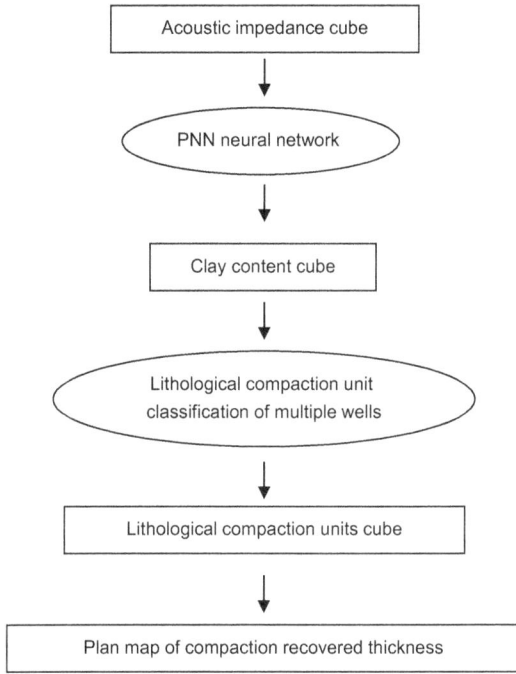

Fig. 6 Flowchart of compaction recovery method based on 3-D clay-content data

The method for characterizing clastic compaction units based on clay content was then applied by assessing integrated well log data for multiple wells, including porosity and clay-content data, to classify lithological compaction units.

Step 4 Calculation of a lithological compaction unit cube

Based on the clay-content range of the classified lithological compaction units (Step 3), a clay-content inversion cube can then be transferred into a lithological compaction unit cube containing a set of compaction units.

Step 5 Calculation of compaction history

Based on the porosity-depth exponential model, the initial porosities and compaction coefficients for the various compaction units, as along with the 3-D lithological compaction unit cube in the depth domain, are integrated to calculate paleo-geomorphic thickness. The programming designed to do this involves two main components.

(1)   During the process of stratal compaction, the rock matrix volume remains constant. The porosity-depth exponential model of $\phi(z) = \phi_0 e^{-cz}$ was applied to calculate H_s for the strata matrix thickness (z_1, z_2 separately represent the top and bottom depth of the present strata; i.e., the top and bottom depth of the corresponding compaction units; ϕ_0, c represent the initial porosity and compaction coefficient of the compaction unit)

$$H_s = z_2 - z_1 - \frac{\phi_0}{c}(e^{-cz_1} - e^{-cz_2}). \quad (2)$$

The strata thickness during the compaction process can be calculated by:

$$H = z_4 - z_3 = H_s + \frac{\phi_0}{c}(e^{-cz_3} - e^{-cz_4}), \quad (3)$$

where z_3, z_4 separately represent the top and bottom depth of the strata during the compaction process; H is the corresponding strata thickness; when z_3 is zero, the H represents the paleo-thickness).

(2)   As the initial porosity and compaction coefficient of each lithological compaction unit are available, they are combined with the top and bottom depths to calculate the thickness of each lithological compaction unit during the deposition process. Finally, the paleo-geomorphic thickness can be recovered in plan view.

4.2 Compaction recovery method based on the plan-view interpolation of compaction correction parameters

The compaction correction parameters of wells were calculated by applying the compaction unit characterization method based on clay content. Then, the compaction correction parameters are interpolated laterally. It is both economical and efficient to develop a compaction recovery thickness map by integrating stratal thickness and the plane map of the compaction parameter. The detailed calculation procedure follows (Fig. 7).

Step 1 Data quality control

The clay-content data come in two types: well log data and interpretations based on well log data. It is necessary to calibrate these two data sets to ensure that their differences can be ignored. The well data should be abandoned if the clay content determined for these two types differs considerably.

Step 2 Well-based lithological compaction unit characterizations

The clastic compaction unit characterization method based on clay content was applied by using quality-controlled porosity and clay-content data to classify lithological compaction units.

Step 3 Well-based compaction correction degree calculation

The compaction correction degree value for each well should be calculated (i.e., compaction corrected thickness/ original strata thickness).

```
┌─────────────────────────────────────────────┐
│        Calibration of data from wire-line log │
│          and well-logging interpretation      │
└─────────────────────────────────────────────┘
                      │
                      ▼
        ╭─────────────────────────────╮
        │  Classification of well-based │
        │   lithological compaction unit │
        ╰─────────────────────────────╯
                      │
                      ▼
┌─────────────────────────────────────────────┐
│     Calculation of compaction correction parameter │
└─────────────────────────────────────────────┘
                      │
                      ▼
            ╭──────────────────────╮
            │  Interpolating method │
            ╰──────────────────────╯
```

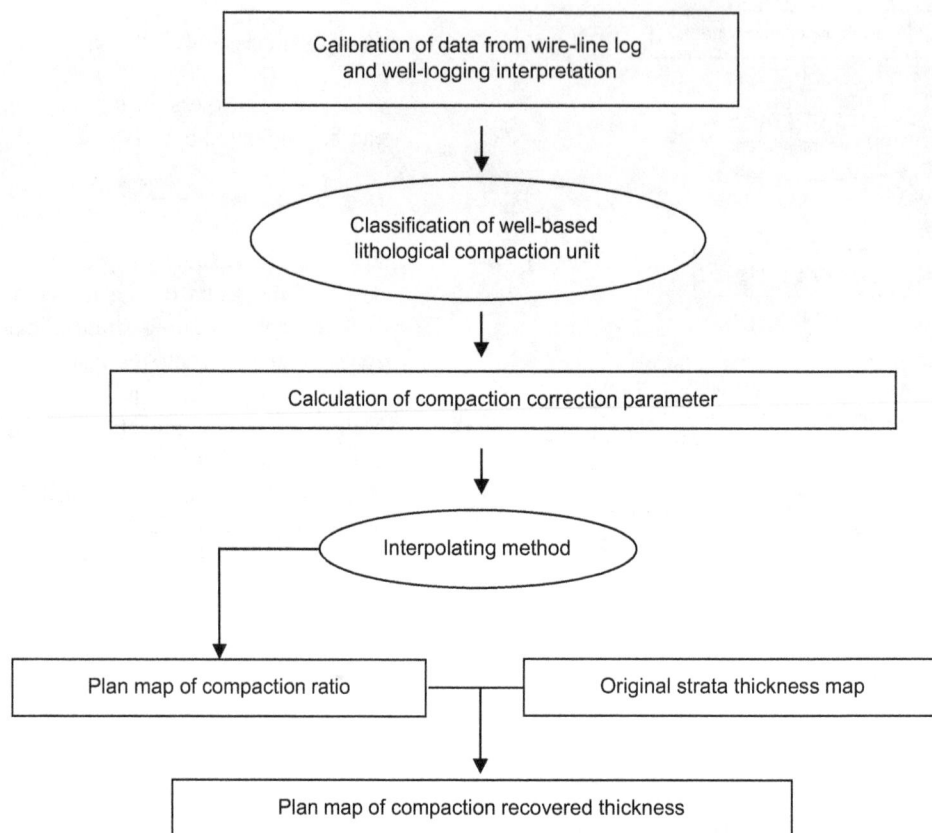

Fig. 7 Flowchart of the compaction recovery method based on a plan-view interpolation of the compaction correction degree parameter

Step 4 Plan-view interpolation

When the compaction recovery parameters for the wells have been obtained, an interpolation method, e.g., Kriging, should be employed to obtain a plan-view distribution map of the compaction correction parameter.

Step 5 Mapping compaction recovered thickness

A time domain thickness map of target horizons can be obtained through seismic data interpretation. The thickness map in the time domain then can be transferred into the depth domain with the use of a time-depth model. By multiplying the compaction correction degree parameters in the map by the thicknesses in the depth domain, a plan map of compaction recovered thicknesses eventually can be obtained.

5 Case studies

5.1 Case study of a compaction recovery method based on 3-D clay-content data

A 3-D seismic survey, covering an area of 100 km² in the Qinan Sag, contains a complete series of clastic Tertiary deposits, including the Kongdian, Shahejie, Dongying, Guantao, and Minghuazhen Formations (from bottom to top). The fracture system within the sag is not developed within the seismic survey, so the influence on porosity evolution attributed to the fracture system largely can be eliminated.

5.1.1 Identification of lithological compaction units within the wells

The porosity data were selected with the aid of a cross-plot of porosity and depth values. The simple principle behind the selection of porosity data is to exclude obviously abnormal values such as those from sections where the pressure was unusually high. The porosity data for a group of wells were jointly assessed by the proposed compaction unit classification method. Figure 8 and Table 2 show the resulting lithological compaction units separated into three groups. As no porosity data were available for mudstone in the study area, lithologies with clay contents ranging from 57% to 100% were regarded as mudstone. According to a previous local study (Gui 2008), the compaction coefficient of mudstone is 0.000686.

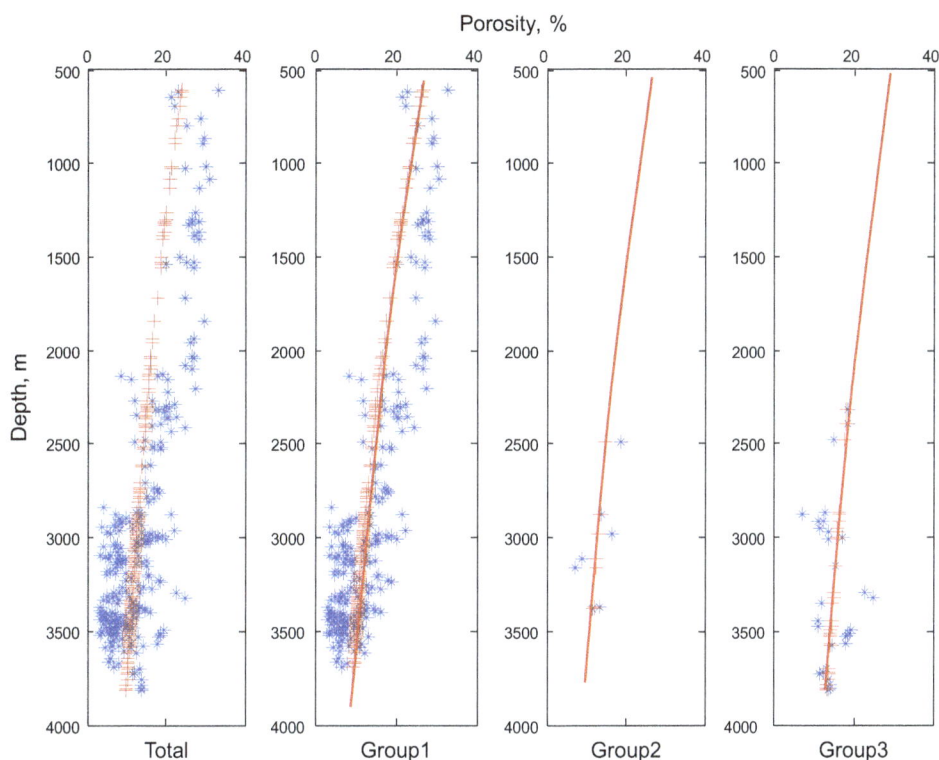

Fig. 8 Automatic classification of the clastic compaction units based on clay content for multiple wells

Table 2 Automatic classification of the clastic compaction units in the study area

	Lower bound of clay content, %	Upper bound of clay content, %	Initial porosity, %	Compaction coefficient	Fitted residual
Group 1	0	35	32.6	0.0003333	23.2
Group 2	35	40.5	24.9	0.0002155	10.3
Group 3	40.5	57	29.3	0.0002049	15.7
Mudstone	57	100	47.1	0.000686	

5.1.2 Calculation of a lithological compaction unit cube in the depth domain

The study region consists of a 3-D post-stack seismic data volume. To calculate the compaction of lithological units, such as the Es_1^z member, for example, the explicit calculation procedures are as follows.

(1) Seismic acoustic impedance inversion was conducted on the targeted Es_1^z member horizon in the seismic data.

(2) Based on well log data (acoustic impedance and sand content), the nonlinear relationship between the sand content and acoustic impedance was obtained by applying a neural network method.

(3) Based on a nonlinear relationship between acoustic impedance and sand content, the sand content within

the 3-D seismic cube was calculated accordingly (Fig. 9a).

(4) According to a 3-D time-depth model for the region, the sand content in the 3-D seismic volume was transformed from the time domain into a 3-D clay-content seismic volume in the depth domain (Fig. 9b).

(5) Finally, by applying the lithology compaction unit information from the Table 2, a 3-D lithology compaction unit seismic cube in the depth domain was derived from the clay-content cube. Figure 9c shows that there are four distinct lithology compaction units.

(6) The compaction correction thickness (Fig. 10) can be calculated based on the related procedure in Step (5).

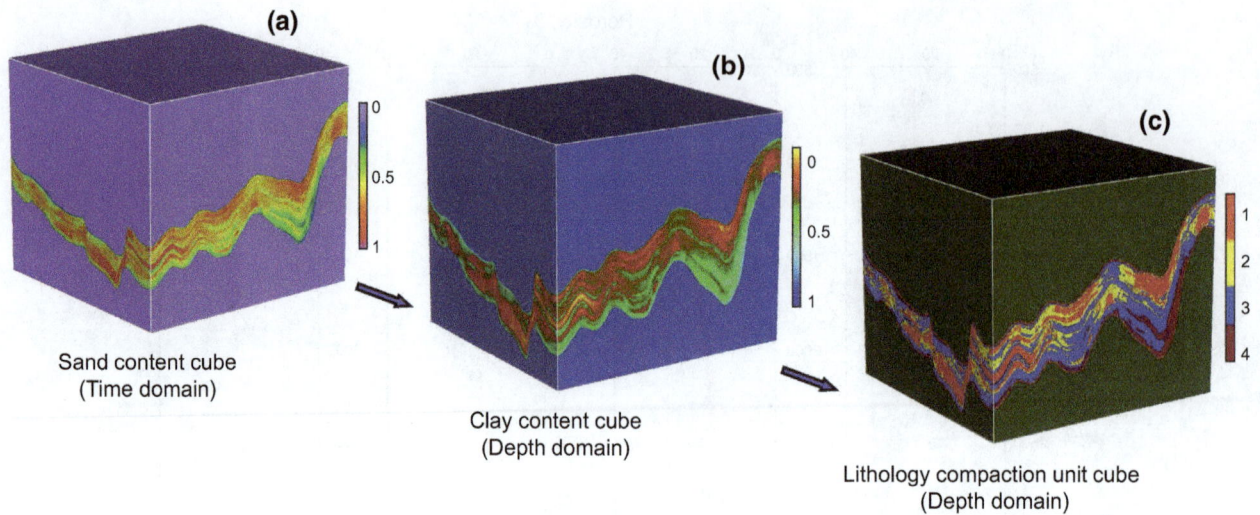

Fig. 9 Calculation of the lithology compaction unit cube for Es_1^z

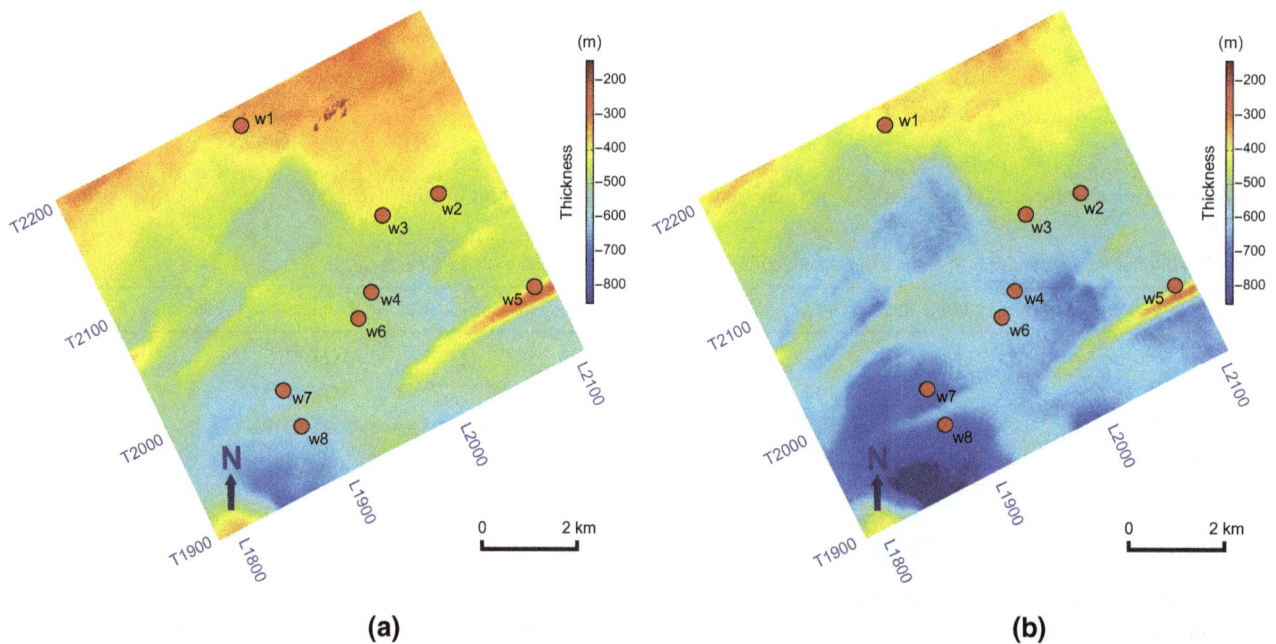

Fig. 10 Plan view of the recovered paleo-geomorphic thickness of the Es_1^z member of the study area (**a** without compaction recovery, **b** after compaction recovery)

5.1.3 Analysis of compaction recovery results

When compaction recovery was applied, the difference in paleo-geomorphic thickness was remarkable. In terms of tectonic geometry, the trend of the post-compaction recovery was almost the same as the original geometry. The inverse phenomenon of tectonic geometry does not occur. The trend in thickness change only presents a slight difference. Relatively high positions become higher, while shallow positions become deeper. Take the geomorphology of W7 and W8, for example. After compaction recovery, the paleo-geomorphology becomes deeper and has steeper slopes. In this case,

the Es_1^z member of the study area does not belong to a region of facies change in which the vertical lithology varies rapidly. Therefore, the "topographic inverse" phenomenon of tectonic geometry would not occur after compaction recovery. Yet as a whole, the test of the method based on 3-D clay-content data has proven its feasibility.

5.2 Case study of the plan-view interpolation-based method

A 400 km^2 3-D seismic survey located in the Qibei Sag contains a complete series of clastic Tertiary deposits,

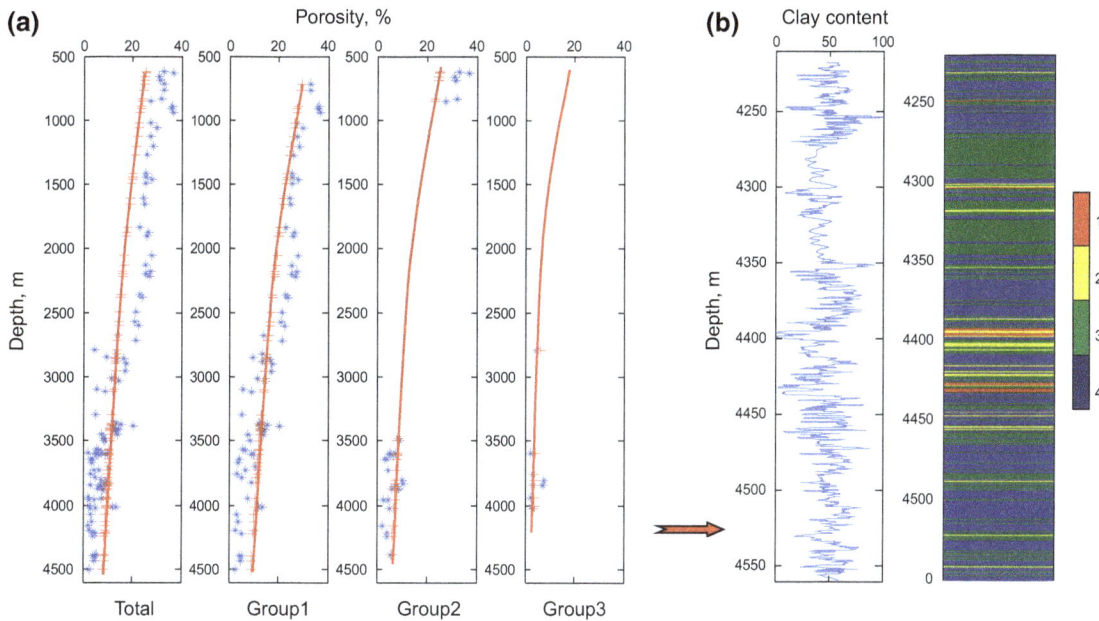

Fig. 11 Calculation of compaction degree for a single well

including the Kongdian, Shahejie, Dongying, Guantao, and Minghuazhen Formations (from bottom to top). Again, the regional fracture system is not developed within the seismic survey, so the influence on porosity evolution attributed to the fracture system largely can be eliminated.

5.2.1 Calculation of the compaction parameter for a single well

Well Binsh1 is used for an example here because the clay-content data from the well log curves match those from interpretations of the well logs. First, the method for characterizing clastic compaction units based on clay content was applied to identify lithological compaction units (Fig. 11a). According to the clay-content range of the classified lithological compaction units, the clay-content curves can be transferred into a lithological compaction unit curve (Fig. 11b). The compaction recovery thicknesses for each of the different compaction units were calculated, and then the stratal thicknesses of target horizons were obtained following compaction corrections. Finally, by dividing the compaction corrected thickness by the current stratal thickness, the compaction correction degree can be obtained.

5.2.2 Calculation and analysis of compaction recovered thickness

In the same way, the compaction correction parameters for other wells were calculated on the basis of quality-controlled data. Then, the plan-view distribution map of compaction correction parameters can be obtained by employing the Kriging interpolation method (Fig. 12b).

The depth domain thickness map of member Es_2 can be obtained through time-depth conversion (Fig. 12a). By multiplying the plan-view distribution map of the compaction correction parameters by the thickness map in the depth domain, a plan-view map of the compaction recovered thickness of the Es_2 member can eventually be calculated (Fig. 12c). Before applying the compaction correction, the area of well Binsh1 exhibited flat topography, similar to the topographic features of well Binsh18 (Fig. 12a). After compaction recovery (Fig. 12c), the area of well Binsh1 was used for a different topographic assessment which obviously was lower than the area of well Binsh18. It can be interpreted that the present strata thicknesses in the area of wells Binsh1 and Wellbinsh18 area are nearly the same, whereas well Binsh1 contains a greater proportion of shale layers than well Binsh18. After the application of the compaction correction, the corrected thickness of well Binsh1 was greater than that of well Binsh18; i.e., the "topography inverse" phenomenon occurs. The location of well Binsh1 can be interpreted as the center of the lake basin, which matches well with regional sedimentary features (Fig. 12d). Also, the paleo-geomorphology corresponds to depositional distribution characteristics. With the aid of the 3-D interpolated compaction correction method that is based on the parameters of the compaction correction, the precision of the compaction correction also has been improved.

(a)

(b)

(c)

(d)

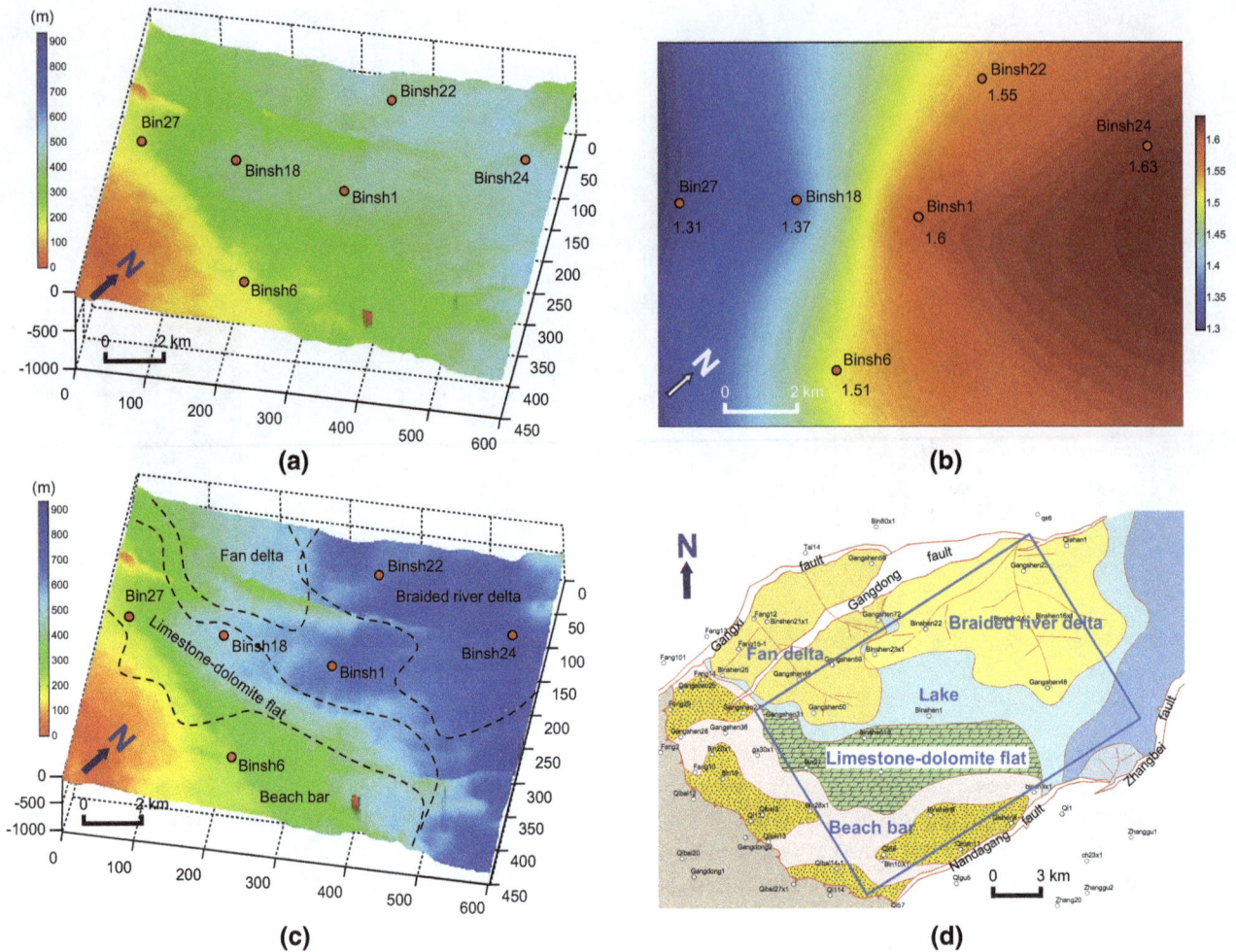

Fig. 12 Plan views of the paleo-geomorphic recovered thickness of the Es_2 formation in the study area (**a** paleo-geomorphic thickness before compaction recovery; **b** plan map of compaction correction degree; **c** paleo-geomorphic thickness after compaction recovery; **d** depositional map of Es_2 formation)

5.3 Comparison

Both of proposed integrated well- and seismic-data compaction recovery methods have merits and disadvantages. The most helpful method can be chosen according to specific seismic or geological conditions and data. When the seismic or geological conditions of a clastic survey are favorable, the precision of the clay-content inversion can be guaranteed. The compaction recovery method, based on 3-D clay-content inversion data, is able to ensure the precision of compaction characterizations that are some distance away from wells. Compared with the other method based on interpolation, this method has its advantages in plan-view propagation with more reliability. Yet, as the impedance inversion, clay-content inversion, and time-depth conversion are all involved in this method, the accumulated errors cannot be ignored. Also, this method is time-consuming and complicated to conduct within an industrial research setting. However, it can be regarded for

reference as a compaction recovery method that integrates well and seismic data.

The plane interpolation of the compaction correction method (based on compaction correction parameters) makes full use of compaction information from wells. When the number of wells is sufficient, the plan compaction characteristics derived from the interpolation method are credible. Also, as few calculation procedures are involved in this method, the accumulated errors are small. Compaction characterizations from wells largely can be retained. On the whole, this method based on the plan-view interpolation of compaction correction parameters is economical and efficient to employ—with obvious value for industrial applications.

6 Conclusions

(1) It is reasonable and necessary to develop a porosity-clay content-depth compaction model.

(2) The research precision of the compaction correction can be effectively improved by applying the method of classifying clastic compaction units based on clay content.

(3) The proposed compaction correction method, based on the plan-view interpolation of the compaction correction parameters, can retain and largely make full use of compaction information from wells. The geomorphological attributes closest to the real paleo-geomorphology can be obtained. The efficiency and feasibility of the process make industrial applications possible.

(4) In addition to mechanical compaction, several other factors, such as the erosion effect, tectonic activity, diagenesis, and abnormal pressure, can have considerable impacts on porosity evolution. Porosity data influenced by these factors should be excluded in the study. Also, a compaction recovery method that can only consider mechanical compaction oversimplifies the compaction process. The precision of the compaction recovery work is inevitably influenced. When more data are involved (e.g., geochemical data), the recovered topography that is closer to the real paleo-geomorphology can possibly be obtained.

Acknowledgements We thank Dr. Tapan Mukerji from the Geophysics Department at Stanford University for his constructive suggestions. We particularly appreciate his great help and guidance.

References

Alberts L, Weltje GJ. Predicting initial porosity as a function of grain-size distribution from simulations of random sphere packs, IAMG Conference, paper. 2001.

Athy LF. Density, porosity, and compaction of sedimentary rocks. AAPG Bull. 1930;14(1):1–24.

Atkins JE, McBride EF. Porosity and packing of Holocene river, dune and beach sands. AAPG Bull. 1992;76(3):339–55.

Beard DC, Weyl PK. Influence of texture on porosity and permeability of unconsolidated sand. AAPG Bull. 1973;57(2):349–69.

Dvorkin J, Gutierrez MA, Grana D. Seismic reflections of rock properties. Cambridge: Cambridge University Press; 2014.

Falvey DA, Middleton MF. Passive continental margins: evidence for a prebreakup deep crustal metamorphic subsidence mechanism. Oceanol Acta. 1981;4:103–14.

Fawad M, Mondol NH, Jahren J. Mechanical compaction and ultrasonic velocity of sands with different texture and mineralogical composition. Geophys Prospect. 2011;59(4):697–720.

Gui BL. The paleo-geomorphologic reconstruction of Es_2 in Zhuangxi area, Bohai Bay Basin. Master of Science thesis. Beijing: China University of Geosciences; 2008. (**in Chinese**)

Harbaugh JW, Watney WL, Rankey EC, et al. Numerical experiments in stratigraphy: recent advances in stratigraphic and sedimentologic computer simulations. SEPM Special Publication. 1999. No. 62.

He JQ, Zhou ZY, Jiang XG. Optimum estimation of the amount of erosion by porosity data. Pet Geol Exp. 2002;24(6):561–4 (**in Chinese**).

Li CL, Kong XY. A new equation of porosity-depth curve. Xinjiang Pet Geol. 2001;22(2):152 (**in Chinese**).

Li SH, Wu CL, Wu JF. A new method for compaction correction. Pet Geol Exp. 2000;22(2):110–3 (**in Chinese**).

Li SH. The compaction correction based on the principle keeping formation grain volume and mass constant. Pet Geol Exp. 2001;23(3):357–60 (**in Chinese**).

Mavko G, Mukerji T, Dvorkin J. The rock physics handbook. Cambridge: Cambridge University Press; 2009.

Pedersen BK, Norda K. Petrophysical evalutation of thin beds: a review of the Thomas–Stieber approach. Course 24034 Report. Norwegian University of Science and Technology. 1999.

Pryor WA. Permeability-porosity patterns and variations in some Holocene sand bodies. AAPG Bull. 1973;57(1):162–89.

Qi JH, Yang Q. A discussion about the method of decomposition correction. Pet Geol Exp. 2001;23(3):351–6 (**in Chinese**).

Ramm M, Bjørlykke K. Porosity/depth trends in reservoir sandstones: assessing the quantitative effects of varying pore-pressure, temperature history and mineralogy, Norwegian shelf data. Clay Miner. 1994;29(4):475–90.

Schon JH. Physical properties of rocks: fundamentals and principles of petrophysics. Amsterdam: Elsevier; 2004.

Syvitski JPR, Bahr DB. Numerical models of marine sediment transport and deposition. Comput Geosci. 2001;27(6):617–8.

Thomas EC, Stieber SJ. The distribution of shale in sandstone and its effect upon porosity. In: Transactions of the 16th auunal logging symposium of the SPWLA, paper T. 1975.

Thomas EC, Stieber SJ. Log derived shale distributions in sandstone and its effect upon porosity, water saturation, and permeability. In: Transactions of the 6th formation evaluation symposium of the canadian well logging society. 1977.

Yang Q, Qi JF. Method of delaminated decompaction correction. Pet Geol Exp. 2003;25(2):206–10 (**in Chinese**).

Yin H. Acoustic velocity and attenuation of rocks: isotropy, intrinsic anisotropy, and stress-induced anisotropy. Ph.D. thesis. Stanford University. 1992.

Zaimy SS, Rasaei MR. Reconstruction of porosity distribution for history matching using genetic algorithm. Pet Sci Technol. 2013;31(11):1145–11158.

The potential of domestic production and imports of oil and gas in China: an energy return on investment perspective

Zhao-Yang Kong[1] · Xiu-Cheng Dong[1] · Qian Shao[2] ·
Xin Wan[3] · Da-Lin Tang[4] · Gui-Xian Liu[1]

Abstract Concerns about China's energy security have escalated because of the country's high dependency on oil and gas imports, so it is necessary to calculate the availability of domestic oil and gas resources and China's ability to obtain foreign energy through trade. In this work, the calculation was done by using the energy return on investment (EROI) method. The results showed that the $EROI_{stnd}$ (i.e., standard EROI) of China's oil and gas extraction decreased from approximately 17.3:1 in 1986 to 8.4:1 in 2003, but it increased to 12.2:1 in 2013. From a company-level perspective, the $EROI_{stnd}$ differed for different companies and was in the range of (8–12):1. The $EROI_{2,d}$ (EROI considering energy outputs after processed and direct energy inputs) for different companies was in the range of (3–7):1. The EROI of imported oil ($EROI_{IO}$) declined from 14.8:1 in 1998 to approximately 4.8:1 in 2014, and the EROI of imported natural gas ($EROI_{ING}$) declined from 16.7:1 in 2009 to 8.6:1 in 2014. In 2015, the $EROI_{IO}$ and $EROI_{ING}$ showed a slight increase due to decreasing import prices. In general, this paper suggests that from a net energy perspective, it has become more difficult for China to obtain oil and gas from both domestic production and imports. China is experiencing an EROI decline, which demonstrates the risk in the use of unsustainable fossil resources.

Keywords EROI · Oil and gas extraction · Imported oil · Imported natural gas · China

1 Introduction

Few issues, if any, are as fundamentally important to industrial societies and their economies as the future oil and gas supplies (Cleveland 2005; Gagnon et al. 2009). Oil and gas provide nearly 60 % of the world's energy (BP 2014). Global food production and most economies rely heavily on oil and gas, and historical restrictions on the availability of oil have had major economic impacts (Munasinghe 2002). China has become the world's largest energy consumer, with consumption increasing from 16.7×10^{12} in 1978 to 99.8×10^{12} MJ in 2012 (National Bureau of Statistics of China, 2014; Fan et al. 2015). Together, oil and gas comprise approximately 25.5 % of the consumption of primary energy resources (Safronov and Sokolov 2014). Due to China's limited domestic production capacity, however, increasing amounts of oil and gas have been imported from counties such as Qatar, Indonesia, Malaysia, Russia, and Australia (Kong et al. 2015). Over the last 7 years, China's dependency on imported oil (IO) has increased by 21.5 % annually, and in 2013, it reached 59 percent of total use (Fig. 1). Moreover, China's dependence on imported natural gas (ING) is also rising (Fig. 1) and, according to the BP Energy Outlook 2030, will reach over 40 percent of total use by 2030 (Kong et al. 2015). Thus, oil and gas security has become an issue that cannot be ignored. If an interruption in energy imports

✉ Xiu-Cheng Dong
  dongxiucheng@cup.edu.cn

[1] School of Business Administration, China University of Petroleum (Beijing), Beijing 102249, China

[2] School of Business, Tianjin University of Finance and Economics, Tianjin 300222, China

[3] Tangshan Iron and Steel Group Co., Ltd, Tangshan 261000, China

[4] China Petroleum Enterprise Association, Beijing 100724, China

Edited by Xiu-Qin Zhu

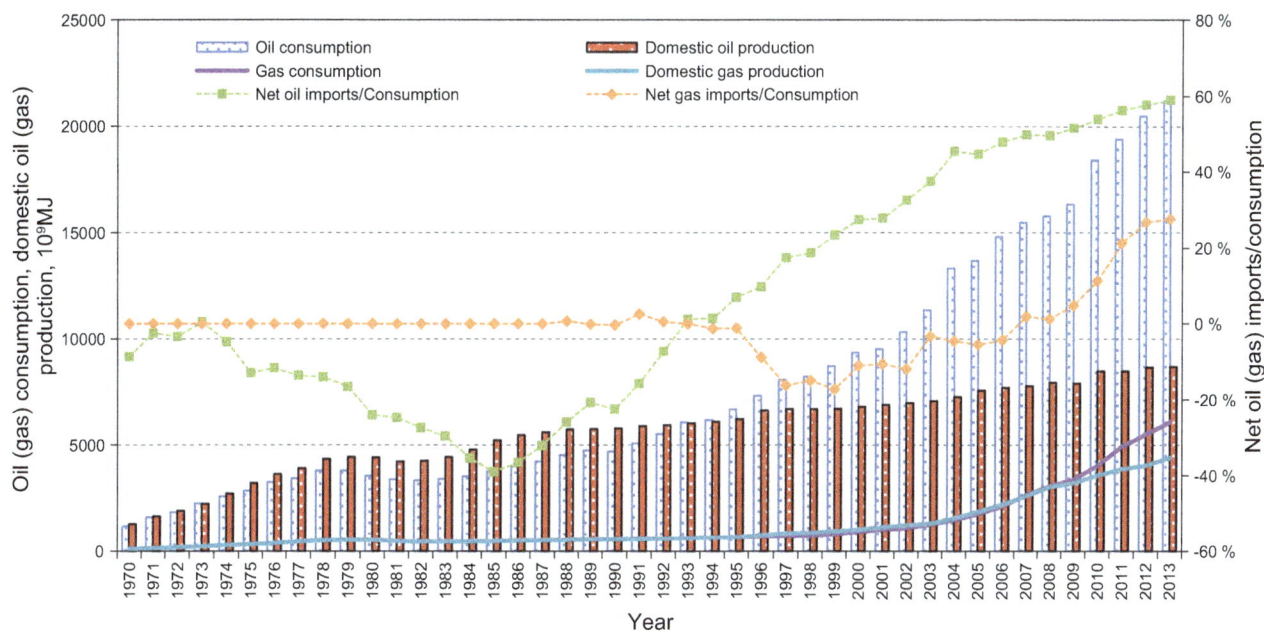

Fig. 1 China's dependence on foreign oil and gas

occurs or the import price increases, China's economy will be seriously affected. For example, as a result of the 1973 oil shock, the world economy endured the hitherto worst recession in postwar history (Kesicki 2010). Therefore, to ensure the country's oil and gas security, it is necessary to calculate the availability of domestic oil and gas resources and China's ability to obtain foreign oil and gas through trade.

The energy return on investment (EROI) is a useful approach for assessing the productive availability of an energy source (Heun and Wit 2012). It is the ratio of energy that is produced by a process to the energy that is consumed in carrying out that process (Gagnon et al. 2009). If the EROI of a fuel is high, then only a small fraction of the energy produced is required to maintain production, and the majority of that produced energy can be used to run the general economy. In contrast, if the EROI is very low, the majority of the energy produced must be used to ensure continued energy production, and very little net energy is available for useful economic work. Thus, high EROI fuels are vital to economic growth and productivity (Gagnon et al. 2009).

Unfortunately, the peer-reviewed literature in this field has paid only minimal attention to the EROI of oil and gas extraction (OGE) in China. To the best of our knowledge, only three papers have examined the EROI of OGE in China. In 2011, Hu et al. (2011) derived an EROI of the Daqing oil field, the largest oil field in China. They estimated that its EROI was 10:1 in 2001 but declined to 6.5:1 in 2009. In 2013, Hu et al. (2013) found that the EROI for China's OGE fluctuated from 12:1 to 14:1 in the mid-1990s

and declined to 10:1 during the period of 2007–2010. In 2014, Xu et al. (2014) forecasted that the Daqing oil field's EROI would continuously decline from 7.3:1 to 4.7:1.

In this paper, we address the EROI of OGE in China not only from an industry-level perspective, but also from a company-level perspective. An analysis of the EROI of oil companies could provide some information about energy inputs and thus serve as a reference for policymakers and investors. Besides, we prefer to analyze EROI not only at the mine mouth (the conventional approach) but also at the refinery, because crude oil cannot be directly used by cars and needs be processed. Of course, oil processing consumes energy, which should also be considered in energy inputs.

Lambert et al. (2014) studied the EROI of imported oil ($EROI_{IO}$) for 12 developing countries, including China. They have found that most developing nations have EROI values below 8–10. As a large gas importer, it is necessary to estimate the EROI of imported natural gas ($EROI_{ING}$) of China. Through this study, we aimed to answer four questions:

(1) What is the EROI of China' oil and gas extraction in 1985–2012?

(2) What are the EROI values of China's four oil companies: CNPC (China National Petroleum Corporation), Sinopec (Sinopec Group), CNOOC (China National Offshore Oil Corporation), and Yanchang (Shanxi Yanchang Petroleum (Group) Co., Ltd.)?

(3) Are the EROI values of these four oil companies lower or higher than the national average value?

(4) What are the $EROI_{IO}$ and $EROI_{ING}$ values of China?

2 EROI methodology and data for China's domestic production of oil and gas

2.1 EROI methodology for domestic production of oil and gas

EROI can broadly be described as the ratio of the energy made available to society through a certain process and the energy cost to implement this process (Cleveland and O'Connor 2011; Lundin 2013). The general equation for EROI is given in Eq. (1):

$$\text{EROI} = \frac{\text{Energy produced (outputs)}}{\text{Energy consumed (inputs)}} \tag{1}$$

The numerator is the sum of all energy produced in a given timeframe, and the denominator is the sum of the energy costs. EROI is typically calculated without discounting for time. Because the numerator and denominator are usually assessed in the same units, the ratio derived is dimensionless and often expressed as EROI: 1 in text (Lundin 2013), e.g., 10:1. This implies that a particular process yields 10 joules on an investment of 1 J.

Some previous EROI analyses have generated a wide variety of results, including apparently conflicting results, when applied to the same energy resource. The reasons for these differences are not limited to intrinsic variations in energy resource quality, extraction technology, and varying geology but also include methodological issues including different boundaries of analysis, different methods used to estimate indirect energy inputs, and issues related to energy quality (Hu et al. 2013).

In order to formalize the analysis of EROI, Mulder and Hagens (2008) established a consistent theoretical framework for EROI analysis that encompasses the various methodologies. Murphy et al. (2011) propose a more explicit two-dimensional framework for EROI analysis that describes three boundaries for energy analysis and five levels of energy inputs, as shown in Table 1.

In Table 1, the numbers "1," "2," and "3" describe the boundary for energy outputs, i.e., where the analysis is terminated (mine mouths, refinery or point of use), while the "d," "i," "aux," "lab," or "env" in subscript refer to the abbreviations for different types of inputs considered: They are direct energy (d) used on site, indirect energy (i) used to purchase material inputs constructed offsite such as steel for sand pipes, embodied energy in the wages of labor (lab), energy afforded by governmental services in the public sector (aux), and energy embodied in environmental costs for assessment (env), respectively.

Because most EROI analyses account for both direct and indirect energy and material inputs, but not for labor or environmental costs, Murphy et al. (2011) deem this boundary to be the standard EROI and assign it the name $\text{EROI}_{\text{stnd}}$. Using the standard calculation, we have the following equation:

$$\text{EROI}_{\text{stnd}} = \frac{E_O}{E_d + E_i} \tag{2}$$

where E_o is all energy outputs, J; E_d, and E_i represent the total input and direct input, J, respectively, of different types of energy. The challenge is that the indirect energy inputs are rarely available as physical energy units. Rather, the data are available in monetary units as, e.g., investments in industrial equipment. Thus, Eq. 3 is used to complete the EROI analysis:

$$\text{EROI}_{\text{stnd}} = \frac{E_O}{E_d + M_i \times E_{\text{ins}}} \tag{3}$$

where M_i represents the indirect inputs in monetary terms and E_{ins} expresses the energy intensity of a dollar input for indirect components.

Other approaches (e.g., including environmental) can be conducted as sensitivity analyses, which will examine how changing variables affect the outcome. If both environmental and indirect costs are considered, the EROI can be expressed as $\text{EROI}_{1,i+\text{env}}$. The critical point is to clarify what is included in the analysis (Murphy et al. 2011).

Table 1 Two-dimensional framework for EROI analysis	Levels for energy inputs	Boundary for energy outputs		
		1. Extraction	2. Processing	3. End-use
	Direct energy and material inputs	$\text{EROI}_{1,d}$	$\text{EROI}_{2,d}$	$\text{EROI}_{3,d}$
	Indirect energy and material inputs	$\text{EROI}_{\text{stnd}}$	$\text{EROI}_{2,i}$	$\text{EROI}_{3,i}$
	Indirect labor consumption	$\text{EROI}_{1,\text{lab}}$	$\text{EROI}_{2,\text{lab}}$	$\text{EROI}_{3,\text{lab}}$
	Auxiliary services consumption	$\text{EROI}_{1,\text{aux}}$	$\text{EROI}_{2,\text{aux}}$	$\text{EROI}_{3,\text{aux}}$
	Environmental	$\text{EROI}_{1,\text{env}}$	$\text{EROI}_{2,\text{env}}$	$\text{EROI}_{3,\text{env}}$

2.2 China's oil and gas extraction data

2.2.1 Energy outputs

The National Bureau of Statistics of China provides data on energy outputs and energy inputs for OGE (Table 2) (National Bureau of Statistics of China 2014). Open access data are available from 1985 onward, so we calculated the $EROI_{OGC}$ from 1985 to 2012. This output was converted to heat units using the values in Table 3. Thus, we can obtain energy output as heat equivalents (Fig. 2).

2.2.2 Energy inputs

Direct energy inputs to the OGE sector mainly include raw coal, crude oil, gasoline, diesel oil, fuel oil, natural gas, and

Table 3 Conversion factors from physical units to MJ

	Conversion factor
Oil	41.8, MJ/kg
Natural gas	38.9, MJ/m^3 or 57.18, MJ/kg
Raw coal	20.9, MJ/kg
Crude oil	41.8, MJ/kg
Gasoline	43.1, MJ/kg
Diesel	42.7, MJ/kg
Fuel oil	41.8, MJ/kg
Kerosene	43.1, MJ/kg
Electricity	3.6, MJ/kWh
Coal equivalent	29.3, MJ/kg

Table 2 Energy output and energy inputs of OGE in China, in physical units (tce is tonnes of coal equivalent)

Year	Energy output		Energy inputs							
	Oil 10^4, metric tons	Gas 10^8, m^3	Raw coal 10^4, metric tons	Oil 10^4, metric tons	Gasoline 10^4, metric tons	Diesel 10^4, metric tons	Fuel oil 10^4, metric tons	Gas 10^8, m^3	Electricity 10^8, kWh	Others 10^4, tce
1985	12,490	129	92	141	31	28	47	34	85	95
1986	13,037	134	96	132	27	41	50	34	98	71
1987	13,392	135	88	120	28	46	79	39	109	62
1988	13,685	139	105	138	35	50	80	40	123	78
1989	13,748	145	121	138	36	57	71	39	131	86
1990	13,831	153	132	141	39	64	130	36	145	92
1991	13,968	154	117	133	38	64	98	32	158	37
1992	14,196	157	140	109	42	68	109	40	171	34
1993	14,400	163	215	141	58	137	157	39	232	75
1994	14,607	167	229	212	52	142	160	45	240	66
1995	15,005	180	220	175	59	146	161	42	259	132
1996	15,729	201	262	174	53	191	105	30	259	61
1997	16,044	223	310	317	36	157	108	38	315	211
1998	16,052	223	205	313	33	103	127	35	298	251
1999	16,000	252	176	331	42	145	144	46	308	325
2000	16,300	272	186	409	45	162	146	50	322	353
2001	16,396	303	162	423	44	177	150	58	356	356
2002	16,700	327	163	448	44	198	142	59	365	369
2003	16,960	350	186	552	39	168	121	62	357	355
2004	17,587	415	188	499	37	185	33	49	363	273
2005	18,135	493	184	504	26	186	26	49	385	269
2006	18,477	586	187	565	29	187	28	55	316	228
2007	18,632	692	179	569	31	198	26	64	311	186
2008	19,044	803	153	696	28	272	37	86	318	180
2009	18,949	853	155	487	25	230	26	89	333	164
2010	20,301	949	157	482	24	186	31	102	348	176
2011	20,288	1027	151	368	22	192	26	96	375	193
2012	20,748	1072	129	463	14	63	12	96	397	153

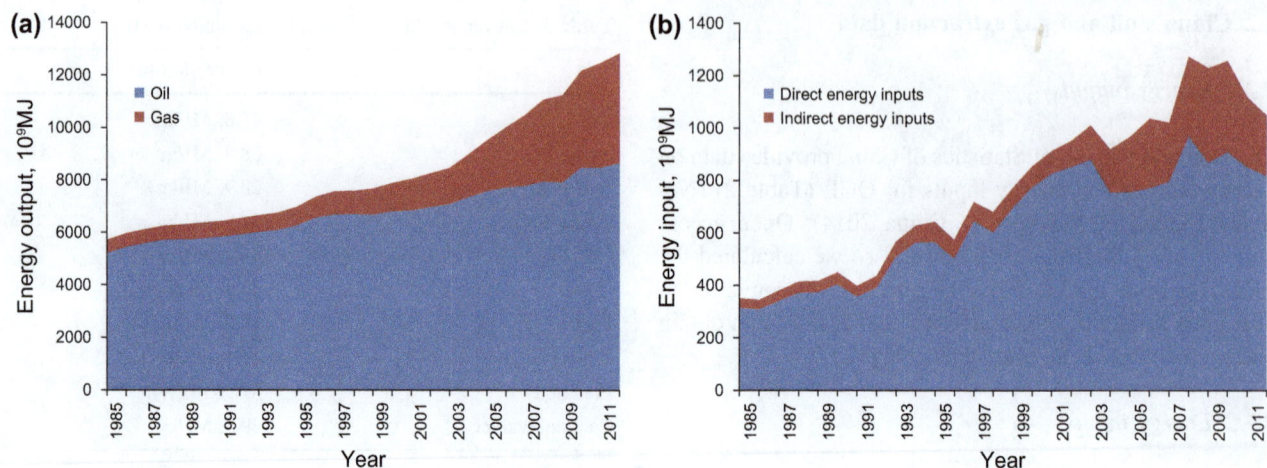

Fig. 2 Energy output and energy inputs for OGE in China

electricity. All of the raw data for direct energy inputs in physical units (Table 2) are also converted into thermal units (Fig. 2) using the conversion factors in Table 3.

We derived indirect energy inputs through multiplying costs by industry energy intensity factors (Table 4), because there is no energy inputs accounting in routine economic data. Indirect monetary costs include the sum of "purchase of equipment and instruments" and "other

expenses" of "investment" in the China Statistic Yearbook database (National Bureau of Statistics of China 2014). China Statistic Yearbook database did not provide the data about indirect monetary costs before 1995. The ratio of indirect energy inputs to total energy inputs between 1995 and 2001 fluctuated around 10 %, so it was assumed to be 10 % before 1995. Thus, the indirect energy inputs for OGE can be obtained, as shown in Fig. 2.

Table 4 Indirect energy inputs of oil and natural gas extraction in China

	Raw data, 10^9, yuan			Energy intensity for industry, MJ/yuan	Indirect energy inputs, 10^9 MJ
	Purchase of equipment and instruction	Other expenses	Total		
1995	4.5	2.2	6.7	11.3	76
1996	5.2	2.3	7.5	9.6	72
1997	5.5	2.6	8.1	8.7	70
1998	6.1	3.2	9.3	8.4	78
1999	6.1	3	9.1	8.2	75
2000	6.9	3.6	10.5	7.6	80
2001	7.7	4.2	11.9	7.2	86
2002	8.6	5.6	14.2	7	99
2003	11	7.5	18.5	7	130
2004	13.6	10.5	24.1	6.8	164
2005	19.8	12.9	32.7	6.4	209
2006	25.1	19.7	44.8	5.9	264
2007	25.9	16.5	42.4	5.3	225
2008	44.4	20.2	64.6	4.7	304
2009	55.2	21.6	76.8	4.7	361
2010	58.4	24	82.4	4.2	346
2011	41.6	28.5	70.1	3.8	266
2012	33.3	28.8	62.1	3.7	230
2013	39.3	27	66.3	3.5	232

2.3 Oil and gas production data from companies

2.3.1 Energy outputs

The volumes of oil and gas extracted by each company (Table 5) can be easily found in their annual reports (CNPC Annual Reports; Sinopec Annual Reports; CNOOC Annual Reports; Yanchang Social Responsibility Reports). For $EROI_{stnd}$, the energy outputs of each company in Table 6 are equal to the volumes (Table 5) multiplied by the conversion factors.

When calculating the $EROI_{2,d}$ of each company, it is found that the volumes of oil extracted are not equal to the volumes of oil processed, due to the existence of oil imports. To calculate the $EROI_{2,d}$ easily, it is assumed the volumes of oil extracted are 4.18×10^4 MJ per tonne. According to Brandt et al. (2013), the energy losses in the refining process are about 7 %. Then, the energy output is 3.89×10^4 MJ per tonne. We assume the energy output of CNPC, Sinopec, and Yanchang is 3.89×10^4 MJ, then that of CNOOC is 2.14×10^4 MJ, because the yield of light oil products for CNOOC is lower about 1.75×10^4 MJ per tonne than that of other companies.

2.3.2 Energy inputs

2.3.2.1 Energy investment in OGE

(1) Direct energy inputs

The CNPC's and Yanchang's total direct energy inputs data in Table 7 for the OGE sector are from the CNPC Statistical Yearbooks (CNPC 2014) and the Yanchang Social Responsibility Reports, respectively. The Sinopec Statistical Yearbook provides data on the units of direct energy inputs for OGE (Table 7) (Sinopec 2014). From the volume of oil and gas extracted by Sinopec (Table 5), the total direct energy inputs of Sinopec can be calculated and shown in Table 8.CNOOC's annual reports published

only the total direct energy inputs for the whole company without a division by sectors of other activities from 2005 to 2009 (Table 7) (CNOOC Annual Reports). From 2005 to 2008, the direct energy inputs were mainly consumed by OGE, while part of the direct energy inputs was used in the oil processing sector in 2009, when CNOOC entered the refining business (CNOOC Annual Reports). For CNOOC, we do not have any data on direct energy inputs separated by processing sector for 2009. However, knowing the volume of oil processed by company CNOOC (Table 9) and the specific direct energy inputs required for processing 1 tonne of oil, the amount of total direct energy inputs in the oil processing sector can be calculated. In this paper, it is assumed that the average energy requirement for oil processing by CNOOC equals that needed for the Huizhou Refinery, the largest refinery owned by CNOOC. In 2009, 0.28×10^4 MJ is needed per tonne of oil processed in the Huizhou Refinery (Gong and Wang 2013). Given this amount, the direct energy inputs for the OGE sector can be calculated by subtracting the direct energy inputs for the processing sector from the total direct energy inputs. After obtaining the total direct energy inputs of the four companies (CNPC, Sinopec, SNOOC, Yanchang), they were converted into thermal units (Table 8).

Table 10 shows the direct energy inputs per tonne of oil extraction. These data are required when calculating the $EROI_{2,d}$ of each company.

(2) Indirect energy inputs

Currently, Chinese companies do not provide enough data on indirect monetary costs for oil and gas extraction. To calculate indirect energy inputs, in this work, it is assumed that the ratio of indirect energy inputs to total energy inputs for each company is equal to that for China's oil and gas extraction. For example, the ratio of indirect energy inputs to total energy inputs is 9.4 % in 2001 for China's oil and gas extraction; then, the ratio for CNPC in 2001 is also

Table 5 Amount of oil and gas extracted by each company (oil: 10^4, tonnes; gas: 10^9, m^3)

	2001	2002	2003	2004	2005	2006	2007	2008	2009	2010	2011	2012	2013
Oil													
CNPC	11,484	11,757	11,695	12,097	12,598	13,471	13,762	13,875	13,745	14,144	14,927	15,188	15,981
Sinopec						4017	4108	4180	4242	4256	4273	4318	4378
CNOCC					3197	3154	3055	3244	3697				
Yanchang											1232	1264	1263
Gas													
CNPC	212	233	263	313	396	480	578	664	738	829	882	935	1039
Sinopec						73	80	83	85	125	146	169	187
CNOCC					70	88	99	105	107				
Yanchang											0.2	2.6	4.7

Table 6 Energy output of OGE for each company, in thermal units (10^9, MJ)

	2001	2002	2003	2004	2005	2006	2007	2008	2009	2010	2011	2012	2013
Oil													
CNPC	4800	4914	4889	5057	5266	5631	5753	5800	5745	5912	6239	6349	6680
Sinopec						1679	1717	1747	1773	1779	1786	1805	1830
CNOCC				1336	1318	1277	1356	1545					
Yanchang											515	528	528
Gas													
CNPC	825	906	1023	1218	1540	1867	2248	2583	2871	3225	3431	3637	4042
Sinopec						284	311	323	331	486	568	657	727
CNOCC				272	342	385	408	416					
Yanchang											1	10	18

Note: In the Oil section, CNOCC values 1336, 1318, 1277, 1356, 1545 align under 2005–2009; in the Gas section, CNOCC values 272, 342, 385, 408, 416 align under 2005–2009.

Table 7 Total direct energy inputs for OGE for each company (10^4, tce)

	2001	2002	2003	2004	2005	2006	2007	2008	2009	2010	2011	2012	2013
CNPC	1840	1851	1905	1960	2106	2191	2410	2443	2467	2492	2517	2542	2568
Sinopec[a]						111.7	104.9	105.1	102.4	100.4	105.9	105.9	105.3
CNOCC[b]					391	413	480	581	746				
Yanchang											147.8	146.6	145

[a] Sinopec's units of direct energy inputs for OGE are kgce/t

[b] The 2009 data for CNOOC refer to the total direct inputs for OGE and oil processing

Table 8 Total direct energy inputs of OGE for each company, in thermal units (10^9, MJ)

	2001	2002	2003	2004	2005	2006	2007	2008	2009	2010	2011	2012	2013
CNPC	539.1	542.3	558.2	574.3	617.1	642.0	706.1	715.8	722.8	730.2	737.5	744.8	752.4
Sinopec						153.5	149.1	152.7	150.9	159.4	174.9	182.8	188.7
CNOOC					114.6	113.1	119.3	123.9	140.1				
Yanchang											43.4	43.1	42.5

Table 9 Amount of oil processed (10^4, tonnes)

	2001	2002	2003	2004	2005	2006	2007	2008	2009	2010	2011	2012	2013
CNPC	8795	8947	9255	10,370	11,061	11,587	12,173	12,530	12,512	13,529	14,484	14,716	14,602
Sinopec						15,651	16,576	17,294	18,824	21,297	21,892	22,309	23,370
CNOCC									2081				
Yanchang											1302	1400	1405

9.4 %. The direct energy inputs for CNPC in 2001 is 539.1×10^9 MJ; then, the indirect energy inputs is 56×10^9 MJ. Indirect energy inputs of each company are in Table 11.

2.3.2.2 Energy investment in oil transportation Oil companies do not provide an explicit accounting of energy consumption during the process of oil transportation. Oil transportation relies on pipelines in China, and the average

Table 10 Direct energy inputs per tonne of oil extraction, in thermal units (10^4, MJ)

	2001	2002	2003	2004	2005	2006	2007	2008	2009	2010	2011	2012	2013
CNPC	0.40	0.39	0.39	0.38	0.38	0.36	0.35	0.37	0.35	0.33	0.32	0.31	0.29
Sinopec							0.33	0.31	0.31	0.30	0.29	0.31	0.31	0.31
CNOOC					0.30	0.28	0.30	0.29	0.30				
Yanchang											0.35	0.33	0.33

Table 11 Indirect energy inputs of each company (10^9, MJ)

	2001	2002	2003	2004	2005	2006	2007	2008	2009	2010	2011	2012	2013
CNPC	56	63	82	124	170	220	199	224	300	277	230	209	211
Sinopec						52	42	48	63	60	54	51	53
CNOOC					32	39	34	39	58				
Yanchang											14	12	12

Table 12 Energy intensity and fuel mix for each transportation mode

	Energy intensity, MJ/tonne-km	Fuel mix and their percentage
Sea tanker	0.023	Residual oil (100 %)
Pipeline: oil	0.3	Residual oil (50 %) and electricity (50 %)
Pipeline: NG	0.372	NG (99 %) and electricity (1 %)

distance from oilfield to oil processing plant is assumed to be 1000 km. The energy intensity by oil pipeline is approximately 0.3 MJ/tonne-km (Table 12) (Ou et al. 2011). Thus, the direct energy input of transporting one tonne of oil for 1000 km is 300 MJ.

2.3.2.3 Energy investment in oil processing The direct energy inputs of CNPC, Sinopec, and CNOOC for processing per tonne of oil in Table 13 are from literature (CNPC 2014; Sinopec 2014; Gong and Wang 2013). For Yanchang, the total direct energy inputs for oil processing from 2011 to 2013 are 60×10^9, 58×10^9, and 57×10^9 MJ, respectively (Yanchang Social Responsibility Reports). Then Yanchang's direct energy input for processing per tonne of oil (Table 13) is equal to its total direct energy inputs for oil processing divided by the amount of oil processed (Table 9). Table 14 shows the direct energy inputs in MJ for processing per tonne of oil.

3 EROI methodology and data for imports of oil and gas

3.1 EROI methodology for imported oil (IO) and imported natural gas (ING)

An economy without enough domestic fossil fuel must import fuel and pay for it with some type of surplus economic activity. The ability to purchase critically required energy depends upon what else the economy can generate to sell it to the world, as well as the fuel required to grow or

produce that material (Hall et al. 2009). In 1986, Kaufmann (1986) derived an explicit method to quantitatively assess the EROI_{IO} (Eq. 4). Because such financial data are usually available, the EROI_{IO} can be derived with a moderate degree of accuracy (Lambert et al. 2014). In 2010, King (2010) developed a metric called the energy intensity ratio (EIR), which is similar to Kaufmann's EROI_{IO}, and calculated it for various industrial fuels in the US over time. His study suggested that the EIR is an easily calculated and effective proxy for the EROI for individual fuels.

$$\text{EROI}_{\text{IO}} = \frac{E_{\text{IO}}}{E_{\text{p,OIL}}} = \frac{E_{\text{OIL}} \times M_{\text{IO}}}{EI_{\text{GDP}} \times P_{\text{OIL}} \times M_{\text{IO}}} \quad (4)$$

where, E_{OIL} is the unit energy content of oil, P_{OIL} is the price of total oil imported, M_{IO} is the amount of oil purchased, EI_{GDP} is the economic intensity of the economy, E_{IO} is the total energy content of the oil purchased, and $E_{\text{p,OIL}}$ is the total energy inputs in the purchasing phase. Usually, IO includes two phases: purchasing and international transportation (Fig. 3). Equation (4) only considers the purchasing phase, while both phases are considered in this paper (Eq. 5).

$$\text{EROI}_{\text{IO}} = \frac{E_{\text{IO}}}{E_{\text{p,OIL}} + E_{\text{t,OIL}}} \quad (5)$$

where $E_{\text{t,OIL}}$ refers to the total energy inputs in the international transportation phase.

Similar to the EROI_{IO} equation, the EROI_{ING} can be calculated as follows:

$$\text{EROI}_{\text{ING}} = \frac{E_{\text{ING}}}{E_{\text{p,NG}} + E_{\text{t,NG}}} \quad (6)$$

Table 13 Direct energy inputs for processing per tonne of oil, in physical units

	2001	2002	2003	2004	2005	2006	2007	2008	2009	2010	2011	2012	2013
CNPC, kgoe	86.4	89.3	83.4	78.7	80.6	78.0	75.6	71.6	67.6	65.5	65.0	64.1	64.0
Sinopec, kgoe						66.9	65.9	63.8	61.3	58.2	57.0	56.2	57.5
CNOCC, kgoe									63.9				
Yanchang, kgce											157.7	143.9	139.5

Table 14 Direct energy inputs for processing per tonne of oil, in thermal units (10^4, MJ)

	2001	2002	2003	2004	2005	2006	2007	2008	2009	2010	2011	2012	2013
CNPC	0.36	0.37	0.35	0.33	0.34	0.33	0.32	0.30	0.28	0.27	0.27	0.27	0.27
Sinopec						0.28	0.28	0.27	0.26	0.24	0.24	0.24	0.24
CNOCC									0.28				
Yanchang											0.46	0.42	0.41

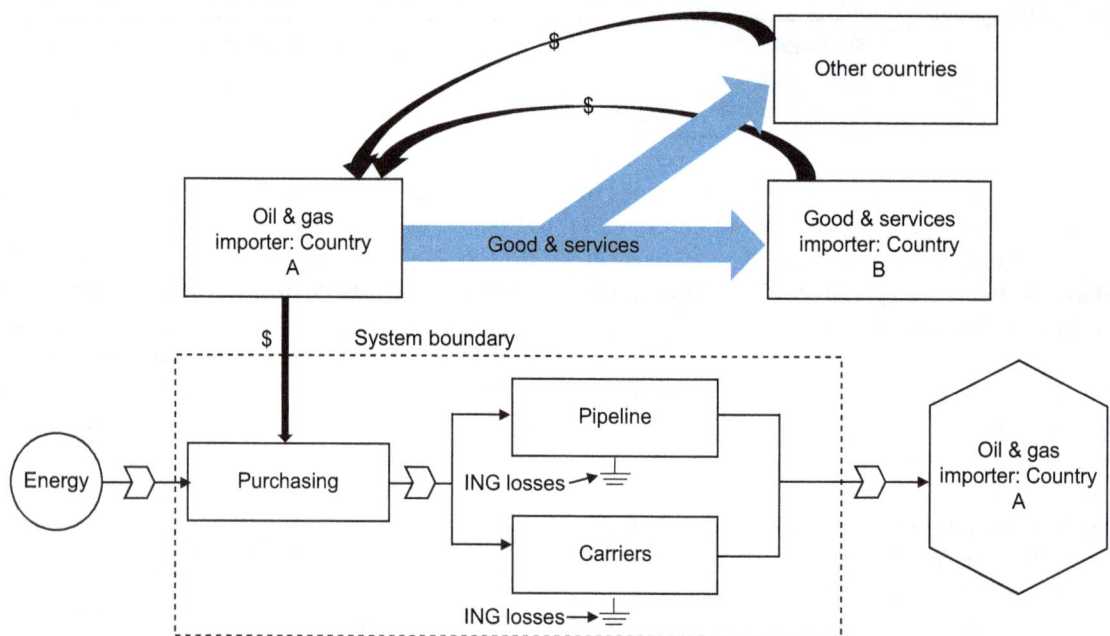

Fig. 3 System boundaries of IO and ING

where E_{ING} is the total energy content of the gas purchased, and $E_{p,NG}$ and $E_{t,NG}$ refer to the total energy inputs in the purchasing phase (equal to the energy required to make the goods exported to pay for the gas) and the international transportation phase, respectively. Unlike coal and oil, which remain almost unchanged after long distance transportation, gas may suffer some losses in transportation (Lin et al. 2010). Therefore, gas losses in transportation will be estimated and excluded from the total energy outputs. The EROI$_{ING}$ can be calculated with Eq. (7).

$$\text{EROI}_{ING} = \frac{E_{ING} - L_{t,NG}}{E_{p,NG} + E_{t,NG}} \tag{7}$$

where $L_{t,NG}$ refers to the gas losses in international transportation.

For gas transportation by pipeline, gas losses usually result from fugitive emissions and flaring. According to the Intergovernmental Panel on Climate Change, gas losses in pipeline transportation can be calculated as follows (Zhang et al. 2013):

$$L_t = M_t \times LR \tag{8}$$

where L_t is the volume of gas losses, kg; M_t is the volume of gas transported, m^3; and LR is the loss rate caused by fugitive emissions and flaring, kg/m^3.

For gas transportation by tanker, gas losses result from boil-off gas (BOG) (Zakaria et al. 2013). Due to heat transfer from the surroundings to cryogenic LNG (liquefied natural gas), LNG is unavoidably vaporized, thus generating BOG in LNG tankers. To reduce the losses caused by

BOG, some technologies are applied to re-liquefy the BOG at the expense of power consumption for liquefaction and the initial cost of the liquefying facilities (Lin et al. 2010). Here, we assume that the recovery rate of BOG is r, so L_t can be calculated using Eq. (9).

$$L_t = B(1 - r) = M_t \left[1 - (1 - BR)^{\frac{D_t}{S \times 24}} \right] (1 - r) \qquad (9)$$

here B is the BOG in transportation by LNG tankers, m^3; M_t is the volume of gas transported, m^3; BR is the boil-off rate, which refers to the percentage of LNG that needs to be boiled off to keep the LNG at the same temperature when heat is added to the LNG fuel ($\times\%/$day, e.g., 0.5 %/day); D_t is the distance of international transportation, km; and S is the speed of the LNG tanker, km/h.

3.2 Data for IO and ING

3.2.1 Energy content of oil and gas purchased

The amounts of China's imported oil and gas (Table 15) are available from Wind Information Co., Ltd (Wind Info). Wind Info Import data cover the period from 1996 to 2015 (2015 data are only the sum of the first seven months). E_{IO} and E_{ING} (Table 16) are equal to these annual import volumes (Table 15) multiplied by the energy content factors in Table 3.

3.2.2 Energy inputs

Energy investment in purchasing: The costs of purchasing oil and gas are from Wind Info (Table 15). The EI_{GDP} (in 2010 constant prices) in Fig. 4 is from National Bureau of Statistics of China (2014) and The People's Bank of China (2014). EI_{GDP} is multiplied by the cost of purchasing oil (gas) to create a time series of $E_{p,OIL}$ ($E_{p,NG}$) (Table 16).

Energy investment in international transportation: The oil imported by tanker and by pipeline and the gas imported by tanker and by pipeline are shown in Fig. 5. Table 12 presents all of the data on the energy intensity and the fuel mix for each transportation mode. We assume that the transport distances for pipelines and tankers are 2000 and 8000 km, respectively. Thus, the $E_{p,OIL}$ and $E_{p,NG}$ in Table 16 can be calculated; these values are equal to the transport distance multiplied by the traffic intensity and then multiplied by the transport volume.

3.2.3 Gas losses in international transportation

The values for transport volume by pipeline and by LNG tanker are shown in Fig. 5. Zhang et al. (2013) found that the loss rate (LR) is 0.607×10^{-3} kg/m^3. The volume of gas losses (L_t) for pipeline transport can be calculated using

Eq. (8). In this paper, we set the distance for LNG transport at 8000 km. According to Lin et al. (2010), for large LNG carriers, the boil-off rate (BR) is usually between 0.05 %/day and 0.1 %/day, where the middle value is 0.075 %/day. Chu (2000) collected data on approximately 108 LNG ships and found the operating speed to be concentrated within the range of 32.4–37 km/h. The average speed is 34.7 km/h, which is the speed used in this paper. Assuming that the recovery rate of BOG (r) is 0.7, the volume of gas losses (L_t) for tanker transport can be calculated using Eq. (9). The $L_{t,NG}$ (Table 16) is the sum of the L_t for pipeline transport and the L_t for tanker transport.

4 Results

4.1 EROI for China's oil and gas extraction

The $EROI_{stnd}$ for China varies from about 8.4:1 to 17.3:1, decreasing from 1985 to 2003 and then increasing again (Fig. 6). Obviously, with the depletion of oil reserves, production becomes more costly. In addition, the growth in energy consumption in turn leads to a decrease in the EROI, which is entirely consistent with the fact that until 2003, China's EROI was continually declining. However,

Table 15 Amounts and costs of IO and ING

Year	The amounts (10^8, kg)		The costs (10^8, dollars)	
	IO	ING	IO	ING
1996	155		21	
1997	340		53	
1998	274		33	
1999	328		53	
2000	640		135	
2001	558		108	
2002	634		116	
2003	865		188	
2004	1141		315	
2005	1232		464	
2006	1385		634	
2007	1534		750	
2008	1733		1247	
2009	1952	55	850	13
2010	2309	119	1302	40
2011	2464	226	1904	104
2012	2635	305	2142	168
2013	2732	378	2128	204
2014	2864	404	2122	225
2015	1759	226	754	101

Table 16 Energy outputs and inputs for IO and ING

Year	E_{IO}	$E_{p,OIL}$	$E_{t,OIL}$	E_{ING}	$L_{t,NG}$	$E_{p,NG}$	$E_{t,NG}$
1996	6479	569	28				
1997	14,212	1317	62				
1998	11,453	761	49				
1999	13,710	1172	59				
2000	26,752	2851	114				
2001	23,324	2176	99				
2002	26,501	2272	110				
2003	36,157	3857	148				
2004	47,694	6818	189				
2005	51,498	9877	202				
2006	57,893	12,774	222				
2007	64,121	13,695	247				
2008	72,439	19,710	288				
2009	81,594	12,730	321	3163	9	189	10
2010	96,516	18,541	379	6833	16	571	37
2011	102,995	25,366	398	12,908	24	1387	100
2012	110,143	26,876	422	17,432	31	2107	145
2013	114,198	25,710	437	21,631	38	2463	181
2014	119,715	25,381	458	23,113	41	2689	189
2015	73,526	8800	282	12,944	22	1180	112

after 2003, the EROI increased from 8.4:1 to 12.2:1. This increase is mainly a result of the increasing EROI for gas extraction. Another factor is that energy savings programs in the industry have been implemented.

4.2 The EROI for oil and gas companies

As shown in Table 17, the $EROI_{stnd}$ for CNPC increased significantly from 9.5:1 in 2001 to 11.1:1 in 2013. This increase in $EROI_{stnd}$ is attributed to the fact that CNPC has made great effort to develop natural gas, which has a higher EROI than oil. From 2001 to 2013, natural gas production has increased from 825×10^9 to 4042×10^9 MJ, which is an average annual increase in 14.2 %. Meanwhile, in the development of natural gas, in 2012, the OGE of the CNPC Changqing Oilfield reached 1800×10^9 MJ (for the first time it produced more oil and gas than the CNPC Daqing Oilfield), making it the largest oil and gas field in China. Moreover, CNPC vigorously developed energy-saving technologies to improve its energy efficiency, which in turn helped to improve its EROI.

For Sinopec's OGE, since 2009, the EROI increased from 9.9:1 in 2009 to 10.6:1 in 2013 (Table 17). The $EROI_{2,d}$ increased from 6.1:1 in 2006 to 6.9:1 in 2010 and remained unchanged at approximately 6.7:1 between 2010 and 2013 (Table 18). Because the information is limited in the public domain, it is impossible to explain accurately all

of the above phenomena in this paper. However, we can see that one reason is that the potential for energy savings was being exhausted, as shown in Fig. 7 (CNPC Annual Reports; Sinopec Annual Reports; CNOOC Annual Reports; Yanchang Social Responsibility Reports).

CNOOC's $EROI_{OGE}$ is higher than that of CNPC, Sinopec, and Yanchang. The main reason may be that compared with onshore oil fields, the degree of development of offshore oil resources is relatively low, and the production cost is relatively cheap. There are two reasons for the low degree of exploration: On the one hand, offshore oil resource development takes longer time than onshore oil development; on the other hand, at present, only CNOOC is engaged in offshore oil exploration and development, whereas CNPC and Sinopec, the two largest oil companies in China, are engaged mainly in onshore oil exploration and development. Currently, because of the lack of data, we cannot provide an answer as to why the CNOOC's $EROI_{OGE}$ between 2005 and 2008 has been slightly fluctuating near approximately 10.9:1. To provide an explanation, we need more detailed data on energy consumption. In terms of refining, in 2009, CNOOC's 12 Mt/a Huizhou Refinery Project Phase I was completed and put into production, marking the CNOOC entrance into the refining business (CNOOC Annual Reports). The Huizhou Refinery Project is a large-scale refinery in China that is especially designed to process high acid heavy crude oil, which requires more energy consumption and has a lower light oil extraction rate, resulting in a lower EROI. As shown in Table 18, the $EROI_{2,d}$ for CNOOC is significantly lower than that of the other companies in 2009.

Yanchang's $EROI_{OGE}$ is lower than that of CNPC, Sinopec, and CNOOC (Table 17), because the Yanchang has been producing oil from depleted fields. In 1905, the Yanchang Petroleum Factory was established, and in 1907, Yanchang drilled the first oil well in mainland China (Zuo 2009). As shown in Fig. 8, the growth rate of crude oil production of Yanchang declines with time; in 2013, it fell 0.1 % compared with that in 2012. However, the energy efficiency of OGE for Yanchang has a trend of slight increase, although the Yanchang's EROI is lower than that of the other companies. This may be attributed to the increased natural gas production and energy efficiency measures. In 2010, Yanchang began to produce natural gas, and the production of natural gas increased from 0.4×10^9 in 2010 to 18×10^9 MJ in 2013 (Fig. 8). In terms of energy efficiency, Yanchang formulated the "Environmental Governance Programme 2011–2013," in which energy conservation was considered an important environmental protection measure. Yanchang's energy savings in 2013 was 2×10^9 MJ, which was slightly higher than in 2011 (Fig. 8) (Yanchang Social Responsibility Reports).

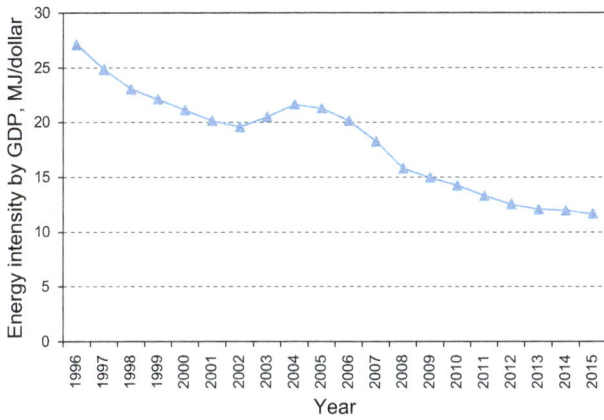

Fig. 4 Energy intensity of GDP

4.3 The EROI for imported oil and imported natural gas

The $EROI_{IO}$ and $EROI_{ING}$ are calculated based on Eqs. (5) and (7), respectively. The $EROI_{IO}$ values in China during "good times" (i.e., the late 1990s) and "bad times" (i.e., 2006–2008) are shown in Fig. 9. The $EROI_{IO}$ shows a peak of approximately 14.8 in 1998 and a value of approximately 8.4 in 2015; overall, the figure presents a fluctuating but declining trend over the entire study period (slope of -0.44, $R^2 = 0.6$) reflecting increasing relative prices of petroleum. The patterns in the EROI values for IO and ING have broadly similar trends during the period from 2009 to 2015. From 2009 to 2012, they show a sharp and sustained decline. From 2012 to 2014, they remained relatively

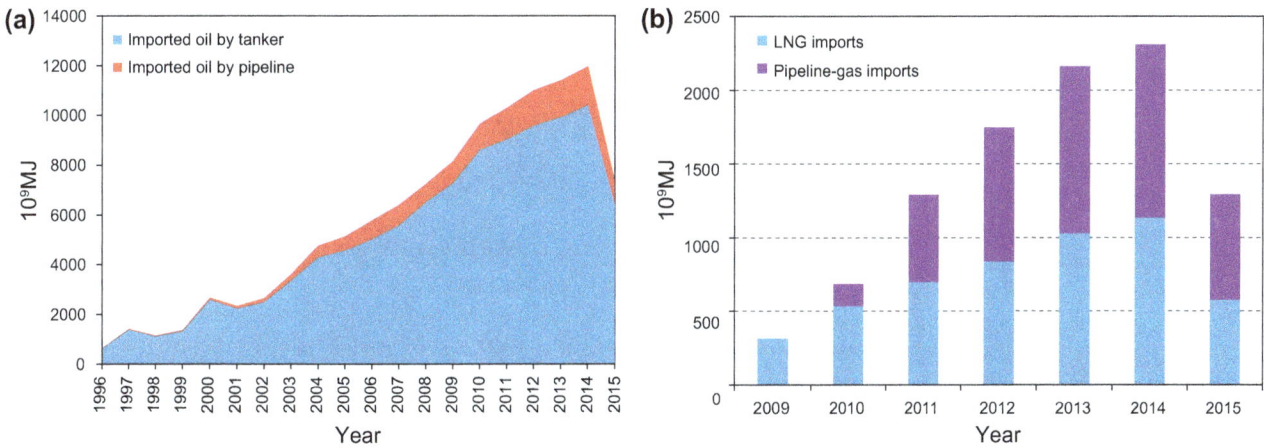

Fig. 5 Amounts of IO and ING by tanker and by pipeline (NB. 2015 data are only for 7 months)

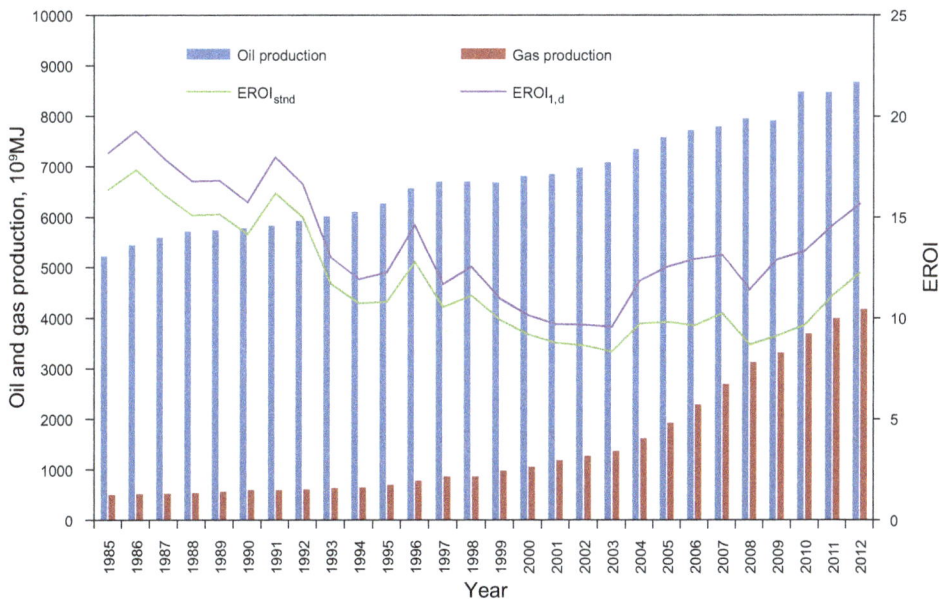

Fig. 6 EROI of China's OGE

Table 17 EROI$_{stnd}$ of each company

	2001	2002	2003	2004	2005	2006	2007	2008	2009	2010	2011	2012	2013
CNPC	9.5	9.6	9.2	9.0	8.6	8.7	9.0	8.7	8.4	9.1	10.0	10.5	11.1
Sinopec						9.5	10.6	10.3	9.9	10.3	10.3	10.5	10.6
CNOOC					11.0	10.9	10.9	10.8	9.9				
Yanchang											9.1	9.8	10.0

Table 18 EROI$_{2,d}$ of each company

	2001	2002	2003	2004	2005	2006	2007	2008	2009	2010	2011	2012	2013
CNPC	4.9	4.9	5.0	5.2	5.2	5.4	5.5	5.6	5.9	6.1	6.3	6.4	6.6
Sinopec						6.1	6.3	6.4	6.6	6.9	6.7	6.7	6.7
CNOOC									3.5				
Yanchang											4.6	5.0	5.1

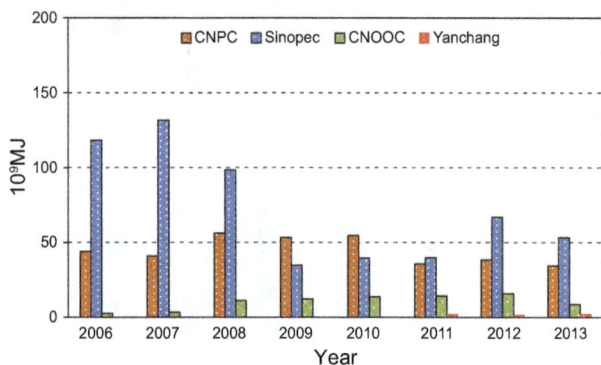

Fig. 7 Amount of energy savings for China's oil companies

stable. In 2015, they began to increase as oil prices again decreased relative to exported commodities.

5 Discussion

5.1 Comparison with previous estimates for China's oil and gas

Hu et al. (2013) showed that the EROI$_{stnd}$ for China's oil and gas extraction fluctuated from 12 to 14:1 in the mid-1990s and declined to 10:1 in the period from 2007 to 2010 (Fig. 10). The EROI$_{stnd}$ trends documented in this paper are

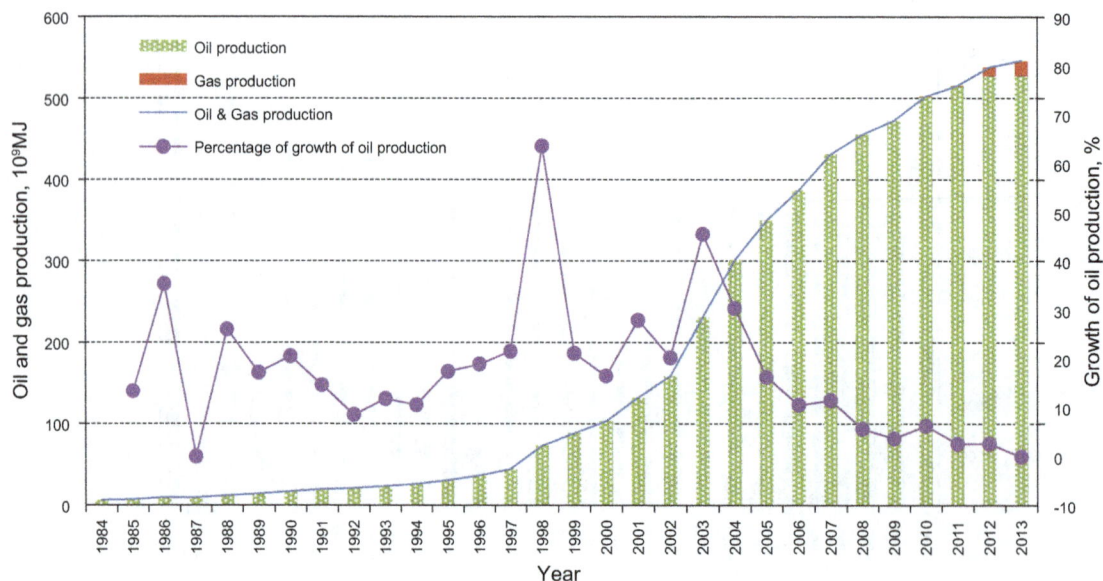

Fig. 8 OGE of Yanchang

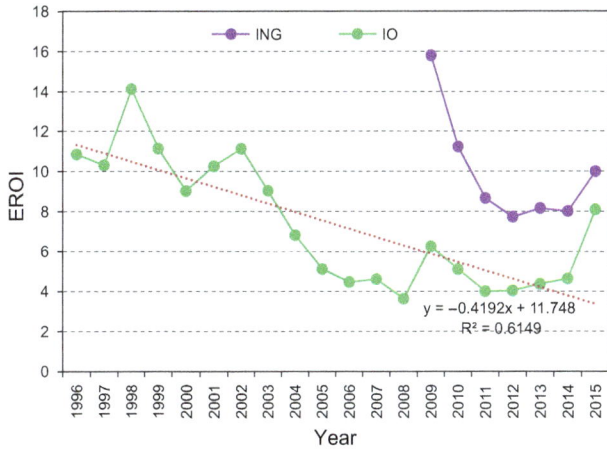

Fig. 9 $EROI_{IO}$ and $EROI_{ING}$ in China

Hu et al. (2011) estimated that the EROI of CNPC's Daqing Oilfield, China's largest, declined continuously from 10:1 in 2001 to 6:1 in 2009. From Fig. 10, we can find that the EROI of the Daqing Oilfield is lower than that of CNPC overall, and the discrepancy continues to increase. There are two reasons for this result. The principal reason is that as Daqing's fields age, they require more energy-intensive techniques, such as high pressure water and polymer injections. The discovery of the Daqing oil field in 1959 made China an oil-rich country (Hu et al. 2011). After 40 years of development, its oil production began to decline in 1998. The production of the Daqing oil field has been decreasing from its peak of 2328×10^9 to 1672×10^9 MJ in 2013. During this period, oil production was maintained, and the water content was mainly controlled by increasing the water pressure beneath the oil and using polymer flooding technology, thus leading to an increase in energy inputs. Daqing's natural gas production in 2013 was 134×10^9 MJ, and compared with 1998, it only increased by 43×10^9 MJ. The other reason is that other CNPC oil fields produced more oil and gas. Of these fields, the most noteworthy is the Changqing Oilfield. Over the period from 1998 to 2013, its oil and gas production increased by 12 per cent and 31 per cent annually, respectively, reaching 1017×10^9 and 1349×10^9 MJ (Fig. 11) (Wang 2004a, b; Zhao and Xiao 2009a, b; Xiao 2014a, b).

similar to those of Hu et al. (2013) with the only difference being that the EROIs in this paper are somewhat lower than theirs, which results from that their study considering only 8 main fuels (natural gas, crude oil, electricity, diesel oil, raw coal, fuel oil, gasoline, and refinery gas), and ignoring some fuels that are used in small amounts such as liquefied petroleum gas, while this paper considers them. The discrepancy between $EROI_{1,d}$ and $EROI_{stnd}$ has been increasing, which suggests that indirect energy inputs are increasing.

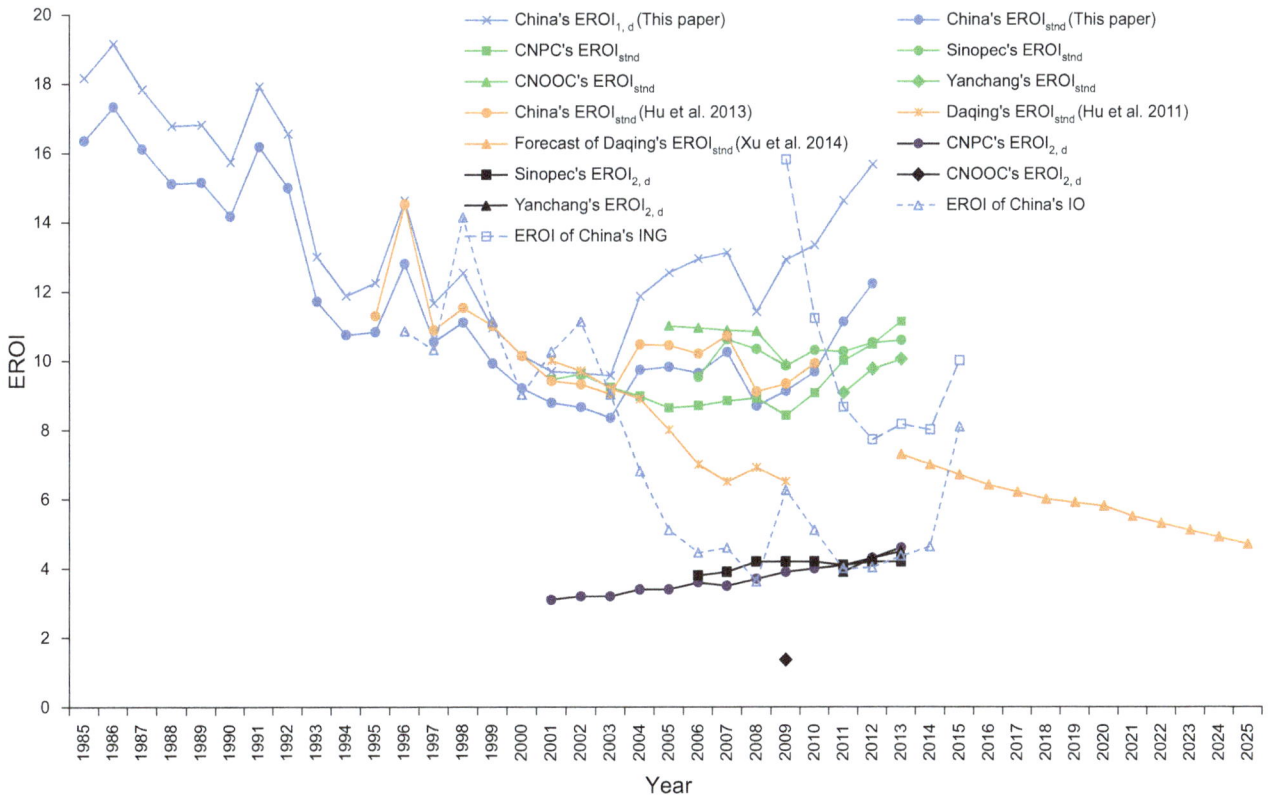

Fig. 10 Comparison with previous estimates of China's oil and gas

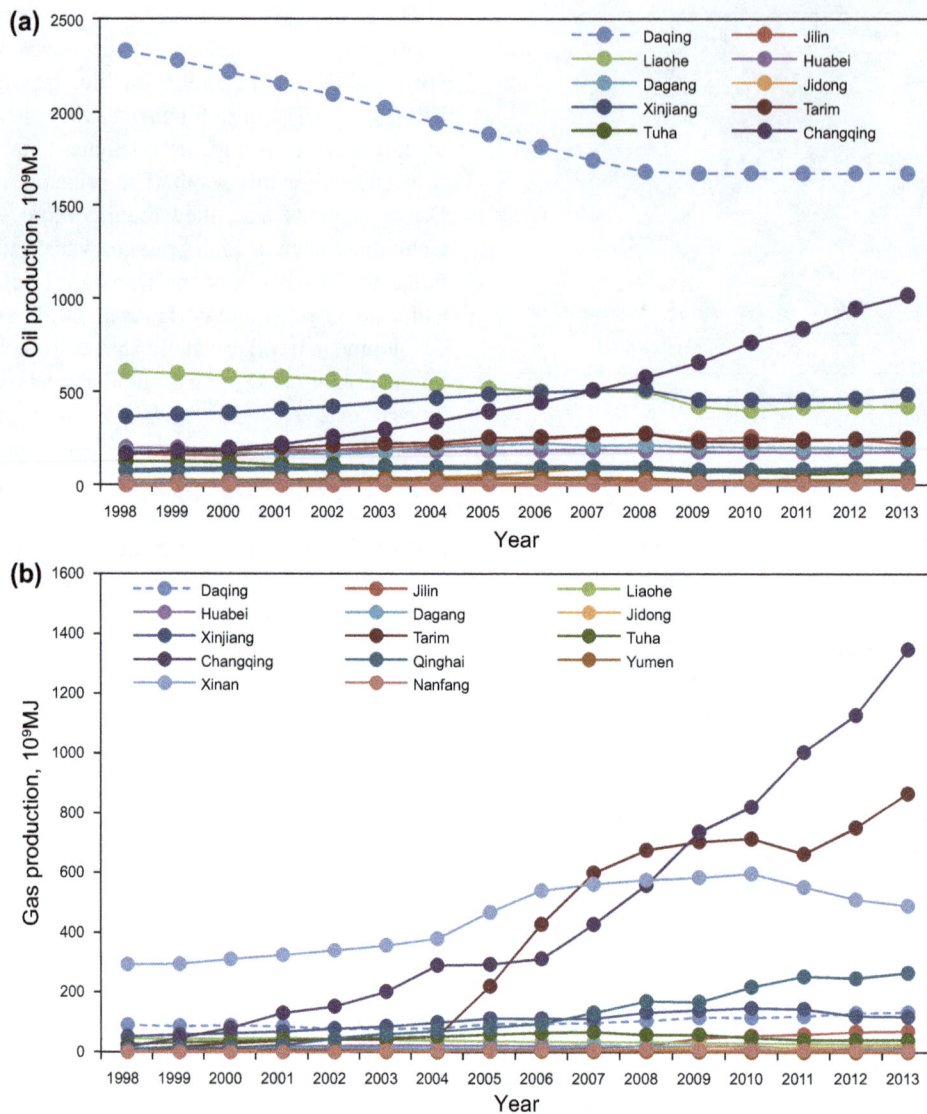

Fig. 11 Oil (*left*) and gas (*right*) production of CNPC's oilfields

Our EROI estimates for IO fall within the range of previously published studies. Lambert et al. (2014) provided an EROI analysis for IO for 12 developing countries, including China. The authors found that most developing nations had $EROI_{IO}$ values below 8–10, which are very similar to our results. They also found that there were two $EROI_{IO}$ peaks for oil imported into China. One peak occurred in 1998, which is the same as our results. Until now, no studies have estimated the $EROI_{ING}$. Our study shows that the $EROI_{ING}$ has also shown a declining trend, similar to the $EROI_{IO}$.

5.2 Policy implications

From 1985 to 2003, the $EROI_{stnd}$ in China has shown a declining trend, while after 2003, the $EROI_{stnd}$ started to increase due the development of NG, which may have a better energy return. Therefore, the government should

take measures to increase gas production. We believe that the most significant measure would be to rationalize the domestic NG pricing mechanisms, which would improve the enthusiasm of enterprises to develop NG. Since the 1990s, the Chinese government has implemented several NG price reforms. The pricing method for NG has evolved from government pricing to a two-track implementation to prices set with government guidance and finally to the current market net back value method (Kong et al. 2015). Although the pricing mechanisms have gradually improved, market-oriented pricing is still not used for China's NG.

In Sect. 5.1, by comparison with $EROI_{stnd}$ and $EROI_{1,d}$, we found that indirect energy inputs have a significant and negative impact on China's EROI. Therefore, to improve the EROI, indirect energy inputs should be controlled. Using the "purchase of equipment and instruments" as an example, on the one hand, oil companies should reduce their amount of equipment and instruments by improving their utilization. On the other hand, equipment manufacturing enterprises should improve their energy efficiency, thereby reducing energy consumption in the equipment production process.

The $EROI_{IO}$ and $EROI_{ING}$ are both continually decreasing. If import prices continue to increase, and hence, the $EROI_{IO}$ declines, the result will likely correspond to lower quality of life indices for China's citizens (Lambert et al. 2014). Therefore, the question that must be asked is, "What opportunities does China have to mitigate the effects of these rising energy prices and the declining EROI of its imported fuel?" Improving the efficiency at which China's economy converts energy and material into marketable goods, and services is one means of improving the country's energy security (Lambert et al. 2014). The other method is to support moderate EROI renewable energy production, which might serve to improve China's net energy balance based on a poor EROI for imported oil and gas through trade.

China is experiencing an EROI decline, which is a risk of unsustainable fossil resource use. Therefore, China must be able to foresee, understand, and plan for changes in its broad energy landscape, particularly during what researchers have characterized as post-peak oil production. Furthermore, sustainability could be effectively addressed by the emergence of a new field: Transition Engineering, which is a new approach to engineering to address the risks of unsustainability so that a vision of a desirable future can be identified and delivered (Transition Engineering). It addresses three key engineering challenges:

Climate: re-engineering systems so they do not cause and are resilient to climate change.

Peak oil: re-engineering systems so they do not depend on fuel; e.g., using 90 % less fuel.

EROI: re-engineering systems so they only use energy with high EROI; e.g., on the order of >10.

Meeting these three challenges could be the way that Chinese society reduces both fossil fuel use and the detrimental social and environmental impacts of industrialization (Krumdieck 2013). Although it is still in its infancy, Transition Engineering should be investigated as a direction for future research by governments, scholars, scientists, and even ordinary people.

6 Conclusion

In this paper, we calculated the EROI for domestic production and imports of oil and gas in China. Our estimates show that the $EROI_{stnd}$ in China decreased from approximately 17.3:1 in 1986 to 8.4:1 in 2003, and it increased to 12.2:1 in

2013. From a company-level perspective, the $EROI_{stnd}$ differs for different companies and was in the range of 8–12:1. Compared with the $EROI_{stnd}$, the $EROI_{2,d}$ declined by 50 %–80 % and was in the range of 3–7:1. The $EROI_{IO}$ declined from 14.8:1 in 1998 to approximately 4.8:1 in 2014, and the $EROI_{ING}$ declined from 16.7:1 to 8.6:1 from 2009 to 2014. In 2015, the $EROI_{IO}$ and $EROI_{ING}$ have shown a slight increase due to decreasing oil and gas prices. In general, this paper suggests that from a net energy perspective, it will become more difficult for China to obtain oil and gas from both domestic production and imports.

Acknowledgments We gratefully acknowledge that this work is supported by the National Natural Science Foundation of China (No. 71273277) and the Philosophy and Social Sciences Major Research Project of the Ministry of Education (No. 11JZD048).

References

BP Statistical Review of World Energy. 2014. http://www.bp.com/en/global/corporate/about-bp/energy-economics/statistical-review-of-world-energy.html. Accessed 10 Jan 2015.

Brandt AR, Englander J, Bharadwaj S. The energy efficiency of oil sands extraction: energy return ratios from 1970 to 2010. Energy. 2013;55:693–702.

Cleveland C. Net energy from the extraction of oil and gas in the United States. Energy. 2005;30:769–82.

Cleveland CJ, O'Connor PA. Energy return on investment (EROI) of oil shale. Sustainability. 2011;3:2307–22.

CNOOC Annual Reports. http://www.cnooc.com.cn/col/col1921/index.html. Accessed 21 Jan 2016.

CNPC Annual Reports. http://www.cnpc.com.cn/cnpc/ndbg/gywm_list.shtml. Accessed 25 July 2016.

Chu LY. Ship Type Evaluation for LNG import transportation of China. Master's thesis. Dalian Maritime University, Dalian, China. 2000 **(in Chinese)**.

Fan J, Wang Q, Sun W. The failure of China's Energy Development Strategy 2050 and its impact on carbon emissions. Renew Sustain Energy Rev. 2015;49:1160–70.

Gagnon N, Hall CAS, Brinker L. A preliminary investigation of energy return on energy investment for global oil and gas production. Energies. 2009;2:490–503.

Gong CB, Wang TY. Analysis of energy consumption of refineries and energy-saving optimization practice. Sino Glob Energy. 2013;18:90–4 **(in Chinese)**.

Heun MK, Wit DM. Energy return on (energy) invested (EROI), oil prices, and energy transitions. Energy Policy. 2012;40:147–58.

Hu Y, Hall CAS, Wang J, et al. Energy return on investment (EROI) of China's conventional fossil fuels: historical and future trends. Energy. 2013;54:352–64.

Hu Y, Feng L, Hall CAS, et al. Analysis of the energy return on investment (EROI) of the huge Daqing oil field in China. Sustainability. 2011;3:2323–38.

Hall CAS, Balogh S, Murphy DJR. What is the minimum EROI that a sustainable society must have? Energies. 2009;2:25–47.

Kong Z, Dong X, Zhou Z. Seasonal imbalances in natural gas imports in major northeast Asian countries: variations, reasons, outlooks and countermeasures. Sustainability. 2015;7:1690–711.

Kesicki F. The third oil price surge-what's different this time? Energy Policy. 2010;38:1596–606.

King CW. Energy intensity ratios as net energy measures of United

States energy production and expenditures. Environ Res Lett. 2010;5:1–10.

Kaufmann R. In energy and resource quality: the ecology of the economic process. New York: Wiley; 1986.

Krumdieck S. Transition engineering: adaptation of complex systems for survival. Int J Sustain Dev. 2013;16:310–21.

Lundin J. EROI of crystalline silicon photovoltaics-variations under different assumptions regarding manufacturing energy inputs and energy output. 2013. http://uu.diva-portal.org/smash/record.jsf?pid=diva2%3A620665&dswid=-3486. Accessed 2 Oct 2014.

Lin W, Zhang N, Gu A. LNG (liquefied natural gas): a necessary part in China's future energy infrastructure. Energy. 2010;35:4383–91.

Lambert JG, Hall CAS, Balogh S, et al. Energy, EROI and quality of life. Energy Policy. 2014;64:153–67.

Murphy DJ, Hall CAS, Dale M, et al. Order from chaos: a preliminary protocol for determining the EROI of fuels. Sustainability. 2011;3:1888–907.

Mulder K, Hagens JN. Energy return on investment: toward a consistent framework. J Hum Environ. 2008;37:74–9.

Munasinghe M. The sustainomics trans-disciplinary meta-framework for making development more sustainable. Int J Sustain Dev. 2002;5:125–82.

National Bureau of Statistics of China. China statistic yearbook 1989–2013. Beijing: China Statistics Press; 2014 (in Chinese).

Ou X, Yan XY, Zhang X. Life-cycle energy consumption and greenhouse gas emissions for electricity generation and supply in China. Appl Energy. 2011;88:289–97.

Safronov A, Sokolov A. Preliminary calculation of the EROI for the production of crude oil and light oil products in Russia. Sustainability. 2014;6:5801–19.

Sinopec Annual Reports. http://www.sinopecgroup.com/group/gsjs/gsbg/. Accessed 20 Aug 2014.

The People's Bank of China. Chinese Exchange Rate. http://www.pbc.gov.cn/publish/english/963/index.html. Accessed 26 Dec 2014.

Transition Engineering. http://www.transitionnetwork.org/initiatives/transition-engineering. Accessed 21 Jan 2016.

Wang XX. Natural gas production in China's oil and gas fields in 1998–2003. Int Pet Econ. 2004a;2:61 (in Chinese).

Wang XX. Crude oil production in China's oil and gas fields in 1998–2003. Int Pet Econ. 2004b;2:60 (in Chinese).

Wind Info (Wind Information Co., Ltd.). http://www.wind.com.cn/En/Default.aspx. Accessed 21 Jan 2016.

Xu B, Feng L, Wei W, et al. A preliminary forecast of the production status of China's Daqing oil field from the perspective of EROI. Sustainability. 2014;6:8262–82.

Xiao L. China's natural gas production in 2008–2013. Int Pet Econ. 2014a;4:93 (in Chinese).

Xiao L. China's crude oil production in 2008–2013. Int Pet Econ. 2014b;4:92 (in Chinese).

Yanchang Social Responsibility Reports. http://www.sxycpc.com/shzr.jsp?urltype=tree.TreeTempUrl&wbtreeid=1735. Accessed 25 Aug 2014.

Zhang B, Chen GQ, Li JS, et al. Methane emissions of energy activities in China 1980–2007. Renew Sustain Energy Rev. 2013;29:11–21.

Zakaria MS, Osman K, Abdullah H. Greenhouse gas reduction by utilization of cold LNG boil-off gas. Procedia Eng. 2013;53:645–9.

Zuo XL. Oil exploration and production in northwest of China during the national government. Master's thesis. Tianjin Normal University. Tianjin, China. 2009 (in Chinese).

Zhao H, Xiao L. China's natural gas production in 2003–2008. Int Pet Econ. 2009a;6:65 (in Chinese).

Zhao H, Xiao L. China's crude oil production in 2003–2008. Int Pet Econ. 2009b;6:66 (in Chinese).

4

Interactions between the fluid and an isolation tool in a pipe: laboratory experiments and numerical simulation

Hong Zhao[1] · Yi-Xin Zhao[2] · Zhi-Hui Ye[3]

Abstract A remote-control tether-less isolation tool is a mechanical device that is normally used in pipelines to block the flow at a given position by transforming a blocking module. In this study, the interactions between the fluid and the plug module of the isolation tool were investigated. Simulations of the plug process and particle image velocimetry measurements were performed to study the flow characteristics. Numerical solutions for the continuity, momentum, and energy equations were obtained by using commercial software based on finite-volume techniques. Box–Behnken design was applied, and response surface methodology (RSM)-based CFD simulation analysis was conducted. The dynamic model in the plug process was built by RSM and used to evaluate the influences of the main mechanical parameters on the pressure during the plug process. The diameter of the isolation tool and the diameter of the plug module have strong influences on the process, and the length of the isolation tool has only a little influence on the plug process.

Keywords Isolation tool · Numerical simulation · Transformation · Blockage · Response surface methodology

✉ Hong Zhao
hzhao_cn@163.com

[1] College of Mechanical and Transportation Engineering, China University of Petroleum, Beijing 102249, China

[2] Department of Mining, China University of Mining and Technology, Beijing 100086, China

[3] College of Petroleum Engineering, China University of Petroleum, Beijing 102249, China

Edited by Yan-Hua Sun

1 Introduction

Pipelines have been used as one of the safest ways to transport oil and gas in industry. When the pipelines do not work effectively, a remote-controlled tether-less isolation tool is used in maintenance to isolate high pressure in pipelines and block the fluid without losing the pressure. Understanding of the interaction between the fluid and the isolation tool at different isolation stages is necessary for engineers to design and perform suitable plug operations.

A literature survey has revealed a few papers discussing the interactions between the isolation tool and the fluid in the pipe. Most of the available studies are mechanical designs or have a commercial basis. Tveit and Aleksandersen (2000) introduced a PSI Smart Plug to isolate high pressure in pipelines and risers. Selden (2009) showed a successful application case of a PSI Smart Plug. The isolation tool is developed from a smart Pipeline Inspection Gauge (PIG) and in-pipe robot in engineering. Dynamic analyses of the PIG model under different conditions were carried out (Nieckele et al. 2001; Yeung and Lima 2002; Xu and Gong 2005; Saeidbakhsh et al. 2009; Lesani et al. 2012; Zeng et al. 2014). Minami and Shoham (1995) developed a pigging model and analyzed PIG transient operations, coupling it with the Taitel simplified transient model. Nguyen et al. (2001a, b, c) proposed a computational scheme to estimate the pigging dynamics. Solghar and Davoudian (2012) investigated the transient PIG motion in natural gas pipelines by basic differential forms of the mass and linear momentum equations and validated it using experimental data. Minami and Shoham (1995) developed a dynamic model considering the length of the pig. In in-pipe robot designs, researchers mainly focused on the mechanical design analysis (Minami and Shoham 1995; Nguyen et al. 2001a, b, c; Ono and Kato 2004; Wang et al.

2010; Zhao et al. 2010; Vahabi et al. 2011; Solghar and Davoudian 2012; Mirshamsi and Rafeeyan 2012, 2015) and control techniques (Roh et al. 2009; Huang et al. 2010).

The main mechanical parameters influencing the plug process are important for analyzing the interactions between the fluid and the isolation tool. The response surface methodology (RSM) is a statistical and mathematical method which is used in engineering modeling (Han et al. 2012; Saravanakumar et al. 2014; Chen et al. 2015; Li et al. 2015; Poompipatpong and Kengpol 2015; Zhang et al. 2015). Song et al. (2014) conducted an optimal design of the internal flushing channel of a drill bit using the response surface methodology (RSM) and CFD simulation and obtained very good results.

To the best of our knowledge, the modeling effects in the plug process are important for the design of the isolation tool between the geometric transformation and its complicated structure. Moreover, the flow characteristics in a pipe during the plug process are also important for suitable operations of the isolation tool, but those have not been studied. There are also some unanswered questions about interactions between the fluid and the isolation tool, including (1) what is the relationship between the flow characteristics and transforming structures of the plug process in a limited space under turbulent flow conditions; (2) which are the main mechanical parameters of the isolation tool influencing the fluid characteristics in the plug process; (3) how does the flow affect the isolation tool in the plug process.

The aim of this paper is to study the plug process interaction between a transformable isolation tool and the fluid in a pipe and to evaluate the influences of the main mechanical parameters. The plug experiments are conducted using particle image velocimetry (PIV) measurements for estimating interactions in the plug process. The modeling of dynamic characteristics is also conducted from a series of CFD simulations by RSM in the plug process. The influences of the main mechanical parameters are discussed from simulation results.

2 Experimental

In order to visualize the plug process of an isolation tool in pipe flow, a Lucite pipe setup with four models in plug stages was designed. The dimensions of the real isolation tool were relatively big, but the test rig was of limited size. The size of experimental models was scaled down to one-tenth of the real size. The practical Reynolds number, Re was 24,925. Particle image velocimetry (PIV) was used to measure the velocity in the pipe, and the measured velocity results were compared with the results from numerical simulations.

2.1 Physical model of the isolation tool

Figure 1 shows a model of the isolation tool. The isolation tool consists of two pressure heads, a bowl, a plug module, and a cylinder module. Normally, an isolation tool is used to plug the flow in a pipe. The plug operation involves the following steps: The right pressure head drives the plug module until the isolation tool is in the designated position. The plug module rapidly expands along the outside edge of the bowl, causing the outside wall of the plug module to adhere to the inner wall of the pipe. Thus, the plug operation is done without losing the pressure in the pipe.

2.2 Geometric deformation of the isolation tool

Figure 2 illustrates two states of the isolation tool experienced. One is the normal state as shown in Fig. 2a. The other is that the plug module is expanded at 99 % (99 % blockage) (see Fig. 2b). Here, d and d_1 are the diameters of the wheel hub and the plug module, respectively. D is the inner diameter of the pipe. The distance from the left boundary to the left end of the isolation tool is L_1. The distance from the right boundary of the pipe to the right end of the isolation tool is L_2. L is the length, while the isolation tool is in the normal state. L_p is the length, while the isolation tool is in the blocking state. L_d is the length of the isolation tool from the left end of the plug module to the rear end of the isolation tool. This value changes with the expanding percentage increased from 0 (the normal state) to 99 % state. L_{p1} and L_{p2} are the lengths from the left boundary to the left end of the isolation tool and from the right boundary to the right end of the isolation tool, respectively. The relationship between the lengths is described by Eq. (1).

$$L_1 + L + L_2 = L_{p1} + L_p + L_{p1} \tag{1}$$

The normal type and four blockage cases were studied as described in Table 1. The expanding percentages ranged from 0 (the normal state) to the 99 % state. The 100 % blockage state cannot be numerically simulated and tested in experiments.

2.3 The preparation of the test model

For practical experiments, four similar structures of the test models of the isolation tool in water were examined. Four test models were created to understand the effects of the geometric deformation of the isolation tool. The plug modules of the test models were geometrically similar to the physical isolation tool. These four models produced 25 %, 50 %, 75 %, and 99 % blockage (as listed in Table 1). For small changes in the length of the test

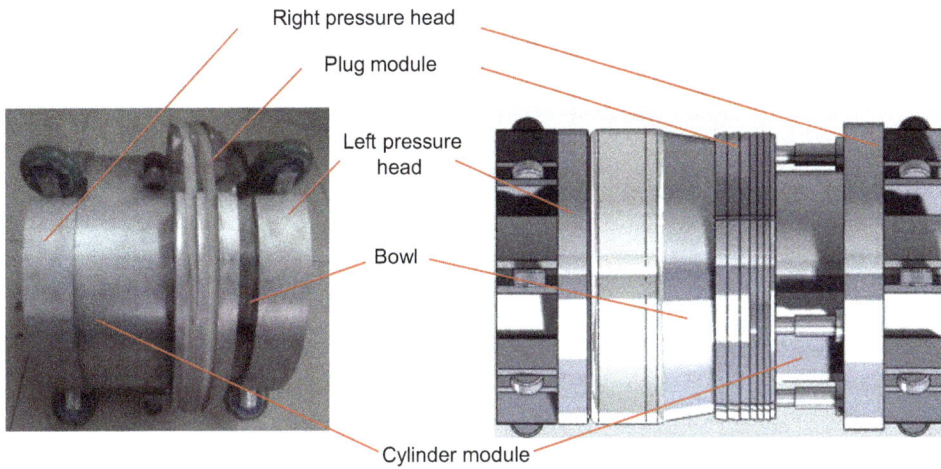

Fig. 1 Physical model of an isolation tool

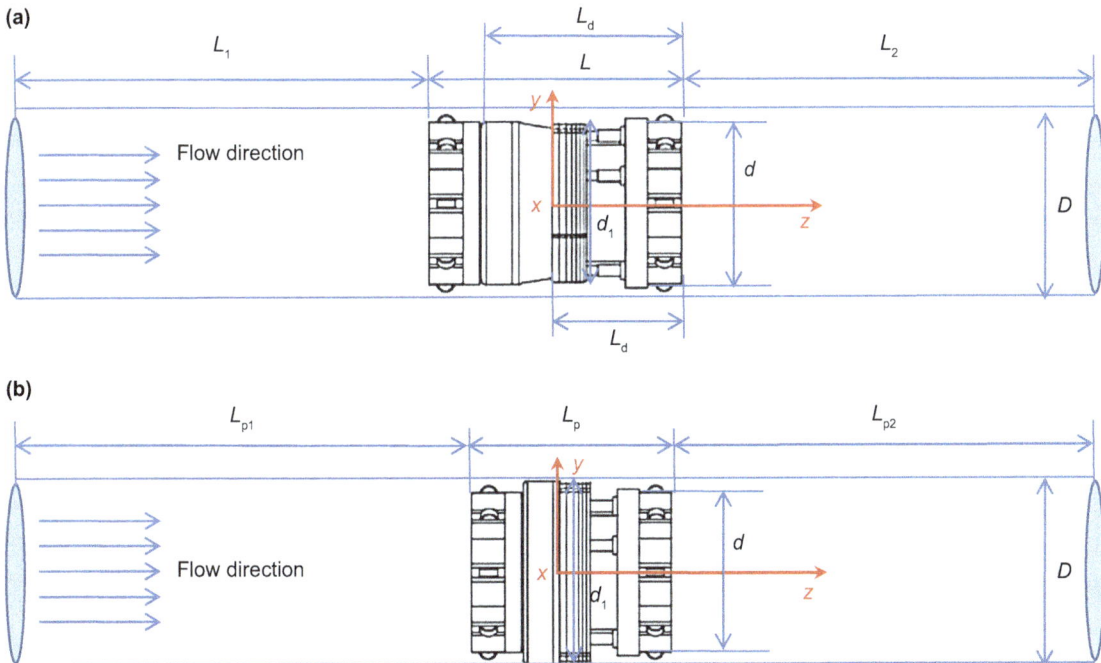

Fig. 2 Deformation models of the isolation tool in a pipe. **a** Normal state. **b** Case 5

Table 1 Studied cases with $L_d = 30$ mm

Case	Blockage percentage	d_1, mm	d_1/D
1	Normal	20.00	0.80
2	25	21.25	0.85
3	50	22.50	0.90
4	75	23.75	0.95
5	99	24.75	0.99

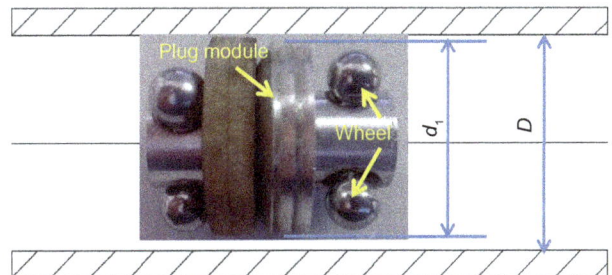

Fig. 3 Test model isolation tool

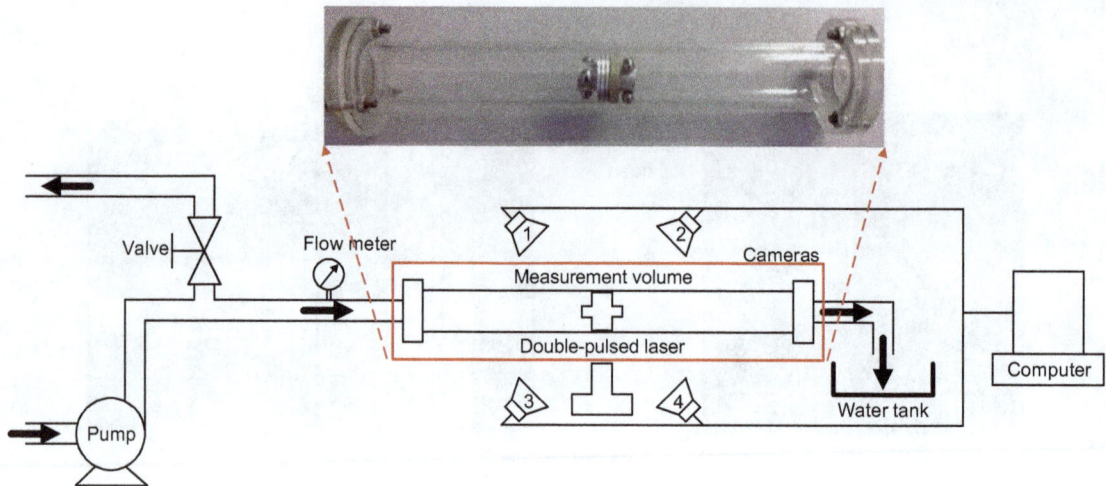

Fig. 4 Test setup of plug processes

Table 2 Test parameters

Inner diameter of the test pipe, mm	Outer diameter of the test pipe, mm	Water density ρ_{water}, kg m^{-3}	Water viscosity μ_{water}, kg ms^{-1}	Inlet velocity w_0, m s^{-1}	Test pressure P_{test}, MPa	Reynolds number Re
25	30	1000	0.001003	1	0.1	24,925

models, the lengths of L_d were all set at 30 mm. The test model with wheels is shown in Fig. 3. The three other types have a similar geometry but a different d_1 values.

A PIV was used to record the particle traces in water, using a camera and a double-pulse laser. The data were then input into a computer to calculate the flow features of particles (hollow glass slivered beads with μm diameter).

The PIV system is shown in Fig. 4. The pipe in this setup is made of Lucite with an inner diameter of 25 mm. A flow meter was connected to the pipe, and the operating conditions were controlled by the pump. The isolation tool was positioned in the middle of the pipe. To ensure that the flow was fully developed, the lengths of the pipe both before and after the isolation tool were in excess of 2 m. Water containing tracer particles was pumped into the test pipe, and then it flowed into the 100-L water tank. PIV measurements were taken at the symmetry plane, that is, at $x = 0$. Detailed measurements of the velocity fields were taken using the PIV system (Dantec Dynamics). The plane under investigation was illuminated by a double-pulsed laser. For comparison, numerical simulations were also carried out at the same flow conditions (as shown in Table 2). According to the real condition of the plug stages, the velocity of the isolation tool was slow down to zero and the isolation tool was set at the fixed position with a thin steel line.

3 Results of PIV experiments

The experiments were carried out at four blockage percentages, namely $d_1/D = 0.85$ (25 % blockage), $d_1/D = 0.9$ (50 % blockage), $d_1/D = 0.95$ (75 % blockage), and $d_1/D = 0.99$ (99 % blockage). The results are shown in Fig. 5. Each figure consists of a geometric graph (sizes are in mm), images captured by a camera installed outside the experimental pipe, and 2D velocity vectors measured by the PIV. The region measured with the PIV (the section indicated by the red square) approximates to the centerline downstream near the bottom of the test model.

The velocity of fluid particles around the test model in the pipe varied considerably and increased from 25 % blockage stage to 99 % blockage stage, as shown in Fig. 5. In Fig. 5, 2D velocity vectors illustrate the flow pattern downstream of the test model as the blockage percentage increased at the symmetry plane. The section indicated by the red square is where the PIV measurements were conducted. For 99 % blockage (Fig. 5d), the velocities of the most of fluid particles in the measurement section decreased significantly and the recirculation structure and flow pattern disappeared. As the graph shows, the length of the vectors represents the velocity, which falls from 25 % blockage stage to 99 % blockage stage. The velocity value was the smallest in the case of 99 % blockage because the flow was almost completely stopped. As proposed by Oztop et al. (2012) for turbulent

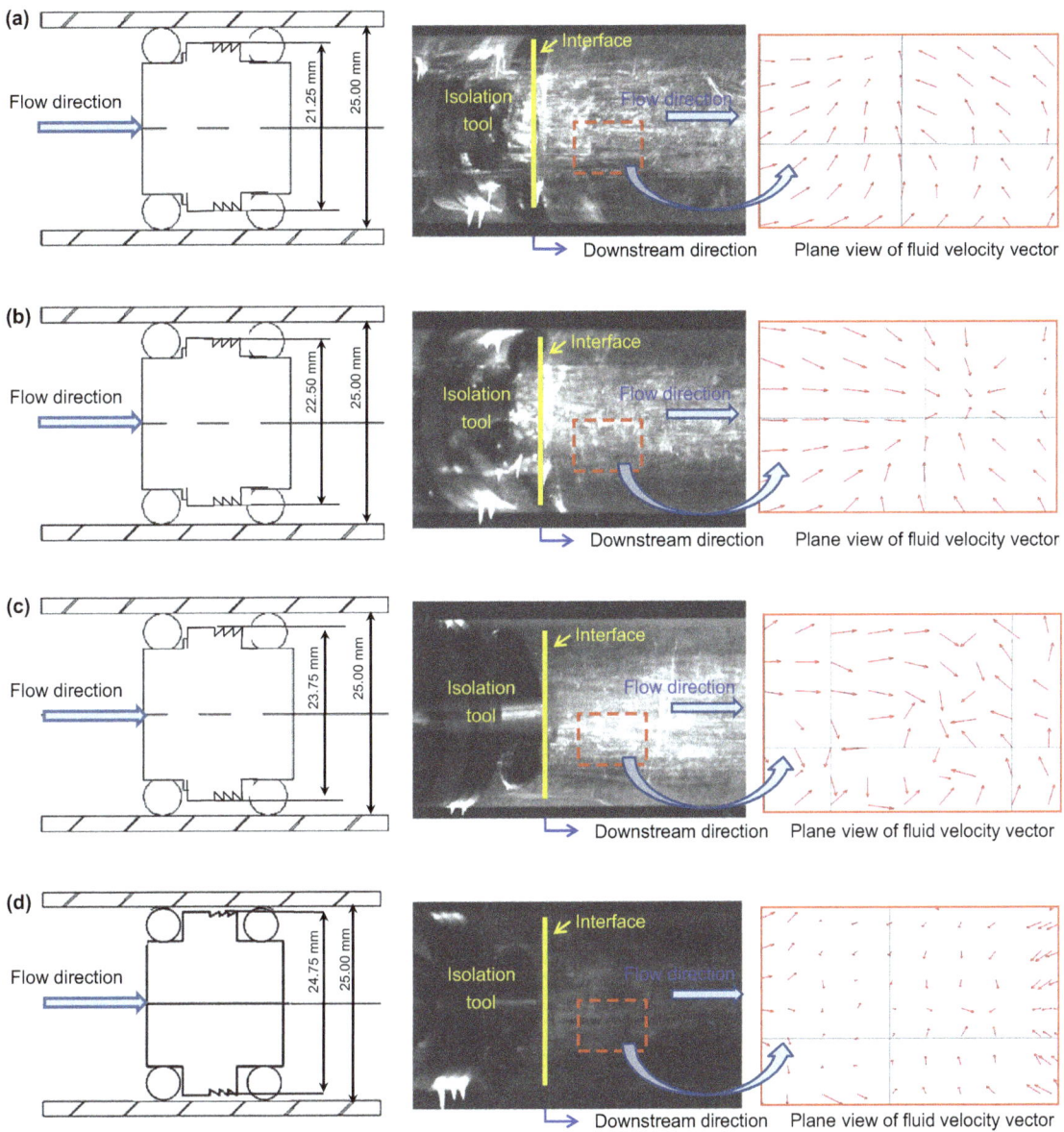

Fig. 5 Experimental velocity vectors for increasing degrees of blockage at the vertical yz plane. **a** 25 % blockage. **b** 50 % blockage. **c** 75 % blockage. **d** 99 % blockage

flow over a double forward-facing step with obstacles, an increase in the step height produced the same distribution of the velocity vector with an increase in the blockage percentage.

As the blockage percentage changed, the recirculation structure and flow pattern varied as well. To analyze the effect of the geometry deformation in the flow field downstream, the velocities along the centerline for different blockage types were obtained from numerical simulations. The mean values of the obtained experimental data are also shown in Fig. 6. Given the limitations of the experiment, the velocity profile at the centerline could only be obtained at the position from

$z = 0.04$ m to $z = 0.044$ m. The values of the velocities show the variation in the obstructed flow for different blockage percentages. At the beginning of the transformation, the velocities dropped quickly. The flow velocities changed rapidly as the transformable isolation tool applied 25 % blockage and 75 % blockage. Furthermore, the velocities became steady at approximately 0.04 m s^{-1} in the 99 % blockage state. Due to the measured data only focusing on a small section, it is basically impossible to consider the main velocity tendency of the flow. A numerical simulation was conducted under experimental conditions, and the characteristics of the flow at different plug processes would be studied for the entire area. From

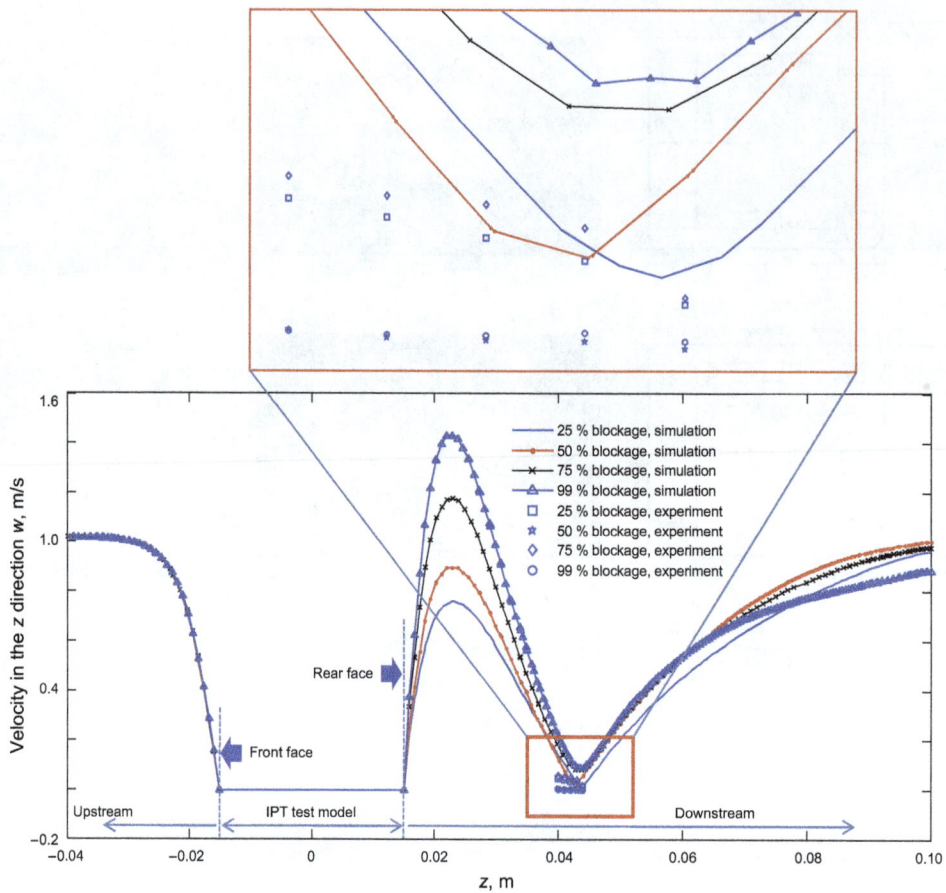

Fig. 6 Experimental and simulated velocities along the z direction at different degrees of blockage

the verification given by Fig. 6, the simulation results can be used to study the effects of the plug process in greater depth.

4 Interaction between the fluid and the isolation tool in the plug process

As mentioned before, the experimental tests had limitations and the numerical simulation was presented to study the interaction between the fluid and the deformable isolation tool in the plug process. The standard k-ε turbulence model was used with Fluent software for the simulation.

4.1 Computational models

The governing equations of mass conservation [Eq. (2)] for fluid flow are described below. In the numerical simulation model, it is assumed that the fluid is fully developed and incompressible under turbulent conditions and no heat transfer occurs. The numerical method is based on the time-marching version of the semi-implicit method for pressure-linked equations consistent (SIMPLEC).

$$\frac{\partial \rho}{\partial t} + \frac{\partial(\rho u)}{\partial x} + \frac{\partial(\rho v)}{\partial y} + \frac{\partial(\rho w)}{\partial z} = 0 \tag{2}$$

where ρ is the fluid density, kg m^{-3}; u is the fluid velocity in the x direction, m s^{-1}; v is the fluid velocity in the y direction, m s^{-1}; w is the fluid velocity in the z direction, m s^{-1}.

The features of the flow field through the isolation tool are as follows: the single phase flow is incompressible, and the fluid velocity is low. Equation (3) depicts the turbulence kinetic energy k, and the equation for the turbulence dissipation rate ε is given as Eq. (4).

$$\rho \frac{\partial k}{\partial t} + \rho v \frac{\partial k}{\partial y} = \frac{\partial}{\partial y}\left[\left(\eta + \frac{\eta_t}{\sigma_k}\right)\frac{\partial k}{\partial y}\right] + \eta_t \frac{\partial u}{\partial y}\left(\frac{\partial u}{\partial y} + \frac{\partial v}{\partial y}\right) - \rho\varepsilon \tag{3}$$

$$\rho \frac{\partial \varepsilon}{\partial t} + \rho w \frac{\partial k}{\partial z} = \frac{\partial}{\partial z}\left[\left(\eta + \frac{\eta_t}{\sigma_k}\right)\frac{\partial \varepsilon}{\partial z}\right] + \frac{c_1\varepsilon}{k}\eta_t \frac{\partial u}{\partial y}\left(\frac{\partial u}{\partial y} + \frac{\partial v}{\partial y}\right) - c_2\rho\frac{\varepsilon^2}{k} \tag{4}$$

where k is the turbulent kinetic energy, m^2 s^{-1}; η is the dynamic viscosity, kg (s m)$^{-1}$; η_t is the turbulence

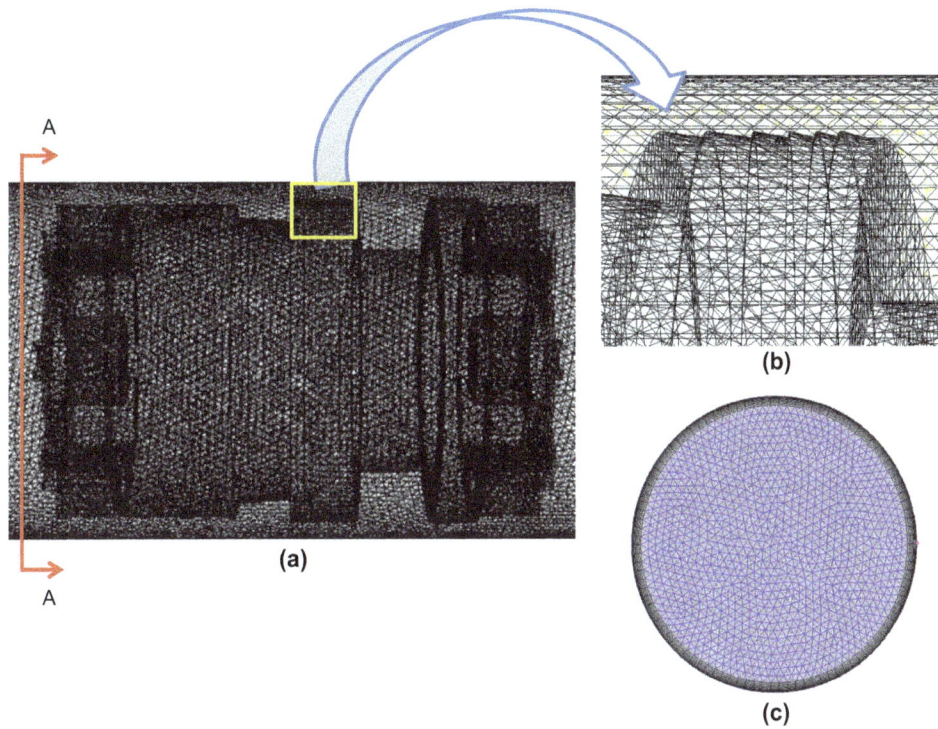

Fig. 7 Mesh model. **a** Three-dimensional mesh models of the isolation tool and the pipe. **b** Mesh between the isolation tool and the wall. **c** Mesh model of the inlet face A–A

Table 3 Level of design factors

Factor	Level 1	Level 2	Level 3
L, mm	30.00	40.00	50.00
d, mm	20.00	22.00	24.00
d_1, mm	20.00	22.37	24.75

viscosity, $\eta_t = c_\mu \rho \frac{k^2}{\varepsilon}$, kg (s m)$^{-1}$; ε is the turbulence dissipation rate, m^2 s^{-1}; c_1 and c_2 are the turbulent dissipation rate coefficients, $c_1 = 1.44$, $c_2 = 1.92$; and the model constants $c_\mu = 0.09$.

4.2 Boundary conditions

To improve the efficiency of calculation, three-dimensional mesh models of the isolation tool and pipe were created, as shown in Fig. 7. A no-slip condition at the pipe walls was assumed. There are ten rows in the boundary condition of the structure of the isolation tool. The mesh areas of the inlet face consisted of triangular cells. A tetrahedral mesh type was applied to the overall model. The whole grid system had 463,904 cells and 89,035 nodes and can be simulated accurately and display clearly. The meshing process was conducted more densely from the boundary of the isolation tool to the flow field. The three-dimensional mesh model and the inlet face are shown in Fig. 7.

Table 4 Design layout and corresponding responses

No.	L, mm	d_1, mm	d, mm	Δp, Pa
1	30.00	20.00	22.00	425
2	50.00	20.00	22.00	400
3	30.00	24.75	22.00	40
4	50.00	24.75	22.00	40
5	30.00	22.38	20.00	20,000
6	50.00	22.38	20.00	13,000
7	30.00	22.38	24.00	2,350,000
8	50.00	22.38	24.00	160,000
9	40.00	20.00	20.00	5000
10	40.00	24.75	20.00	13,000

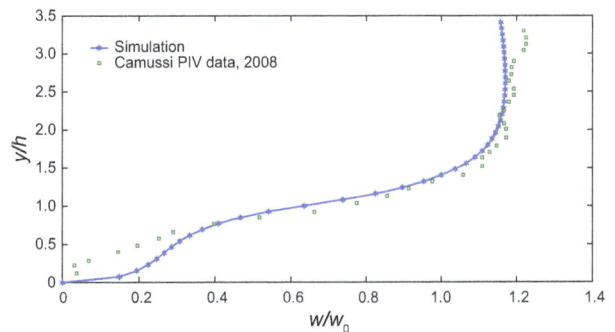

Fig. 8 A comparison of the velocity profile obtained by Camussi et al. (2008) with the simulated velocity profile

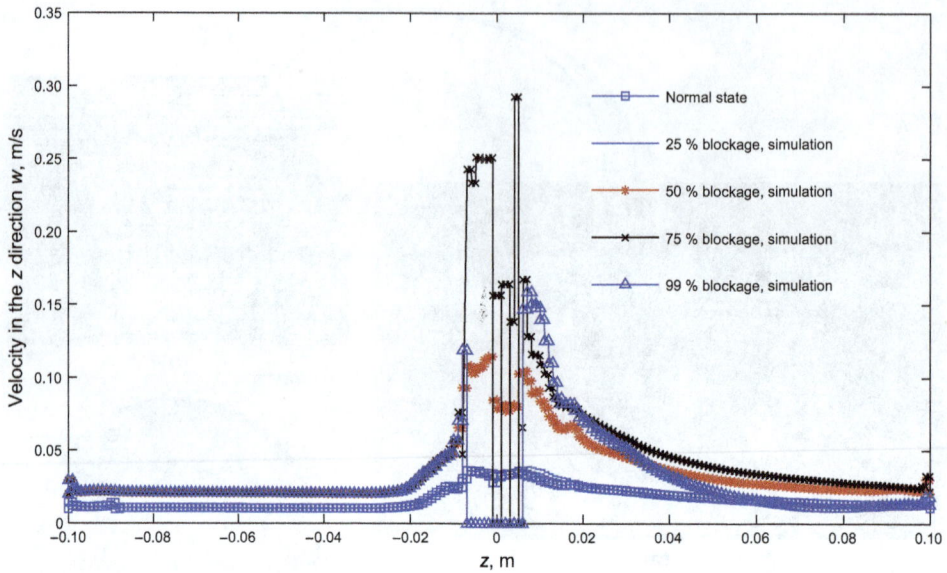

Fig. 9 Numerical streamlines along the pipe wall for increasing degrees of blockage in the vertical yz plane

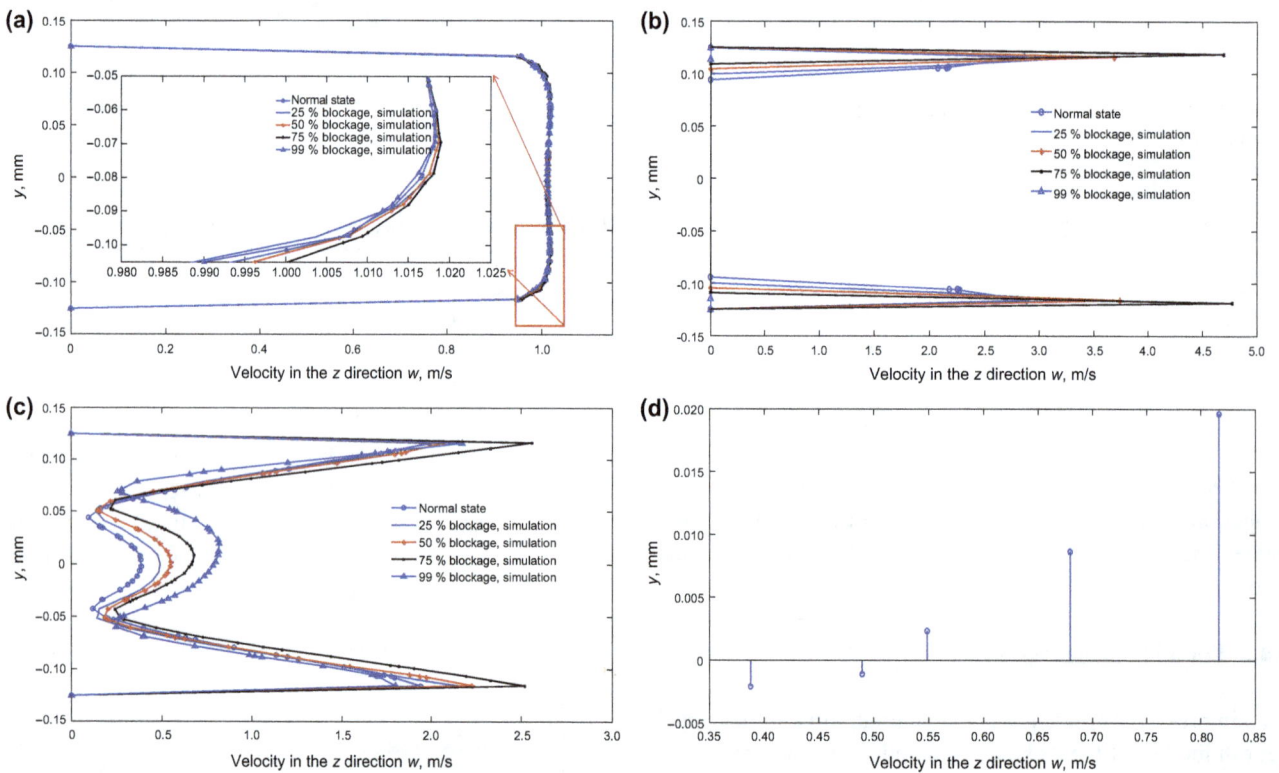

Fig. 10 Velocities of fluid particles in different planes. **a** Upstream velocity in the plane ($x = 0$, $z = -0.032$ m). **b** Velocity of the flow between the plug module and the pipe wall in the plane ($x = 0$, $z = 0$ m). **c** Downstream velocity in the plane ($x = 0$, $z = 0.032$ m).

d Peak recirculation velocity under different conditions in the plane ($z = 0.032$ m)

4.3 Box–Behnken simulation design

The Box–Behnken design is a response surface methodology design, and it is effective to identify regression model coefficients. In each block, a number of factors are put through all combinations for the factorial design, while the other factors are kept at the central values. Wu et al. (2012) conducted an optimal

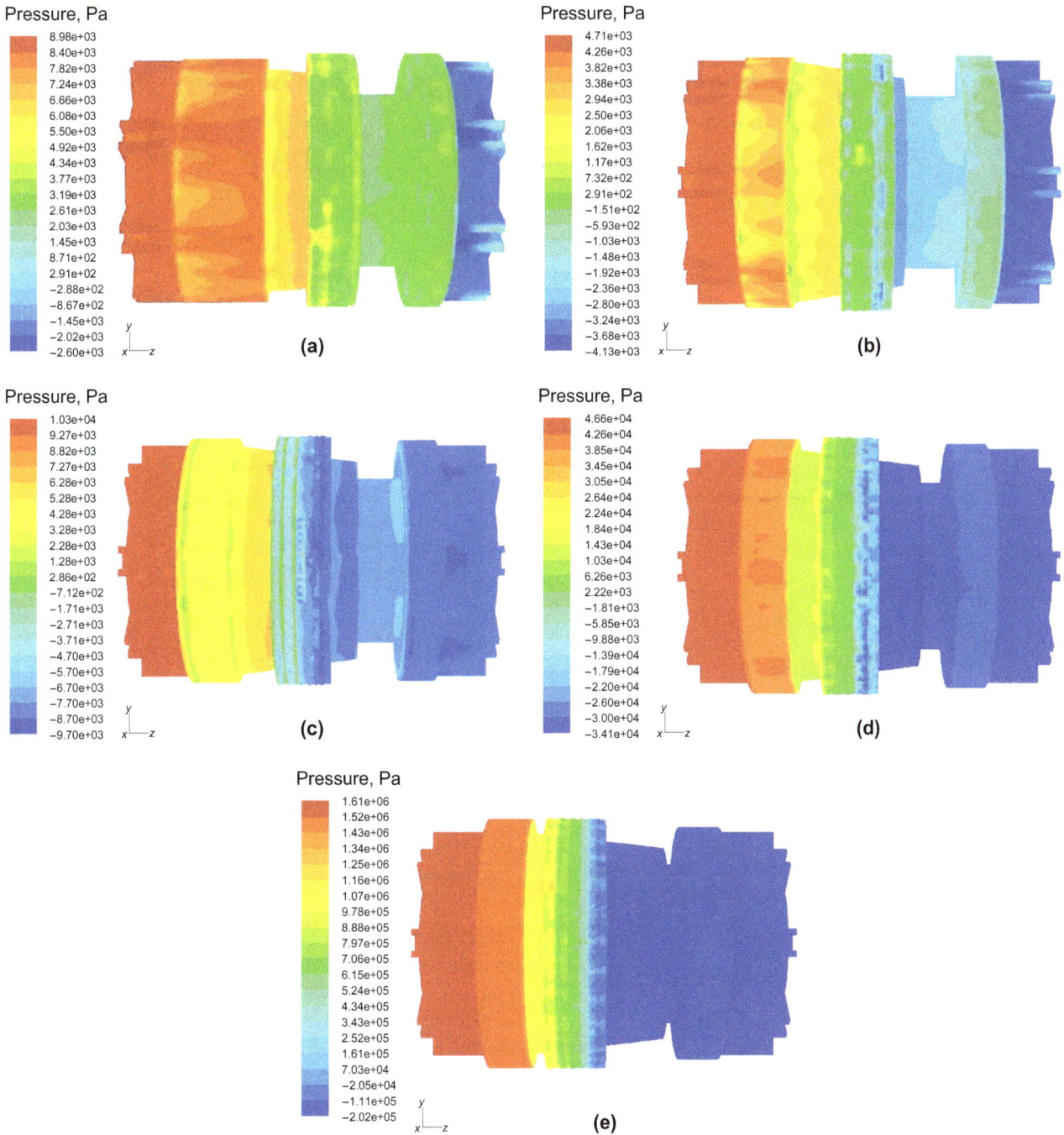

Fig. 11 Pressure contours on the isolation tool. **a** Normal condition. **b** 25 % blockage. **c** 50 % blockage. **d** 75 % blockage. **e** 99 % blockage

design for the foam cup molding process with the Box–Behnken design and obtained very good results. In this study, three factors are selected to evaluate their influences on the pressure drop between upstream and downstream of the isolation tool. Leontini and Thompson (2013) discussed the geometrical effects, and it is important to study the effects of the length (L), diameter of the plug module (d_1), and the diameter of the pressure

head (d). Three factors were influential parameters, and the level three was selected as shown in Table 3. Then, the 3-factor Box–Behnken design coordinates are listed in Table 4.

CFD simulations were conducted using the experimental design. The results for the pressure drop over the isolation tool, Δp, are listed in Table 4. The resulting design combinations are also listed in Table 4.

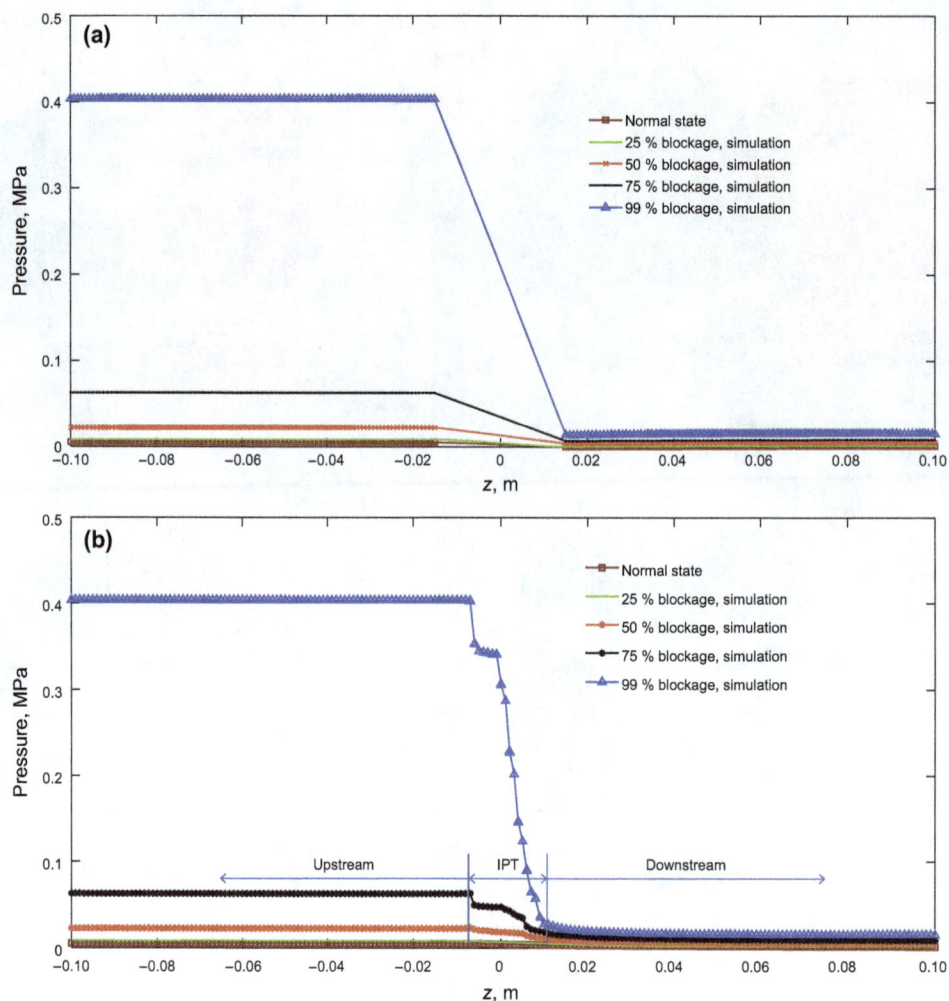

Fig. 12 Pressure distribution along different positions with increasing blockage percentages in the vertical yz plane. **a** Pressure at the centerline. **b** Pressure along the pipe wall

4.4 Numerical results

4.4.1 Validation of the numerical model

The model was validated by the normalized axial velocity profiles from previous research (Camussi et al. 2008). Computations were performed for Reynolds number $Re = 8800$. In Fig. 8, the normalized velocity profile is in a good agreement with PIV results of Camussi et al. (2008), where w/w_0 is the velocity ratio profiles and y/h is a position to downstream of the step (w is the fluid velocity in the z direction; w_0 is the inlet velocity; y is coordinate in the y axis; h is the height of step).

4.4.2 Effects on velocities between flow and the isolation tool in the plug process

Figure 6 shows the centerline velocity of flow for each degree of blockage. The velocity was measured at the symmetry plane, and the plots show both the experimental

values (symbols) and numerical data (symbol lines). The upstream velocity of the test model appears to be steady state and remains almost the same regardless of the degree of blockage, approximating to the inlet velocity. However, the downstream velocity changes rapidly as the degree of blockage changes and a significant change appears at the rear end of the model. The peak flow velocity increases with an increase in the degree of blockage. The same phenomenon was found at high Reynolds numbers (Yoshioka et al. 2001) in flow over backward-facing steps. The velocity changes considerably in the region near the rear face of the model, leading to recirculation.

The velocity of the fluid between the model and the wall (as shown in Fig. 9) reached a maximum value when the blockage percentage approached 75 %. Subsequently, the velocity dropped quickly when the degree of backflow recirculation reached 50 % blockage. The experimental data exhibited the same trend as the simulation results, thus confirming the existence of low velocities and the

Table 5 Pressure drop and difference

Blockage percentage, %	Pressure drop at the centerline, Pa	Pressure drop along the pipe wall, Pa	Pressure difference, Pa
0 (normal)	4991.9	2808	2183.9
25	8420.5	5724	2696.5
50	18,758	15,270	3488
75	55,983	38,200	17,783
99	388,570	381,300	7270

appearance of recirculation around the isolation tool. Subsequently, the velocity behind the isolation tool became steady without any large fluctuations downstream.

To understand the behavior of the velocity of flow through the isolation tool and the effects of the deformation of the isolation tool, Fig. 10 presents transverse velocity profiles at the upstream location ($x = 0$, $z = -0.032$ m), the symmetry plane ($x = 0$, $z = 0$ m), and the downstream location ($x = 0$, $z = 0.032$ m). Figure 10a shows the velocity profiles at the upstream location for four degrees of blockage. The velocity in this figure is in a good agreement with previous observations in which the upstream velocity is steady for each degree of blockage, being almost the same as the inlet velocity. In the region near the wall, however, the velocity fluctuates slightly, indicating the effect of the deformation of the isolation tool. Then, for the symmetry profile ($x = 0$, $z = 0$) (shown in Fig. 10b), the velocity is higher near the pipe wall because the pressure drop increases rapidly with the degree of blockage. For 75 % blockage, the velocity is the highest in this region. The pressure in this region increases due to the deformation of the isolation tool, so the flow velocity increases sharply. When the degree of blockage is 99 %, the velocity is zero in this region as few particles are detected in the fluid. Figure 10c shows that the velocity downstream begins to fluctuate and recirculation appears. In the downstream section, the velocity near the wall increases with the degree of deformation, with the velocity peaking for the 75 % blockage. The flow velocity decreases near the pipe wall for the 99 % blockage, and the deformation of the isolation tool has almost completely obstructed the pipe. The fluctuation around the centerline changed remarkably, however. The peak velocity of recirculation exhibits asymmetry, together with an upward trend (as shown in Fig. 10d).

4.4.3 Interaction model and analysis of main mechanical parameters

Figure 11 shows the pressure contour on the isolation tool from 0 (normal state) to 99 % blockage. It can be seen that the left pressure head experiences the highest pressure and the pressure applied on the plug module is not so high. The

Table 6 Estimated regression coefficients

Term	Model coefficient
b_0	$-1.877\mathrm{e}+008$
b_1	$1.954\mathrm{e}+006$
b_2	$1.554\mathrm{e}+007$
b_3	$6.221\mathrm{e}+006$
b_4	$-24{,}251.236$
b_5	$-2.862\mathrm{e}+005$
b_6	$40{,}743.593$
b_7	-703.750
b_8	$-83{,}440.157$
b_9	$-6.566\mathrm{e}+005$
b_{10}	1043.005
b_{11}	$13{,}489.196$

pressure applied on the right pressure head is low, but most of the modules, except for the wheel and the wheel hub, are exhibited in a high negative pressure. A negative pressure is created by the complicated structure and a sudden expansion of flow which is not obstructed by the rear end of the isolation tool. To better understand the pressure variation with the geometric transformation, the pressures at the centerline and along the pipe wall are shown in Fig. 12, respectively. Figure 12a shows that all of the upstream pressures are higher than those downstream. The upstream pressure increases with the degree of blockage, especially between 75 % and 99 % blockage. The upstream and downstream pressures are stable and do not fluctuate. The pressures along the pipe wall (Fig. 12b) are different from the pressures shown in Fig. 12a in terms of the isolation tool location. It can be seen that only a small amount of fluctuation appears as the degree of blockage increases. Table 5 shows the pressure drop between the upstream and downstream areas at the centerline and along the pipe wall. The pressure drop at the centerline is higher than that along the pipe wall. The pressure difference is computed in Table 5. This shows that the pressure difference is not linear with the blockage increasing and the highest is 17,783 Pa for the 75 % blockage stages, rather than for the 99 % blockage.

Analysis of variance (ANOVA) and response surface analysis were used to determine the statistical significance of the model. The adequacy of the model was predicted

Table 7 ANOVA table for Δp

Source	DF	Sum of squares	Mean square	F value	p value	Percentage contribution C, %
Model	11	9.268e+11	8.425e+10	230.44	0.0004	
A-L	1	1.098e+9	1.098e+9	3.00	0.1815	0.12
B-d1	1	2.458e+11	2.458e+11	672.32	0.0001	10.78
C-d	1	3.491e+10	3.491e+10	95.49	0.0023	3.76
AB	1	156.25	156.25	4.274e-7	0.9995	0
AC	1	7.924e+8	7.924e+8	2.17	0.2374	0
BC	1	2.538e+11	2.538e+11	694.19	0.0001	27.35
A^2	1	3.084e+10	3.084e+10	84.36	0.0027	3.32
B^2	1	1.291e+10	1.291e+10	35.30	0.0095	27.44
C^2	1	9.807e+10	9.807e+10	268.23	0.0005	1.39
A^2B	1	1.227e+11	1.227e+11	335.66	0.0004	13.22
B^2C	1	4.631e+10	4.631e+10	126.67	0.0015	4.99
Residual error	3	1.097e+9	3.656e+8			
Lack of fit	1	1.097e+9	1.097e+9			
Pure error	2	0	0			
Total	14	9.279e+11				

$R^2 = 99.88$ %

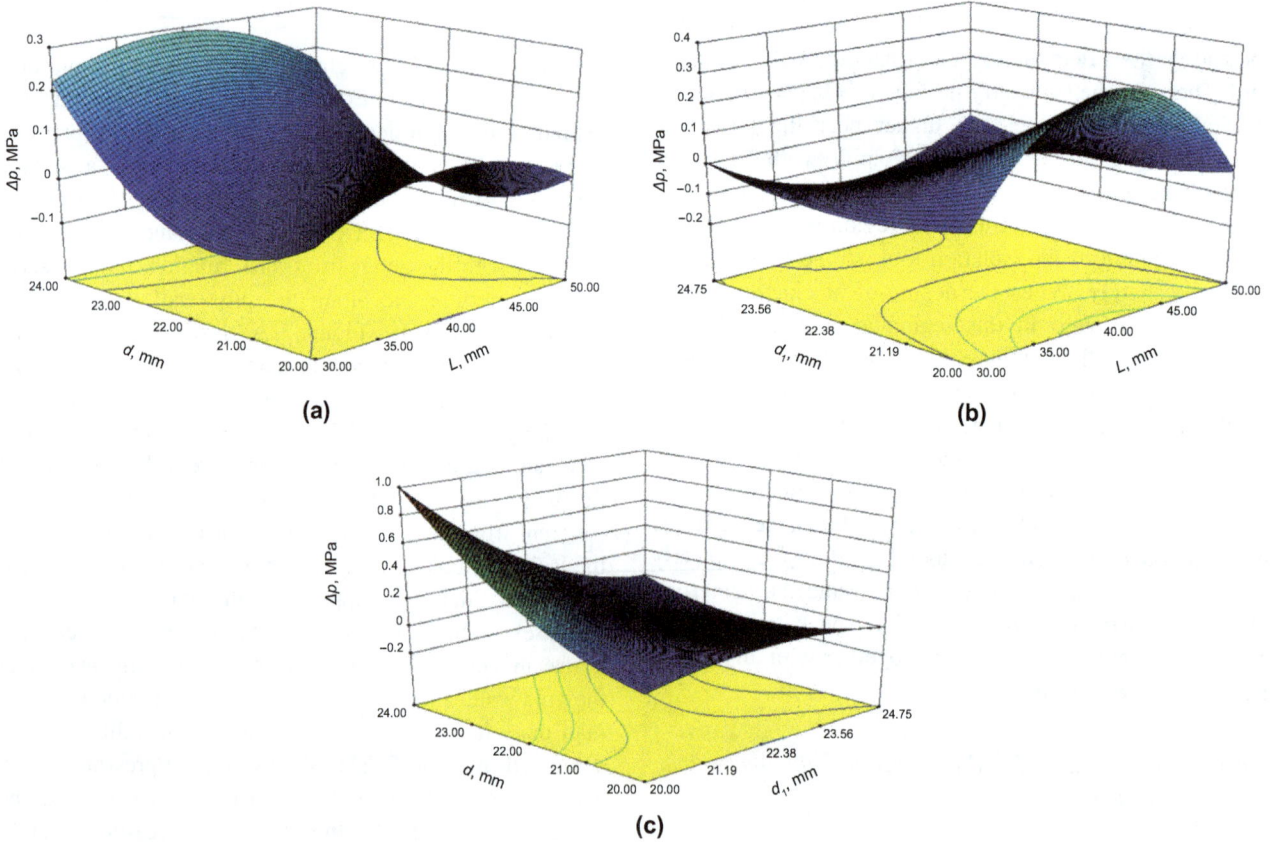

(a)

(b)

(c)

Fig. 13 Surface plots between the pressure drop Δp and three parameters. **a** d and L with $d_1 = 22.38$ mm. **b** d_1 and L with $d = 22$ mm. **c** d_1 and d with $L = 40$ mm

with the ANOVA ($P < 0.05$) and regression analysis (R^2). The relationship between the response and independent variables was demonstrated using a response surface plot.

The second-order regression model of Δp is shown in Eq. (5), and estimated regression coefficients are listed in Table 6.

$$\Delta p = b_0 + b_1 L + b_2 d_1 + b_3 d + b_4 L^2 + b_5 d_1^2 + b_6 d^2 + b_7 (Ld)$$
$$+ b_8 (Ld_1) + b_9 (dd_1) + b_{10} (L^2 d_1) + b_{11} (d_1^2 d) \quad (5)$$

The result of ANOVA (Table 7) reveals the relationships between mechanical parameters and the pressure drop Δp over the isolation tool. The indices in Table 7 illustrate the sum of squares, mean squares, the degrees of freedom (DF), F value, and probability (p value), as well as the percentage contribution (C). The low p value indicates that the regression model can predict Δp from the designed factors with 99.88 % confidence. It is clear that d_1 greatly influences Δp which has 10.78 % of the contribution from the results of ANOVA. The parameter d has a contribution of 3.76 %. The parameter L has a minimum influence among the single parameters with 0.12 % contributions. The coupling terms like BC($d_1 d$), B^2(d_1^2), A^{2}B($L^2 d_1$) are dominant 27.35 %, 27.44 %, and 13.22 % with contributions. Three-dimensional surface plots between Δp and three parameters are shown in Fig. 13 with help of the Minitab software package. Figure 13 shows the surface plots of the pressure drop Δp against two variable factors and one fixed variable. In Fig. 13a, the fixed variable d_1 was held at the minimum value. The surface plot shows that the pressure drop Δp increases with the values of d. The value L becomes greater, and the pressure drop Δp shows nonlinear relations with it. In Fig. 13b, the fixed variable d was held at the minimum value. The surface plot shows that the greater the values of d_1, the greater the pressure drop Δp is. The value L becomes greater, and the pressure drop Δp shows nonlinear relations with it. In Fig. 13c, the fixed variable L was held at the middle value. The surface plot shows that the greater the values of d_1 and d, the greater the pressure drop Δp is.

5 Conclusions

In this study, interactions between the turbulent flow and the isolation tool in the blocking process were investigated with experimental tests and numerical simulations. A comparison between the experimental results and numerical simulation indicates that the standard k-ε model can be used to predict the flow characteristics in the plug process. The influences of the main mechanical parameters were evaluated, and the regression model can be used to predict

the relationships between the pressure drop (Δp) and the three main mechanical parameters of the isolation tool in the plug process. The major results can be summarized as follows:

(1) The downstream pressure and velocity changed more rapidly than those upstream, while the pressure drop (Δp) changed considerably.
(2) The pressure drop was found to change as the degree of blockage increased, with the variation being the greatest for 75 % blockage. The regression model between the pressure drop and mechanical main parameters was obtained.
(3) The diameter d_1 greatly influences Δp which has 10.78 % of the contribution and the parameter d has a contribution of 3.76 %. The parameter L has a minimum influence among the single parameters with 0.12 % contribution.

Note that this study provides the way to analyze the interactions between the flow and the isolation tool in the plug process, and the regression model was built to evaluate the influences of the main parameters. It is pointed that the plug process of the isolation tool is a slow dynamic process. Further study will focus on the transient plug process of the isolation tool in numerical simulations and experimental tests.

Acknowledgments This work was financially supported by the National Natural Science Foundation of China (Grant No. 51575528), the Scientific Research Foundation of the Education Ministry for Returned Chinese Scholars (China), the State Key Laboratory for Coal Resources and Safe Mining, China University of Mining and Technology (No. SKLCRSM10KFB04), and the Science Foundation of China University of Petroleum, Beijing (No. YXQN-2014-02).

References

Camussi R, Felli M, Pereira F, et al. Statistical properties of wall pressure fluctuations over a forward-facing step. Phys Fluids. 2008;20(7):75–113. doi:10.1063/1.2959172.

Chen P, Chen Y, Pan C, et al. Parameter optimization of micro milling brass mold inserts for micro channels with Taguchi method. Int J Precis Eng Manuf. 2015;16(4):647–51. doi:10.1007/s12541-015-0086-1.

Han B, Wang Z, Zhao H. Strain-based design for buried pipelines subjected to landslides. Pet Sci. 2012;9(2):236–41. doi:10.1007/s12182-012-0204-y.

Huang HP, Yan JL, Cheng TH. Development and fuzzy control of a pipe inspection robot. IEEE Trans Ind Electron. 2010;57(3):1088–95. doi:10.1109/TIE.2009.2031671.

Lesani M, Rafeeyan M, Sohankar A. Dynamic analysis of small pig through two and three dimensional liquid pipeline. J Appl Fluid Mech. 2012;5(2):75–83.

Leontini JS, Thompson MC. Vortex-induced vibrations of a diamond cross-section: sensitivity to corner sharpness. J Fluid Struct. 2013;39:371–90. doi:10.1016/j.jfluidstructs.2013.01.002.

Li SE, Park J, Lim J, et al. Design and control of a passive magnetic levitation carrier system. Int J Precis Eng Manuf. 2015;16(4):693–700. doi:10.1007/s12541-015-0092-3.

Minami K, Shoham O. Pigging dynamics in two-phase flow pipelines: experiment and modeling. SPE Prod Facil. 1995;10(4):225–32. doi:10.2118/26568-PA.

Mirshamsi M, Rafeeyan M. Speed control of pipeline pig using QFT method. Oil Gas Sci Technol. 2012;67(4):693–701. doi:10.2516/ogst/2012008.

Mirshamsi M, Rafeeyan M. Dynamic analysis of pig through two and three dimensional gas pipeline. J Appl Fluid Mech. 2015;8(1):43–54. doi:10.1016/j.jngse.2015.02.004.

Nguyen TT, Hui RY, Yong WR, et al. Speed control of pig bypass flow in natural gas pipeline. In: International symposium on industrial electronics, June 12–16, Pusan, Korea; 2001a. doi:10.1109/ISIE.2001.931581.

Nguyen TT, Kim SB, Yoo HR, et al. Modeling and simulation for pig flow control in natural gas pipeline. J Mech Sci Technol. 2001;15(8):1165–73. doi:10.1007/BF03185671.

Nguyen, TT, Kim DK, Rho YW, et al. Dynamic modeling and its analysis for PIG flow through curved section in natural gas pipeline. In: International symposium on computational intelligence in robotics and automation, July 29–August 1, Banff, Alberta, Canada; 2001c. doi:10.1109/CIRA.2001.1013250.

Nieckele AO, Braga AMB, Azevedo LFA. Transient pig motion trough gas and liquid pipelines. ASME J Energy Resour. 2001;12(3):260–9. doi:10.1115/1.1413466.

Ono M, Kato S. A study of an earthworm type inspection robot movable in long pipes. Int J Adv Robot Syst. 2004;7(1):85–90. doi:10.5772/7248.

Oztop HF, Mushatet KS, Yilmaz I. Analysis of turbulent flow and heat transfer over a double forward facing step with obstacles. Int Commun Heat Mass Transf. 2012;39:1395–403. doi:10.1016/j.icheatmasstransfer.2012.07.011.

Poompipatpong C, Kengpol A. Design of a decision support methodology using response surface for torque comparison: an empirical study on an engine fueled with waste plastic pyrolysis oil. Energy. 2015;82:850–6. doi:10.1016/j.energy.2015.01.095.

Roh S, Kim DW, Lee J, et al. In-pipe robot based on selective drive mechanism. Int J Control Autom Syst. 2009;7(1):105–12. doi:10.1007/s12555-009-0113-z.

Saeidbakhsh M, Rafeeyan M, Ziaei-rad S. Dynamic analysis of small pigs in space pipelines. Oil Gas Sci Technol. 2009;64(2):155–64. doi:10.2516/ogst:2008046.

Saravanakumar D, Mohan B, Muthuramalingam T. Application of response surface methodology on finding influencing parameters

in servo pneumatic system. Measurement. 2014;54(8):40–50. doi:10.1016/j.measurement.2014.04.017.

Selden RA. Innovative solution for emergency repair of a deep water riser. In: The offshore technology conference, 4–7 May, Houston, Texas, USA; 2009. doi:10.4043/20154-MS.

Song C, Kwon K, Park J, et al. Optimum design of internal flushing channel of drill bit using RSM and CFD simulation. J Precis Eng Manuf. 2014;15(6):1041–50. doi:10.1007/s12541-014-0434-6.

Solghar AA, Davoudian M. Analysis of transient PIG motion in natural gas pipeline. Mech Ind. 2012;13(5):293–300. doi:10.1051/meca/2012039.

Tveit E, Aleksandersen O. Remote controlled (Tether-Less) high pressure isolation system. In: SPE Asia Pacific oil and gas conference and exhibition, Brisbane, Australia; 2000. doi:10.2118/64513-MS.16-18.

Vahabi M, Mehdizadeh E, Kabganian M, et al. Modelling of a novel in-pipe micro robot design with IPMClegs. J Syst Control Eng. 2011;225(I1):63–73. doi:10.1243/09596518JSCE1042.

Wang ZW, Cao QX, Luan N, et al. Development of an autonomous in-pipe robot for offshore pipeline maintenance. Ind Robot Int J. 2010;37(2):177–84. doi:10.1108/01439911011018957.

Wu L, Yick K, Ng S, et al. Parametric design and process parameter optimization for bra cup molding via response surface methodology. Expert Syst Appl. 2012;39:162–71. doi:10.1016/j.eswa.2011.07.003.

Xu XX, Gong J. Pigging simulation for horizontal gas-condensate pipelines with low-liquid loading. J Pet Sci Eng. 2005;48:272–80. doi:10.1016/j.petrol.2005.06.005.

Yeung HC, Lima PCR. Modeling of pig assisted production methods. J Energy Resour Technol Trans ASME. 2002;124(3):8–13. doi:10.1115/1.1446474.

Yoshioka S, Obi S, Masuda S. Turbulence statistics of periodically perturbed separated flow over a backward-facing step. Int J Heat Fluid Flow. 2001;22(4):393–401. doi:10.1016/S0142-727X(01)00079-0.

Zeng D, Deng K, Lin Y. Theoretical and experimental study of the thermal strength of anticorrosive lined steel pipes. Pet Sci. 2014;11(3):417–23. doi:10.1007/s12182-014-0356-z.

Zhang L, Long Z, Cai J, et al. Multi-objective optimization design of a connection frame in macro-micro motion platform. Appl Soft Comput. 2015;32:369–82. doi:10.1016/j.asoc.2015.03.044.

Zhao B, Li C, Zhang J, et al. The isolation technology of oil and gas pipeline in China. In: International offshore and polar engineering conference, June 20–25, Beijing, China; 2010.

5

Formation mechanisms and sequence response of authigenic grain-coating chlorite: evidence from the Upper Triassic Xujiahe Formation in the southern Sichuan Basin, China

Yu Yu[1,2] · Liang-Biao Lin[1,2] · Jian Gao[1,2]

Abstract Authigenic grain-coating chlorite is widely distributed in the clastic rocks of many sedimentary basins around the world. These iron minerals were mainly derived from flocculent precipitates formed when rivers flow into the ocean, especially in deltaic environments with high hydrodynamic conditions. At the same time, sandstone sequences with grain-coating chlorites also tend to have relatively high glauconite and pyrite content. EPMA composition analysis shows that glauconites with "high Al and low Fe" content indicate slightly to semi-saline marine environments with weak alkaline and weakly reducing conditions. By analyzing the chlorite-containing sandstone bodies of the southern Sichuan Xujiahe Formation, this study found that chlorite was mainly distributed in sedimentary microfacies, including underwater distributary channels, distributary channels, shallow lake sandstone dams, and mouth bars. Chlorite had a tendency to form in the upper parts of sandstone bodies with signs of increased base level, representing the influence of marine (lacustrine) transgression. This is believed to be influenced by megamonsoons in the Middle and Upper Yangtze Region during the Late Triassic Epoch. During periods of abundant precipitation, river discharges increased and more Fe particulates flowed into the ocean (lake). In the meantime, increases or decreases in lake level were only affected by precipitation for short periods of time. The sedimentary environment shifted from weakly oxidizing to weak alkaline, weakly reducing conditions as sea level increased, and Fe-rich minerals as authigenic chlorite and glauconite began to form and deposit.

Keywords Sichuan Basin · Xujiahe Formation · Grain-coating chlorite · Glauconite · Pyrite

1 Introduction

Grain-coating chlorite, also known as pore-lining chlorite, is widely distributed in many sedimentary basins around the world. Studies have been performed on many sedimentary formations such as the Upper Triassic Yanchang Formation in the Ordos Basin (Zhang et al. 2011; Ding et al. 2013; Zhang et al. 2013; Zhu et al. 2015), the Upper Jurassic in the Songliao Basin (Zeng 1996), Xujiahe Formation in the Sichuan Basin (Liu et al. 2009; Huang et al. 2010; Zou et al. 2013), Middle Jurassic Shaximiao Formation in the Sichuan Basin (Lü et al. 2015), Jurassic Sangonghe Formation in the Junggar Basin (Xi et al. 2015), Upper Cretaceous in the Santos Basin, Brazil (Bahlis and Luiz 2013), and the Lower Vicksburg Formation in Southern Texas (Grigsby 2001). However, the clastic sequences which developed grain-coating chlorite in these sedimentary basins have varying characteristics. For example, the Yanchang Formation in the Ordos Basin was deposited in a continental sedimentary environment, the Upper Jurassic of the Songliao Basin is from a marine sedimentary environment, and the Xujiahe Formation in the Sichuan Basin is in a hybrid marine–continental sedimentary environment. However, all of these sandstones were deposited in high salinity environments. Huang et al.

✉ Liang-Biao Lin
linliangbiao08@cdut.cn

[1] State Key Laboratory of Oil and Gas Reservoir Geology and Exploitation, Chengdu University of Technology, Chengdu 610059, Sichuan, China

[2] Institute of Sedimentary Geology, Chengdu University of Technology, Chengdu 610059, Sichuan, China

Edited by Jie Hao

(2004) also thought chlorite formed in slightly saline water or semi-saline marine environments and characterized by successive ocean fluids.

Many studies performed on grain-coating chlorite mainly focus on the genetic mechanism of chlorite and its effects on reservoir porosity preservation. Bloch et al. (2002), Huang et al. (2004), and Zhu et al. (2004) thought grain-coating chlorite was beneficial to reservoir porosity preservation. Grain-coating chlorite which formed during the early diagenetic stage could enhance compaction resistance and inhibit quartz overgrowth cementation. Quartz overgrowths are considered one of the key factors responsible for reservoir densification (Zhu et al. 2009; Gu et al. 2014). However, Liu et al. (1998) believed primary porosity was reduced because of grain-coating chlorite which would block pore throats and reduce permeability. As for the study of chlorite (mainly referring to chlorite which precipitates from pore waters and forms early in diagenesis), study of the genetic mechanism got relatively consistent results. Grain-coating chlorite formation was thought to be mainly controlled by sedimentary facies and formed in high hydrodynamic conditions and weak alkaline environments with abundant Fe and Al sources (Huang et al. 2004). Sandstone sequences with well-developed chlorite were found in sedimentary facies conditions described above, including underwater distributary channels and distributary channels (Ehrenberg 1993; Pen et al. 2009; Okwese et al. 2012; Sun et al. 2014). However, there is no clear answer to why authigenic grain-coating chlorite tends to form in such conditions. Is there any connection between chlorite and the sedimentary environment conditions? What factors control the distribution and development of chlorite besides the sedimentary facies? Solving these problems would not only enhance the scientific understanding of authigenic grain-coating chlorite, but also provide data to search for sandstone sequences containing authigenic grain-coating chlorite.

Authigenic chlorite is widely distributed in Xujiahe Formation sandstones. Chlorite can be found in the main reservoir intervals (Xu 2 Member, Xu 4 Member, Xu 6 Member). The majority of authigenic chlorite in the Xujiahe sandstones is found as grain coatings. This study examines samples from the Upper Triassic Xujiahe Formation sandstones taken from the southern Sichuan Basin. The study analyzes the occurrence and microcharacteristics of grain-coating chlorite through thin section, SEM, and EPMA analysis. The study also examines the characteristics of paragenetic minerals (glauconite, pyrite) and the vertical distribution patterns of authigenic chlorite in various sandstone bodies to study the formation environment, genetic mechanism, and other factors which affected the distribution of authigenic grain-coating chlorite.

2 Geological setting

Petroleum and gas exploration in the southern Sichuan Basin began in the 1950s, but the Upper Triassic Xujiahe Formation had not been widely explored until recently. In recent years, exploration of Danfengchang and Guanyinchang (which had seven industrial gas wells drilled by the end of 2013 with a cumulative gas production of 1.63×10^8 m^3) had discovered a batch of commercial gas reservoirs (Zhu et al. 2014; Zhang et al. 2014). These discoveries show the Xujiahe Formation has excellent hydrocarbon potential.

The southern Sichuan Basin can be divided into the western, central, southwestern, and southeastern Sichuan low and steep structural areas (Lin et al. 2015). The Upper Triassic was the key period for Sichuan Basin sedimentary development, where the basin experienced transformation from a marine to a continental sedimentary environment (Lin et al. 2006) as marine facies to transitional facies to continental facies. The sedimentary systems of the Xujiahe Formation include alluvial fans, marine (lacustrine) deltas, and lacustrine environments. The Xujiahe Formation strata are divided into six members. The Xu 2 Member (T_3x^2), Xu 4 Member (T_3x^4), and Xu 6 Member (T_3x^6) are the main reservoir intervals in the study area, and their lithology consists primarily of fine-medium sandstone. The lithology of the Xu 1 Member (T_3x^1), Xu 3 Member (T_3x^3), and Xu 5 Member (T_3x^5) is mainly black shale and mudstone with gray silt stone and sandstone. T_3x^1 and T_3x^2 are absent in the center and south of the basin (Fig. 1).

3 Characteristics of grain-coating chlorite

3.1 Occurrences and formation time

Chlorite is a complicated, layered aluminosilicate mineral that is composed primarily of Fe, Mg, Si, and Al. The chemical formula is as follows:

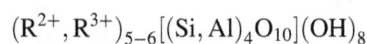

$$(R^{2+}, R^{3+})_{5-6}[(Si, Al)_4 O_{10}](OH)_8$$

$$R^{2+} = Mg, Fe, Mn, Ni$$

$$R^{3+} = Al, Fe, Cr, Mn$$

Chlorite group minerals have a stratified structure, consisting of alternating interlayers of talc and brucite. The majority of chlorites are monoclinic, though there are some triclinic polymorphs. Chlorite crystals mainly form either hexagonal sheets or tabular patterns, though a few are barrel shaped. The cross section of chlorite is either hexagonal or irregular flake shaped, while the longitudinal section of chlorite crystals is rectangular (Chang 2006).

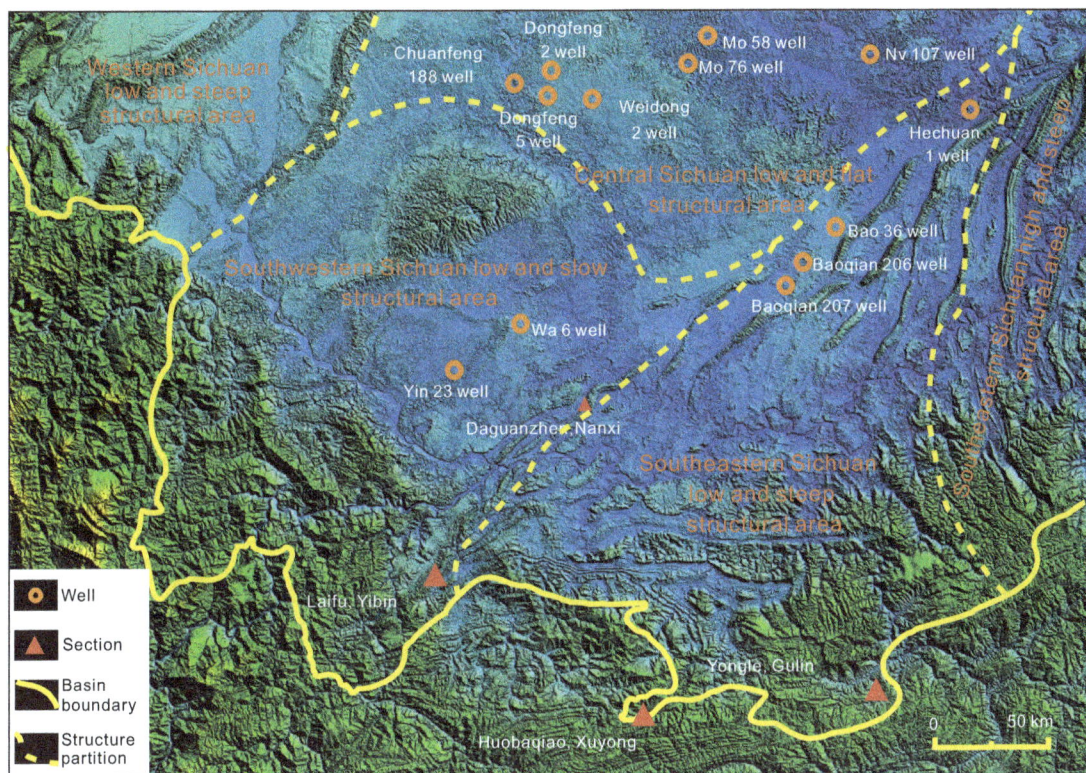

Fig. 1 Geological setting of study area, including well locations, basin boundaries, and structure delineations

Thin section and SEM analysis found that almost all authigenic chlorite grew around the grains (Fig. 2). Chlorite growth could clearly be observed on grain surfaces as scaly sheets in SEM images (Fig. 2e, f).

Authigenic chlorite in Xujiahe Formation sandstones has the following characteristics: (1) Because chlorite had already developed between grain contacts, it can be concluded authigenic chlorite formed on grain surfaces before mechanical compaction; (2) authigenic chlorite growths were found on the surface of quartz overgrowths and silica cement which filled in intergranular pores, indicating that authigenic chlorite may have formed after silica cementation; (3) chlorite was not found in intergranular dissolved pores or mouldpores, while chlorite has been found on the surface of feldspar grains with dissolved pores, suggesting authigenic chlorite formed before dissolution. These characteristics indicate authigenic chlorite growth in the Xujiahe Formation occurred in multiple phases. The first stage took place during syngenesis–early diagenesis A stage. Authigenic chlorite had already coated the grains before clastic grain contacts developed. This is also the formation stage of most authigenic chlorite in the study area. The second stage took place after silica cementation but before dissolution. Authigenic chlorite growths could be seen on the surface of silica cement, but chlorite was not found in intragranular dissolved pores. This suggests authigenic chlorite at this stage formed before aluminosilicate

minerals were dissolved by organic acids. According to "China's Dividing Standard for Diagenetic Stage of Clastic Rocks (SY/T5477-2003)," the second stage of chlorite growth occurred around the end of early diagenesis A stage to the beginning of middle diagenesis A stage.

3.2 Source material

The formation process of authigenic chlorite in clastic rocks is still widely debated. Some researchers believe authigenic chlorite transformed from Fe-rich grain coating (Grigsby 2001; Lanson et al. 2002). Billault et al. (2003) thought authigenic chlorite precipitated from pore water. Clay minerals which transform into chlorite leave honeycomb-like structures which were not found in Xujiahe Formation sandstones. Most authigenic chlorite rims have nearly equal thicknesses, and most grain-coating chlorite formed during syngenesis–early diagenesis A stage. These characteristics suggest authigenic chlorite in the Xujiahe Formation sandstones mainly precipitated from pore waters and coat other grains.

Sandstones from the Xu 2 and Xu 4 Formation Members with grain-coating chlorite were selected for EPMA testing. The analyses were performed in the State Key Laboratory of Oil and Gas Reservoir Geology and Exploitation, Chengdu University of Technology (Table 1). Xujiahe Formation chlorite is mainly composed of FeO, SiO_2,

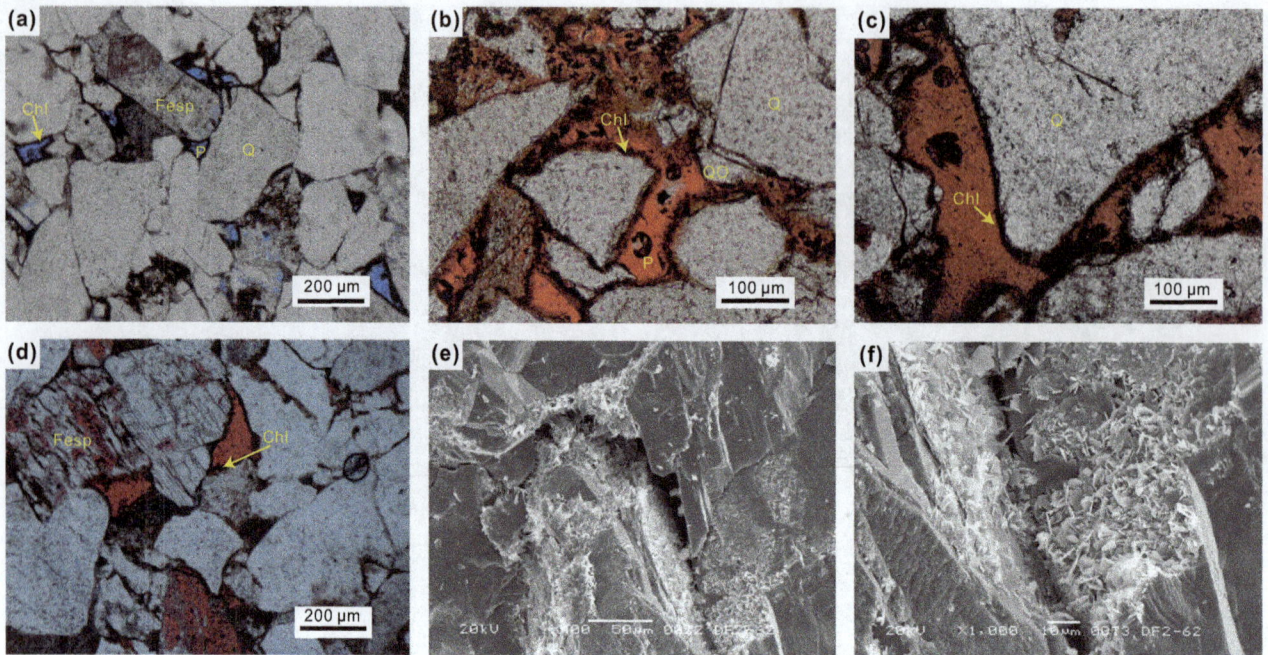

Fig. 2 Microphotographs of sandstone with grain-coating chlorite of the Xujiahe Formation in the southern Sichuan Basin. Figure descriptions including well numbers and formation are listed as follows: **a** authigenic chlorite coating grain, Lu 107 well, Xu 2 Member, PPL. **b** Authigenic chlorite growth on silica cement in top right corner, Hechuan 1 well, Xu 2 Member, PPL. **c** Authigenic chlorite growth on silica cement, Hechuan 1 well, Xu 2 Member, PPL. **d** Authigenic chlorite growth on feldspar surface with dissolved pores in grains, Bao 36 well, Xu 4 Member, PPL. **e** Scaly sheet chlorite sticking to the surfaces of quartz grains, Dongfeng 2 well, Xu 4 Member. **f** Zoom in **e**, Dongfeng 2 well, Xu 4 Member. Major minerals are abbreviated as *Q* quartz, *QO* quartz overgrowth, *Chl* chlorite, *P* pore, *Fesp* feldspar

Table 1 Primary mineral components of authigenic grain-coating chlorite in Xujiahe Formation sandstones ($\times 10^{-2}$)

Age	FeO	Al_2O_3	MgO	SiO_2	K_2O	CaO
Xu 2 Member	50.35	10.91	0.85	27.99	1.39	0.54
Xu 4 Member	52.56	8.20	1.61	35.46	1.14	0.729

Al_2O_3, and other oxides. Fe is the most important chemical component of authigenic chlorite.

Previous studies showed that iron ions in clastic rock chlorite were mainly derived from alteration of iron-rich debris (Billault et al. 2003; Bloch S et al. 2002). As rivers flow into the ocean, ferruginous components formed flocculent precipitates and become a source of chlorite (Grigsby 2001) and filled mudstones with iron and magnesium ion-rich fluids (Hilier et al. 1996). After thin section analysis, study area debris was identified mainly as sedimentary and metamorphic rock debris, with minor amounts of iron-rich volcanic rock debris. Therefore, the study rules out the theory that Fe was derived from the alteration of iron-rich debris. The Xu 1, Xu 3, and Xu 5 Members of the Sichuan Basin are composed primarily of mudstone and argillaceous siltstone. During periods of sea

(lake) expansion, iron ion could be provided for reservoirs such as the Xu 2 and Xu 4 Members. Authigenic grain-coating chlorite formed during these periods of rising base level. As mudstones were filled with iron- and magnesium-rich fluids, magnesium ions reacted with iron-rich grain coatings to form grain-coating chlorite. In conclusion, the iron content of grain-coating chlorite in mudstone did not precipitate from pore water.

The relative content of authigenic chlorite in different sedimentary microfacies was determined from thin section analysis, and the results are shown in Fig. 3. Authigenic chlorite distribution was controlled by the sedimentary environment and mainly developed in underwater distributary channels, distributary channel, shallow lake sandstone dam, mouth bar facies, etc. The first three environments belong to delta sedimentary facies, and mouth bars belong to shore-shallow lake sedimentary facies. However, evidence of suspended boulder-sized particles was found in the fluvial environments, which require highly hydrodynamic conditions.

Therefore, this study proposes that the iron component of authigenic chlorite was mainly derived from flocculent precipitates formed when rivers flow into the ocean. The difference between electrolyte composition and the colloidal properties of river water and sea water (lake water)

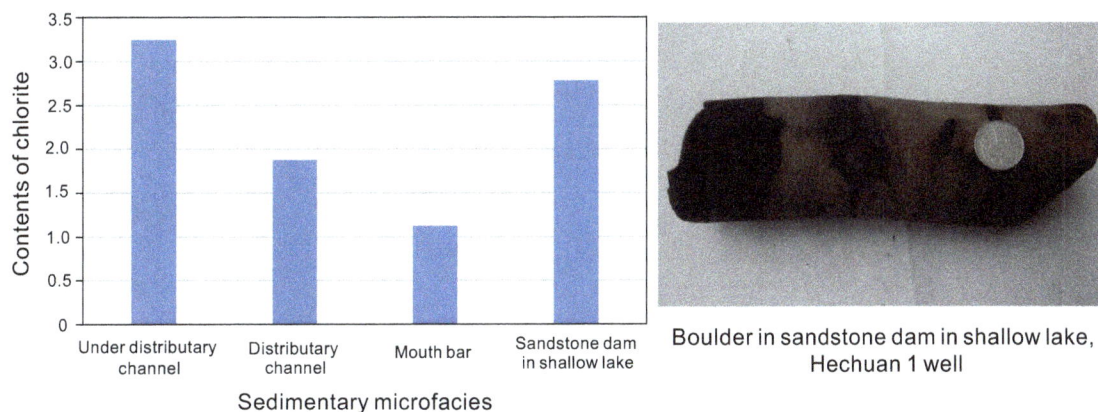

Fig. 3 Comparison of grain-coating chlorite content in different sedimentary microfacies, including: underwater distributary channels, distributary channels, mouth bars, and shallow lake sandstone dams

resulted in flocculation and the formation of iron-rich precipitates (Huang et al. 2004; Ehrenberg 1993; Yao et al. 2011). Gould et al. (2010) indicated primary grain-coating chlorite was formed by ferruginous flocculent precipitates above the sulfate-reduction zone. The formation of primary grain-coating chlorite is similar to the formation process of oolitic lime, which requires highly hydrodynamic conditions and for grains to constantly adsorb flocculent precipitates from the water (Yao et al. 2011). Therefore, grain-coating chlorites are most likely to form in highly hydrodynamic environments, such as underwater distributary channel sandstone bodies.

3.3 Paragenetic mineral characteristics

Based on thin section analysis, this study finds that sandstone sequences with grain-coating chlorite tend to have relatively well-developed glauconite and pyrite. Pyrite is mostly granular with an intergranular or zonal distribution (Fig. 4a–c). Semi-saline water and iron-rich conditions are favorable for pyrite formation, and the appearance of pyrite in the study section indicates a reducing environment.

Glauconite is a sedimentary mineral with the molecular formula: $K_{1-x}\{(Al,Fe)_2[Al_{1-x}Si_{3+x}O_{10}](OH)_2\}$. Glauconite is thought to be an indicator mineral of marine sedimentary environments, especially transgressive facies, and formed in warm shallow oceans with abundant organic matter over long periods of time (Chang 2006). Glauconite in the Xujiahe Formation sandstones mostly appears emerald and light green under the microscope, brighter than chlorite, with irregular clumps and a granular appearance (Fig. 4d–f). Glauconite tends to have a soft, gel-like texture after deposition, which makes it easy for the intrusion of other grains and the formation of irregular shapes.

The glauconite was analyzed by EPMA (Table 2). The main chemical components included FeO, SiO_2 and Al_2O_3 and were comparable to the composition of modern glauconite (Chen 1994). Glauconite in Xujiahe Formation sandstones tends to exhibit characteristics of "high Al, low Fe, and high K content". Potassium (K) content is the key indicator of glauconite maturity. Primary glauconite is indicated when K_2O content is less than 4 %, K_2O between 4 % and 6 % indicates low maturity glauconite, K_2O between 6 % and 8 % indicates moderately mature glauconite, while K_2O values greater than 8 % indicate highly mature glauconite (Odin and Matter 1981). Previous studies show that Fe content in glauconite would be replaced by Al during diagenesis. At the same time, K would be absorbed from pore waters which results in an increase of Al_2O_3 and K_2O, while FeO content decreases. Marine sedimentary glauconite exhibits "high Al, low Fe" (Chen 1994). In conclusion, the Xujiahe Formation glauconites exhibit marine sedimentary glauconite characteristics.

3.4 Relationship of grain-coating chlorite and reservoir porosity

Based on identification of cast thin sections, the study found that the Xujiahe Formation reservoir space is composed of primary intergranular pores, solution pores, and microfractures. Analysis of the relationship between authigenic grain-coating chlorite content and surface porosity, primary intergranular pore porosity, solution porosity, and sample porosity (Fig. 5) shows that the correlation between chlorite content to surface porosity and sample porosity is very poor, while the correlation with primary intergranular porosity is positive, and the correlation with solution porosity is negative.

The results of this study indicate authigenic grain-coating chlorite has a good positive correlation with primary intergranular porosity, a conclusion shared with the studies by Sun et al. (2012, 2014) and Xiang et al. (2016). Many researchers thought grain-coating chlorite was

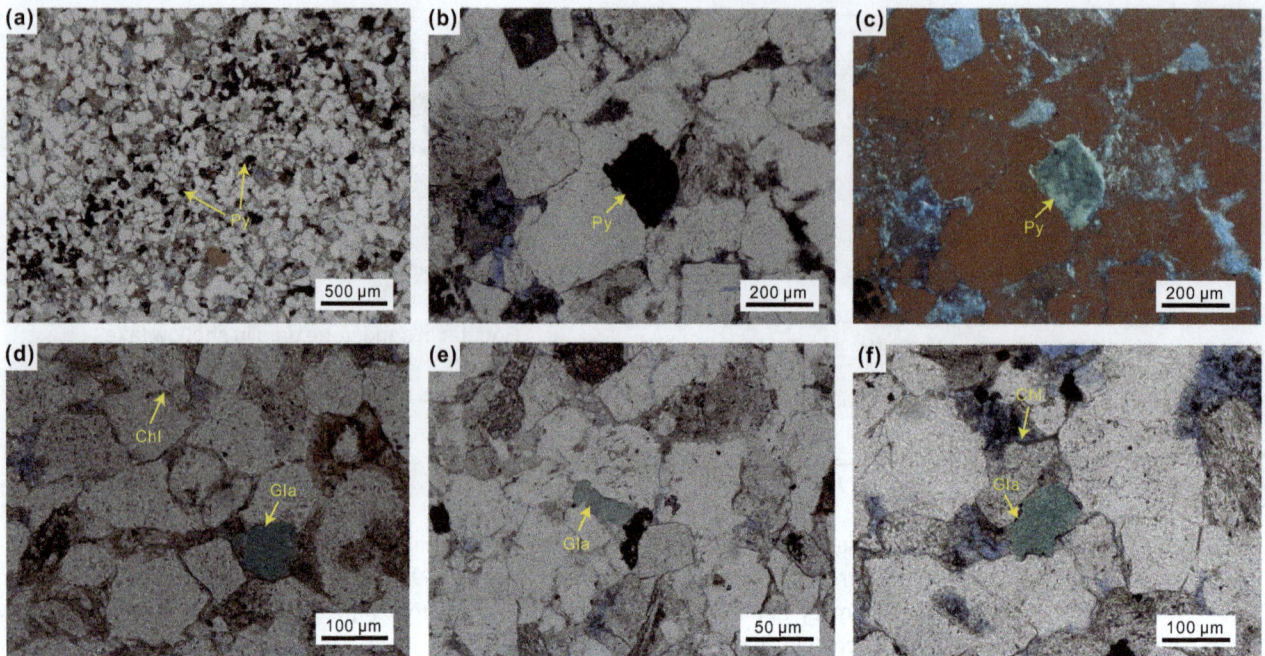

Fig. 4 Characteristics of pyrite and glauconite in Xujiahe Formation sandstones. **a** Black granular pyrite with zonal distribution, Weidong 2 well, Xu 2 Member, PPL. **b** Black granular pyrite, Weidong 2 well, Xu 2 Member, PPL. **c** Reflected light thin section of image **b**. **d** Irregular, granular glauconite with grain-coating chlorite on the surface, Weidong 2 well, Xu 2 Member, PPL. **e** Irregular granular glauconite, Mo 76 well, Xu 2 Member, PPL. **f** Granular glauconite, grain-coating chlorite developed on the grain surface, Weidong 2 well, Xu 2 Member, PPL. *Py* Pyrite, *Gla* glauconite, *Chl* chlorite

Table 2 Main chemical components of glauconite in Xujiahe Formation sandstones ($\times 10^{-2}$)

Components	Na$_2$O	SiO$_2$	FeO	CaO	Al$_2$O$_3$	MnO	K$_2$O	MgO
Study area	0.06	49.68	12.5	0.23	18.68	0.02	7.9	3.37
Modern[a]	0.43	44.20	25.07	2.43	5.08	0.07	3.28	5.84

[a] Data are cited from Chen (1994)

beneficial to reservoir porosity preservation, especially for primary intergranular pores. However, some researchers believed good-quality reservoirs develop as a result of good primary intergranular porosity, and the existence of grain-coating chlorite is only indicative of primary intergranular porosity (Xiang et al. 2016). Based on the correlation of chlorite content with surface porosity and sample porosity in this study, it was concluded that the development of chlorite did not play a positive role in the reservoir porosity preservation and it may have influenced dissolution, to a certain extent. This may be attributed to two reasons: (1) As primary intergranular porosity increased with chlorite content, acidic fluids flowed into the reservoir, and the increased pore space diluted the acidic fluids, reducing the effects of dissolution; (2) it is rare to find authigenic chlorite grains in areas which have not undergone dissolution; therefore, the increased chlorite content may cause a reduction in solution porosity. Further study is needed to discover the effects of authigenic grain-coating

chlorite on reservoir porosity preservation. However, the effects of chlorite development on sandstone reservoir properties are not disputed.

4 Characteristics of sand bodies with grain-coating chlorite

Sand bodies with authigenic grain-coating chlorite in the study area may be divided into four types of sedimentary microfacies: underwater distributary channels, distributary channel, shallow lake sandstone dams, and mouth bars.

4.1 Sandstone dam in shallow lake

Figure 6 shows the characteristics of a sandstone dam in the shallow lake environment of the Hechuan 1 well in the Xu 2 Member in the study area. Tabular cross-bedding is developed at the bottom of the sand body, and there are

(a)

(b)

(c)

(d)

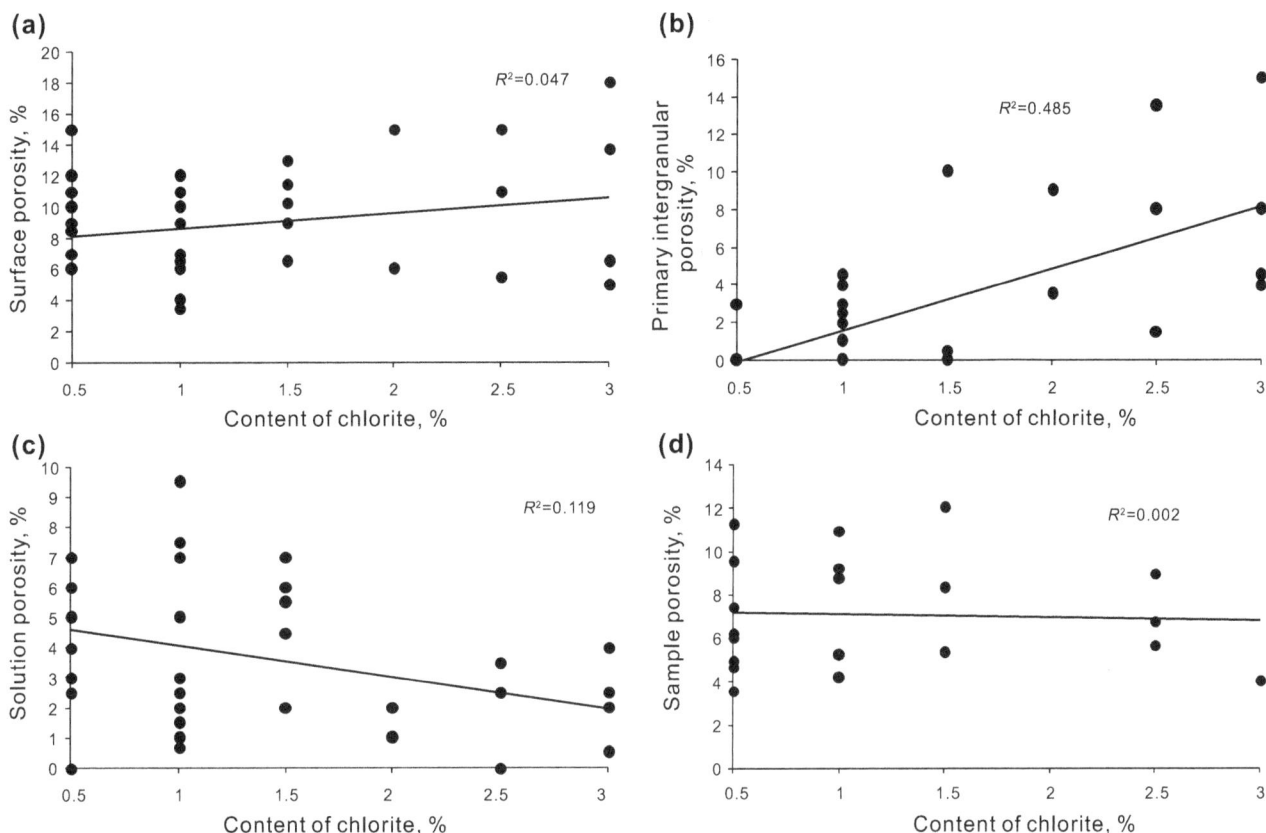

Fig. 5 Correlation of authigenic grain-coating chlorite content with: **a** surface porosity; **b** primary intergranular porosity; **c** solution porosity; and **d** sample porosity. The results indicate chlorite content has a positive correlation with primary intergranular porosity and surface porosity, a negative correlation with solution porosity, and a poor negative correlation with sample porosity

boulder-sized grains in contact with the lower shale. The development of upward progradation in the sand body can be considered as the result of the rising base-level cycle. Thin section analysis shows that grain-coating chlorite was poorly developed at the bottom of sand bodies with poor sorting (Fig. 6a). Higher muddy matrix content also results in poor porosity. However, muddy matrix content is lower in the upper sand bodies as shown in Fig. 6b, c. The upper sand bodies also have well-developed grain-coating chlorite with good porosity. The upper sand bodies also have intergranular pores conducive to the formation of good reservoirs.

4.2 Distributary channel

Figure 7 shows a sand body which developed in a distributary channel and swamp with a burial depth of 2100–2200 m in the Bao 36 well and the Xu 4 Member. The sand body is divided into two base-level rising cycles because there are two gravel horizons at 2200 and 2155 m. Two sandstone horizons dominated by cementation of authigenic chlorite developed at depths of 2120 and 2155 m. The upper horizon with developed chlorite also contains high amounts of silica cement which filled in

intergranular pores (Fig. 7a, b). As a result, the porosity of the lower horizon is higher than that of the upper horizon. The high content of silica cement in the upper horizon is related to the upper covered mudstone or shale, as during diagenesis, montmorillonite would be converted into illite and release large amounts of SiO_2. Fluids containing SiO_2 move to the lower layers and cause the precipitation of silica cements in intergranular pores.

4.3 Underwater distributary channel

The sand body shown in Fig. 8 is buried at a depth ranging from 2050 to 2080 m in the Xu 4 Member in the Bao 36 well. The sedimentary facies of the sand body include distributary channels and swamp, underwater distributary channels, and peat flats. Grain-coating chlorite in the sand body mainly developed in underwater distributary channels.

4.4 Mouth bar

Figure 9 shows a lithologic log of the Xu 2 Member in the Weidong 2 well from 2140 to 2157 m. The sedimentary microfacies were divided into underwater distributary channel, mouth bar, and peat flat. Grain-coating chlorite

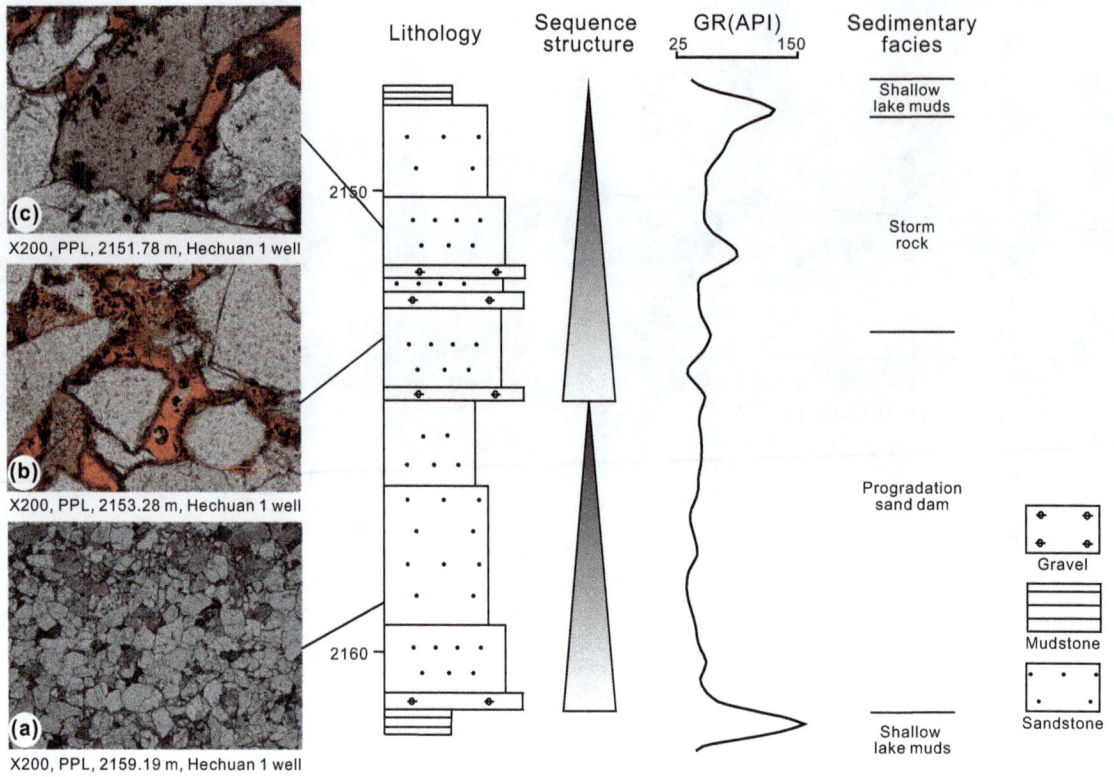

Fig. 6 Lithologic log and characteristics of shallow lake sandstone dam samples taken from the Hechuan 1 well. Photomicrographs of samples taken from the Hechuan 1 well at depths of **a** 2159.19 m; **b** 2153.28 m; and **c** 2151.78 m

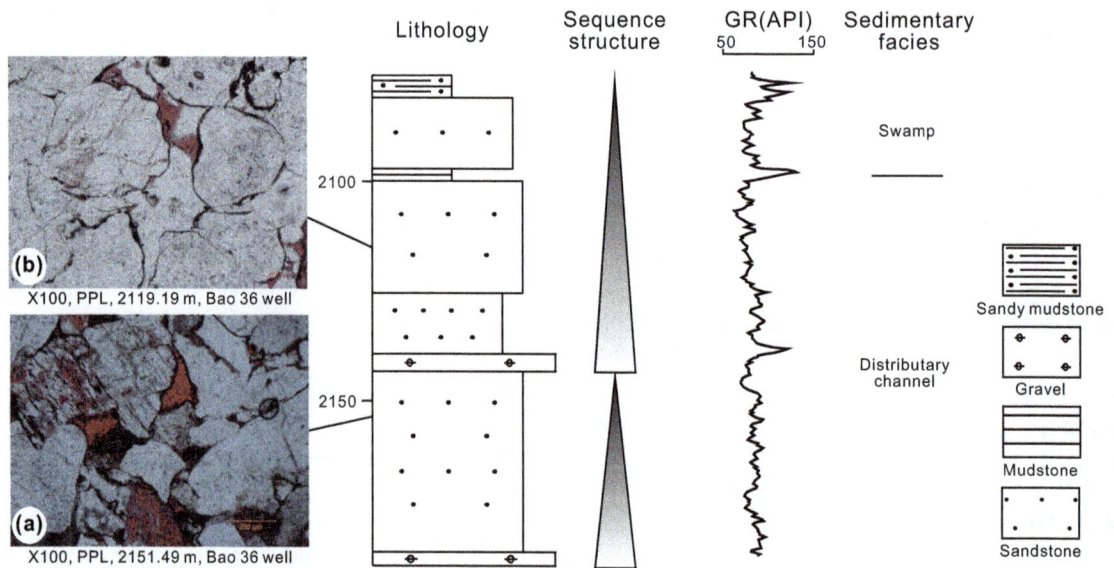

Fig. 7 Lithology and characteristics of distributary channel sand bodies taken from Bao 36 well samples. Photomicrographs of samples from depths of **a** 2151.49 m and **b** 2119.19 m

mainly developed in the upper part of the mouth bar sand body. Primary porosity rarely formed due to strong compaction, although considerable secondary porosity formed due to dissolution (Fig. 10).

Based on the four types of chlorite-containing sand microfacies, grain-coating chlorite was found almost exclusively in the upper part of sandstone bodies, which indicates rising sea level. Base level is an abstract surface

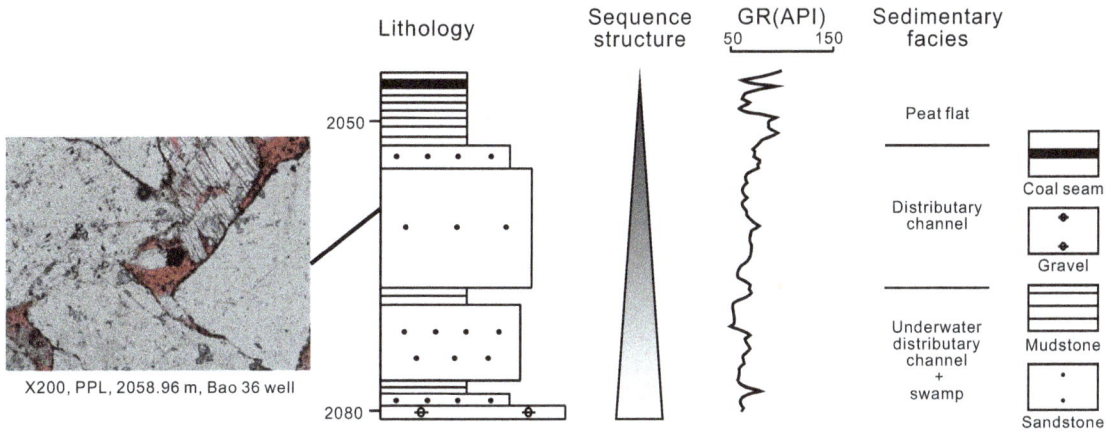

Fig. 8 Characteristics of underwater distributary channel

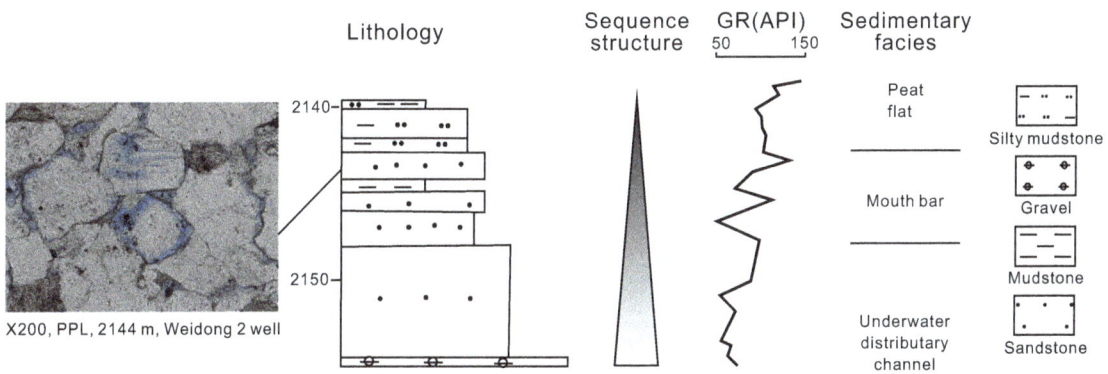

Fig. 9 Mouth bar characteristics

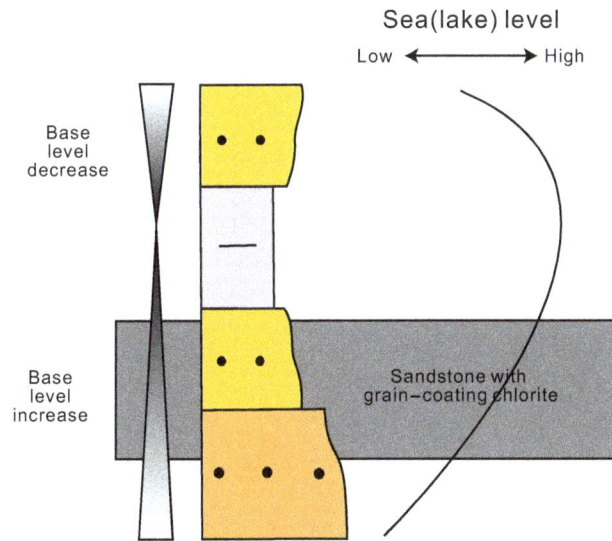

Fig. 10 Relationship between developed grain-coating chlorite horizon and sequence

that is equivalent to continuous, wave-like movement on the earth's surface with a slight declination to the basin. Its location, direction of motion, and change in elevation vary

with time (Deng et al. 2000). For simplicity, base level is usually equivalent to sea (lake) level. An increase in base level is indicative of marine (lacustrine) transgression.

5 Formation model of grain-coating chlorite in the Xujiahe Formation

The evidence shown above indicates that grain-coating chlorite formed in high hydrodynamic, weak alkaline, weakly reducing, and iron-rich conditions. By analyzing authigenic grain-coating chlorite horizons, the study found that while chlorite mainly formed at the base of the sand bodies, it experienced a large increase in content, especially in the upper part of the sandstone bodies. What causes it?

Previous studies show that the Sichuan Basin shifted from a marine to a continental sedimentary environment. Because of the influence by the An'xian movement in the Xu 4, the ancient Longmen Mountain thrust belt was uplifted and transformed the Sichuan Basin into a terrestrial sedimentary environment (Lin et al. 2006). Although

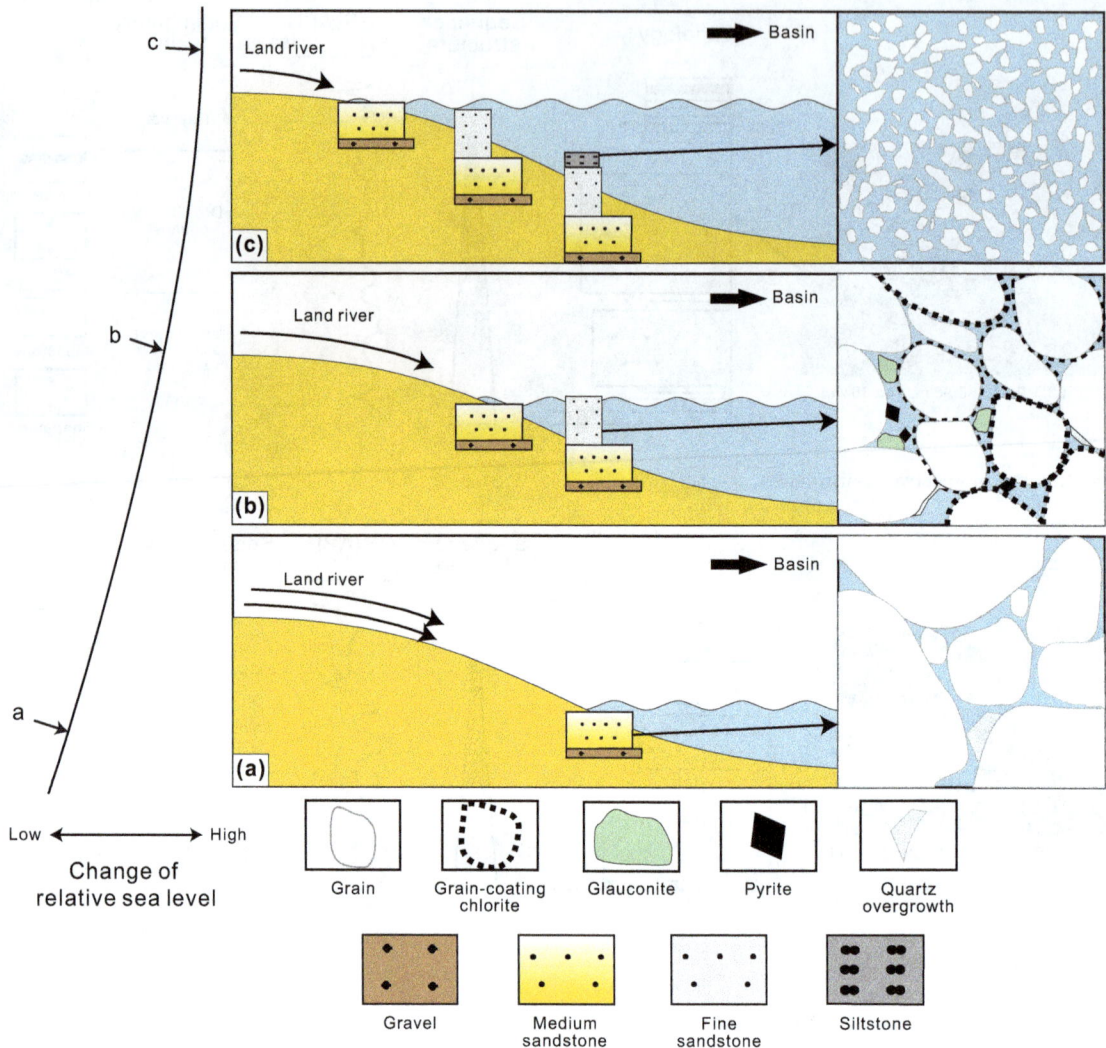

Fig. 11 Formation model of grain-coating chlorite

the Sichuan Basin was already in a continental sedimentary environment during Xu 4 deposition, the lake water was still slightly saline to semi-saline as the basin was just cut off from the Paleo-Tethys sea. Because of the inherited properties of ocean water, the Xu 4 lake water had semi-saline characteristics and could be distinguished from Xu 2 sea water in terms of salinity, pH, and other characteristics.

Grain-coating chlorite, glauconite, and pyrite are all iron-rich and can form authigenically. Based on findings of this study, Fe was thought to be derived from the flocculent precipitates formed when rivers flowed into the ocean. The sand bodies of the Xujiahe Formation typically indicate increasing base level and generally contain gravel. The bottom layers usually include river rejuvenation surfaces or river erosion surfaces and generally contain gravel which overlaps with the lower black mudstone or shale, indicating when base level started to increase. Decreasing base level exposes previously deposited sediments to the surface. The

influx of fresh water from rivers would reduce pH and salinity, where the environment could not reach the conditions for the deposition of grain-coating chlorite or glauconite. As base level rose, the pH and salinity of sea (lake) water began to rise and the sedimentary environment shifted from weakly oxidizing to weakly reducing. When the pH became weakly alkaline, the sedimentary environment became weakly reducing. Pyrite began to deposit, while chlorite and glauconite were deposited as grain coating and crystal grains, respectively (Fig. 11b).

However, grain-coating chlorite is rarely observed in study area sand bodies which developed during periods of decreasing base level. The shift in base level may be related to megamonsoons during the Carnian-Raleigh of Late Triassic Epoch (Shi et al. 2010) in the Middle and Upper Yangtze regions of the Sichuan Basin. The monsoon climate resulted in uneven rainfall distribution and seasonal humidity. During periods of increased rainfall, river discharges increased and more Fe particulates were deposited. Water was relatively more saturated in Fe, and Fe-rich minerals such as chlorite became easier to deposit. The rising base level resulted in the formation of additional sand body cycles because of increased river discharges. During the even longer seasonal dry periods, river flow decreased and less Fe particulates were deposited. This resulted in less grain-coating chlorite and the sand bodies formed reflected decreased base level. This model assumes lake level would only be affected by rainfall while ignoring the influence of tectonic activity.

Based on this chlorite formation model, authigenic chlorite-containing sandstones can be identified with greater certainty. During periods of high water level, areas which once deposited authigenic chlorite began to deposit siltstone due to rising base level. When the values of pH and oxidation–reduction potential (ORP) reached the conditions necessary for authigenic chlorite deposition in shallow water environments, grain-coating chlorite began to deposit in sand bodies.

6 Conclusions

Authigenic chlorite growth in the southern Sichuan Basin Xujiahe Formation was found to have occurred in multiple phases and mostly formed during syngenesis–early diagenesis A stage based on thin section and SEM analysis of grain-coating chlorite. Ferruginous minerals were derived from flocculent precipitates which formed as rivers flowed into the ocean, and chlorite formed in weak alkaline, weakly reducing environments with high hydrodynamic conditions. The sedimentary microfacies could be divided into underwater distributary channel, distributary channel, shallow lake sandstone dam, mouth bar, etc., based on analysis of authigenic chlorite-containing sandstone bodies from the southern Sichuan Xujiahe Formation. Chlorite had a tendency to form in the upper part of sandstone bodies, indicating increased base level. The increase in base level may result from lacustrine transgression related to Late Triassic megamonsoons in the Middle and Upper Yangtze Region. During periods of abundant precipitation, river discharges increased and more Fe particulates were brought to the lake. Base level would increase or decrease over short periods in response to precipitation. As the sedimentary environment shifted from weakly oxidizing to weak alkaline, weakly reducing because of increased base level, Fe-rich minerals such as authigenic chlorite and glauconite started to form and deposit.

Acknowledgments This study is financially supported by the National Science and Technology Major Project of the Ministry of Science and Technology of China (Nos. 2011ZX05002-004-006HZ, 2016ZX05002-004-010).

References

Bahlis AB, Luiz F. Origin and impact of authigenic chlorite in the Upper Cretaceous sandstone reservoirs of the Santos Basin, eastern Brazil. Pet Geosci. 2013;19(2):185–99.

Billault V, Beaufort D, Baronnet A, et al. A nanopetrographic and textural study of grain-coating chlorites in sandstone reservoirs. Clay Miner. 2003;38(3):315–28.

Bloch S, Lander RH, Bonnell L. Anomalously high porosity and permeability in deeply buried sandstone reservoirs: origin and predictability. AAPG Bull. 2002;86(2):301–28.

Chang LH. Identification manual of transparent minerals in thin-section. Beijing: Geological Publishing House. 2006. 1–235; (in Chinese).

Chen LR. Evolution of authigenic glauconite in early diagenesis. Chin Sci Bull. 1994;39(18):1550–3.

Deng HW, Wang HL, Ning N. Sediment volume partition principle: theory basis for high-resolution sequence tratigraphy. Earth Sci Front. 2000;7(4):305–13 **(in Chinese)**.

Ding XQ, Han MM, Zhang SN. The role of provenance in the diagenesis of siliciclastic reservoirs in the Upper Triassic Yanchang Formation, Ordos Basin, China. Pet Sci. 2013;10(2):149–60.

Ehrenberg SN. Preservation of anomalously high porosity in deeply buried sandstones by grain-coating chlorite: examples from the Norwegian continental shelf. AAPG Bull. 1993;77(7):1260–86.

Gould K, Pe-Piper G, Piper D. Relationship of diagenetic chlorite rims to depositional facies in Lower Cretaceous reservoir sandstones of the Scotian Basin. Sedimentology. 2010;57(2):587–610.

Grigsby JD. Origin and growth mechanism of authigenic chlorite in sandstones of the Lower Vicksburg Formation, South Texas. J Sediment Res. 2001;71(1):27–36.

Gu N, Tian JC, Zhang X, et al. Densification mechanism analysis of sandstone densification within Xujiahe Formation in low-step structure in southern Sichuan Basin. J Northeast Pet Univ. 2014;38(5):7–14 **(in Chinese)**.

Hilier S, Fallick AE, Matter A. Origin of pore-lining chlorite in the aeolian Rotliegend of northern Germany. Clay Miner. 1996;31(2):153–71.

Huang SJ, Xie LW, Zhang M, et al. Formation mechanism of authigenic chlorite and relation to preservation of porosity in nonmarine Triassic reservoir sandstones, Ordos Basin and Sichuan Basin, China. J Chengdu Univ Technol Sci Technol Ed. 2004;31(3):273–81 **(in Chinese)**.

Huang J, Zhu RK, Hou DJ, et al. Influences of depositional environment and sequence stratigraphy on secondary porosity development: a case of the Xujiahe Formation clastic reservoir in the central Sichuan Basin. Pet Explor Dev. 2010;37(2):158–66 **(in Chinese)**.

Lanson B, Beaufort D, Berger G, et al. Authigenic kaolin and illitic minerals during burial diagenesis of sandstones: a review. Clay Miner. 2002;37(1):1–22.

Lin LB, Chen HD, Zhai CB, et al. Sandstone compositions and paleogeographic evolution of the Upper Triassic Xujiahe Formation in the Western Sichuan Basin, China. Pet Geol Exp. 2006;28(6):511–7 **(in Chinese)**.

Lin LB, Yu Y, Gao J, et al. Main control factors of Xujiahe Formation sandstone reservoir in South Sichuan, China. J Chengdu Univ Technol Sci Technol Ed. 2015;42(4):400–9 **(in Chinese)**

Liu JK, Peng J, Liu JJ, et al. Pore-preserving mechanism of chlorite rims in tight sandstone—an example from the T_3x Formation of Baojie area in the transitional zone from the central to southern Sichuan Basin. Oil Gas Geol. 2009;30(1):53–60 (**in Chinese**).

Liu LY, Qu ZH, Sun W, et al. Properties of clay mineral of clastic rock in Shanshan oil field, Xinjiang. J Northwest Univ Nat Sci Ed. 1998;28(5):77–80 (**in Chinese**).

Lü ZX, Ye SJ, Yang X, et al. Quantification and timing of porosity evolution in tight sand gas reservoirs: an example from the Middle Jurassic Shaximiao Formation, western Sichuan, China. Pet Sci. 2015;12(2):207–17.

Odin GS, Matter A. De glauconarium origine. Sedimentology. 1981;28:611–41.

Okwese AC, Pe-Piper G, Piper D. Controls on regional variability in marine pore-water diagenesis below the seafloor in Upper Jurassic—Lower Cretaceous prodeltaic sandstone and shales, Scotian Basin, Eastern Canada. Mar Pet Geol. 2012;29(1):175–91.

Pen J, Liu JK, Wang Y, et al. Origin and controlling factors of chlorite coatings—an example from the reservoir of T_3x Group of the Baojie area, Sichuan Basin, China. Pet Sci. 2009;6(4):376–82.

Shi ZQ, Zeng DY, Xiong ZJ, et al. Sedimentary records of Triassic megamonsoon in Upper Yangtze Area. Bull Mineral Pet Geochem. 2010;29(2):164–72 (**in Chinese**).

Sun QL, Sun HS, Jia B, et al. Genesis of chlorites and its relationship with high-quality reservoirs in the Xujiahe Formation tight sandstones, western Sichuan depression. Oil Gas Geol. 2012;33(5):751–7 (**in Chinese**).

Sun ZX, Sun ZL, Yao J, et al. Porosity preservation due to authigenic chlorite coatings in deeply buried Upper Triassic Xujiahe Formation sandstones, Sichuan Basin, western China. J Pet Geol. 2014;37(3):251–68.

Xi KL, Cao YC, Wang YZ, et al. Diagenesis and porosity-permeability evolution of low permeability reservoirs: a case study of Jurassic Sangonghe Formation in Block 1, central Junggar Basin, NW China. Pet Explor De. 2015;42(4):434–43 (**in Chinese**).

Xiang F, Feng Q, Zhang DY, et al. Further study of chlorite rim in sandstone: evidences from Yanchang Formation in Zhenjing area, Ordos Basin, China. J Chengdu Univ Technol Sci Technol Ed. 2016;43(1):59–67 (**in Chinese**).

Yao JL, Wang Q, Zhang R, et al. Forming mechanism and their environmental implications of chlorite-coatings in Chang 6 sandstone (Upper Triassic) of Hua-Qing Area, Ordos Basin. Acta Sedimentol Sin. 2011;29(1):72–9 (**in Chinese**).

Zeng W. Diagenesis and reservoir distribution of Upper Jurassic Series in Zhangqiang Hollow. J Southwest Pet Inst. 1996;18(4):11–6 (**in Chinese**).

Zhang JZ, Chen SJ, Xiao Y, et al. Characteristics of the Chang 8 tight sandstone reservoirs and their genesis in Huaqing area, Ordos Basin. Oil Gas Geol. 2013;34(5):679–84 (**in Chinese**).

Zhang X, Lin CM, Chen ZY. Characteristics of chlorite minerals from Upper Triassic Yanchang Formation in the Zhenjing Area, Ordos Basin. Acta Geol Sin. 2011;85(10):1659–71 (**in Chinese**).

Zhang X, Tian JC, Du BQ, et al. Matching between sandstone tightening and hydrocarbon accumulation of the Xujiahe Formation, Guanyinchang area in southern Sichuan Basin. Oil Gas Geol. 2014;35(2):231–7 (**in Chinese**).

Zhu HH, Zhong DK, Yao JL, et al. Alkaline diagenesis and its effects on reservoir porosity: a case study of Upper Triassic Chang 7 tight sandstones in Ordos Basin, NW China. Pet Explor Dev. 2015;42(1):51–9 (**in Chinese**).

Zhu HH, Zhong DK, Zhang YX, et al. Pore types and controlling factors on porosity and permeability of Upper Triassic Xujiahe tight sandstone reservoir in Southern Sichuan Basin. Oil Gas Geol. 2014;35(1):65–76 (**in Chinese**).

Zhu P, Huang SJ, Li DM, et al. Effect and protection of chlorite on clastic reservoir rocks. J Chengdu Univ Technol Sci Technol Ed. 2004;31(2):153–6 (**in Chinese**).

Zhu RK, Zou CN, Zhang N, et al. Diagenetic fluids evolution and genetic mechanism of tight sandstone gas reservoirs in Upper Triassic Xujiahe Formation in Sichuan Basin, China. Sci China Ser D. 2009;39(3):327–39 (**in Chinese**).

Zou CN, Gong YJ, Tao SZ, et al. Geological characteristics and accumulation mechanisms of the "continuous" tight gas reservoirs of the Xu2 Member in the middle-south transition region, Sichuan Basin, China. Pet Sci. 2013;10(2):171–82.

Sedimentary characteristics and processes of the Paleogene Dainan Formation in the Gaoyou Depression, North Jiangsu Basin, eastern China

Xia Zhang[1] · Chun-Ming Lin[1] · Yong Yin[2] · Ni Zhang[1] · Jian Zhou[1] ·
Yu-Rui Liu[3]

Abstract In this paper, the type, vertical evolution, and distribution pattern of sedimentary facies of the Paleogene Dainan Formation in the Gaoyou Depression of the North Jiangsu Basin are studied in detail. Results show that fan delta, delta, nearshore subaqueous fan, and lacustrine facies developed during the Dainan Formation period and their distribution pattern was mainly controlled by tectonics and paleogeography. The fan delta and nearshore subaqueous fan facies predominantly occur in the southern steep slope region where fault-induced subsidence is thought to have created substantial accommodation, whereas the delta facies are distributed on the northern gentle slope which is thought to have experienced less subsidence. Finally, the lacustrine facies is shown to have developed in the center of the depression, as well as on the flanks of the fan delta, delta, and nearshore subaqueous fan facies. Vertically, the Dainan Formation represents an integrated transgressive–regressive cycle, with the E_2d_1 being the transgressive sequence and the E_2d_2 being the regressive sequence. This distribution model of sedimentary facies plays an important role in predicting favorable reservoir belts for the Dainan Formation in the Gaoyou Depression and similar areas. In the Gaoyou Depression, sandstones of the subaqueous distributary channels in the fan delta and the subaqueous branch channels in the delta are characterized by physical properties favorable for reservoir formation.

Keywords Sedimentary facies · Distribution pattern · Sedimentary evolution · Dainan Formation · Gaoyou Depression · North Jiangsu Basin

1 Introduction

Lacustrine rift basins are distributed widely in eastern China. About 300 Mesozoic–Cenozoic rift basins cover a total area of approximately 2×10^6 km^2. These depressions occur as one of the most important petroliferous basin types in China, and have therefore been the focus of exploration for subtle reservoirs (Xian et al. 2007; Wang et al. 2014; Jiang et al. 2014). The sedimentary systems developed in these rift basins in eastern China, such as the Bohai Bay Basin, the southern part of the North China Basin, the Erlian Basin, and the Ural Basin, tend to form favorable lithologic or structural-lithologic reservoirs, even in the conglomerates and/or sandy conglomerates of the nearshore subaqueous fans that they host (Sui 2003; Zhao et al. 2011; Cao et al. 2014; Zhang et al. 2014a). The North Jiangsu Basin is one of the richest regions for oil and gas in eastern China, given its thick and wide distribution of Mesozoic–Cenozoic strata. The Paleogene Dainan Formation is one of the most productive reservoir intervals in the Gaoyou Depression, North Jiangsu Basin (Qiu et al. 2006). The production of most major oil fields in the Dainan Formation is now in decline; thus, a precise description of the sedimentary facies of these reservoir sandstones is greatly needed. To date, studies of the Dainan Formation have focused primarily on the paleontological, sequence stratigraphic, and structural

✉ Chun-Ming Lin
cmlin@nju.edu.cn

[1] State Key Laboratory for Mineral Deposits Research, School of Earth Sciences and Engineering, Nanjing University, Nanjing 210023, Jiangsu, China

[2] School of Geographic and Oceanographic Sciences, Nanjing University, Nanjing 210023, Jiangsu, China

[3] Institute of Geological Sciences, Jiangsu Oilfield Branch Company, SINOPEC, Yangzhou 225009, Jiangsu, China

Edited by Jie Hao

compartmentalization of the basin (Dong 1999; Lu 2000; Zhang et al. 2005; Pang and Cao 2005; Zhu et al. 2013; Chen et al. 2015). There have also been studies of the sedimentology (Chen and Wu 2006; Zhang et al. 2007; Xia et al. 2008; Ji et al. 2012; Zhao et al. 2015) and provenance (Zhou et al. 2010; Zhang et al. 2014b) of the basin fill. However, the regional distribution pattern and processes of sedimentary facies have yet to be understood at a sufficiently high temporal and spatial resolution. Such detail is crucial for reliable predictions of depression-scale sedimentary architecture within and/or between individual oil fields. The objectives of this study are to (1) describe the characteristics, spatial distribution, and evolution processes of sedimentary facies of the Dainan Formation, and (2) reconstruct the sedimentary system and model of the Gaoyou Depression, which may have broad implications for other similar rift basins.

2 Geological setting

The North Jiangsu Basin is a large Mesozoic–Cenozoic fault-depressed basin with the basement being composed of Proterozoic metamorphic rock and Early Mesozoic carbonate, turbidite, and clastic rocks (Shu et al. 2005). It is located east of the Lower Yangtze Plate covering an area of approximately 35×10^3 km^2 (Fig. 1a), and it can be divided into four east-westward oriented tectonic units: the Dongtai Depression, Jianhu Uplift, Yanfu Depression, and the Binhai Uplift (Qiu et al. 2006; Fig. 1b).

The Gaoyou Depression is located in the central Dongtai Depression with an area of about 2.7×10^3 km^2. It is characteristic of a dustpan-like depression (Chen 2001; Zeng 2007; Zhu et al. 2013; Fig. 1c, d) resulting from the differential subsidence of fault blocks during the Yizheng Movement in the Late Cretaceous and the Wubao Movement in the Late Paleocene (Chen 2001). The Gaoyou Depression is bounded to the south by the Zhenwu fault belt (separating it from the Tongyang uplift), and links to the Zheduo low uplift through a slope in the north (Fig. 1c). The western and eastern boundaries are the Lingtangqiao low uplift and the Wubao low uplift, respectively (Fig. 1c). Due to the influence of Indian and Pacific plate movements, there are three groups of fractures (ENE, NE, and NW orientations) developed in the Gaoyou Depression, with those oriented ENE dominant. These ENE faults (Zhen 1, Zhen 2, and Hanliu faults) separate the Gaoyou Depression into three ENE trending sections from south to north: southern step-fault zone, central deep depression zone, and northern slope zone (Qiu et al. 2006; Chen 2001; Fig. 1c). The central deep depression zone can be further divided into three subdepressions from west to east: Shaobo, Fanchuan, and Liuwushe (Fig. 1e).

The Mesozoic–Cenozoic sedimentary thickness in the Gaoyou Depression can reach up to 7000 m. Of this, the Dainan Formation (E_2d) has a thickness of approximately 1500 m and has been one of the most productive reservoir intervals in the Gaoyou Depression over the last 30 years, hosting over 15 oil–gas fields containing about 4.1×10^8 tons of recoverable oil. The E_2d lies between the overlying Funing Formation (E_1f) and the underlying Sanduo Formation (E_2s) (Table 1), and can be divided into two members in the ascending order: 1st member (E_2d_1) and 2nd member (E_2d_2).

3 Sedimentary characteristics and facies

Four sedimentary facies (fan delta, delta, nearshore subaqueous fan, and lacustrine) have been identified within the Dainan Formation in the Gaoyou Depression based on the variations in lithology, sedimentary structures, and vertical successions.

3.1 Fan delta

Fan deltas occur mainly in the southern steep slope of the Gaoyou Depression, with three subfacies: fan delta plain, fan delta front, and profan delta. The fan delta plain subfacies is the subaerial part of the fan delta and contains distributary channels and back swamps. It is comparable with high-energy gravel-rich braided river facies (Blair and McPherson 1994; Lin et al. 2003; Krézsek et al. 2010). Distributary channels, which are the dominant microfacies of the fan delta plain, consist of gray or mottled conglomerate, gray conglomeratic sandstone, and coarse sandstone. The conglomerate gravels are common poorly sorted, subangular to subrounded in shape, and randomly distributed in a matrix of fine- to coarse-grained sands which indicate proximal deposition. Also, they have complex compositions, which include siliceous rocks, phyllite, limestone, mud pebbles and gypsum, and have diameters ranging from 1 to 8 cm. The structureless conglomerates overlie basal scour surfaces and progressively change upwards into parallel-bedded conglomeratic sandstones and coarse-grained sandstones. These characteristics suggest deposition from waning high-density flows. The spontaneous potential logs (SP) are jagged with low to moderate amplitudes. The brown mudstones and silty mudstones are interpreted as deposits of a back-swamp environment.

Fan delta front subfacies consist primarily of subaqueous distributary channels and interchannels, and subordinate mouth bar and sand sheet deposits (Fig. 2). Subaqueous distributary channel microfacies are characterized by light gray conglomeratic sandstones, and gray to

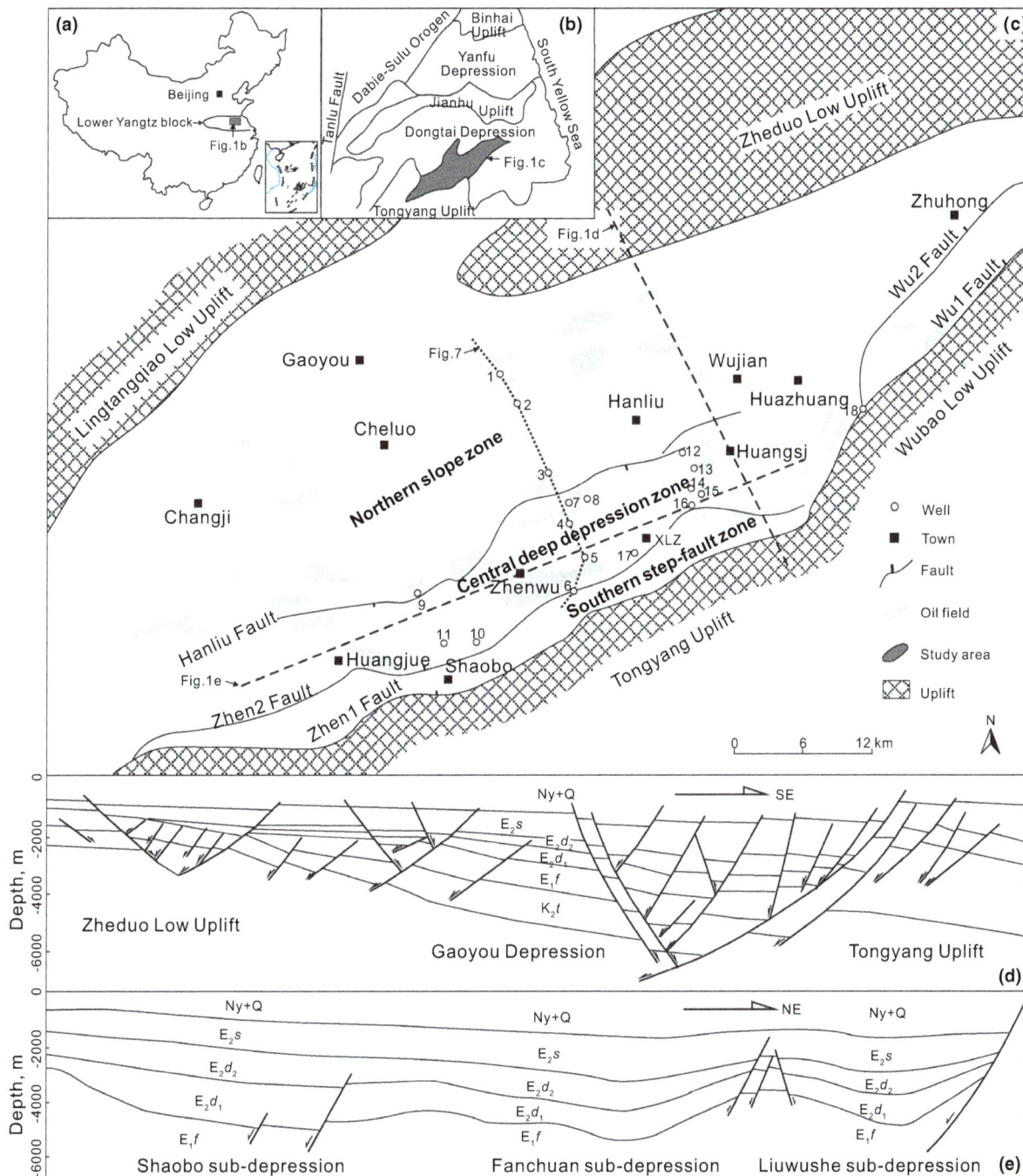

Fig. 1 **a** Location of the North Jiangsu Basin. **b** Simplified geological map of the North Jiangsu Basin and the location of the Gaoyou Depression. **c** Simplified geological map of the Gaoyou Depression. **d, e** the tectonic profiles derived from the seismogeological interpretation (see locations in **c**). Arabic numerals represent locations of the wells used in this paper. *1* Fa 1; *2* Jia 4; *3* Yong 20; *4* Yong 14; *5* Cao 20; *6* Xu 27; *7* Yong 7; *8* Yong 16; *9* Lian 7; *10* Shao 9; *11* Shao 6; *12* Fu 16; *13* Fu 35; *14* Fu 5; *15* Fu 44; *16* Fu 23; *17* Xiao 3; *18* Zhou 52; K$_2$t: Taizhou Formation; E$_1$f: Funing Formation; E$_2$d$_1$: the 1st member of the Dainan Formation; E$_2$d$_2$: the 2nd member of the Dainan Formation; E$_2$s: Sanduo Formation; Ny: Yancheng Formation; Q: Quaternary; XLZ: Xiaoliuzhuang

brown fine sandstones and siltstones which exhibit an upward-fining trend (Fig. 2). From bottom to top, sedimentary structures include a scour surface (Fig. 3a), graded bedding, tabular cross bedding (Fig. 3b), parallel bedding

(Fig. 3b), climbing-ripple cross stratification (Fig. 3c), wavy bedding, and convolute bedding (Fig. 3d). Furthermore, mud pebbles are pervasively present above the scour surface, having diameters ranging from 0.5 to 2.5 cm, and

Table 1 Stratigraphic division of the Paleogene Dainan Formation in the Gaoyou Depression

Stratigraphy				Maximum thickness m	Lithology	Lithology description	Geologic event
Paleogene	Eocene	Dainan Formation	Sanduo Formation	700		Light brownish-gray sandstone and gray black mudstone	Zhenwu event
			E_2d_2 $E_2d_2{}^1$	150		Gray, dark brown mudstone, silty mudstone, muddy siltstone interbedded with gray siltstone and fine sandstone	
			$E_2d_2{}^2$	200			
			$E_2d_2{}^3$	150		Brown, dark brown mudstone intercalated with light brown, light gray siltstone, fine sandstone, fining upward, mainly mudstone	
			$E_2d_2{}^4$	200			
			$E_2d_2{}^5$	250		Light brown, light gray siltstone, fine sandstone, silty-fine sandstone interbedded with gray, brown, purple red mudstone, silty mudstone	
			E_2d_1 $E_2d_1{}^1$	200		Dark, dark gray, purple red mudstone intercalated with dark gray silty mudstone, light gray sandstone	
			$E_2d_1{}^2$	300		Brown, gray silty-fine sandstone, siltstone, unequal-thickness interbedded with brown, dark purple red mudstone	
			$E_2d_1{}^3$	400		Gray fine sandstone, silty-fine sandstone, conglomeratic sandstone unequal-thickness interbedded with dark gray, gray, dark purple red mudstone, silty mudstone	Wubao event
			Funing Formation	500		Dark gray, gray black mudstone and shale	

Mudstone	Silty mudstone	Muddy siltstone	Siltstone	Silty-fine sandstone	Fine sandstone	Conglomeratic sandstone

their abundance and grain size become progressively lower and smaller, respectively, towards the top (Fig. 3a). Single subaqueous distributary channels are 5–10 m thick, but they can amalgamate and superimpose upon one another with resultant thicknesses reaching more than 50 m. SP curves display an obvious negative anomaly. Subaqueous distributary interchannel microfacies are composed of gray silty mudstones, and purple red, brown, dark gray mudstones, which together are occasionally intercalated with muddy siltstones. Horizontal, wavy, and lenticular beddings are also present, commonly having bioturbation and

abundant plant remains. Mudstones of subaqueous distributary interchannels usually display significant scour and can even be completely removed by successive high-discharge events. Such mudstones are commonly laminated as indicated by the SP curves close to the shale line and resistivity log (R) curves displaying a low-magnitude jagged pattern (Fig. 2). Mouth bar microfacies mainly contain gray to brown siltstone and fine-grained sandstone with a thickness of 4–6 m. These exhibit an upward-coarsening succession, as shown by the funnel-shaped SP curve. Cross, parallel, and wavy beddings are common. The sandstones

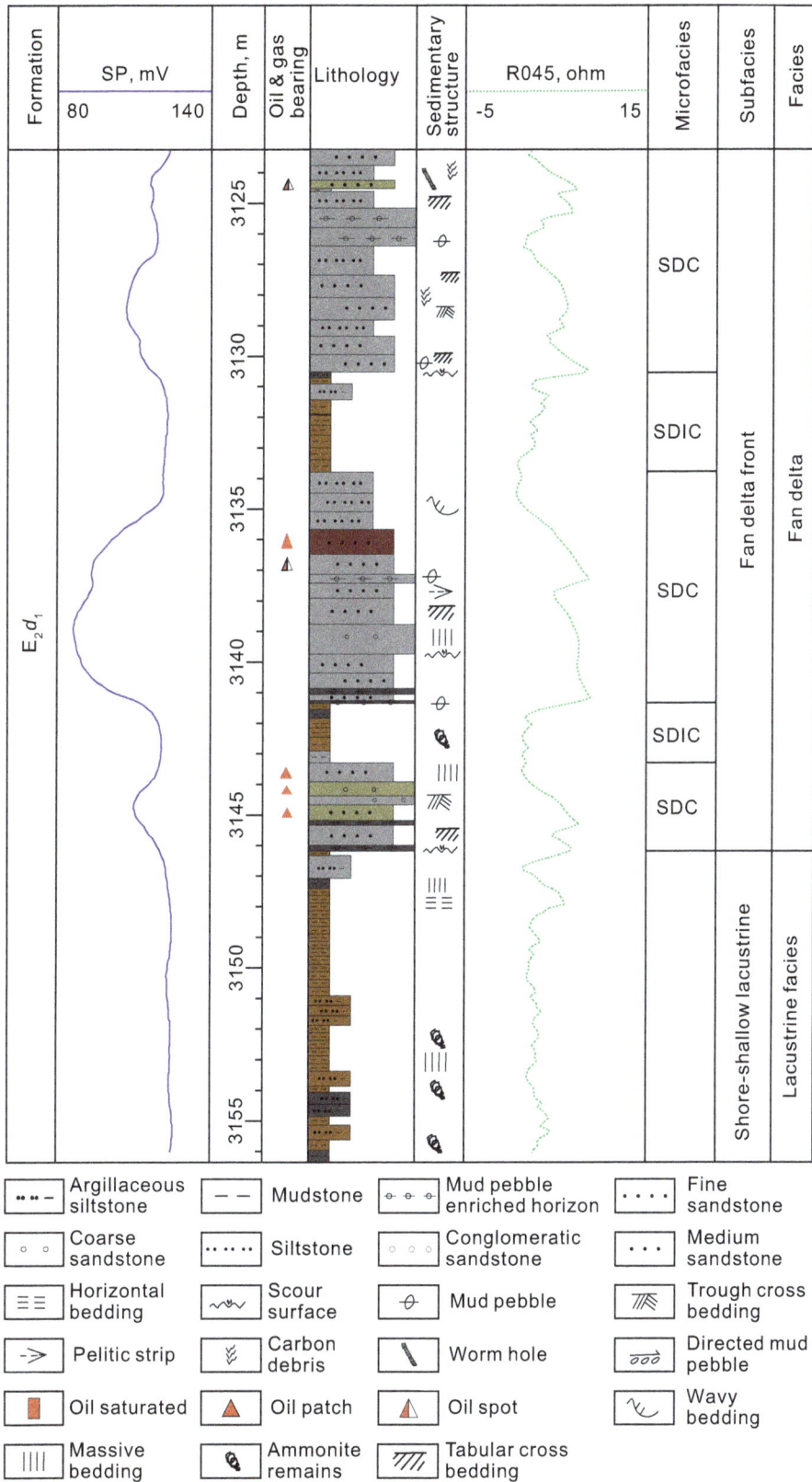

Fig. 2 Sedimentary characteristics of the fan delta for the Dainan Formation in Well Fu 35 of the Gaoyou Depression (see location in Fig. 1c). *SP* spontaneous potential curve; *R045* 0.45 m potential resistivity curve. *SDC* Subaqueous distributary channel; *SDIC* Subaqueous distributary interchannel

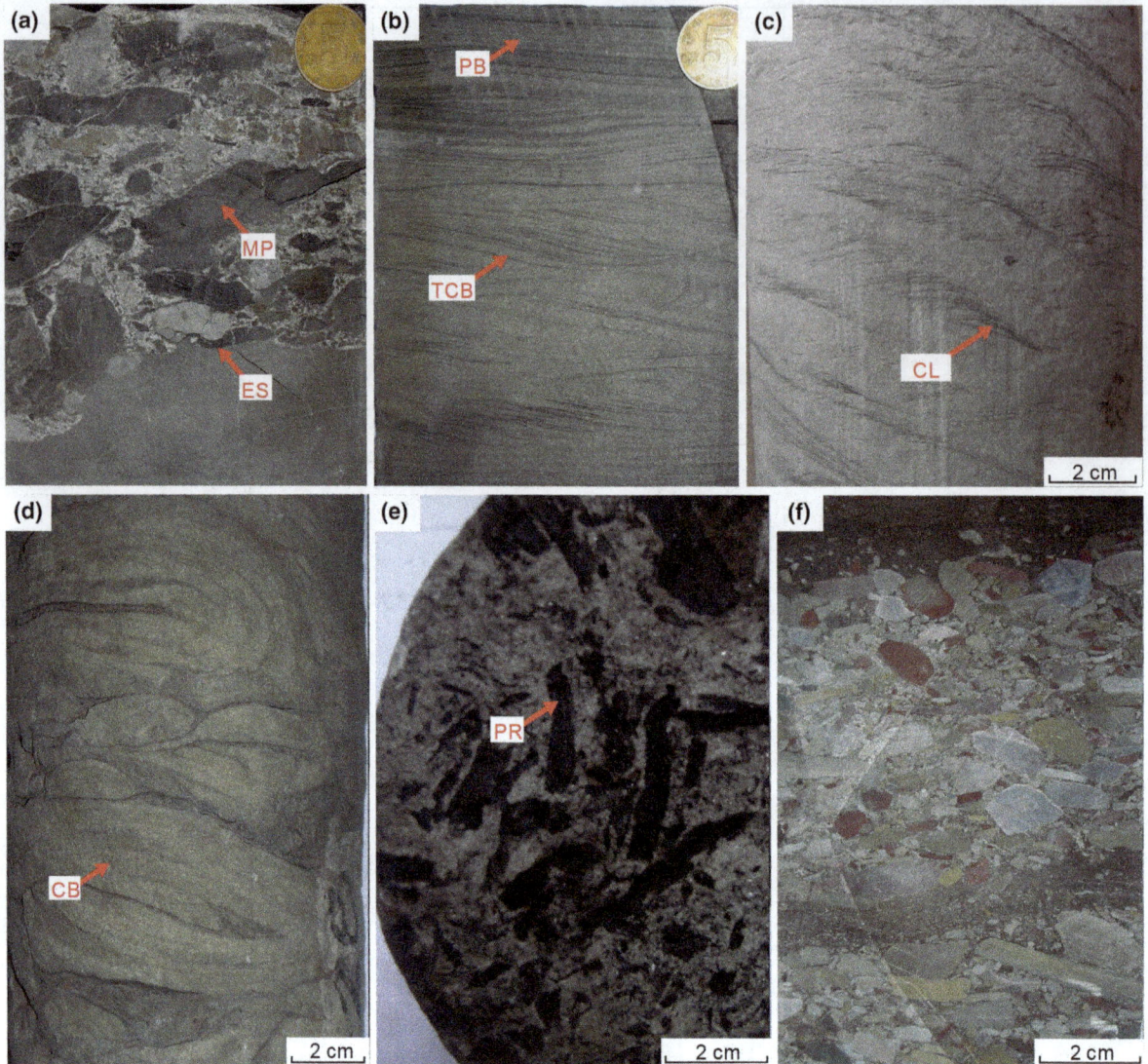

Fig. 3 Typical sedimentary structures of the Dainan Formation in the Gaoyou Depression. Subaqueous distributary channel of the fan delta front. **a** Erosional base surface (ES) and mud pebbles (MP), Well Fu 23, 2933.7 m; **b** Tabular cross bedding (TCB) and parallel bedding (PB), Well Fu 23, 3094.6 m; **c** Climbing-ripple cross stratification (CL), Well Zhou 52, 1655.8 m; and **d** Convolute bedding (CB), Well Cao 20, 3251.3 m. **e** Plant remains (PR) from the subaqueous interchannel in the delta front, Well Fu 35, 3126.8 m. **f** Structureless mottled conglomerate from the inner fan of the nearshore subaqueous fan, Well Shao 9, 2284.5 m. The coins in **a**, **b** are 2 cm in diameter. See Fig. 1c for well locations

of subaqueous distributary channels and mouth bars were vulnerable to being reworked by wave processes (Coleman 1988; Johnson and Levell 1995; Hoy and Ridgway 2003), forming a thin-bedded, widely distributed sand sheet in the distal part of the fan delta front. Sand sheets consist mainly of siltstone and muddy siltstone and display an intimate association with the mudstones of the shallow lacustrine facies (finger-like pattern in the SP curve). The thickness of individual sand sheets is 2–3 m.

Profan delta subfacies mainly consist of grayish brown mudstones with sand strips and masses. Horizontal bedding is most common, with wavy and lenticular beddings being

less common. The SP curve is relatively straight, while the *R* curve displays a low-amplitude jagged pattern.

3.2 Delta

Delta facies occur predominantly in the northern gentle slope of the Gaoyou Depression, and are marked by fine-grained sandstones with less conglomerate and more mudstone compared to the fan deltas (Table 2). Seismic profiles occur as parallel to subparallel reflection configurations. The deltas can be divided into three distinct yet genetically related subfacies: delta plain, delta front, and

Table 2 Comparisons among the fan delta, delta, and nearshore subaqueous fan facies of the Paleogene Dainan Formation

	Fan delta	Delta	Nearshore subaqueous fan
Distribution location	Step-fault zone in the southern steep slope	Northern gentle slope zone	The single fault zone in the southern steep slope
Sedimentary characteristics	Composed primarily of relatively coarse-grained, moderately sorted, and grain-supported conglomerate, conglomeratic sandstone, and coarse sandstone, with some siltstone and mudstone Cumulative grain-size distribution curve presents two sections. Contain sedimentary structures formed by tractive currents	Composed mainly of relatively fine-grained and well-sorted fine to coarse sandstones, subordinate siltstone, and mudstone Cumulative grain-size distribution curve presents two sections Contain sedimentary structures formed by tractive currents	Composed of poorly sorted and coarsest-grained sedimentary rocks, like matrix-supported conglomerate, indicating the strongest hydrodynamic force Cumulative grain-size distribution curve is similar to gravity flow. Contain sedimentary structures formed by gravity flow
Microfacies type	Microfacies type is monotonous, and consists mainly of subaqueous distributary channels and interchannels of the fan delta front. Fan delta plain subfacies are not well developed	Delta plain and delta front subfacies are widely developed, with the latter dominant. Furthermore, the delta front is typified by various microfacies, including subaqueous branch channel, subaqueous branch interchannel, branching mouth bar, and sand sheet	Dominated by inner fan and middle fan subfacies, with the former including one or a few main channels and the latter involving braided channels and channel bays
Distribution pattern	Small-scaled lobe shape in plane view	Large-scaled lobe shape in plane view	Small-scaled lobe shape in plane view
Logging curve property	SP curves show prominent negative anomalies, and are slightly jagged with low to moderate amplitudes. R curves are moderately jagged with medium amplitudes. These curves could show bell or cylinder shapes	SP curves show prominent negative anomalies, with moderate to high amplitudes and significant negative values. SP curves show pronounced bell, cylinder, and funnel shapes. R curves are moderately jagged with relatively low values and amplitudes	SP curves show prominent negative anomalies, and are slightly serrated with low to moderate amplitudes. R values are extremely high, with the curves being seriously jagged
Seismic facies characteristics	Wedge-shaped foreset with large thickness	Parallel and subparallel sheet seismic reflections	Wedge- and/or mound-shaped seismic configurations

prodelta, with the delta front as the majority of those found in the Gaoyou Depression. The delta plain constitutes the subaerial part of the delta and mainly consists of branch channels and branch interchannels. The basal sections of branch channels are erosionally based and are typified by medium- to coarse-grained sandstones with scattered mud pebbles. They can be structureless, or contain pervasive trough-cross and parallel beddings. The upper portion of the branch-channel succession is primarily composed of fine-grained sandstones and siltstones (with occasional mudstones), containing wavy bedding and climbing-ripple cross stratification. The thickness of individual successions ranges from 6 to 8 m.

The delta front includes subaqueous branch channels, subaqueous branch interchannels, branching mouth bars, and sheet sands (Fig. 4). Subaqueous branch channels are erosionally based, and are mainly composed of grayish brown fine-grained sandstones and siltstones, with many rounded mud pebbles at the base. Sedimentary structures comprise graded bedding, parallel bedding, cross bedding, climbing-ripple cross stratification, and horizontal bedding from bottom to top. In general, subaqueous branch channels represent an upward-fining succession with the corresponding SP curve characteristically bell-shaped (Fig. 4). The thickness of individual successions is 5–7 m. Subaqueous branch interchannels are located between adjacent subaqueous branch channels and consist of brownish-gray mudstones and muddy siltstones. Horizontal and wavy beddings, plant remains (Fig. 3e), and bioturbations are common.

Branching mouth bars usually occur as an upward-coarsening succession with grayish brown muddy siltstones at the bottom, and siltstones and fine-grained sandstones towards the top. Parallel, wavy, and cross beddings are pervasive (Fig. 4). The thickness of individual successions is 3–5 m. Sheet sands are located in the distal part of the delta front, and are mainly composed of thin-bedded fine-grained sandstone and siltstone (Fig. 4) with wavy, cross, and horizontal beddings involved. The thickness of individual successions is commonly 2–3 m. Prodelta subfacies mainly consist of gray-brown mudstones and shales intercalated with thin-bedded siltstones. The SP curve displays a finger-like pattern within the low-magnitude range.

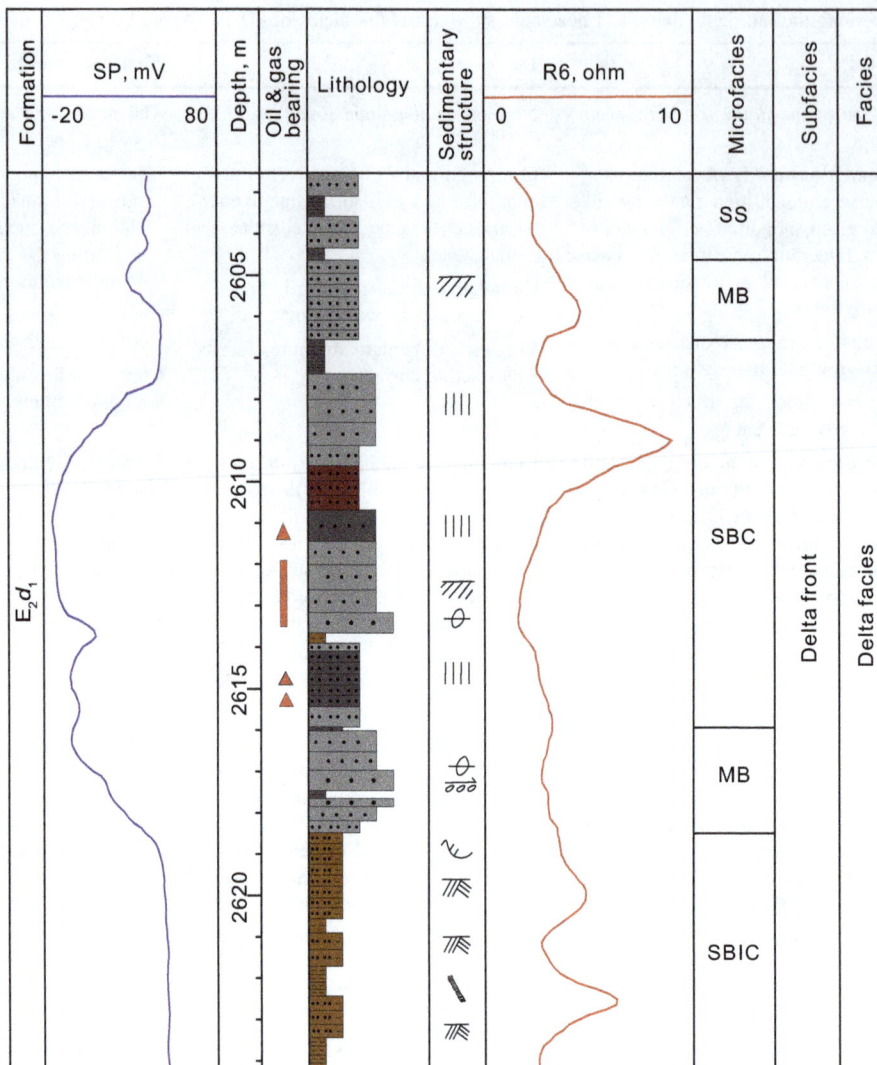

Fig. 4 Sedimentary characteristics of the delta for the Dainan Formation in Well Yong 16 (see location in Fig. 1c) of the Gaoyou Depression. *SP* Spontaneous potential curve; *R6* 6 m bottom gradient resistivity curve; *SS* Sand sheet; *MB* Branching mouth bar; *SBC* Subaqueous branch channel; *SBIC* Subaqueous branch interchannel. See legends in Fig. 2

3.3 Nearshore subaqueous fan

The term "nearshore subaqueous fan" refers to a coarse-grained fan that lacks a subaerial component. It develops where an alluvial river or fan debouches directly into excessively deep coastal waters (Colella and Prior 1993), and is derived from gravity flow (Zhang and Tian 1999; Table 2). Deposits of these systems have been commonly found in Mesozoic–Cenozoic rift basins of eastern China (Zhang and Shen 1991; Zhou et al. 1991; Zhang and Tian 1999).

In the study area, nearshore subaqueous fans developed in the Shaobo, Xiaoliuzhuang, and Zhouzhuang areas during the E_2d_1 period, and these can be further divided into three microfacies: inner fan, middle fan, and outer fan.

The inner fan subfacies is characterized by one or a few main channels which can be described as undercompensated incised valleys (Liu 2003). The main channel consists of poorly sorted mottled conglomerates and sandy conglomerates that have complex compositions and are commonly matrix-supported and/or grain-supported. Those that are matrix-supported are interpreted to have been deposited by debris flows, characterized by gravels floating randomly in a fine-grained matrix (appearing structureless) (Fig. 3f). Resultant successions have sharp boundaries or scour surfaces at their base and load structures developed towards the underlying unit. Grain-supported conglomerates, on the other hand, are thought to have been deposited by high-density turbidity currents. Deposits of such turbidity currents (turbidites) are typified by normal and reverse graded bedding, occasional crude cross bedding, and slump deformation structures. The corresponding SP curve of such successions displays a jagged bell- or cylinder-shaped pattern.

The middle fan facies includes braided channels and interchannels, with deposits of the former constituting the majority of the succession (Fig. 5). The braided channels are typified by gray and grayish-white sandy conglomerates and conglomeratic sandstones with low compositional maturity and moderate textural maturity. Clasts constituting these conglomerates involve limestone and quartz fragments with diameters of 10–20 mm. Scour surfaces and flute casts are common at the bottom of the braided channel successions. Graded, parallel, and cross beddings are most common. The fine-grained sediments formed in the interchannels are easily washed away (Walker 1978). The corresponding SP curve displays a bell-shaped or jagged cylinder-shaped pattern (Fig. 5).

The outer fan is located in the seaward extremity of the nearshore subaqueous fan and consists mainly of dark mudstones intercalated with siltstones, muddy siltstones, and locally thinly bedded fine-grained sandstones. Sedimentary structures mainly include wavy and horizontal beddings, reflecting a relatively lower flow regime and quiet environment. The corresponding SP curve is flat with low amplitudes.

Compared to the deltas and fan deltas, the nearshore subaqueous fan is characterized by the strongest hydrodynamic force and poor sorting (Table 2), which is clearly reflected in the probability cumulative grain-size distribution curves (Fig. 6). Figure 6 shows that (1) the grain-size range of saltation components for nearshore subaqueous fans is between -1.0 and $3.0\ \Phi$, whereas those of fan delta and delta are 0.5–3.5 and 1.0–$4.0\ \Phi$, respectively, which indicates that the hydrodynamic force of nearshore subaqueous fans is the strongest (Lin et al. 2005); (2) the slope of saltation components for delta, fan delta, and nearshore subaqueous fan is about $71°$, $65°$, and $52°$, respectively, of which the latter is the smallest, suggesting poor sorting; (3) the abundance of the suspension component for the

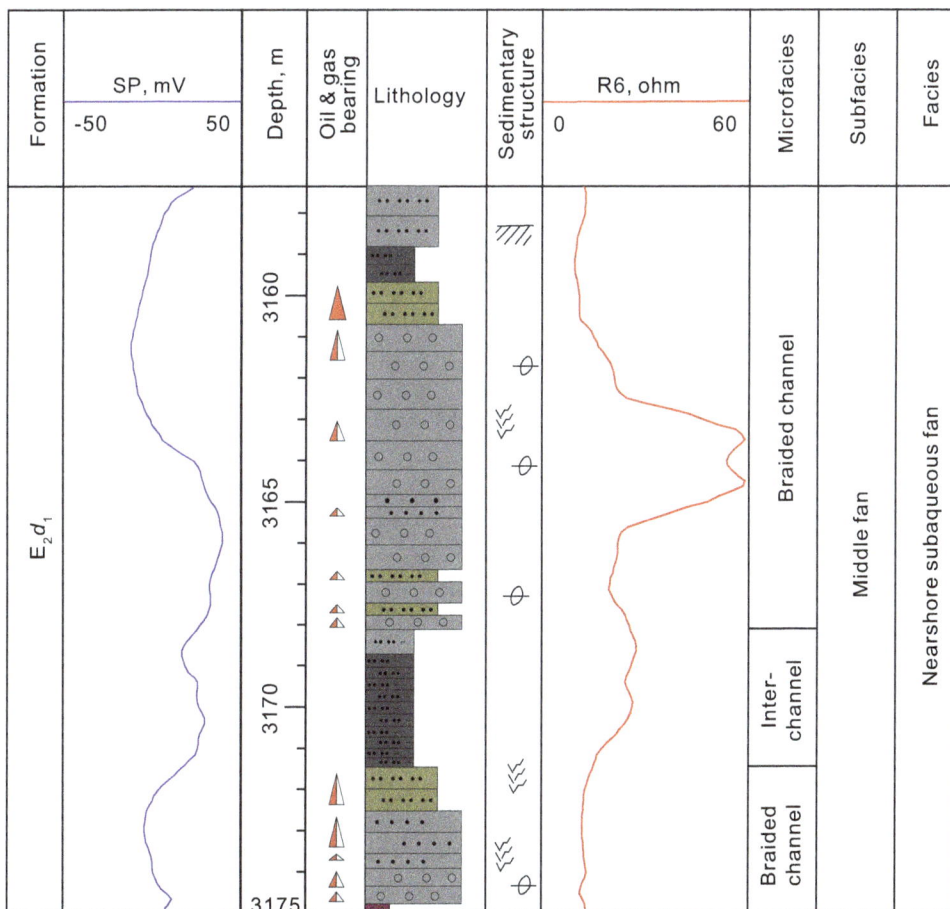

Fig. 5 Sedimentary characteristics of the nearshore subaqueous fan for the Dainan Formation in Well Xiao 3 (see location in Fig. 1c) of the Gaoyou Depression. *SP* Spontaneous potential curve; *R6* 6 m bottom gradient resistivity curve. See legends in Fig. 2

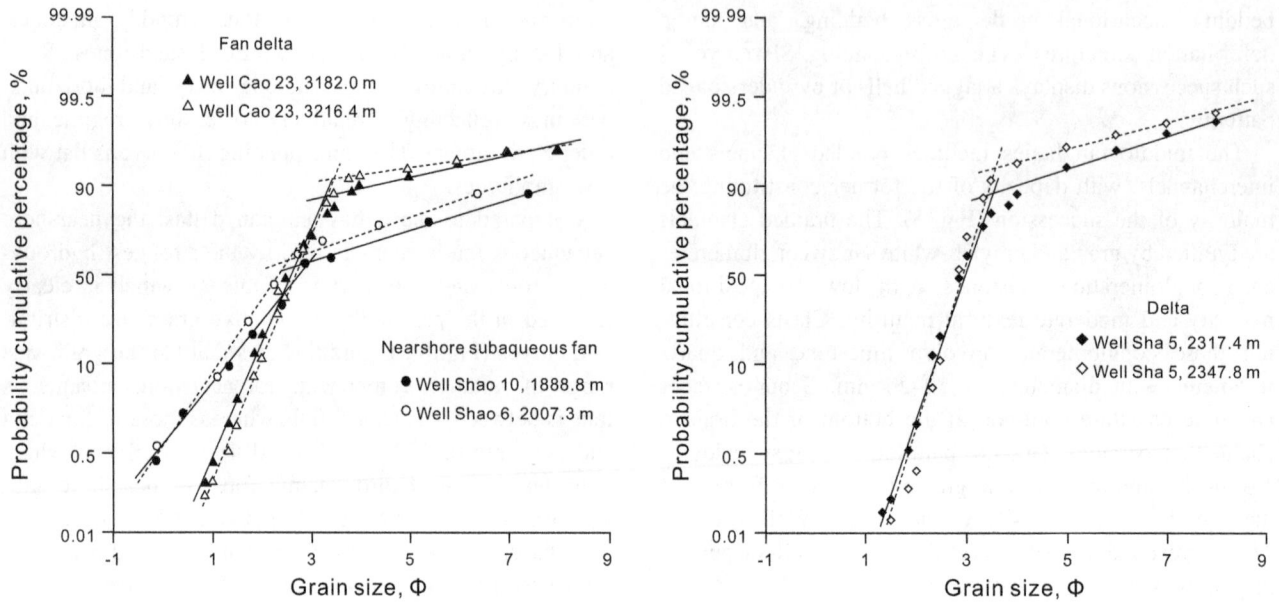

Fig. 6 Probability cumulative grain-size distribution curves of the fan delta, delta, and nearshore subaqueous fan in the Dainan Formation. See Fig. 1c for the well locations

nearshore subaqueous fans is the greatest, generally more than 30 %, supporting a graywacke classification.

3.4 Lacustrine facies

Lacustrine facies mainly occur at the center of the Gaoyou Depression, as well as on the flanks of the fan deltas, deltas, and nearshore subaqueous fans. Shore-shallow lacustrine and semideep lacustrine subfacies are identified in the study area. The nature of the shore-shallow lacustrine subfacies is controlled by the provenance and hydrodynamic force. If the provenance is typified by gravel and sand, then gravelly and/or sandy lacustrine beaches form. However, if the terrain of the lacustrine beach is gentle, the hydrodynamic force will be weak, and supplied sediment will consist mainly of mud, allowing mudflats to form (Lin et al. 2003). In the study area, the shore-shallow lacustrine subfacies comprise mainly of siltstones and mudstones with a variety of colors: brownish-gray, dark-purple, and dark-brown (Fig. 2). Horizontal and wavy beddings, bioturbation (especially vertical worm burrows), and plant remains are common. The corresponding SP curve is linear and low in amplitude, while the R curve is jagged having low to moderate amplitudes (Fig. 2). Semideep lacustrine facies are located under the fair-weather wave base, i.e., a generally anoxic environment. Sedimentary rocks are mainly composed of dark mudstones with high organic matter contents, with horizontal and lenticular beddings dominant.

4 Sedimentary distribution and processes

The sedimentary succession of the Dainan Formation in the Gaoyou Depression exhibits a complete transgressive–regressive cycle with sediment grain sizes displaying a coarse–fine-coarse pattern in the ascending order. As a result, there are two major sedimentological periods for the Dainan Formation: E_2d_1 and E_2d_2. E_2d_1 consists of three stages: $E_2d_1^3$, $E_2d_1^2$, and $E_2d_1^1$, and E_2d_2 is composed of five stages: $E_2d_2^5$, $E_2d_2^4$, $E_2d_2^3$, $E_2d_2^2$, and $E_2d_2^1$ (Table 1).

4.1 Sedimentary period of E_2d_1

During the E_2d_1 period, the Dainan Formation began to form and overlay the Funing Formation by an unconformity which had resulted from Wubao Movement. Deposition during the E_2d_1 is thought to have been in phase with the uplifting of the basement and the development of northeastern faults in the Gaoyou Depression (syndepositional). The strata are thick in the south and gradually thin towards the north. The subsidence center was located in the Shaobo and Fanchuan subdepressions with strata thickness up to 900 m. The fan delta, delta, nearshore subaqueous fan, and lacustrine facies developed within this period, representing a transgressive succession with the grain size of clastic particles fining upwards and the relative thickness of sandstones reducing gradually upwards (Fig. 7).

In $E_2d_1^3$, the strong movements of the Zhen 2 and Hanliu faults controlled and limited the distribution of sediments (Chen 2001), so that the Zhen 2 fault acted as a southern

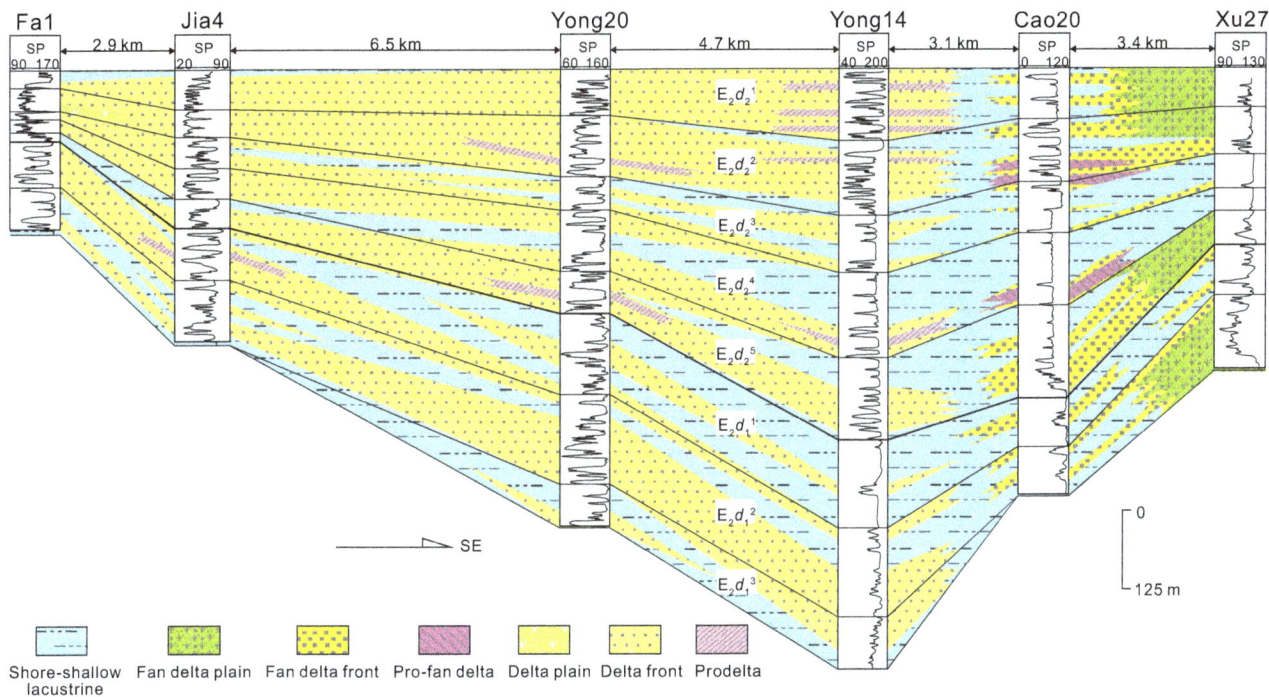

Fig. 7 Cross-sectional profile showing the vertical evolution and distribution pattern of the different sedimentary facies of the Dainan Formation in the Gaoyou Depression. The Dainan Formation presents as an integrated transgressive–regressive cycle. See Fig. 1c for the location. *SP* Spontaneous potential curve (mV)

boundary and the Hanliu fault formed a fault-step zone as the northern margin. The subsidence center is located in the Shaobo subdepression with stratum thickness up to 400 m (Fig. 8a). The total sandstone thickness is greater in the east than in the west, reaching up to 60 m in the Fumin area (generally 20–40 m) (Fig. 8b). Four fan deltas (Huangjue (HJ), Zhenwu–Caozhuang (ZC), Fumin (FM), Zhouzhuang (ZZ)) and a nearshore subaqueous fan (Shaobo (SB)) developed along the Zhen 2 fault. Four small-scale and independent deltas involving the western part of Lianmengzhuang (LLZ), and eastern parts of Lianmengzhuang (RLZ), Yong'an (YA) and Fumin–Huazhuang (FH) were formed on the gentle northern slope (Fig. 9a). Shore-shallow lacustrine facies occurred mainly in the center of the depression (Fig. 9a).

In $E_2d_1^2$, the sedimentary area extended with the boundary crossing the Zhen 2 and Hanliu faults because of the small-scale increase of lake water and decrease in tectonic activity (Fig. 9b). The stratigraphic overlap in the northern part of the Gaoyou Depression is easily observed in the seismic profile. The subsidence center was also located in the Shaobo and Fanchuan subdepressions with a thickness of 300 m (Fig. 8c). The gross sandstone thickness is greater in the east than towards the west, with maximum thickness reaching 80 m in the Fumin and Yong'an areas (20–40 m in general) (Fig. 8d). Compared

to the $E_2d_1^3$, the range and scale of the $E_2d_1^2$ period fan delta, delta deposits, and nearshore subaqueous fan sediments expanded due to augmented accommodation and sufficient sediment supply. Four fan deltas (HJ, ZC, FM, and ZZ) and three nearshore subaqueous fans (SB, Xiaoliuzhuang (XLZ), and ZZ) developed along the steep southern slope zone (Fig. 9b). Also, the four deltas of $E_2d_1^3$ in the northern slope of the Gaoyou Depression merged to form one larger delta, and the Majiazui (MJZ) area started receiving sediment in this stage, resulting in the formation of two detached deltas (Fig. 9b). Shore-shallow lacustrine deposits mainly accumulated in the center of the depression, as well as on the flanks of the fan deltas, deltas, and nearshore subaqueous fans (Fig. 9b).

During the $E_2d_1^1$ period, the lake transgression reached a maximum and the lateral extent of deposition continued to expand with the "five high-conductivity" dark mudstones representing the sedimentary boundary of the Gaoyou Depression. The subsidence center was still located in the Shaobo, Fanchuan, and Liuwushe subdepressions with strata thickness of 200 m (Fig. 8e). The gross sandstone thickness is generally 20 m, but can be locally greater, such as in the Fumin, Yong'an, and Shanian areas, where thicknesses can reach up to 40 m (Fig. 8f). Compared to the $E_2d_1^3$ and $E_2d_1^2$, the scales of fan deltas, deltas and nearshore subaqueous fans of the $E_2d_1^1$ period were reduced

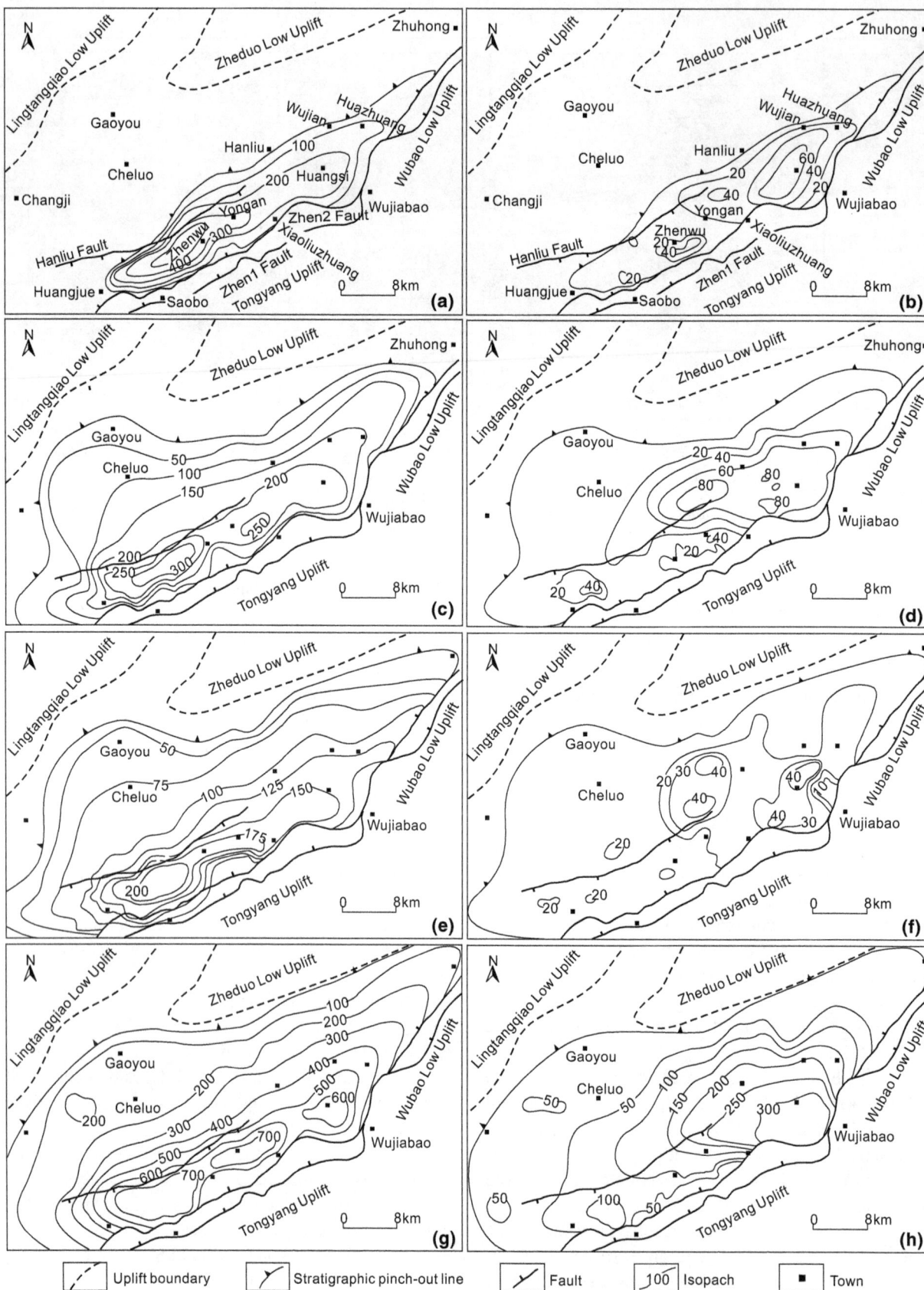

◄**Fig. 8** Isopach maps of the strata (*left side*) and gross sandstone layers (*right side*) in different stages for the Dainan Formation, Gaoyou Depression. **a**, **b** The $E_2d_1^3$ stage; **c**, **d**: the $E_2d_1^2$ stage; **e**, **f** the $E_2d_1^1$ stage; **g**, **h** the E_2d_2 stage

and these moved back towards the lakeshore. The ZC and ZZ fan deltas of $E_2d_1^2$ were each replaced with two detached and small-scale fan deltas. In the northern slope, the delta was still a unified delta as that of $E_2d_1^2$. The Wazhuang (WZ) area started receiving sediment in this stage occurring as the WZ delta. The ZZ nearshore subaqueous fan was substituted by shore-shallow lacustrine subfacies. Semideep lacustrine subfacies developed in the Shaobo, Fanchuan, and Liuwushe subdepressions, and the shore-shallow lacustrine subfacies deposited mainly on the flanks of fan deltas, deltas, and nearshore subaqueous fans (Fig. 9c).

4.2 Sedimentary period of E_2d_2

During the E_2d_2 period, the Gaoyou Depression was characterized by weak movement of faults and basement uplift, resulted in a shallowing water depth, a decreasing slope gradient, and disappearance of nearshore subaqueous fans in the southern slope (Figs. 9d, 10). The stratum thickness is still thick in the south and thins out towards the north (Fig. 8g). Subsidence centers were located in the Shaobo and Fanchuan subdepressions with strata thicknesses up to 700 m (Fig. 8g). The sandstones were primarily deposited in the eastern Gaoyou Depression, including the Shanian and Yong'an areas of the northern slope, as well as the Fumin and Fanchuan regions of the southern slope, with thicknesses commonly reaching 100–300 m. These thicknesses generally thin out towards the western Gaoyou Depression to approximately <50 m, with only a few areas reaching 100 m (Fig. 8h). In general, the sedimentary framework of E_2d_2 occurs as a regressive succession composed of a second-order transgressive–regressive cycle, with the grain size and thickness of the sandstone displaying a coarse–fine-coarse and thick-thin-thick trend upwards, respectively (Fig. 7).

During the $E_2d_2^5$ depositional period, the dispersal of sediment reduced and water depth became shallower compared to $E_2d_1^1$. Due to the sufficient supply of sediment, the fan deltas in the steep southern slope and the deltas in the northern gentle slope prograded into the center of the depression with their lateral extent amplified. In the Fumin area, the fan delta front and delta front converged (Fig. 9d). The Zhenwu and Caozhuang fan deltas of the $E_2d_1^1$ period also joined together as one unified fan delta. Semideep lacustrine subfacies were replaced by shore-shallow lacustrine subfacies (Fig. 9d).

In $E_2d_2^4$, the lateral extent of sediment deposition extended and water depth increased compared to $E_2d_2^5$. Together, this resulted in the retrogradation of the fan deltas and deltas with the scale of such systems reduced. The fan delta fronts and delta fronts, however, still converged in the Fumin area. The Zhenwu–Caozhuang and Shaobo fan deltas and Lianmengzhuang–Yong'an–Fumin delta of the $E_2d_2^5$ period were each replaced by two detached and small-scale fan deltas and deltas. Shore-shallow lacustrine subfacies developed in the center of the depression and on the flanks of deltas/fan deltas (Fig. 9e).

In $E_2d_2^3$, the water depth continued to increase, which resulted in the persistent retrogradation of the fan deltas and deltas. The fan delta front and delta front separated in the Fumin area. The Lianmengzhuang and Yong'an–Fumin delta of the $E_2d_2^4$ period converged into a unified delta, and the two separated fan deltas in the Shaobo area also joined together. Shore-shallow lacustrine subfacies developed in the center of the depression and on the flanks of deltas/fan deltas (Fig. 9f).

In $E_2d_2^2$, the distribution pattern of sedimentary facies is similar to that of the $E_2d_2^3$ period; however, the deltas and fan deltas prograded into the center of the depression with the scale of lateral deposition increased due to the shallow water depth and sufficient supply of sediment (Fig. 9g).

In $E_2d_2^1$, the deltas and fan deltas continued to prograde into the center of depressions with the scales increased due to the shallow water depth and sufficient supply of sediment. In the Fumin area, the fan delta front and delta front met and blended together again. The Huangjue and Shaobo fan deltas also merged together to form a unified fan delta (Fig. 9h). In addition, during the late stage of the $E_2d_2^1$ period, the Gaoyou Depression uplifted as a result of Zhenwu Movement leading to the denudation of Dainan Formation which was unconformably overlain by the Sanduo Formation (Table 1).

5 Sedimentary architecture and implications

Continental rift basin sediment filling patterns are mainly controlled by tectonics (Lin et al. 2001), and subordinate lake-level fluctuations and sediment supply (Yu et al. 2007). Tectonics primarily determines the type of sedimentary facies present and the associated spatial distribution pattern. The southern slope of the Gaoyou Depression was steep and narrow such that the increased rate of accommodation creation, triggered by tectonically induced subsidence (fault movements), exceeded the rate of sediment supply (A > S). As a result, small-scale and coarse-grained fan deltas and nearshore subaqueous fans preferentially developed (Fig. 11). For instance, in step-fault zones such as the Huangjue and Fumin areas, the slope was

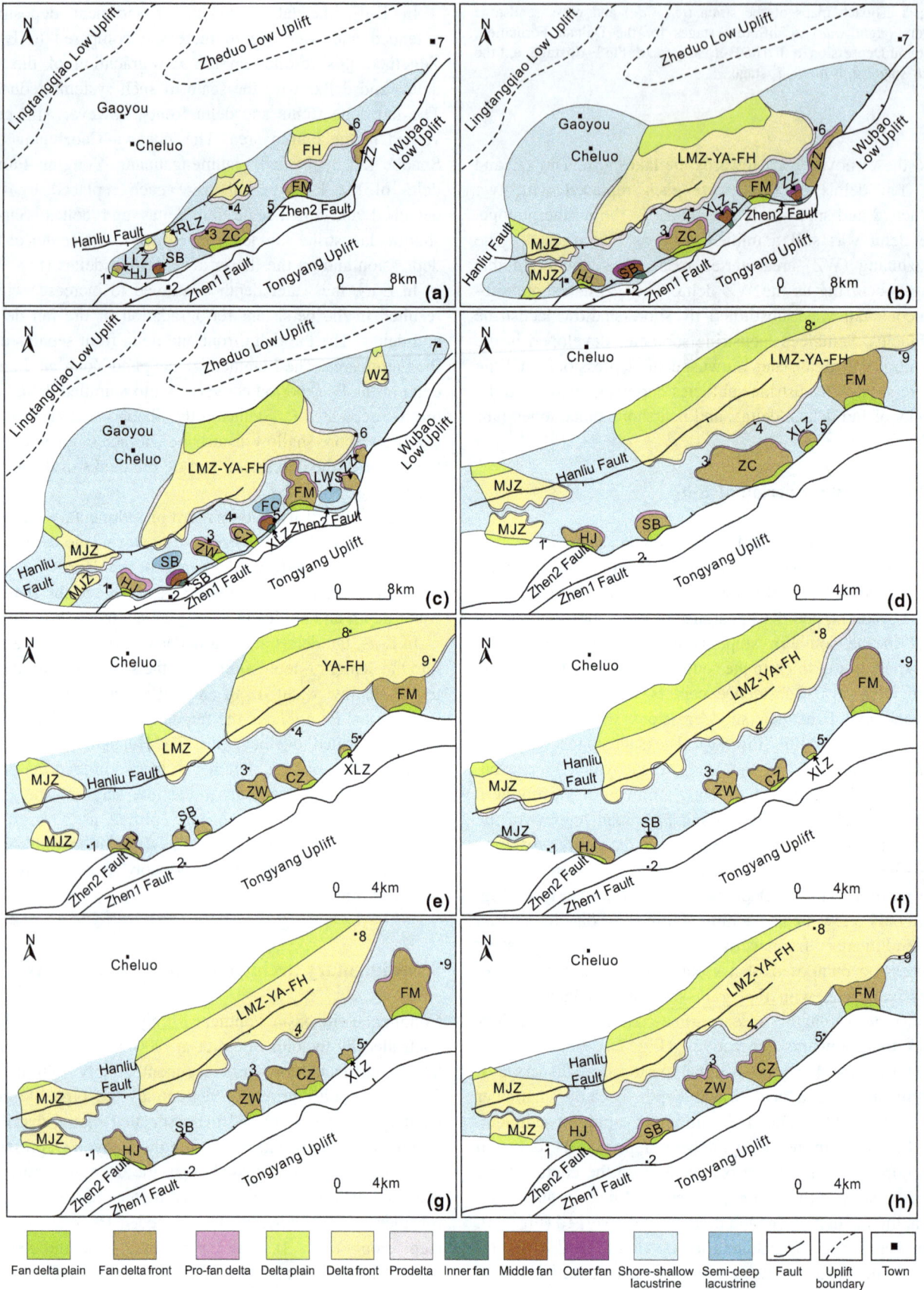

Fan delta plain | Fan delta front | Pro-fan delta | Delta plain | Delta front | Prodelta | Inner fan | Middle fan | Outer fan | Shore-shallow lacustrine | Semi-deep lacustrine | Fault | Uplift boundary | Town

◀Fig. 9 Diagrams showing the distribution pattern of the different sedimentary facies in plan view for different stages during the development of the Dainan Formation, Gaoyou Depression. **a** $E_2d_1^3$ stage; **b** $E_2d_1^2$ stage; **c** $E_2d_1^1$ stage; **d** $E_2d_2^5$ stage; **e** $E_2d_2^4$ stage; **f** $E_2d_2^3$ stage; **g** $E_2d_2^2$ stage; **h** $E_2d_2^1$ stage. *FM* Fumin; *ZZ* Zhouzhuang; *YA* Yong'an; *FH* Fumin–Huazhuang; *RLZ* Right part of the Lianmengzhuang; *LLZ* Left part of the Lianmengzhuang; *HJ* Huangjue; *SB* Shaobo; *ZC* Zhenwu–Caozhuang; *MJZ* Majiazui; *XLZ* Xiaoliuzhuang; *WZ* Wazhuang; *ZW* Zhenwu; *CZ* Caozhuang; *LMZ-YA-FH* Lianmengzhuang–Yong'an–Fumin–Huazhuang; *FC* Fanchuan; *LWS* Liuwushe; *LMZ* Lianmengzhuang; *YA-FH* Yong'an–Fumin–Huazhuang. Arabic numbers show the town locations in the study area, *1* Huangjue; *2* Shaobo; *3* Zhenwu; *4* Yong'an; *5* Xiaoliuzhuang; *6* Huazhuang; *7* Zhuhong; *8* Hanliu; *9* Huangsi

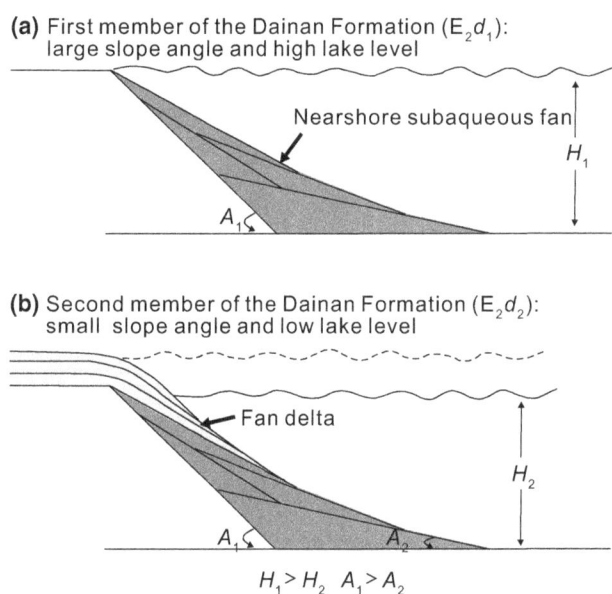

(a) First member of the Dainan Formation (E_2d_1): large slope angle and high lake level

Nearshore subaqueous fan

H_1

A_1

(b) Second member of the Dainan Formation (E_2d_2): small slope angle and low lake level

Fan delta

H_2

A_1 A_2

$H_1 > H_2$ $A_1 > A_2$

Fig. 10 Schematic map showing how the slope gradient and lake level control the formation of the nearshore subaqueous fan and fan delta. H_1 represents the water depth; A_1 indicates the slope angle

relatively gentle enough to allow for the development of fan deltas. In the monofaulted zone, like the Shaobo area, however the slope was sufficiently steep to form nearshore subaqueous fans (Figs. 10, 11). The northern slope was broad and gentle characterized by decreased subsidence as a result of reduced movement of faults, therefore favoring the generation of large-scale fine-grained deltas (Fig. 11). Lake-level fluctuations and sediment supply modulated the distribution pattern and scale of sand bodies by modifying the interrelationship between the rates of accommodation space creation and sediment supply. The rate of sediment supply was able to keep pace with, or exceeded, the increased rate of accommodation space creation. Numerous studies have shown that lake-level fluctuations can cause a shift in the depocenter, which results in the deposition of a wide range of sedimentary facies in the same area of the

basin through each transgressive–regressive cycle as shown in Fig. 7 (Posamentier et al. 1988; Hoy and Ridgway 2003).

Economically important reservoirs in the Gaoyou Depression consist predominantly of deltaic and fan deltaic sandstones which are mainly distributed along the margins of depressions. Additionally, the reservoir quality of sandstones in the subaqueous branch channels of the deltas is generally better than that of sandstones in the subaqueous distributary channels of the fan deltas. Porosity of the former ranges from 10 % to 30 %, and the permeability ranges from 1 to 100 mD. The porosity and permeability of the latter, however, are 10 %–20 % and <1 mD, respectively. The sandstones of nearshore subaqueous fans represent a second reservoir type, consisting mainly of thick-bedded, turbiditic channel-fill sandstones. Some turbiditic channel sandstones have been proven to be important oil reservoirs in the Gaoyou and other depressions (Zhang and Tian 1999; Gao et al. 2009). Thus, a comprehensive understanding of the vertical evolution and distribution patterns of the sedimentary facies of the Dainan Formation in the Gaoyou Depression is of significance in predicting optimal reservoir targets for exploration and exploitation.

6 Conclusions

The Dainan Formation in the Gaoyou Depression was generated during two major sedimentation periods (E_2d_1 and E_2d_2), involving four main sedimentary facies, which include fan delta, delta, nearshore subaqueous fan, and lacustrine facies. In addition, the nearshore subaqueous fan facies were absent during the E_2d_2 period due to the weak movement of faults, shallowing of the water depth, and reduction of the slope gradient. Fan delta and nearshore subaqueous fan facies are distributed predominantly in the southern steep slope, whereas deltaic facies occur in the northern gentle slope. The lacustrine facies are present in the center of the depression and on the flanks of the three facies above. Vertically, the Dainan Formation exhibits an integrated transgressive–regressive cycle with the grain size and relative thickness of sandstones displaying a coarse–fine–coarse and thick–thin–thick trend upwards, respectively. This sedimentary framework and distribution patterns of facies are thought to have been controlled primarily by tectonics, and less by lake level and sediment supply. This study provides a valuable model for the exploration and exploitation of oil and gas in the study area, as the sandstones of the subaqueous distributary channel and subaqueous branch channel facies have favorable physical properties for major lithologic reservoir targets.

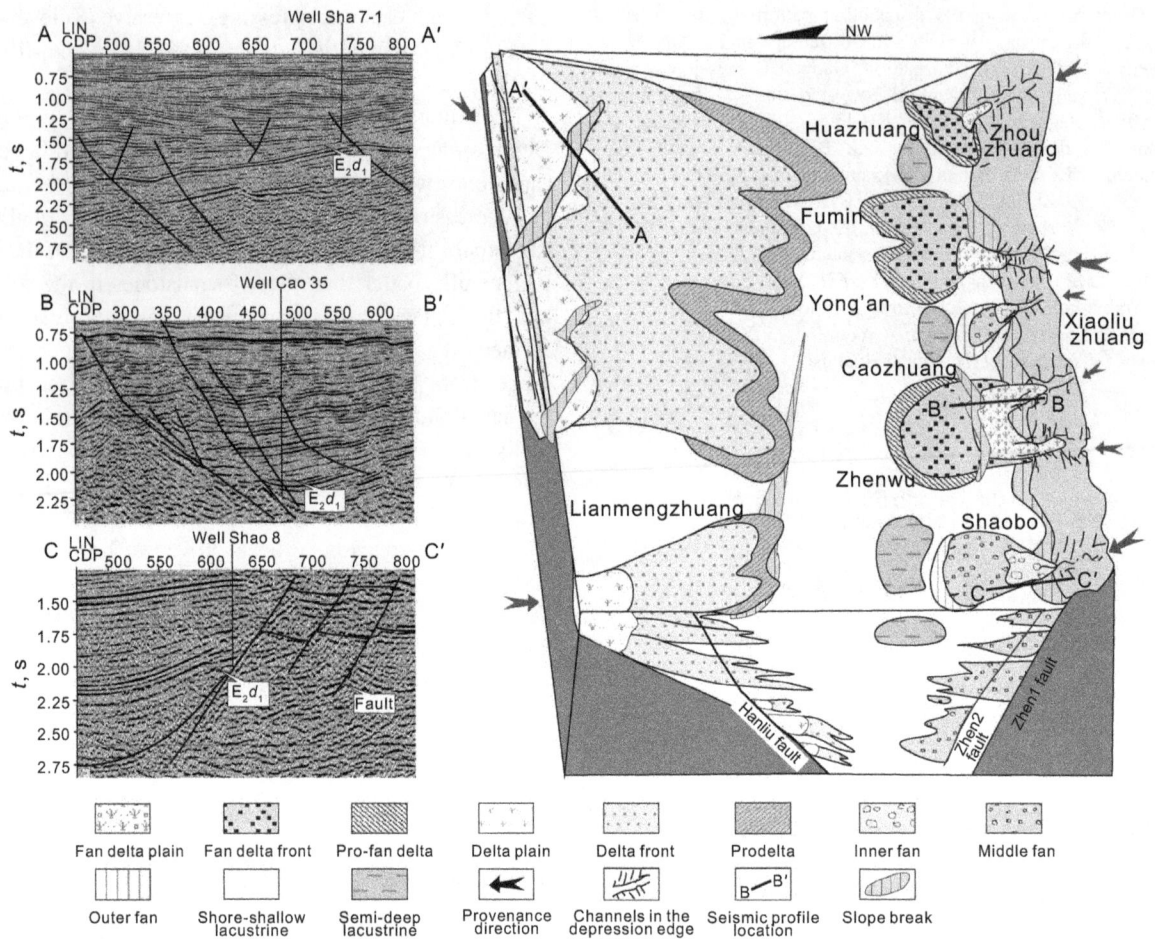

Fig. 11 The sedimentary architecture for the first member of the Dainan Formation (E_2d_1) in the Gaoyou Depression. Fan deltas and nearshore subaqueous fans are distributed in the southern steep slope, and deltas in the northern gentle slope. Lacustrine facies developed in the center of the depression and on the flanks of the fan deltas, deltas, and nearshore subaqueous fans

Acknowledgments This research was financially supported by the National Natural Science Foundation of China (Grants Nos. 41272124 and 41402092), Natural Science Foundation (Youth Science Fund Project) of Jiangsu Province (BK20140604), the Fundamental Research Funds for the Central Universities (20620140386), and the State Key Laboratory for Mineral Deposits Research of Nanjing University (Grant No. ZZKT-201321). We thank X.D. Yue, Y.L. Li, Z.P. Zhang, Y.L. Yao, and L.K. Gao for their helpful discussions, and assistance in field and core observations, and the laboratory work. Especial thanks are given to Y.J. Ma and Q.D. Liu of Jiangsu Oilfield Branch Company, SINOPEC for their invaluable support. Special thanks should be extended to the Petroleum Science editors and anonymous reviewers for their constructive suggestions and comments, and to D.T. Canas of Queen's University, Canada for checking the English presentation.

References

Blair TC, McPherson JG. Alluvial fans and their natural distinction from rivers based on morphology, hydraulic processes, sedimentary processes, and facies assemblages. J Sediment Res. 1994;64a:450–89.

Cao YC, Yuan GH, Li XY, et al. Characteristics and origin of abnormally high porosity zones in buried Paleogene clastic reservoirs in the Shengtuo area, Dongying Sag, East China. Pet Sci. 2014;11(3):346–62.

Chen AD. Dynamic mechanism of formation of dustpan subsidence, Northern Jiangsu. Geol J Chin Univ. 2001;7(4):408–18 (**in Chinese**).

Chen QH, Wu L, Zhou YC. Hydrocarbon accumulation conditions and modes in the area around Shabo sub-sag, Gaoyou Sag. J Chin Unic Min Technol. 2015;44(2):282–91 (**in Chinese**).

Chen ZR, Wu JY. Early deposition feature of Dainan Formation in the west of Gaoyou Depression and its relation to oil and gas. Small Hydrocarbon Reserv. 2006;11(2):11–4 (**in Chinese**).

Colella A, Prior D. Coarse-grained deltas. Spec Publ 10 of the IAS. Oxford: Blackwell; 1993. p. 29–168.

Coleman JM. Dynamic changes and processes in the Mississippi River delta. GSA Bull. 1988;100(7):999–1015.

Dong RX. Evolution of paleontology and sedimentary environment in Dainan-Sanduo Formation of the Tertiary Gaoyou Depression. J Tongji Univ. 1999;27(3):366–70 (**in Chinese**).

Gao ZY, Guo HL, Zhu RK, et al. Sedimentary response of different fan types to the Paleogene-Neogene basin transformation in the

Kuqa Depression, Tarim Basin, Xinjiang Province. Acta Geol Sin. 2009;83(2):411–24.

Hoy RG, Ridgway KD. Sedimentology and sequence stratigraphy of fan-delta and river-delta deposystems, Pennsylvanian Minturn Formation, Colorado. AAPG Bull. 2003;87(7):1169–91.

Ji YL, Li QS, Wang Y, et al. Fan delta sedimentary system and facies models of Dainan Formation of Paleogene in Gaoyou Sag. J Earth Sci Environ. 2012;34(1):9–19 (**in Chinese**).

Jiang ZX, Liang SY, Zhang YF, et al. Sedimentary hydrodynamic study of sand bodies in the upper subsection of the 4th Member of the Paleogene Shahejie Formation in the eastern Dongying Depression, China. Pet Sci. 2014;11(2):189–99.

Johnson HD, Levell BK. Sedimentology of a transgressive, estuarine sand complex: the Lower Cretaceous Woburn Sands (Lower Greensand), southern England. In: Plint AG, editor. Sedimentary Facies Analysis. Spec Publ of the IAS, vol. 22. Oxford: Blackwell; 1995. p. 17–46.

Krézsek C, Filipescu S, Silye L, et al. Miocene facies associations and sedimentary evolution of the Southern Transylvanian Basin (Romania): implications for hydrocarbon exploration. Mar Pet Geol. 2010;27(1):191–214.

Lin CM, Li GY, Zhuo HC, et al. Sedimentary facies of incised valley fillings of the Late Quaternary in Hangzhou Bay area and shallow biogenic gas exploration. J Palaeogeogr. 2005;7(1):12–24 (**in Chinese**).

Lin CM, Song N, Mu R, et al. Sedimentary facies and evolution of Late Cretaceous in the Yanfu Depression from Jiangsu Province. Acta Sedimentol Sin. 2003;21(4):553–9 (**in Chinese**).

Lin CS, Kenneth E, Li ST, et al. Sequence architecture, depositional systems, and controls on development of lacustrine basin fills in part of the Erlian Basin, Northeast China. AAPG Bull. 2001;85(11):2017–43.

Liu ZJ. Lacus subaqueous fan sedimentary characteristics and influence factors: a case study of Shuangyang Formation in Moliqing fault subsidence of Yitong Basin. Acta Sedimentol Sin. 2003;21(1):148–54 (**in Chinese**).

Lu HM. Continental sequence stratigraphy study of Gaoyou Sag in Subei Basin. Fault-block Oil Gas Field. 2000;7:18–22 (**in Chinese**).

Pang JM, Cao B. Origin and exploration practice of E_2d concealed oil and gas accumulation in Gaoyou Depression. Offshore Oil. 2005;25(3):7–13 (**in Chinese**).

Posamentier HW, Jervey MT, Vail PR. Eustatic controls on clastic deposition I—conceptual framework. In: Wilgus CK, Hastings BS, Kendall CGStC, et al., editors. Sea-level changes: an integrated approach, vol. 42. Tulsa: SEPM Spec Publ.; 1988. p. 109–24.

Qiu XM, Liu YR, Fu Q. Sequence stratigraphy and sedimentary evolution of cretaceous to tertiary in Subei Basin. Beijing: Geological Publishing House; 2006. p. 17–21 (**in Chinese**).

Shu LS, Wang B, Wang LS, et al. Analysis of northern Jiangsu prototype basin from Late Cretaceous to Neogene. Geol J Chin Univ. 2005;11(4):534–43 (**in Chinese**).

Sui FG. Characteristics of reservoiring dynamic on the sand-conglomerate fan bodies in the steep-slope belt of continental fault basin: a case study on Dongying Depression. Oil Gas Geol. 2003;24(4):335–40 (**in Chinese**).

Walker RG. Deep-water sandstone facies and ancient submarine fans: models for exploration for stratigraphic traps. AAPG Bull. 1978;62:932–66.

Wang YZ, Cao YC, Ma BB, et al. Mechanism of diagenetic trap formation in nearshore subaqueous fans on steep rift lacustrine basin slopes: a case study from the Shahejie Formation on the north slope of the Minfeng Subsag, Bohai Basin, China. Pet Sci. 2014;11(4):481–94.

Xia LJ, Wu XY, Mao SL, et al. Seismic prediction of nearshore subaqueous fan in Shaobo, Gaoyou Depression. Prog Explor Geophys. 2008;31(3):212–8 (**in Chinese**).

Xian BZ, Wang YS, Zhou TQ, et al. Distribution and controlling factors of glutinite bodies in the actic region of a rift basin: an example from Chezhen Sag, Bohai Bay Basin. Pet Explor Dev. 2007;34(4):429–36 (**in Chinese**).

Yu XH, Jiang H, Li SL, et al. Depositional filling models and controlling factors on Mesozoic and Cenozoic fault basins of terrestrial facies in eastern China: a case study of Dongying Sag of Jiyang Depression. Lithol Reserv. 2007;19(1):39–45 (**in Chinese**).

Zeng P. Synthetic interpretation of the G78 profile and tectonic characteristics of North Jiangsu Basin. J Oil Gas Technol. 2007;29(3):82–6 (**in Chinese**).

Zhang JL, Shen F. Characteristics of nearshore subaqueous fan reservoir in Damoguaihe Formation, Wuerxun Depression. Acta Pet Sin. 1991;12(3):25–35 (**in Chinese**).

Zhang M, Tian JC. The nomenclature, sedimentary characteristics and reservoir potential of nearshore subaqueous fans. Sediment Facies Palaeogeogr. 1999;19(4):42–52 (**in Chinese**).

Zhang Q, Zhu XM, Steel RJ, et al. Variation and mechanisms of clastic reservoir quality in the Paleogene Shahejie Formation of the Dongying Sag, Bohai Bay Basin, China. Pet Sci. 2014a;11(2):200–10.

Zhang N, Lin CM, Zhang X. Petrographic and geochemical characteristics of the Paleogene sedimentary rocks from the North Jiangsu Basin, East China: implications for provenance and tectonic setting. Miner Pet. 2014b;108(4):571–88.

Zhang XB, Zheng RC, Zhang SN. The tectonic-sedimentary system of the Dainan Formation in Majiazui-Lianmenzhuang region, Gaoyou Sag. Pet Geol Oilfield Dev Daqing. 2007;26(1):13–7 (**in Chinese**).

Zhang XL, Zhu XM, Zhong DK, et al. Study on sedimentary facies and their correlations with subtle traps of the Dainan Formation, Paleogene in Gaoyou Sag, Subei Basin. J Palaeogeogr. 2005;7(2):207–18 (**in Chinese**).

Zhao DN, Zhu XM, Liang B, et al. Seismic sedimentology of the Palaeogene Dainan Formation in the deep sag zone of Gaoyou Depression, Jiangsu Province. Geol J Chin Univ. 2015;21(2):336–45 (**in Chinese**).

Zhao XZ, Jin FM, Wang Q, et al. Hydrocarbon accumulation principles in troughs within faulted depressions and their significance in exploration. Pet Sci. 2011;8:1–10.

Zhou J, Lin CM, Li YL, et al. Provenance analysis of Dainan Formation (Paleogene) of Majiazui in Gaoyou Sag, Subei Basin. Acta Sedimentol Sin. 2010;28(6):1117–28 (**in Chinese**).

Zhou SX, Wright VP, Platt NH, et al. Lacustrine sedimentary systems and hydrocarbon. Beijing: Science Press; 1991. p. 203 (**in Chinese**).

Zhu G, Jiang QQ, Piao XF, et al. Role of basement faults in faulting system development of a rift basin: an example from the Gaoyou Depression in southern Subei Basin. Acta Geol Sin. 2013;87(4):441–52 (**in Chinese**).

7

Regularized least-squares migration of simultaneous-source seismic data with adaptive singular spectrum analysis

Chuang Li[1] · Jian-Ping Huang[1] · Zhen-Chun Li[1] · Rong-Rong Wang[2]

Abstract Simultaneous-source acquisition has been recognized as an economic and efficient acquisition method, but the direct imaging of the simultaneous-source data produces migration artifacts because of the interference of adjacent sources. To overcome this problem, we propose the regularized least-squares reverse time migration method (RLSRTM) using the singular spectrum analysis technique that imposes sparseness constraints on the inverted model. Additionally, the difference spectrum theory of singular values is presented so that RLSRTM can be implemented adaptively to eliminate the migration artifacts. With numerical tests on a flat layer model and a Marmousi model, we validate the superior imaging quality, efficiency and convergence of RLSRTM compared with LSRTM when dealing with simultaneous-source data, incomplete data and noisy data.

Keywords Least-squares migration · Adaptive singular spectrum analysis · Regularization · Blended data

1 Introduction

A fundamental factor considered in seismic data acquisition is efficiency. Simultaneous-source acquisition uses simultaneous shooting of two or more sources, resulting in the advantages of high efficiency and allowing denser source sampling and wider azimuths (Beasley 2008; Hampson et al. 2008). However, simultaneous shooting also produces blended data. There are mainly two ways to deal with simultaneous-source data. One is deblending the data (Mahdad et al. 2011; Chen et al. 2014; Chen 2015, 2016; Gan et al. 2016a; Zu et al. 2016) and then processing the deblended data with conventional methods. The other way is imaging the simultaneous-source data directly without separation (Tang and Biondi 2009; Berkhout et al. 2012; Chen et al. 2015). The velocity analysis of simultaneous-source data can also be implemented directly to obtain a precise velocity model in the common-midpoint domain (Gan et al. 2016b). The second approach has the advantage of high computational efficiency, but it suffers from migration artifacts because of the interference of adjacent sources.

Least-squares migration (LSM) is able to suppress the migration artifacts and produce high-quality images (Nemeth et al. 1999; Tang and Biondi 2009; Dai and Schuster 2013; Li et al. 2014, 2015a; Liu and Li 2015; Huang et al. 2013, 2015a). However, the computational cost of LSM is high as it is solved by gradient-based optimization schemes (Huang et al. 2015b; Huang and Zhou 2015; Li et al. 2016a). The computational efficiency and imaging quality can be improved by incorporating some sort of regularization into the LSM (Wang et al. 2009; Liu et al. 2013; Wang 2013; Li et al. 2015b; Lu et al. 2015). Structural constraint is an effective approach which can attenuate the migration artifacts while preserving the information of subsurface structures. Within angle-domain common-image gathers, Kuehl and Sacchi (2003) propose to use a smoothing operator along the ray parameter axis to suppress migration artifacts. This approach can also be implemented with structure-preserving constraints to improve the migration results (Wang and Sacchi 2009).

✉ Jian-Ping Huang
jphuang@upc.edu.cn

[1] School of Geosciences, China University of Petroleum, Qingdao 266580, Shandong, China

[2] Hisense (Shandong) Refrigerator Co. Ltd, Hisense, Qingdao 266580, Shandong, China

Edited by Jie Hao

The angle-domain common-image gathers need more computation and storage, so Xue et al. (2015) employ structure-enhancing filtering (Liu et al. 2010; Swindeman and Fomel 2015) as a shaping regularization operator for effectively removing noise. The structure-enhancing filter is also used as a preconditioning operator that updates the image only along prominent dips (Chen et al. 2015; Dutta and Schuster 2015), but the success of this approach significantly depends on the estimated dips.

Motivated by the excellent denoising performance of singular spectrum analysis (SSA) (Sacchi 2009; Oropeza and Sacchi 2010, 2011; Huang et al. 2014), we propose to incorporate a regularization term using SSA into least-squares reverse time migration (LSRTM) that eliminates migration artifacts caused by simultaneous-source data, incomplete data and noisy data. In order to make the SSA more efficient for large models, we divide large inverted images into several subsections by small spatial windows. Another problem of SSA is the difficulty to properly truncate singular values. The singular values are always selected manually by some criterion, for example, the number of linear events in the analysis window (Oropeza and Sacchi 2011). So we introduce the difference spectrum theory for adaptively determining the proper number of useful components.

In this paper, we first derive the forward modeling and migration operator of simultaneous-source data, and then present the theory of regularized least-squares reverse time migration (RLSRTM). The numerical tests on a flat layer model and a Marmousi model were carried out to compare RTM, LSRTM and RLSRTM when dealing with simultaneous-source data, incomplete data and noisy data. The numerical tests demonstrate the validity and superiority of the proposed method.

2 Method

2.1 Modeling and migration of simultaneous-source data

The forward modeling operator of simultaneous-source data is first derived according to the single shot forward modeling. The relation between the observed seismic data without blending and the reflectivity model can be expressed as

$$
\begin{bmatrix} \mathbf{d}_1 \\ \mathbf{d}_2 \\ \vdots \\ \mathbf{d}_N \end{bmatrix} = \begin{bmatrix} \mathbf{L}_1 \\ \mathbf{L}_2 \\ \vdots \\ \mathbf{L}_N \end{bmatrix} \mathbf{m}
\tag{1}
$$

where $\mathbf{d}_i$ and $\mathbf{L}_i$ denote the observed data and forward modeling operator related to the ith shot; $\mathbf{m}$ denotes the reflectivity model. In LSRTM, the forward modeling operator is a linear operator with the Born approximation (Dai et al. 2012).

Two or more sources are excited simultaneously in the simultaneous-source acquisition. Assuming there are n super shots in a two-dimensional survey and each super shot consists of k sources, the blended seismic data can be expressed as

$$
\begin{bmatrix} \mathbf{D}_1 \\ \mathbf{D}_2 \\ \vdots \\ \mathbf{D}_n \end{bmatrix} = \begin{bmatrix} \sum_{i=1}^{k} \mathbf{d}_{1,i} \\ \sum_{i=1}^{k} \mathbf{d}_{2,i} \\ \vdots \\ \sum_{i=1}^{k} \mathbf{d}_{n,i} \end{bmatrix} = \begin{bmatrix} \sum_{i=1}^{k} \mathbf{L}_{1,i} \\ \sum_{i=1}^{k} \mathbf{L}_{2,i} \\ \vdots \\ \sum_{i=1}^{k} \mathbf{L}_{n,i} \end{bmatrix} \mathbf{m}
\tag{2}
$$

where $\mathbf{D}_j$ represents the jth super shot while $\mathbf{L}_{j,i}$ represents the demigration (forward modeling) operator corresponding to the ith source in the jth super shot.

The sources in the simultaneous-source acquisition can be generated either completely simultaneous or nearly simultaneous. The nearly simultaneous shooting method is distinguished from the completely simultaneous shooting method by a nonzero time-delay between adjacent sources. Introducing the time-shifting matrix into Eq. (2), we get the forwarding modeling operator of the nearly simultaneous-source shooting method,

$$
\begin{bmatrix} \mathbf{D}_1 \\ \mathbf{D}_2 \\ \vdots \\ \mathbf{D}_n \end{bmatrix} = \begin{bmatrix} \sum_{i=1}^{k} \tau_{1,i}\mathbf{d}_{1,i} \\ \sum_{i=1}^{k} \tau_{2,i}\mathbf{d}_{2,i} \\ \vdots \\ \sum_{i=1}^{k} \tau_{n,i}\mathbf{d}_{n,i} \end{bmatrix} = \begin{bmatrix} \sum_{i=1}^{k} \tau_{1,i}\mathbf{L}_{1,i} \\ \sum_{i=1}^{k} \tau_{2,i}\mathbf{L}_{2,i} \\ \vdots \\ \sum_{i=1}^{k} \tau_{n,i}\mathbf{L}_{n,i} \end{bmatrix} \mathbf{m}
\tag{3}
$$

where $\tau_{j,i}$ denotes the time-shifting matrix corresponding to the ith source in the jth super shot. Equations (3) and (2) become equivalent when $\tau_{j,i}$ equals to a unit matrix, which represents the completely simultaneous shooting method.

Then, we rewrite the forward modeling of the simultaneous-source data with a simplified form,

$$
\mathbf{D} = \mathbf{S}\mathbf{m}
\tag{4}
$$

where $\mathbf{S}$ denotes the forward modeling operator of the simultaneous-source data.

The adjoint of the forward modeling operator can be written as,

$$
\begin{aligned}
\mathbf{S}^{\mathrm{T}} &= \left[\mathbf{S}_1^{\mathrm{T}}, \mathbf{S}_2^{\mathrm{T}}, \ldots, \mathbf{S}_n^{\mathrm{T}}\right] \\
&= \left[\sum_{i=1}^{k} \mathbf{L}_{1,i}^{\mathrm{T}}\tau_{1,i}^{\mathrm{T}}, \sum_{i=1}^{k} \mathbf{L}_{2,i}^{\mathrm{T}}\tau_{2,i}^{\mathrm{T}}, \ldots, \sum_{i=1}^{k} \mathbf{L}_{n,i}^{\mathrm{T}}\tau_{n,i}^{\mathrm{T}}\right]
\end{aligned} \tag{5}
$$

where the superscript T denotes the conjugate transpose operator.

So the RTM operator of the simultaneous-source data is

$$
\begin{aligned}
\mathbf{m}_{\mathrm{mig}} &= \left[\mathbf{S}_1^{\mathrm{T}}, \mathbf{S}_2^{\mathrm{T}}, \ldots, \mathbf{S}_n^{\mathrm{T}}\right]
\begin{bmatrix} \mathbf{D}_1 \\ \mathbf{D}_2 \\ \vdots \\ \mathbf{D}_n \end{bmatrix} \\
&= \sum_{l=1}^{n} \mathbf{S}_l^{\mathrm{T}}\mathbf{D}_l = \sum_{l=1}^{n}\sum_{i=1}^{k}\sum_{j=1}^{k} \mathbf{L}_{l,i}^{\mathrm{T}}\tau_{l,i}^{\mathrm{T}}\tau_{l,j}\mathbf{d}_{l,j} \\
&= \sum_{l=1}^{n}\sum_{i=1}^{k} \mathbf{L}_{l,i}^{\mathrm{T}}\tau_{l,i}^{\mathrm{T}}\tau_{l,i}\mathbf{d}_{l,i} + \sum_{l=1}^{n}\sum_{i\neq j}^{k}\sum_{j=1}^{k} \mathbf{L}_{l,i}^{\mathrm{T}}\tau_{l,i}^{\mathrm{T}}\tau_{l,j}\mathbf{d}_{l,j}
\end{aligned}
$$

$$\tag{6}$$

where $\mathbf{m}_{\mathrm{mig}}$ denotes the migration result of the simultaneous-source data. The first term in Eq. (6) is the image of subsurface structures while the second term is the cross-term noise.

2.2 RLSRTM using SSA

LSRTM can produce high-quality and high signal-to-noise ratio (SNR) images by iteratively updating the migration results close to the real reflectivity model. On the basis of the construction of the forward modeling and the migration operator of the simultaneous-source data, the misfit function of RLSRTM can be written as,

$$
J(\mathbf{m}) = \frac{1}{2}\sum_{i=1}^{n} \|\mathbf{S}_i\mathbf{m} - \mathbf{D}_i\|^2 + \frac{\lambda}{2}\mathbf{R}(\mathbf{m}) \tag{7}
$$

where λ denotes the regularization parameter which controls the tradeoff between the data term residual and the regularization term. The regularization parameter can be evaluated from the L-curve whose corner is used as a suitable regularization parameter (Rezghi and Hosseini 2009). However, this approach needs to compute the inverse problem several times to plot the L-curve, so it is too expensive to be practical for LSRTM. We propose that an a priori λ is selected to keep the ratio of the data term gradient to the regularization term gradient γ a fixed value and $0 < \gamma < 1$. Since the data residual will decrease with an increase in iteration, the regularization parameter should be dynamic to prevent oversize regularization. $\mathbf{R}(\mathbf{m})$ represents the regularizer that imposes constraints on the solution $\mathbf{m}$. These constraints are used to ensure that $\mathbf{m}$ should be sparse or the reflectors in $\mathbf{m}$ should be sharp. Here, we

assume that $\mathbf{R}(\mathbf{m}) = \|\mathbf{Wm}\|^2$ is the weighted reflectivity model while the weighting matrix $\mathbf{W}$ would preserve the interfaces of subsurface structures and eliminate the noise, then Eq. (7) can be rewritten as,

$$
J(\mathbf{m}) = \frac{1}{2}\sum_{i=1}^{n} \|\mathbf{S}_i\mathbf{m} - \mathbf{D}_i\|^2 + \frac{\lambda}{2}\|\mathbf{Wm}\|^2 \tag{8}
$$

In this paper, we define $\mathbf{Wm}$ as the SSA denoising of $\mathbf{m}$. Generally, the seismic signals have better coherence compared with the noise, so the noise in the migration results can be eliminated by SSA (Sacchi 2009; Oropeza 2010). The gradient to solve Eq. (8) is,

$$
\frac{\partial J(\mathbf{m})}{\partial \mathbf{m}} = \sum_{i=1}^{n} \mathbf{S}_i^{\mathrm{T}}(\mathbf{S}_i\mathbf{m} - \mathbf{D}) + \lambda \mathbf{W}^{\mathrm{T}}\mathbf{Wm} \tag{9}
$$

Both RLSRTM and LSRTM are performed iteratively using the preconditioned conjugate-gradient algorithm (Nemeth et al. 1999). Two preconditioners, illumination compensation (Plessix and Mulder 2004; Li et al. 2016b) and high-pass filtering (Li et al. 2016b), are employed to improve the migration results.

2.3 Adaptive SSA denoising

The basic assumption made by SSA can be summarized in a few words. If the seismic data consist of a complex events, the associated Hankel matrix of the data is a matrix of rank a (Hua 1992). When the data contain noise, the rank of the Hankel matrix will increase. So the denoising problem of seismic records can be attributed to the rank reduction issues of the Hankel matrix (Sacchi 2009; Oropeza 2010). SSA denoising can be implemented with the following steps (Sacchi 2009; Oropeza 2010). First, apply Fourier transform to the inverted image,

$$
M(x,k) = \frac{1}{2\pi}\int_{-\infty}^{+\infty} m(x,z)e^{-ikz}dz \tag{10}
$$

where $m(x,z)$ denotes the imaging results.

Denote $M_k = [M_1, M_2, M_3, \ldots, M_{N_x}]^{\mathrm{T}}$ as a spatial vector of a given wavenumber k of the signal. The vector can be organized in a Hankel matrix,

$$
\mathbf{M} = \begin{pmatrix}
M_1 & M_2 & \cdots & M_{K_x} \\
M_2 & M_3 & \cdots & M_{K_x+1} \\
\vdots & \vdots & \ddots & \vdots \\
M_{L_x} & M_{L_x+1} & \cdots & M_{N_x}
\end{pmatrix} \tag{11}
$$

where N_x represents the number of traces of the imaging results, and L_x and K_x are selected to make the Hankel matrix approximately square. Here, $L_x = N_x/2 + 1$, $K_x = N_x - L_x + 1$.

Then, apply singular value decomposition (SVD) to the Hankel matrix,

$$\mathbf{M} = \mathbf{U}\boldsymbol{\sigma}\mathbf{V}^{\mathrm{T}} \qquad (12)$$

where $\boldsymbol{\sigma}$, $\mathbf{U}$, $\mathbf{V}$ denotes the singular values matrix and singular vectors associated with the Hankel matrix.

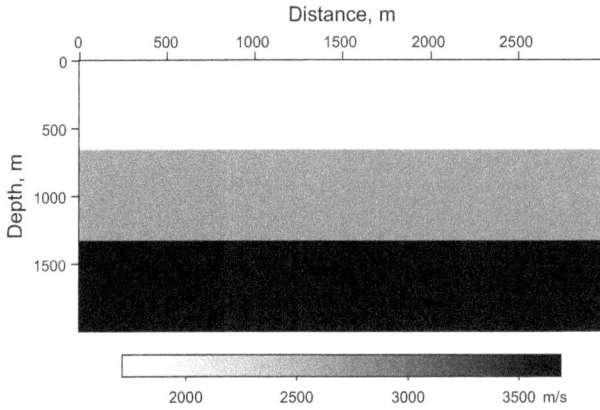

Fig. 1 Velocity of the flat layer model

A key problem of SSA is the difficulty to properly truncate singular values. In this paper, we introduce the difference spectrum theory which can effectively reflect the difference of singular values of the useful components and noise. Assuming the diagonal components of the singular values matrix are denoted by $(\sigma_1, \sigma_2, \sigma_3, \ldots, \sigma_j)$, the difference spectrum of singular values is defined as,

$$\begin{aligned} \mathbf{B} &= (b_1, b_2, \ldots, b_{j-1}) \\ b_i &= \sigma_i - \sigma_{i+1}, \quad i = 1, 2, \ldots, j-1 \end{aligned} \qquad (13)$$

The difference spectrum reflects the changes of two adjacent singular values, and the peak position in the difference spectrum refers to the abrupt change point of singular values. For a noise-free migration image containing a complex events, the associated Hankel matrix of the data is a matrix of rank a, and the peak of the difference spectrum will exist at the ath point. Compared with the useful signals, the noise always has worse coherence and even smaller amplitude, thus corresponds to smaller singular values. In this case, the peak of the difference spectrum could be an effective indicator to preserve effective

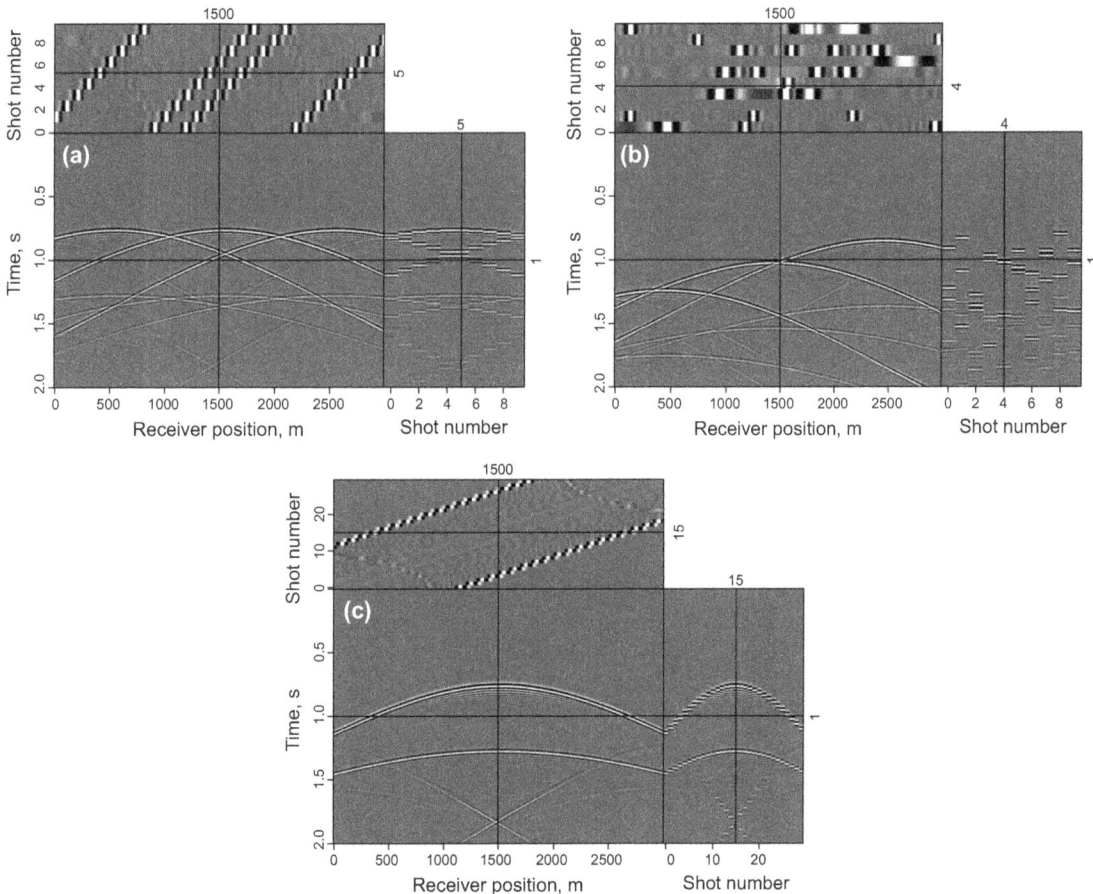

Fig. 2 Synthetic data for the flat layer model: **a** simultaneous-source data with completely simultaneous shooting; **b** simultaneous-source data with nearly simultaneous shooting; and **c** common-shot data without blending

signals while maximizing noise attenuation. The criterion is same as using the numbers of linear events to truncate singular values (Oropeza 2010), but we implement it adaptively without human intervention. However, the difference spectrum may exhibit more than one peak value when the events are curved or the inverted image is complex, because the singular value components of the useful signals are dispersed. In order to minimize this problem, adaptive SSA denoising must be applied using windows in space. In short windows, it is possible to consider a curved

event as linear. And, if multiple peaks cannot be avoided, we will use the last peak point for the consideration of preserving effective signals. Some examples of the adaptive SSA denoising are shown in the next section to test its validity.

If the peak value of the difference spectrum is b_a, the first a largest singular values are intercepted to reconstruct the Hankel matrix,

$$\mathbf{M}_a = \mathbf{U}_a \boldsymbol{\sigma}_a \mathbf{V}_a^{\mathrm{T}} \tag{14}$$

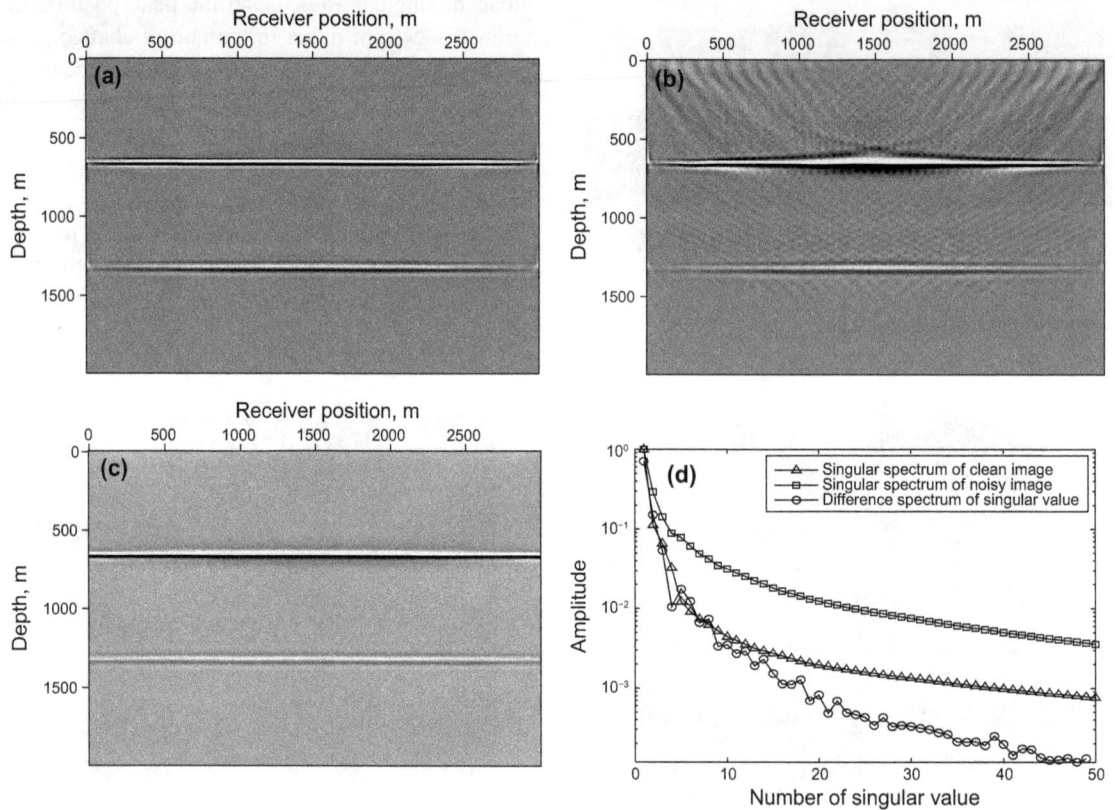

Fig. 3 Synthetic test of adaptive SSA denoising with the flat layer model: **a** LSRTM result of the common-shot data in Fig. 2c; **b** RTM result of simultaneous-source data in Fig. 2a; **c** denoising result of **b**; **d** singular spectrum curves and its difference spectrum curve. Notice that the singular spectrum curves and the difference spectrum curves are plotted in semilogarithmic coordinates

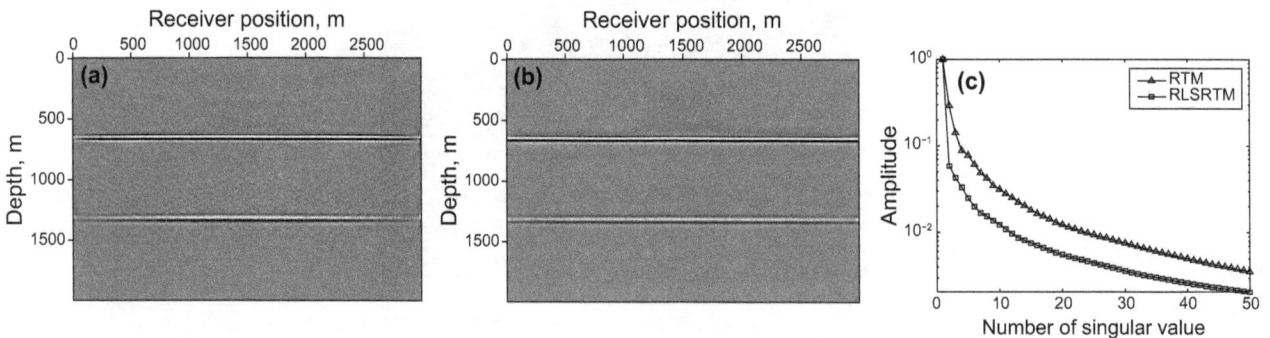

Fig. 4 Imaging results of completely simultaneous-source data: **a** LSRTM result with 40 iterations; **b** RLSRTM result with 25 iterations; **c** singular spectrum curves from RTM result and RLSRTM result

Once the rank reduced Hankel matrix is obtained, the next step entails reconstructing the inverted image by averaging components of the Hankel matrix across its anti-diagonals (Sacchi 2009). Finally, we apply an inverse Fourier transfer to the denoised data.

We should emphasize that SSA denoising needs additional computation, but the computational cost of SSA denoising is negligible compared with the cost of LSRTM. Even for a large size Hankel matrix, such as three-dimensional cases, it has been proven that dividing the data into small cubes and adopting the randomized singular value decomposition (RSVD) to perform the SVD can significantly improve the computational efficiency (Rokhlin et al. 2009; Oropeza and Sacchi 2010, 2011). In this paper, we focus on LSRTM for two-dimensional cases, so SVD is used in the following simulations.

3 Examples

3.1 Flat layer model

In this section, an imaging test of a flat layer model is implemented to demonstrate the validity of the proposed

method and make a comparison between the completely simultaneous shooting method and the nearly simultaneous shooting method. In this example, 10 super shots are recorded by 300 receivers with a 10 m receiver interval. Each super shot contains three sources with a 100 m source interval. The real velocity shown in Fig. 1 is smoothed to be the migration velocity. The data simulated by the completely simultaneous shooting method and nearly simultaneous shooting method are shown in Fig. 2a, b. It

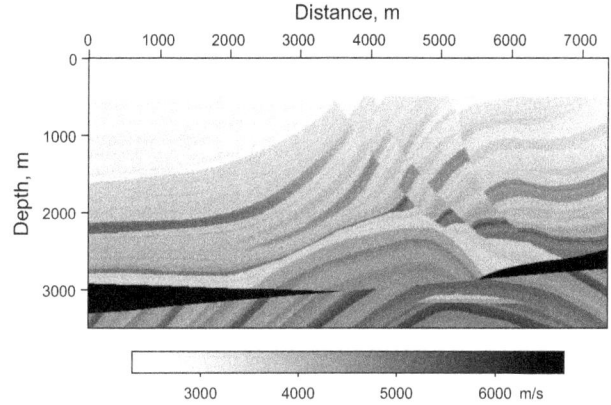

Fig. 6 Velocity of Marmousi model

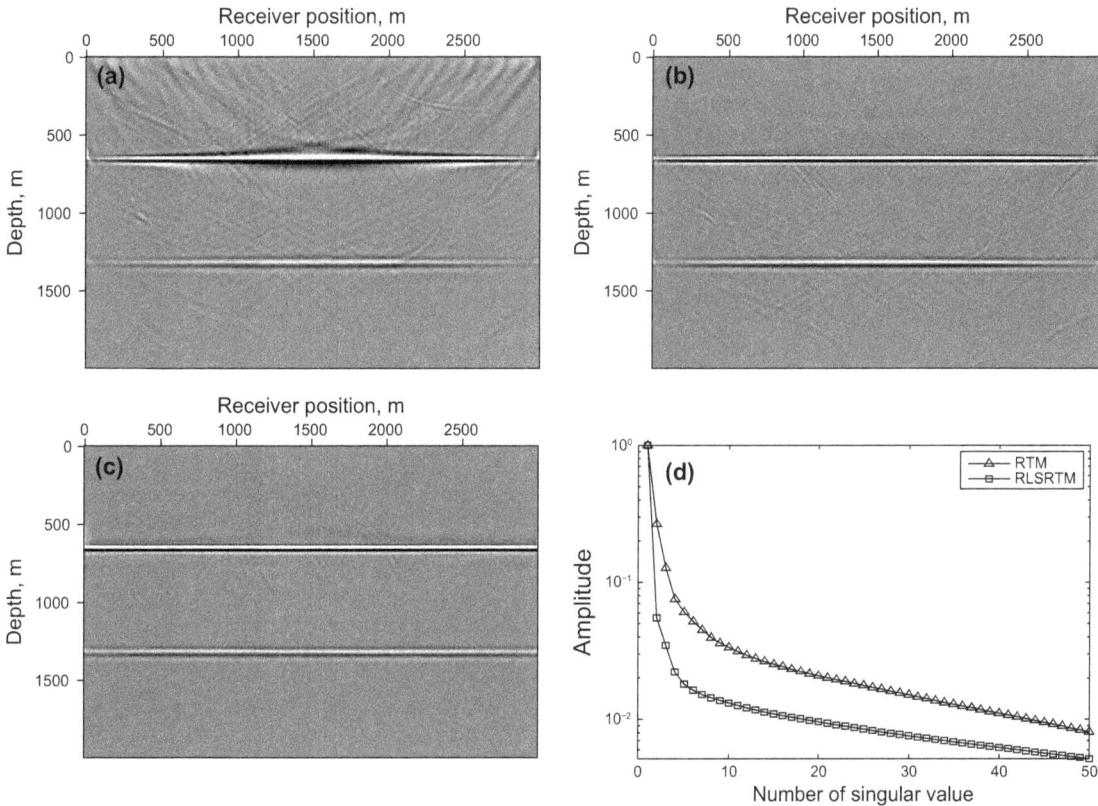

Fig. 5 Imaging results of nearly simultaneous-source data: a RTM result; b LSRTM result with 40 iterations; c RLSRTM result with 25 iterations; d singular spectrum curves from RTM result and RLSRTM result

can be seen that every super shot is blended with three single shots while there is a small time-delay between each single shot in Fig. 2b.

Figure 3 shows the synthetic test of adaptive SSA denoising with the flat layer model. Figure 3a is a clean imaging result of the flat layer model, which is obtained by

LSRTM of the common-shot data without blending (shown in Fig. 2c), Fig. 3b is the RTM result of simultaneous-source data (shown in Fig. 2a), and the denoising result of Fig. 3b is shown in Fig. 3c. Figure 3d shows the singular spectrum curves and the difference spectrum curve, which are plotted in semilogarithmic coordinates. All the singular

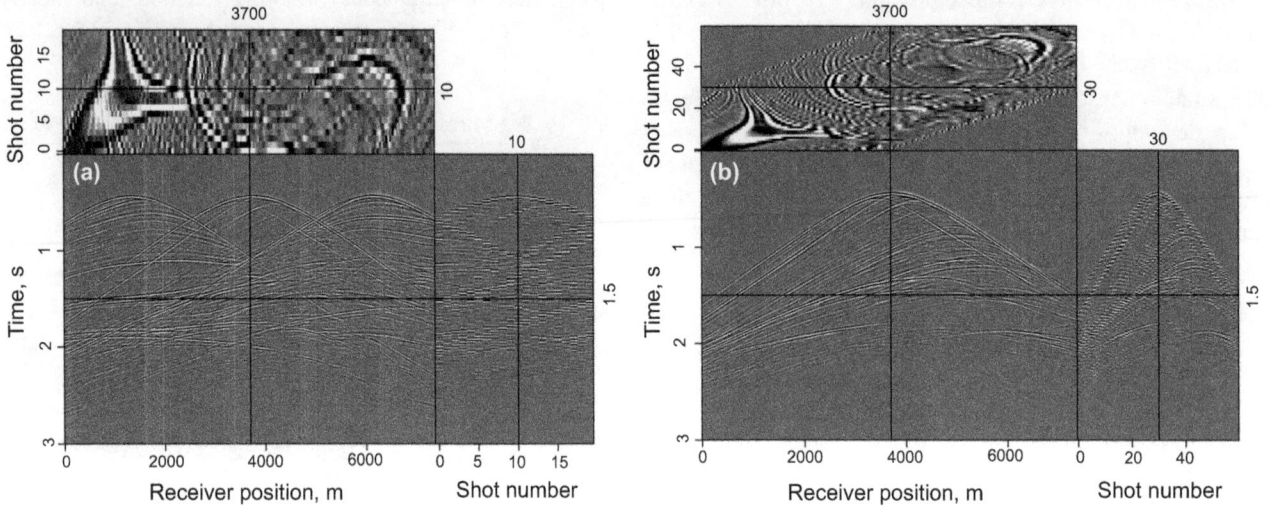

Fig. 7 Synthetic data for the Marmousi model: **a** simultaneous-source data with completely simultaneous shooting method; **b** common-shot data without blending

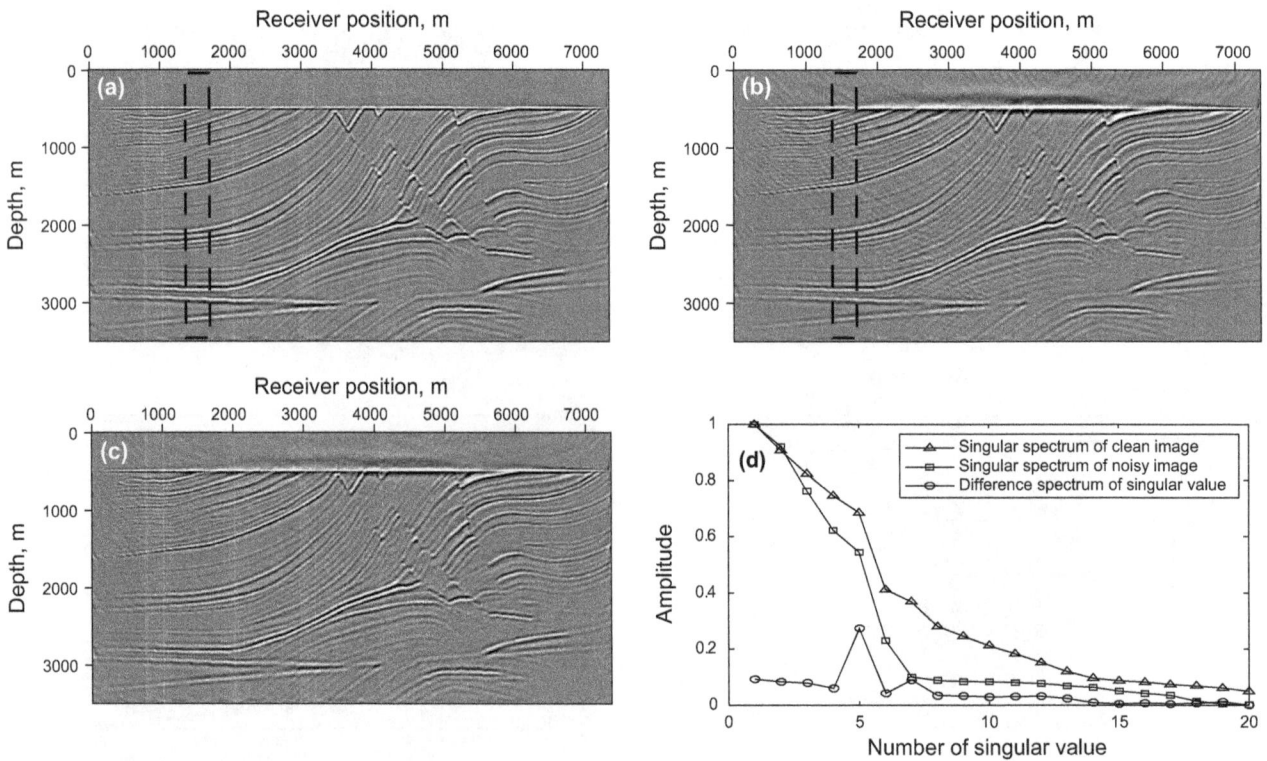

Fig. 8 Synthetic test of adaptive SSA denoising with the Marmousi model: **a** LSRTM result of the common-shot data in Fig. 7b; **b** RTM result of the simultaneous-source data in Fig. 7a; **c** denoising result of **b**; **d** singular spectrum curves and the difference spectrum curve from the imaging results marked in the *black rectangle area*

spectrum curves in this paper are normalized by the first singular value. The useful signals mainly distribute in the first singular value component and the peak of the difference spectrum also exits in the first point. So the first singular value and its corresponding singular vector are used to recover the denoised data. It can be seen that the noise is effectively eliminated by adaptive SSA denoising.

The imaging results of completely simultaneous-source data are shown in Fig. 4. Figure 4a shows the LSRTM result with 40 iterations which still suffers from some migration artifacts. Figure 4b shows the RLSRTM image with 25 iterations which exhibits higher quality image with less noise. The singular spectrum curves are plotted in Fig. 4c. The singular spectrum curve of RLSRTM result is

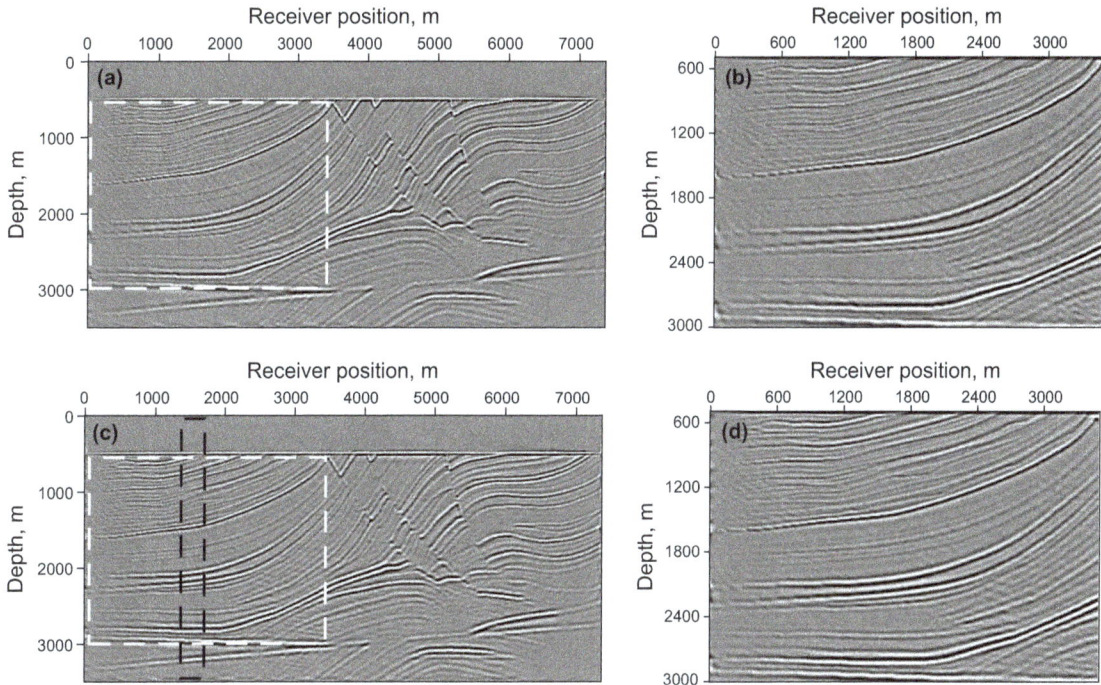

Fig. 9 Imaging results of the simultaneous-source data: **a** LSRTM result with 40 iterations and **b** its zoom view; **c** RLSRTM result with 25 iterations and **d** its zoom view. The zoomed area is highlighted by the *white box*

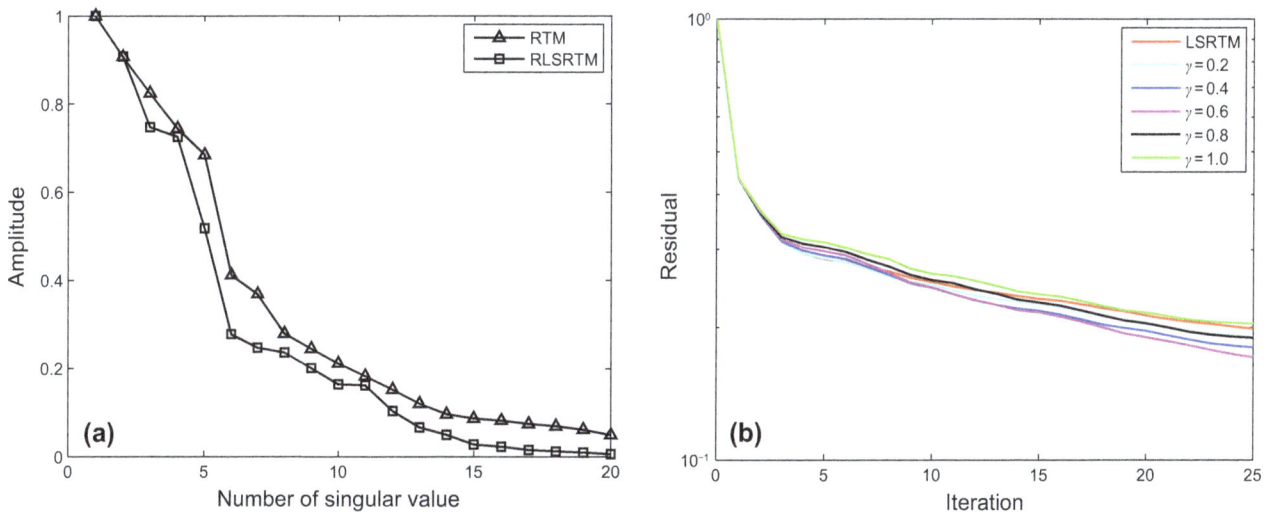

Fig. 10 Singular spectrum curves and data residual convergence curves from the imaging results from simultaneous-source data: **a** singular spectrum curves from RTM and RLSRTM results; **b** data residual convergence curves with different regularization parameters. We only plot the singular spectrum curves from the imaging results marked by the *black rectangle* in Figs. 8a and 9c

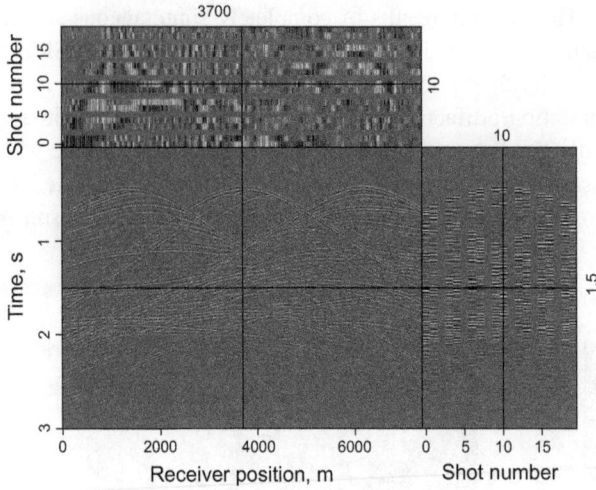

Fig. 11 Synthetic data with 60% of the traces missing for the Marmousi model

more focused on the first point compared with RTM result, indicating that the noise is less. Figure 5 shows the imaging results of nearly simultaneous-source data in which we see similar results compared with Fig. 4. But the cross-term artifacts in Fig. 5 are a little weaker than those in Fig. 4, because stacking the migration results from different super shots can suppress the cross-term artifacts more effectively when the time-delay between adjacent sources is not zero. During the tests, the computer CPU was an Intel(R) Xeon(R) E5-2650 v2 @ 2.60 GHz and the running time of the serial program for LSRTM and RLSRTM with one iteration is 555 and 559 s, respectively.

Figures 4 and 5 demonstrate that (1) direct imaging of simultaneous-source data will introduce migration artifacts which are related to the time-delay between adjacent sources; (2) LSRTM and RLSRTM can suppress the migration artifacts and compensate for unbalanced

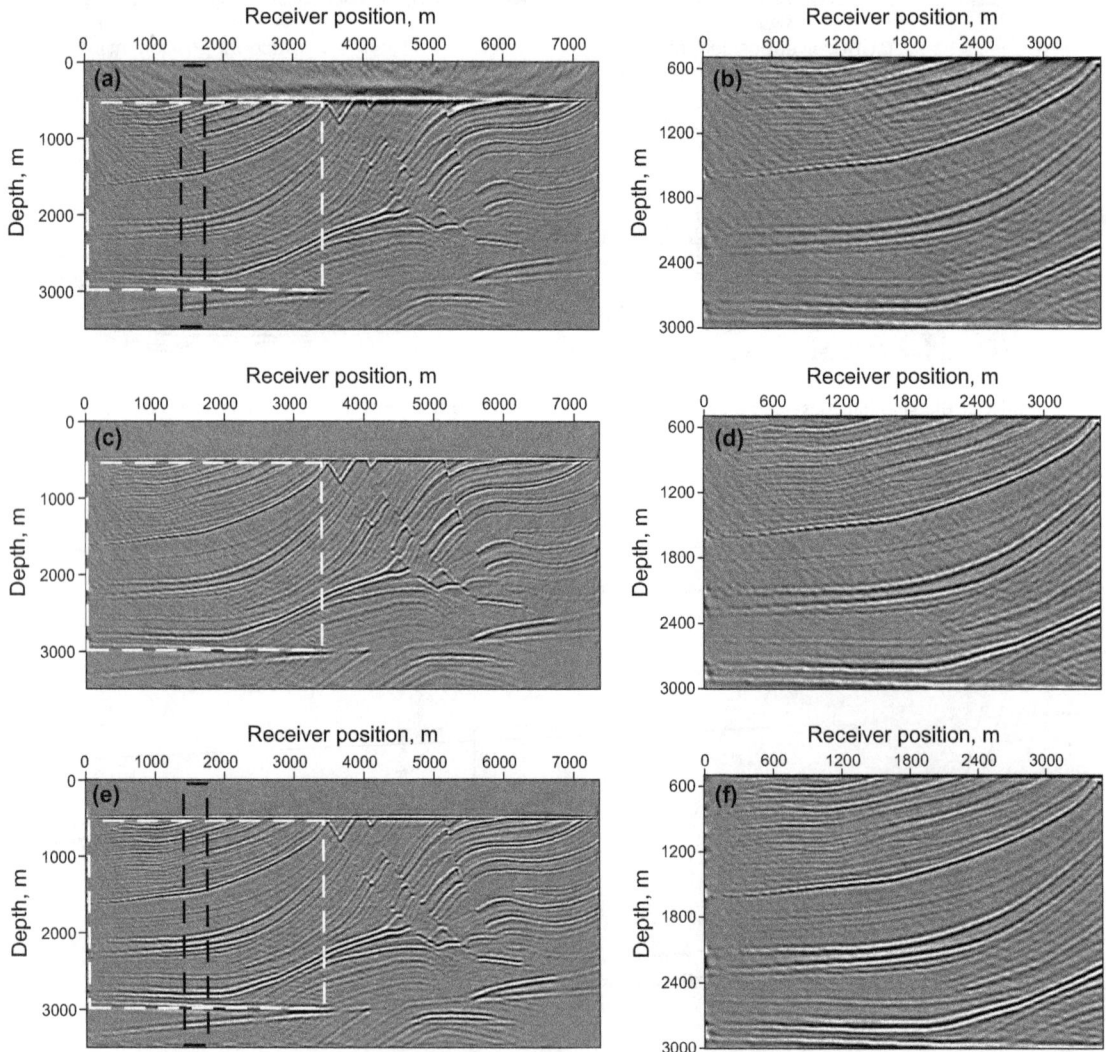

Fig. 12 Imaging results of the incomplete data: **a** RTM result and **b** its zoom view; **c** LSRTM result with 40 iterations and **d** its zoom view; **e** RLSRTM result with 25 iterations and **f** its zoom view. The zoomed area is highlighted by the *white box*

illumination in the RTM image, but RLSRTM produces better images more efficiently compared with LSRTM.

3.2 Marmousi model

We used a more realistic Marmousi model to test the proposed method (shown in Fig. 6). In this example, 20 super shots are simulated by firing three sources at the same time in each shot. The sources are distributed evenly with a 120 m source interval. The shot data shown in Fig. 7a are recorded by 737 receivers with a 10 m receiver interval. The real velocity shown in Fig. 6 is smoothed to be the migration velocity.

We first test adaptive SSA denoising with the synthetic data of the Marmousi model, and the result is shown in Fig. 8. Figure 8a is the LSRTM result of the common-shot data without blending (shown in Fig. 7b), Fig. 8b is the RTM result of simultaneous-source data (shown in Fig. 7a), which contains obvious migration artifacts, and the SSA denoising result of Fig. 8b using spatial windows is shown in Fig. 8c. Thirty windows are selected to cover the entire image in space with 3500 m depth, overlapping every 10 traces. Figure 8d shows the singular spectrum curves and the difference spectrum curve from the imaging results marked by the black rectangle area. From the comparison of the singular spectrum of the clean image and the noisy image, it is clear that the useful signals mainly distribute in the first five singular value components while the noise mainly increases the scale of small singular value components. Thus, truncating the first five singular value components can preserve effective signals and suppress noise. As shown in Fig. 8c, most of the noise is suppressed

after applying adaptive SSA denoising to each windowed image, but there is still some noise left on the image.

An LSRTM image with 40 iterations and its zoom view are shown in Fig. 9a, b, which have higher imaging quality than RTM result, but still contain migration artifacts. Figure 9c, d shows the RLSRTM image with 25 iterations and its zoom view. The imaging quality of RLSRTM is comparable to LSRTM, but the noise in RLSRTM result is a little less. In the test, the running time of the serial program for LSRTM and RLSRTM with one iteration is 9094 and 9105 s, respectively. The singular spectrum curves from RTM and RLSRTM results marked by the black rectangle area in Figs. 8a and 9c are shown in Fig. 10a, in which the singular spectrum curves of RLSRTM result are closer to

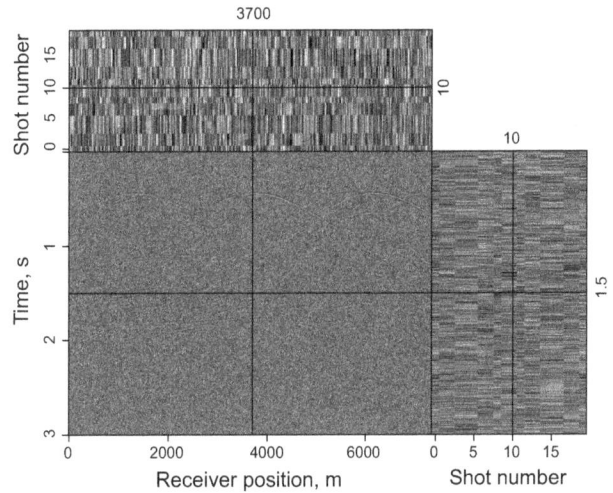

Fig. 14 Synthetic data with random noise for the Marmousi model

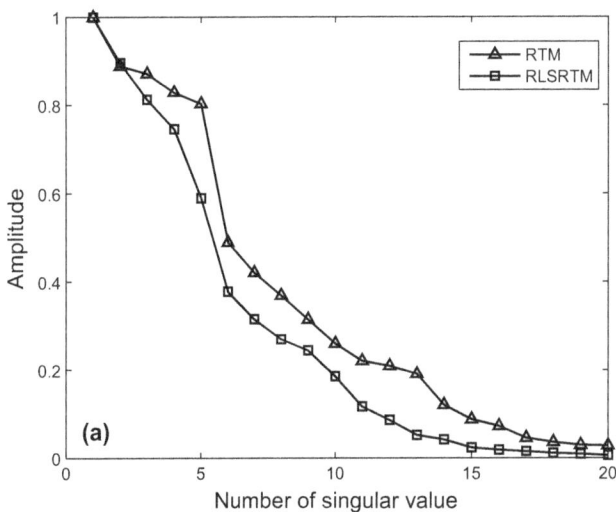

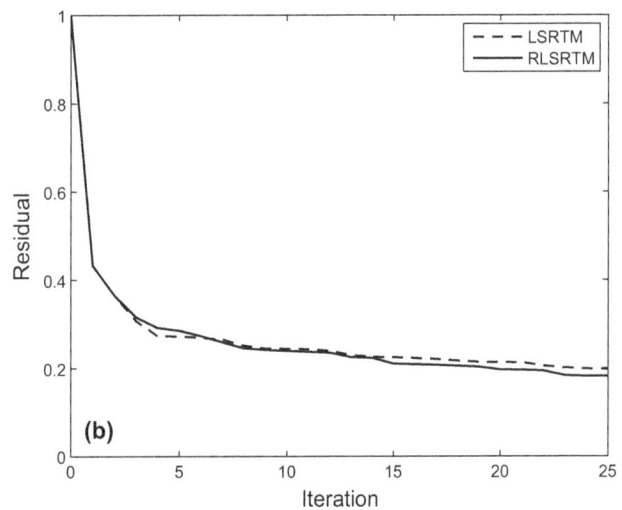

Fig. 13 Singular spectrum curves and data residual convergence curves from the imaging results of the incomplete data: **a** singular spectrum curves from RTM and RLSRTM results; **b** data residual convergence curves. We only plot the singular spectrum curves from the imaging results marked by the *black rectangle* in Fig. 12a, e

the singular spectrum curve of the clean image shown in Fig. 8d. In order to compare the convergence of LSRTM and RLSRTM, we present the data residual convergence curves for the simultaneous-source data with different regularization parameters in Fig. 10b. The convergence curves are plotted in semilogarithmic coordinates so that we can see the differences between LSRTM and RLSRTM more clearly. We notice that RLSRTM exhibits a faster convergence rate than LSRTM in the majority of cases. Only when $\gamma = 1$ are the convergence rates of LSRTM and RLSRTM similar.

Figure 11 shows synthetic data for the Marmousi model with 60% of the data missing. The RTM result of incomplete data shown in Fig. 12a, b contains more severe migration artifacts compared with the RTM result of the complete data. From the comparison of the results of LSRTM and RLSRTM in Fig. 12c–f, we draw the conclusion that LSRTM and RLSRTM can eliminate the migration

artifacts caused by the incomplete data, while RLSRTM is more efficient in attenuating the migration artifacts compared with LSRTM. Figure 13a shows the singular spectrum curves from the RTM and RLSRTM results marked by the black rectangle area in Fig. 12a, e. The singular spectrum curve of RLSRTM result is more focused in the first few points than the singular spectrum curve of RTM result, indicating that RLSRTM result contains less noise. The data residual convergence curves for incomplete data are presented in Fig. 13b, which shows that both the data residuals of LSRTM and RLSRTM decrease fast and the convergence of RLSRTM goes a little faster.

Finally, an imaging test of noisy simultaneous-source data is presented. The noisy data in Fig. 14 are obtained by adding Gaussian noise into the simultaneous-source data with Eq. (15),

$$D_{obs}(x, t) = D(x, t) + \delta \cdot \text{rand(size}(D)) \quad (15)$$

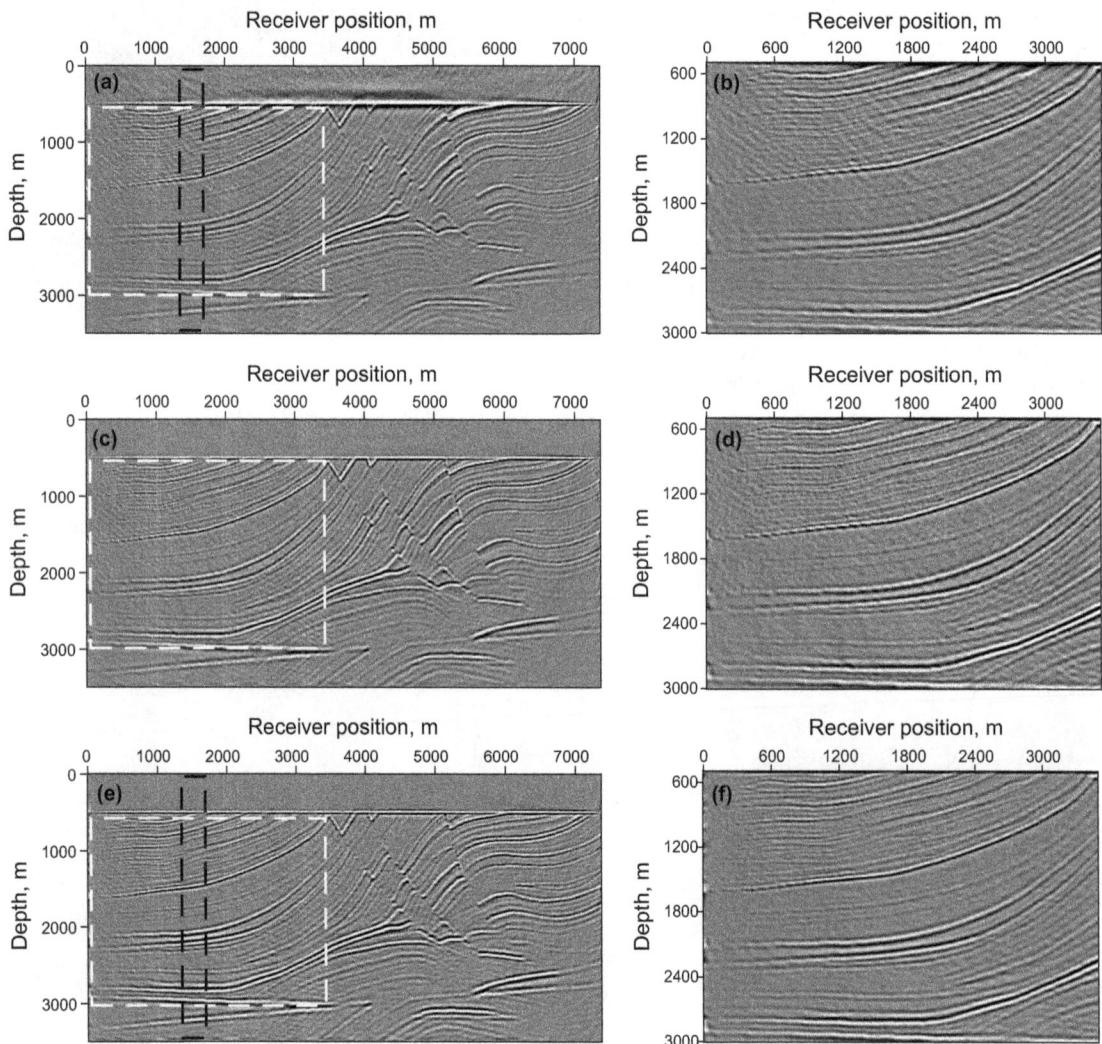

Fig. 15 Imaging results of the noisy data: **a** RTM result and **b** its zoom view; **c** LSRTM result with 25 iterations and **d** its zoom view; **e** RLSRTM result with 25 iterations and **f** its zoom view. The zoomed area is highlighted by the *white box*

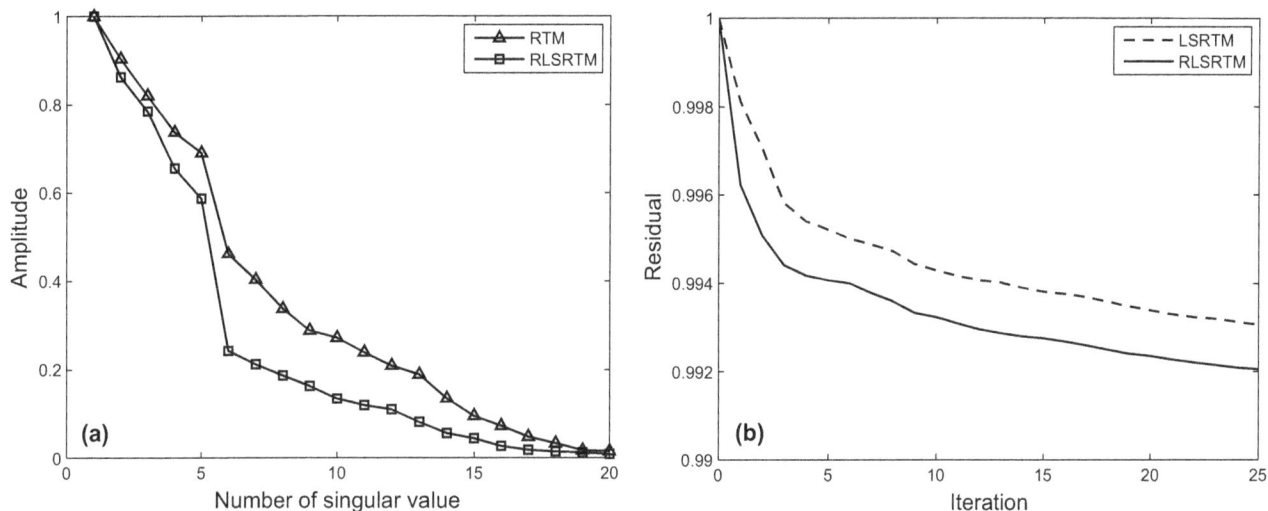

Fig. 16 Singular spectrum curves and data residual convergence curves from the imaging results of noisy data: **a** singular spectrum curves of RTM and RLSRTM results; **b** data residual convergence curves. We only plot the singular spectrum curves from the imaging results marked by the *black rectangle* in Fig. 15a, e

where $\mathrm{rand}(\mathrm{size}(D))$ denotes the Gaussian noise, δ denotes the noise level. In this example, we want to test the robustness of the proposed method when the observed data contain strong noise, with $0.1 < \delta < 1.0$. The imaging results of noisy data are shown in Fig. 15. As shown in Fig. 15a, b, the Gaussian noise in the observed data also introduces slight random noise in the RTM images. The random noise cannot be suppressed but enhanced in the LSRTM image with 25 iterations, because the Gaussian noise cannot be predicted by the forward modeling operator, and will always remain in the data residual. However, the result of RSLRTM with 25 iterations in Fig. 15e, f exhibits less noise compared with the results of RTM and LSRTM. This demonstrates that RLSRTM is effective in producing high SNR images when the observed data suffer from severe Gaussian noise. Figure 16a shows the singular spectrum curves from the RTM and RLSRTM results marked by the black rectangle area in Fig. 15a, e. It is clear that the singular spectrum curve of RTM result is more dispersed because of the influence of the noise. Figure 16b shows the data residual convergence curves for noisy data. We find that the data residual of LSRTM and RLSRTM cannot be converged to below 0.9 because the severe noise will always remain in the data residual.

4 Conclusions

We have proposed the regularized least-squares reverse time migration method using the adaptive SSA technique to solve the direct imaging problems of simultaneous-source data, incomplete data and noisy data. Difference

spectrum theory is presented to implement SSA denoising adaptively. It is important to note that adaptive SSA denoising must be applied using spatial windows for better results when the underground structures are complex. The numerical tests on a flat layer model and a Marmousi model indicate that RLSRTM is able to eliminate migration artifacts efficiently and exhibits superior imaging quality and convergence compared with RTM and LSRTM.

This work can be easily extended to three-dimensional cases. We suggest that dividing the data into small cubes and adopting the RSVD (Rokhlin et al. 2009; Oropeza and Sacchi 2010, 2011) to perform the SVD could be useful to avoid the low computational efficiency problems of a huge Hankel matrix. In addition, the damped multichannel singular spectrum analysis (Huang et al. 2016) can attenuate more noise than traditional SSA. It can help improve the performance when there is random noise in the blended data. Our next work will take these methods into consideration.

Acknowledgements The authors thank the jointly financial support from the National Natural Science Foundation of China (Grant Nos. 41104069, 41274124), National Key Basic Research Program of China (973 Program) (Grant No. 2014CB239006), National Science and Technology Major Project (Grant No. 2011ZX05014-001-008), the Open Foundation of SINOPEC Key Laboratory of Geophysics (Grant No. 33550006-15-FW2099-0033) and the Fundamental Research Funds for the Central Universities (Grant No. 16CX06046A).

References

Beasley CJ. A new look at marine simultaneous sources. Lead Edge. 2008;27(7):914–7. doi:10.1190/1.2954033.

Berkhout AJ, Verschuur DJ, Blacquière G. Illumination properties and imaging promises of blended, multiple-scattering seismic data: a tutorial. Geophys Prospect. 2012;60(4):713–32. doi:10.1111/j.1365-2478.2012.01081.x.

Chen Y, Fomel S, Hu J. Iterative deblending of simultaneous-source seismic data using seislet-domain shaping regularization. Geophysics. 2014;79(5):V179–89. doi:10.1190/geo2013-0449.1.

Chen Y. Iterative deblending with multiple constraints based on shaping regularization. IEEE Geosci Remote Sens Lett. 2015;12(11):2247–51. doi:10.1109/lgrs.2015.2463815.

Chen Y, Yuan J, Zu S, et al. Seismic imaging of simultaneous-source data using constrained least-squares reverse time migration. J Appl Geophys. 2015;114:32–5. doi:10.1016/j.jappgeo.2015.01.004.

Chen Y. Noise attenuation of seismic data from simultaneous-source acquisition. Austin: The University of Texas at Austin; 2016.

Dai W, Fowler P, Schuster GT. Multi-source least-squares reverse time migration. Geophys Prospect. 2012;60(4):681–95. doi:10.1111/j.1365-2478.2012.01092.x.

Dai W, Schuster GT. Plane-wave least-squares reverse-time migration. Geophysics. 2013;78(4):S165–77. doi:10.1190/geo2012-0377.1.

Dutta G, Schuster GT. Sparse least-squares reverse time migration using seislets. 2015 SEG Annual Meeting. Society of Exploration Geophysicists. 2015; 4232–37. doi: 10.1190/segam2015-5869595.1.

Gan S, Wang S, Chen Y, et al. Simultaneous-source separation using iterative seislet-frame thresholding. IEEE Geosci Remote Sens Lett. 2016a;13(2):197–201. doi:10.1109/LGRS.2015.2505319.

Gan S, Wang S, Chen Y, et al. Velocity analysis of simultaneous-source data using high-resolution semblance-coping with the strong noise. Geophys J Int. 2016b;204(2):768–79. doi:10.1093/gji/ggv484.

Hampson G, Stefani J, Herkenhoff F. Acquisition using simultaneous sources. Lead Edge. 2008;27(7):918–23. doi:10.1190/1.2954034.

Hua Y. Estimating two-dimensional frequencies by matrix enhancement and matrix pencil. IEEE Trans Signal Process. 1992;40(9):2267–80. doi:10.1109/icassp.1991.150104.

Huang JP, Li C, Li GL, et al. Simultaneous seismic data de-noising and regularization method based on singular spectrum analysis. Prog Geophys. 2014;29(4):1666–71. doi:10.6038/pg20140423 (in Chinese).

Huang JP, Li C, Li QY, et al. Least-squares reverse time migration with static plane-wave encoding. Chin J Geophys. 2015a;58(6):2046–56. doi:10.6038/cjg20150619 (in Chinese).

Huang JP, Li C, Wang RR, et al. Plane-wave least-squares reverse time migration for rugged topography. J Earth Sci. 2015b;26(4):471–80. doi:10.1007/s12583-015-0556-5.

Huang JP, Li ZC, Kong X, et al. A study of least squares migration imaging method for fractured-type carbonate reservoir. Chin J Geophys. 2013;56(5):1716–25. doi:10.6038/cjg20130529 (in Chinese).

Huang W, Zhou HW. Least-squares seismic inversion with stochastic conjugate gradient method. J Earth Sci. 2015;26(4):463–70. doi:10.1007/s12583-015-0553-8.

Huang W, Wang R, Chen Y, et al. Damped multichannel singular spectrum analysis for 3D random noise attenuation. Geophysics. 2016;81(4):V261–70. doi:10.1190/geo2015-0264.1.

Kuehl H, Sacchi MD. Least-squares wave-equation migration for AVP/AVA inversion. Geophysics. 2003;68(1):262–73. doi:10.1190/1.1543212.

Li C, Huang JP, Li ZC, et al. Analysis on the encoding strategies of plane-wave least-square reverse time migration. Geophys Prospect Pet. 2015a;54(5):592–601. doi:10.3969/j.issn.1000-1441.2015.05.012 (in Chinese).

Li C, Huang JP, Li ZC, et al. Singular spectrum constraint and its application to least-squares migration. 2015 SEG Annual Meeting. Society of Exploration Geophysicists. 2015; 4259–4263, 10.1190/segam2015-5888941.1.

Li C, Huang JP, Li ZC, et al. Plane-wave least-square reverse time migration with encoding strategies. J Seism Explor. 2016a;25(2):177–97.

Li C, Huang JP, Li ZC, et al. Preconditioned least-squares reverse time migration. Oil Geophys Prospect. 2016b;51(3):513–20. doi:10.13810/j.cnki.issn.1000-7210.2016.03.013 (in Chinese).

Li ZC, Guo ZB, Tian K. Least-squares reverse time migration in visco-acoustic medium. Chin J Geophys. 2014;57(1):214–28. doi:10.6038/cjg20140118 (in Chinese).

Liu Y, Fomel S, Liu G. Nonlinear structure-enhancing filtering using plane-wave prediction. Geophys Prospect. 2010;58(3):415–27. doi:10.1111/j.1365-2478.2009.00840.x.

Liu YJ, Li ZC. Least squares reverse time migration with extended imaging condition. Chin J Geophys. 2015;58(10):3771–82. doi:10.6038/cjg20151027 (in Chinese).

Liu YJ, Li ZC, Wu D, et al. The research on local slope constrained least squares migration. Chin J Geophys. 2013;56(3):1003–11. doi:10.6038/cjg20130328 (in Chinese).

Lu X, Han LG, Zhang P, et al. Direct imaging method of multi-source blended data based on total variation. Chin J Geophys. 2015;58(9):3335–45. doi:10.6038/cjg20150926 (in Chinese).

Mahdad A, Doulgeris P, Blacquiere G. Separation of blended data by iterative estimation and subtraction of blending interference noise. Geophysics. 2011;76(3):Q9–17. doi:10.1190/1.3556597.

Nemeth T, Wu C, Schuster GT. Least-squares migration of incomplete reflection data. Geophysics. 1999;64(1):208–21. doi:10.1190/1.1444517.

Oropeza V. The singular spectrum analysis method and its application to seismic data denoising and reconstruction. Edmonton: University of Alberta; 2010.

Oropeza V, Sacchi M. A randomized SVD for multichannel singular spectrum analysis (MSSA) noise attenuation. SEG Annu Meet. 2010;2010:3539–44. doi:10.1190/1.3513584.

Oropeza V, Sacchi M. Simultaneous seismic data denoising and reconstruction via multichannel singular spectrum analysis. Geophysics. 2011;76(3):V25–32. doi:10.1190/1.3552706.

Plessix R, Mulder W. Frequency-domain finite-difference amplitude-preserving migration. Geophys J Int. 2004;157(3):975–87. doi:10.1111/j.1365-246X.2004.02282.x.

Rezghi M, Hosseini SM. A new variant of L-curve for Tikhonov regularization. J Comput Appl Math. 2009;231(2):914–24. doi:10.1016/j.cam.2009.05.016.

Rokhlin V, Szlam A, Tygert M. A randomized algorithm for principal component analysis. SIAM J Matrix Anal Appl. 2009;31(3):1100–24. doi:10.1137/080736417.

Sacchi M. FX singular spectrum analysis. CSPG CSEG CWLS Convention. 2009;392–95.

Swindeman R, Fomel S. Seismic data interpolation using plane-wave shaping regularization. 85th Annual International Meeting, SEG, Expanded Abstracts. 2015;3853–58. doi: 10.1190/segam2015-5926183.1.

Tang Y, Biondi B. Least-squares migration/inversion of blended data. 79th Annual International Meeting, SEG Expanded Abstracts. 2009;2859–63. doi: 10.1190/1.3255444.

Wang J, Sacchi MD Structure constrained least-squares migration. 79th Annual International Meeting, SEG Expanded Abstracts. 2009;2763–67. doi: 10.1190/1.3255423.

Wang YF, Yang CC, Duan QI. On iterative regularization methods for migration deconvolution and inversion in seismic imaging. Chin J Geophys. 2009;52(6):1615–24. doi:10.3969/j.issn.0001-5733.2009.06.024 **(in Chinese)**.

Wang YF. Comparison of interferometric migration and preconditioned regularizing least squares migration inversion methods in seismic imaging. Chin J Geophys. 2013;56(1):230–8. doi:10.6038/cjg20130123 **(in Chinese)**.

Xue ZG, Chen YK, Fomel S, et al. Seismic imaging of incomplete data and simultaneous-source data using least-squares reverse time migration with shaping regularization. Geophysics. 2015;81(1):S11–20. doi:10.1190/geo2014-0524.1.

Zu S, Zhou H, Chen Y, et al. A periodically varying code for improving deblending of simultaneous sources in marine acquisition. Geophysics. 2016;81(3):213–25. doi:10.1190/geo2015-0447.1.

8

The remarkable effect of organic salts on 1,3,5-trioxane synthesis

Liu-Yi Yin[1] · Yu-Feng Hu[1] · Hai-Yan Wang[1]

Abstract The effects of organic salts on 1,3,5-trioxane synthesis were investigated through batch reaction and continuous production experiments. The organic salts used include sodium methanesulfonate (CH_3NaO_3S), sodium benzenesulfonate ($C_6H_5NaO_3S$), sodium 4-methylbenzenesulfonate ($C_7H_7NaO_3S$), and sodium 3-nitrobenzene sulfonate ($C_6H_4NNaO_5S$). It was shown that the effects of organic salts on the yield of 1,3,5-trioxane in reaction solution and distillate follow the order $CH_3NaO_3S <$ $C_6H_5NaO_3S < C_7H_7NaO_3S < C_6H_4NNaO_5S$, which is inversely related to the charge density of the anions of the organic salts. In comparison with Cl^--based salts such as magnesium chloride, organic salts have the advantages of less formic acid generation and low corrosion. Studies on water activity revealed that the effect of organic salts on the activity of water was quite small at low concentration of organic salts. UV–visible spectroscopy and vapor–liquid equilibrium experiments were performed to uncover the mechanisms that govern such effects. The results showed that the effect of organic salts on the yield of 1,3,5-trioxane relies primarily on their ability to increase the catalytic activity of sulfuric acid and increase the relative volatilities of 1,3,5-trioxane and water and of 1,3,5-trioxane and oligomers.

Keywords 1,3,5-Trioxane · Organic salt · Salt effect · Hammett function · Relative volatility

1 Introduction

1,3,5-Trioxane is attracting increasing attention as an alternative starting material to formaldehyde solution for preparing anhydrous formaldehyde which is used to manufacture disinfectant agents, acetal resins, bonding materials, pesticides, molding materials, antibacterial agents, etc. (Augé and Gil 2002). In particular, acetal resins are being more widely used in areas where metals were traditional, because of their superior chemical stability, mechanical strength, and plasticity (Schweitzer et al. 1959; Koch and Lindvig 1959). Accordingly, development of a more practical and economic process for production of 1,3,5-trioxane is important due to the rapid expansion of acetal resin production (Masamoto et al. 2000; Grützner et al. 2007).

In the commercial process, 1,3,5-trioxane is obtained by heating aqueous formaldehyde in the presence sulfuric acid (Masamoto et al. 2000). Although sulfuric acid is the most generally used catalyst due to its low price and the mature processing route, it has shortcomings as well. Among the more troublesome of these are the by-products such as formic acid and methyl formate. The formation of large amounts of by-products will result in a decrease in catalytic activity and of the selectivity of 1,3,5-trioxane formation, subsequently giving rise to an increase in the energy consumption during 1,3,5-trioxane synthesis. Specifically, the formic acid by-product can cause serious corrosion to equipment (Li et al. 2015). Previous works (Cui 1990; Masamoto et al. 2000) report that acidic ion exchange resin and heteropolyacids have the potential to be used as the

✉ Yu-Feng Hu
huyf3581@sina.com

[1] State Key Laboratory of Heavy Oil Processing and High Pressure Fluid Phase Behavior and Property Research Laboratory, China University of Petroleum, Beijing 102249, China

Edited by Xiu-Qin Zhu

catalyst for 1,3,5-trioxane synthesis. Nevertheless, the reaction conditions (e.g., high formaldehyde concentration (60 wt%) and large amount of catalyst (50 wt%) make it difficult to use in pilot plant trials (Masamoto et al. 2000; Guan et al. 2005; Liu and Li 1982). Although acidic ionic liquids (ILs) have been successfully used in a pilot plant scale for the preparation of 1,3,5-trioxane (Song et al. 2012), the high price of ILs catalyst is a problem (Hu et al. 2015; Li et al. 2012). Therefore, development of a new catalytic system (sulfuric acid + assistant reagent) is the key for synthesis of 1,3,5-trioxane.

The salt effect is always a research area of interest in chemical synthesis and separation. More recently, the remarkable role of halide salt additives in the Negishi reaction involving aryl zinc reagents has been reported (McCann and Organ 2014). It is ranked as one of the world's top ten science and technology improvements in 2014 by Chemical & Engineering News. In our previous work, a positive effect of inorganic salts has been found for the synthesis of 1,3,5-trioxane (Yin et al. 2015), and intensive studies have been carried out to explore the salt effect of various types of inorganic salt on the formation of 1,3,5-trioxane (Yin et al. 2015). The results shows that higher yields of 1,3,5-trioxane can be obtained in reaction solution by using the Cl⁻-based salts with smaller cation radius. It can be concluded that the enhancement in the yield of 1,3,5-trioxane is attributed to hydration which can decrease the water activity. Though these investigations covered important aspects of 1,3,5-trioxane synthesis, there is still something to be further explored, since the increase in the acid value by the addition of Cl⁻-based salts is still very remarkable compared to that observed for sulfuric acid alone. In addition, the presence of Cl⁻-based salts and the HCl formed may cause corrosion (Bao 2007). Therefore, in the present work we investigated the influence of organic salts on 1,3,5-trioxane synthesis by batch reaction and continuous production experiments. In addition, UV–visible spectroscopy and vapor–liquid equilibrium experiments were also performed for the first time to investigate the mechanisms that control the effects of organic salts on the yield of 1,3,5-trioxane in the reaction solution and in the distillate from the reaction.

2 Experimental

2.1 1,3,5-trioxane synthesis

The 50 wt% aqueous formaldehyde solution used in experiments was prepared by concentrating ∼37 wt% aqueous formaldehyde solution. The organic salts, analytical grade sodium methanesulfonate (CH_3NaO_3S), sodium benzenesulfonate ($C_6H_5NaO_3S$), sodium 4-methylbenzenesulfonate

($C_7H_7NaO_3S$), and sodium 3-nitrobenzenesulfonate (C_6H_4-NNaO_5S$), supplied by Aladdin Industrial Corporation (Shanghai, China), were used without further purification. The synthesis procedures for batch reaction experiments and continuous production were similar to those we used previously (Yin et al. 2015). The 1,3,5-trioxane concentrations in the reaction mixture and distillate were analyzed by gas chromatography, applying the internal standard procedure, and the acid value was determined by acid–base titration using a potentiometric titrimeter (Leici ZDJ-5, Shanghai INESA Scientific Instrument Co. Ltd., China).

2.2 UV–visible spectroscopy

The acidic scales of the sulfuric acid solution containing organic salt were measured by UV–visible spectra with basic indicators according to the procedure reported by Thomazeau et al. (2003), Xing et al. (2007). UV–visible spectroscopy analysis was conducted on a Shimadzu 2550 UV–visible spectrophotometer (Shimadzu Corporation, Japan).

2.3 Vapor–liquid equilibrium (VLE) experiments

In the present VLE experiments, a modified Othmer still was used as described by Morrison et al. (1990). The Othmer still was operated at atmospheric pressure (101.3 kPa). In each experimental run, 50 wt% formaldehyde solution containing 3 wt% 1,3,5-trioxane and 0.2 mol L⁻¹ organic salts was added to the still which was heated in an oil bath. A steady state was usually reached after 1 h of operation. Then samples of the coexisting phases of the quaternary system (formaldehyde + 1,3,5-trioxane + organic salt + water) were taken from vapor and liquid sampling ports and analyzed. Thus, the relative volatility can by defined by

$$\alpha_2 = \frac{y_2/x_2}{y_1/x_1} \tag{1}$$

where x_i and y_i are, respectively, the liquid- and vapor-phase compositions of species at equilibrium.

3 Results and discussion

The reproducibility of the batch reaction experiments has been studied, and the result was highly satisfactory (Yin et al. 2015).

Therefore, the effects of organic salts on the yield of 1,3,5-trioxane and the formation of by-product(s) in the reaction solution were investigated by batch experiments. Figure 1 shows the results obtained by adding various organic salts.

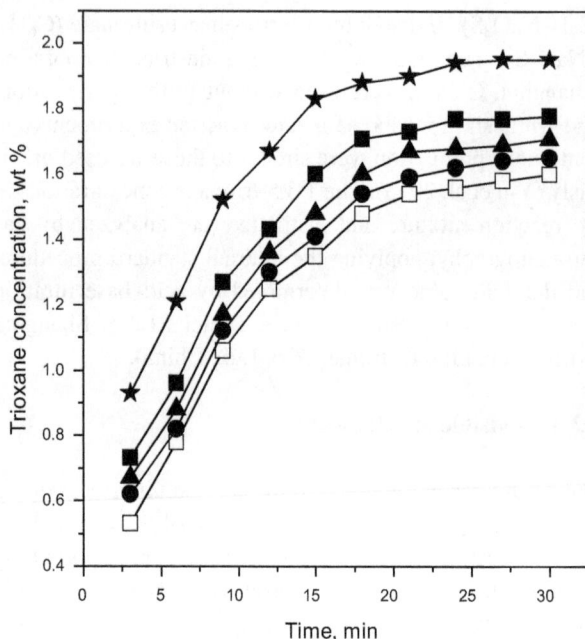

Fig. 1 Effects of organic salts on the yield of 1,3,5-trioxane in the reaction solution (50 wt% formaldehyde–0.4 mol L^{-1} H$_2$SO$_4$–0.2 mol L^{-1} organic salt). *Filled circle* CH$_3$NaO$_3$S; *filled triangle* C$_6$H$_5$NaO$_3$S; *filled square* C$_7$H$_7$NaO$_3$S; *filled star* C$_6$H$_4$NNaO$_5$S; *open square* salt-free

The changes in 1,3,5-trioxane concentration shown in Fig. 1 indicated that the addition of CH$_3$NaO$_3$S, C$_6$H$_5$-NaO$_3$S, C$_7$H$_7$NaO$_3$S, or C$_6$H$_4$NNaO$_5$S remarkably increased the 1,3,5-trioxane concentration in the reaction solution (formaldehyde–H$_2$SO$_4$–organic salt) in comparison with the results obtained in the salt-free system (formaldehyde–H$_2$SO$_4$).

To quantitatively represent the salt effect on the yield of 1,3,5-trioxane in the reaction solution, the rate constant (k) was determined by fitting 1,3,5-trioxane concentration in the reaction solution (c_p) as a function of reaction time according to the reaction dynamics model Eq. (2) (Cui 1990):

$$dc_p/dt = k_1 c_A^3 - k_2 c_p \tag{2}$$

where k_1 and k_2 are, respectively, the rate constants for the forward reaction and the reverse reaction and c_A is the formaldehyde concentration in reaction solution.

Consequently, the effect of the organic salts mentioned in Table 1 on 1,3,5-trioxane formation can be quantitatively represented by the ratio k_{salt}/k, where k_{salt} and k are the rate constants for the salt-containing solution and for the salt-free solution, respectively, and the results are shown in Table 1, which include the ratio k_{salt}/k for inorganic salt MgCl$_2$ for the sake of comparison.

The values of ratio k_{salt}/k shown in Table 1 indicate that the influence of organic salts on the formation of 1,3,5-

Table 1 Effect of the organic salts on the formation of 1,3,5-trioxane catalyzed by sulfuric acid at 98 °C

Salt	$k_{1,salt}/k_1$	$k_{2,salt}/k_2$
CH$_3$NaO$_3$S	1.09	1.07
C$_6$H$_5$NaO$_3$S	1.18	1.12
C$_7$H$_7$NaO$_3$S	1.32	1.21
C$_6$H$_4$NNaO$_5$S	1.59	1.39
MgCl$_2$	1.64	1.41

trioxane follows the order of CH$_3$NaO$_3$S < C$_6$H$_5$NaO$_3$S < C$_7$H$_7$NaO$_3$S < C$_6$H$_4$NNaO$_5$S < MgCl$_2$, which is inversely related to the charge density of the anions of the salts (the value of the acid dissociation constant pK_a decreases in the order of CH$_4$O$_3$S > C$_6$H$_6$O$_3$S > C$_7$H$_8$O$_3$S > C$_6$H$_5$NO$_5$S > HCl) (Guthrie 1978).

Although the influence of MgCl$_2$ on 1,3,5-trioxane formation was greater than that exerted by C$_6$H$_4$NNaO$_5$S, the acid value in solution of 50 wt% formaldehyde–0.4 mol L^{-1} H$_2$SO$_4$–0.2 mol L^{-1} C$_6$H$_4$NNaO$_5$S was considerably smaller than that in the solution of 50 wt% formaldehyde–0.4 mol L^{-1} H$_2$SO$_4$–0.2 mol L^{-1} MgCl$_2$), as is demonstrated in Fig. 2. These comparisons revealed that C$_6$H$_4$NNaO$_5$S had superior advantages over inorganic salts like MgCl$_2$ in 1,3,5-trioxane synthesis.

The control mechanisms for the effect of the organic salts on 1,3,5-trioxane formation in the reaction solutions are very complex. According to the mechanism proposed in

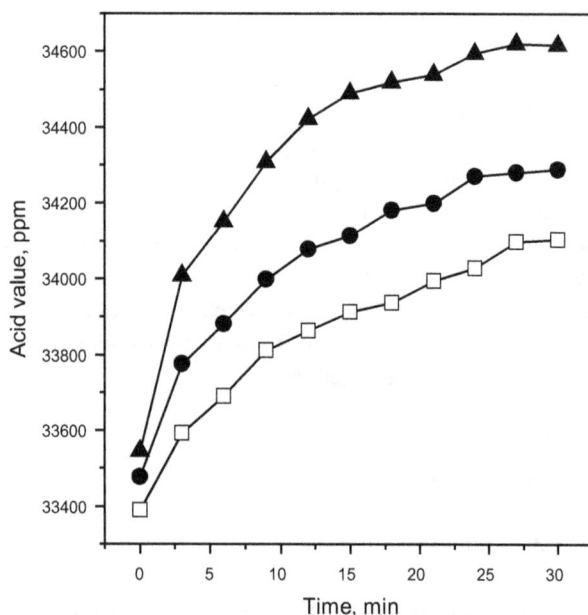

Fig. 2 Effect of salts on the change in the acid value of the reaction solution (50 wt% formaldehyde–0.4 mol L^{-1} H$_2$SO$_4$–0.2 mol L^{-1} salt). *Filled triangle* MgCl$_2$; *filled circle* C$_6$H$_4$NNaO$_5$S; *open square* salt-free

Table 2 Activity of water in the binary solution (organic salt–H_2O) at 25 °C

Organic salts, 0.2 mol kg^{-1}	Osmotic coefficients	Activity of water
CH_3SO_3Na	0.925	0.993
$C_6H_5SO_3Na$	0.932	0.993
$C_7H_7SO_3Na$	0.908	0.993
$C_6H_4NNaO_5S$	0.938	0.993

our previous paper (Yin et al. 2015), the addition of a salt to the reaction solution would decrease the activity of water in the solution and therefore favor the forward reaction of Eq. (3).

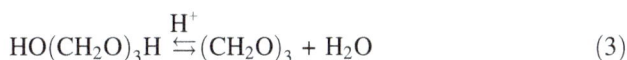

$$HO(CH_2O)_3H \overset{H^+}{\leftrightarrows} (CH_2O)_3 + H_2O \qquad (3)$$

The influence of organic salt on the water activity of the reaction solution can be reflected by the calculated value of activity of water for binary system (organic salt + water) at 25 °C, which is shown in Table 2 (Walter and Wu 1972; Bonner 1981).

Table 2 indicates that these organic salts do not decrease the water activity of the reaction solution (formaldehyde–H_2SO_4–organic salt) at such low concentration of organic salts. However, the addition of these salts can significantly increase 1,3,5-trioxane concentration in the reaction solution. For example, for the same water activity of 0.993, the concentration of 1,3,5-trioxane at $t_r = 30$ min is 1.65 wt% by $C_6H_5NaO_3S$, while 1.93 wt% by $C_6H_4NNaO_5S$. These results mean that the salt effect on the yield of 1,3,5-trioxane in the reaction solution cannot be solely understood by their ability to decrease the activity of water in the reaction solution.

UV–visible experiments were performed to further uncover the mechanism that controls the effect of salts on the formation of 1,3,5-trioxane. The typical spectra are shown in Fig. 3, which discloses that the absorbance of the unprotonated form of the indicator 2-nitroaniline is weak in salt-containing solution compared to the sample of the indicator in salt-free solution.

By taking the total unprotonated form of the indicator as the initial reference, the Hammett function (H_0) for sulfuric acid solution containing various salts was calculated via the $[I]/[IH^+]$ ratio determined from the measured absorbances (A_{max}) (see Table 3). According to Fig. 3 and Table 3, the addition of organic salts to the sulfuric acid solution decreased the Hammett function of the solution (i.e., increased catalytic activity of sulfuric acid). The Hammett function for sulfuric acid solution containing various salts increased in the order of $CH_3NaO_3S < C_6H_5NaO_3S < C_7H_7NaO_3S < C_6H_4NNaO_5S < MgCl_2$, which is consistent with the order for the effect of organic salts on 1,3,5-trioxane concentration in the reaction solution (see

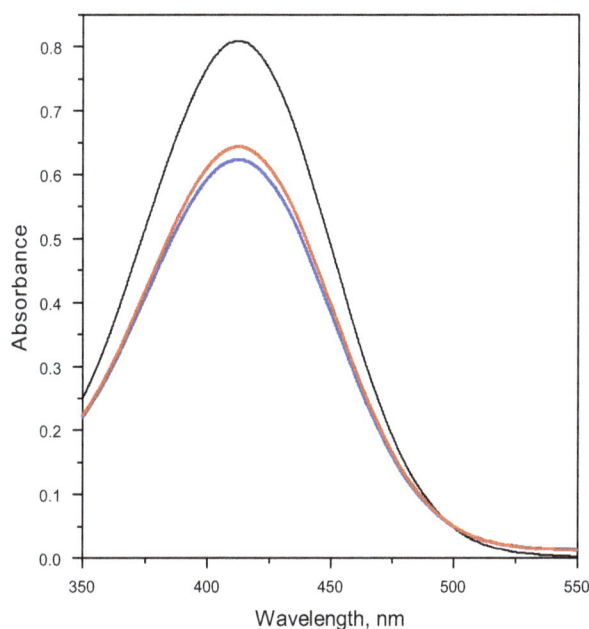

Table 3 Calculation of the Hammett function in sulfuric acid solution containing various salts

Sulfuric acid and organic salt	A_{max}	$[I]^b$, %	$[IH^+]^c$, %	H_0^d
[a]	0.809	100	0	
H_2SO_4	0.643	0.795	0.205	0.299
$H_2SO_4 + CH_3NaO_3S$	0.636	0.786	0.214	0.275
$H_2SO_4 + C_6H_5NaO_3S$	0.632	0.781	0.219	0.262
$H_2SO_4 + C_7H_7NaO_3S$	0.630	0.779	0.221	0.257
$H_2SO_4 + C_6H_4NNaO_5S$	0.628	0.776	0.224	0.250
$H_2SO_4 + MgCl_2$	0.627	0.775	0.225	0.247

[a] The total unprotonated form of the indicator (when no acid and salts are added to the solution of indicator)

[b, c] [I] and [IH$^+$] are the molar concentrations of, respectively, the unprotonated and protonated form of 2-nitroaniline indicator in aqueous solution

[d] $H_0 = pK(I)_{aq} + \log([I]/[IH^+])$, where $pK(I)_{aq}$ is the pK_a value of the indicator referred to an aqueous solution

Fig. 3 Absorption spectra of 2-nitroaniline for various solutions. *Black curve* the total unprotonated form of the indicator; *red curve* sulfuric acid solution; *blue curve* sulfuric acid solution containing CH_3NaO_3S

Table 4 Effect of the added salt on the conversion and time/space yield for 1,3,5-trioxane production

Acid and salt	Conversion[a], %	Time/space yield[b], g h^{-1} L^{-1}
H_2SO_4	19.00	46.28
H_2SO_4 + CH_3NaO_3S	20.54	48.36
H_2SO_4 + $C_6H_5NaO_3S$	22.50	51.04
H_2SO_4 + $C_7H_7NaO_3S$	24.11	54.43
H_2SO_4 + $C_6H_4NNaO_5S$	27.71	58.72
H_2SO_4 + $MgCl_2$	28.69	60.38

[a] The percent of 1,3,5-trioxane converted

[b] The amount of 1,3,5-trioxane formed per hour in 1 l of solution

Fig. 1) and the order for the value of $(k_{1,salt}/k_1)/(k_{2,salt}/k_2)$ (= 1.02, 1.05, 1.09, 1.14, 1.16, respectively, see Table 1). In other words, the effect of these salts on the yield of 1,3,5-trioxane relies primarily on their ability to increase the catalytic activity of sulfuric acid solution, as demonstrated in Table 3.

For the continuous production of 1,3,5-trioxane, the results are summarized in Table 4, which also include the results for $MgCl_2$ for the sake of comparison.

The values of conversion and time/space yield shown in Table 4 indicate that the addition of the organic salts increased the yield of 1,3,5-trioxane in the reaction distillate. Furthermore, such positive effects increased in the order of CH_3NaO_3S < $C_6H_5NaO_3S$ < $C_7H_7NaO_3S$ < $C_6H_4NNaO_5S$ < $MgCl_2$. In the continuous production process, the formed 1,3,5-trioxane was removed from the system by means of azeotropic distillation with water to enhance the separation. Thus, the influences of the organic salts on the conversion and time/space yield can be attributed to many more influencing factors. In addition to their effect on the yield of 1,3,5-trioxane in the reaction solution as discussed above, the vapor–liquid phase equilibrium experiments revealed that the organic salts can also increase the relative volatility of 1,3,5-trioxane and water and of 1,3,5-trioxane and formaldehyde, which is shown in Table 5. Relative volatility is related to the interactions

between 1,3,5-trioxane and the coexisting species. Due to the lack of the -OH group(s) in the molecular structure of 1,3,5-trioxane, the interactions between 1,3,5-trioxane and the coexisting ions of the salt are considerably smaller than those between water (or formaldehyde) and these ions. Therefore, 1,3,5-trioxane shows a very high volatility compared to formaldehyde and water.

4 Conclusions

The effects of organic salts on the yield of 1,3,5-trioxane in the reaction solution were investigated by batch reaction experiments. The results showed that the addition of CH_3NaO_3S, $C_6H_5NaO_3S$, $C_7H_7NaO_3S$, and $C_6H_4NNaO_5S$ to the reaction solution could considerably increase the yield of 1,3,5-trioxane in the reaction solution. The effect of these salts increased in the order of CH_3NaO_3S < $C_6H_5NaO_3S$ < $C_7H_7NaO_3S$ < $C_6H_4NNaO_5S$, which is inversely related to the charge density of the anions of the corresponding salts. The mechanisms that control such effects were systematically studied through calculation of the activity of water and experiments of acidity measurement. The calculated water activity indicated that these organic salts had almost no effect on the activity of water of the reaction solution. The results of acidity measurement experiments showed that the Hammett function for sulfuric acid solution containing various salts increased in the order of CH_3NaO_3S < $C_6H_5NaO_3S$ < $C_7H_7NaO_3S$ < $C_6H_4NNaO_5S$ < $MgCl_2$, which agreed well with the order for the effect of organic salts on 1,3,5-trioxane concentration in the reaction solution.

The influence of organic salts mentioned above on the yield of 1,3,5-trioxane in the reaction distillate was also investigated by continuous production experiments. These salts showed a positive effect on the yield of 1,3,5-trioxane in the distillate, and their effect increased in the order of CH_3NaO_3S < $C_6H_5NaO_3S$ < $C_7H_7NaO_3S$ < $C_6H_4NNaO_5S$. Such an effect is related to the increase in relative volatility, which is verified by vapor–liquid phase

Table 5 Effect of the organic salts on the relative volatility (α) of 1,3,5-trioxane (2) and water (4) and of 1,3,5-trioxane (2) and formaldehyde (1) in the system [formaldehyde (1)–1,3,5-trioxane (2)–organic salt (3)–water (4)]

Organic salt, 0.2 mol kg^{-1}	α_{24}	a_{21}
[a]	4.0506	6.1711
CH_3NaO_3S	4.4763	7.3562
$C_6H_5NaO_3S$	4.5411	7.4864
$C_7H_7NaO_3S$	4.6782	7.6566
$C_6H_4NNaO_5S$	4.7836	7.8072

Formaldehyde concentration is 50 wt%, 1,3,5-trioxane concentration 3 wt%, temperature 98 °C

[a] (formaldehyde–1,3,5-trioxane–water) system

experiments for the system (formaldehyde–1,3,5-trioxane–organic salt–water).

Acknowledgments We acknowledge the National Natural Science Foundation of China (21576285 and 21276271) and Science Foundation of China University of Petroleum, Beijing (qzdx-2011-01), for financial support.

References

Augé J, Gil R. A convenient solvent-free preparation of 1,3,5-trioxanes. Tetrahedron Lett. 2002;43(44):7919–20.

Bonner OD. Study of methanesulfonates and trifluoromethanesulfonates evidence for hydrogen bonding to the trifluoro group. J Am Chem Soc. 1981;103(12):3262–5.

Bao QN. The relationship between chloride ion and the corrosion of stainless steel in cooling water system. Ind Water Treat. 2007;27(7):1.

Cui YL. Kinetics study of synthesis of trioxane from formaldehyde with A-15 catalyst. Petrochem Petrochem Technol. 1990;19(4):214–8.

Guan J, Lin L, Zeng CY. Trioxane synthesis from formaldehyde over the supported PW12/AC catalyst. Nat Gas Chem Ind. 2005;30(4):19–22.

Grützner T, Hasse H, Lang N, Siegert M, Ströfer E. Development of a new industrial process for trioxane production. Chem Eng Sci. 2007;62(18):5613–20.

Guthrie JP. Hydrolysis of esters of oxy acids: pK_a values for strong acids; Brønsted relationship for attack of water at methyl; free energies of hydrolysis of esters of oxy acids; and a linear relationship between free energy of hydrolysis and pK_a holding over a range of 20 pK units. Can J Chem. 1978;56:2342–54.

Hu JJ, Zhao DS, Li JJ, Zhai JH, Hu TT. Synthesis and catalytic esterification performance of caprolactam task-specific ionic liquids. Chin J Org Chem. 2015;35(8):1773–80.

Koch TA, Lindvig PE. Molecular structure of high molecular weight acetal resins. J Appl Polym Sci. 1959;1(2):164–8.

Li Z, Chen J, Xia CG. Advances in industrial application of ionic liquids. Chem Ind Eng Prog. 2012;31(10):2113–82.

Li Z, Luo CB, Yu XM, Liu W, Liu Q, Yang B. Corrosion behavior of carbon steel in high temperature and high pressure formic acid environment. Corros Prot. 2015;36(6):540–6.

Liu HH, Li CL. A study of side reactions in synthesis of trioxymethylene with Anlberlyst-15. Chem Word. 1982;2:41–4.

Morrison JF, Baker JC, Meredith HC III, Newman KE, Waiter TD, Massle JD, Perry RL, Cummings PT. Experimental measurement of vapor-liquid equilibrium in alcohol/water/salt Systems. J Chem Eng Data. 1990;35(4):395.

Masamoto J, Hamanaka K, Yoshida K, Nagahara H, Kagawa K, Iwaisako T, Komaki H. Synthesis of trioxane using heteropolyacids as catalyst. Angew Chem Int Ed. 2000;39(12):2102–4.

McCann LC, Organ MG. On the remarkably different role of salt in the cross-coupling of arylzincs from that seen with alkylzincs. Angew Chem Int Ed. 2014;53(17):4386–9.

Schweitzer CE, Macdonald RN, Punderson JO. Thermally stable high molecular weight polyoxymethylenes. J Appl Polym Sci. 1959;1(2):158–63.

Song HY, Chen J, Xia CG, Li Z. Novel acidic ionic liquids as efficient and recyclable catalysts for the cyclotrimerization of aldehydes. Synth Commun. 2012;42(2):266–73.

Thomazeau C, Olivier-Bourbigou H, Magna L, Luts S, Gilbert B. Determination of an acidic scale in room temperature ionic liquids. J Am Chem Soc. 2003;125(18):5264–5.

Walter JH, Wu YC. Osmotic coefficients and mean activity coefficients of uni-univalent electrolytes in water at 25 °C. J Phys Chem Ref Data. 1972;1(4):1074–100.

Xing HB, Wang T, Zhou ZH, Dai YY. The sulfonic acid-functionalized ionic liquids with pyridinium cations: acidities and their acidity–catalytic activity relationships. J Mol Catal A Chem. 2007;264:53–9.

Yin LY, Hu YF, Zhang XM, Qi JG, Ma WT. The salt effect on the yields of trioxane in reaction solution and in distillate. RSC Adv. 2015;5(47):37697–702.

Effects of ultrasonic waves on carbon dioxide solubility in brine at different pressures and temperatures

Hossein Hamidi[1] · Erfan Mohammadian[2] · Amin Sharifi Haddad[1] ·
Roozbeh Rafati[1] · Amin Azdarpour[3] · Panteha Ghahri[1] · Adi Putra Pradana[1] ·
Bastian Andoni[1] · Chingis Akhmetov[1]

Abstract The adverse impacts of CO_2 emission on the global warming highlight the importance of carbon capture and storage technology and geological storage of CO_2 under solubility trapping mechanisms. Enhancing the solubility of CO_2 in formation water has always been the focus of research in the area of CO_2 sequestration. Ultrasound techniques are one of the environmentally friendly methods that use high-intensity acoustic waves to improve gas solubility in liquids. Ultrasonic waves can alter the properties of different phases that lead to chemical reactions and provide a means to increase the solubility of CO_2 in connate water. In this study, we investigated the effects of ultrasound on the solubility of CO_2 in connate water under different conditions of pressure, temperature, and salinity. The results showed that the solubility of CO_2 was improved with increasing pressure under ultrasonic effects. However, the solubility of CO_2 was inversely proportional to the increase in brine salinity and temperature. Therefore, it was concluded that the solubility of CO_2 might be enhanced in the presence of ultrasound.

Keywords Carbon dioxide · CO_2 sequestration · Ultrasound · High-frequency waves · Solubility

✉ Hossein Hamidi
  hossein.hamidi@abdn.ac.uk

[1] School of Engineering, King's College, University of Aberdeen, Aberdeen AB24 3UE, UK

[2] Faculty of Chemical Engineering, University Technology MARA, 40450 UiTM Shah Alam, Malaysia

[3] Department of Petroleum Engineering, Marvdasht Branch, Islamic Azad University, Marvdasht, Iran

Edited by Xiu-Qin Zhu

1 Introduction

Excessive emission of carbon dioxide (CO_2) into the atmosphere is one of the environmental challenges which causes critical effects such as sea level rise, melting of arctic ice, and increase in the earth's temperature, which is called the phenomenon of global warming. Concerns about global warming and challenges of CO_2 emission reduction highlight the need to develop effective and economical means for CO_2 sequestration (Brooks 1950; Metz et al. 2005; Bachu 2000; Bryant 2007; Gibson-Poole et al. 2007; Sengul 2006). In subsurface CO_2 sequestration, a large amount of CO_2 is injected into deep aquifers. Thus, CO_2 can be stored permanently to reduce its net emissions into the atmosphere (Bachu 2000; Reichle et al. 1999).

Nguyen (2003) suggested four methods for geological storage of CO_2: (1) utilization of CO_2 for enhanced oil recovery (EOR) processes, (2) use of CO_2 to improve recovery from coal-bed methane, (3) injection of CO_2 into depleted oil and gas reservoirs, and (4) injection of CO_2 into deep saline aquifers. Among these methods, saline aquifers are most anticipated due to their considerable storage capacity and extensive distribution around the world (Metz et al. 2005; Bachu 2000; Schrag 2007).

Geological storage of CO_2 in saline aquifers would preferentially occur under supercritical conditions. It depends on the contribution of several CO_2-trapping mechanisms such as (1) physical trapping of a CO_2 plume, (2) solubility trapping, which is the dissolution of CO_2 compounds in the formation fluids (hydrocarbon and brine), (3) hydrodynamic trapping, which is basically quite similar to non-wet trapping of oil within individual pores that are not connected, and (4) mineral trapping (or mineralization), which is defined as formation of carbonates due to the reactions of CO_2, brine and reservoir fluids (Paul et al. 2010).

Among the aforementioned mechanisms by which CO_2 is rendered immobile, solubility and mineralization are the ones that could be manipulated. Therefore, one could expect the improvement in the efficiency of CO_2 sequestration though an increase in the amount of CO_2 sequestrated by these two mechanisms. The mineralization of CO_2 is a very slow process, and despite the efforts for its expedition, it requires at least few hundred years to be accomplished (Azdarpour et al. 2015). The solubility mechanism, on the other hand, is much more rapid than the mineralization process. Therefore, the aim of this research is to increase the solubility of CO_2 in NaCl solution using an unconventional method of ultrasound waves.

At low pressures, there are a large number of data sets on the solubility of CO_2 in brine (Carroll et al. 1991; Zheng et al. 1997). However, due to the fact that in CO_2 sequestration process, CO_2 is injected as a supercritical fluid, high-pressure investigation and analysis are needed. Wiebe and Gaddy (1939) have conducted one of the first studies on the solubility of CO_2 in water under high-pressure conditions. Thereafter, other measurement methods were used to investigate the solubility of CO_2 in water at different pressure and temperature conditions (King et al. 1992; Bamberger et al. 2000). To date, the solubility of CO_2 has been studied by many researchers over a wide range of salinity, temperature and pressure conditions (Rumpf et al. 1994; Koschel et al. 2006; Li et al. 2004; El-Maghraby et al. 2012; Mao et al. 2013; Shedid and Adel 2013; among others).

Two widely used techniques for measuring the solubility of CO_2 are the volumetric technique and the combination technique (Peper and Dohrn 2012; Yan et al. 2011; Tong et al. 2013). Both of these techniques are slow; consequently, studies have been performed to improve and accelerate solubility measurements through these methods (El-Maghraby et al. 2012). Portier and Rochelle (2005) for the first time introduced the potentiometric titration technique to measure the solubility of CO_2. However, the experimental results achieved by this technique were limited to a particular formation brine, with a limited range of salinity, pressure, and temperature conditions. Most recently, Mohammadian et al. (2015) reported the solubility of CO_2 in brine at different salinity, pressure, and temperature conditions, using the potentiometric titration method. They found that increasing pressure enhances the solubility of CO_2 in distilled water and brine, and as the temperature increases, regardless of brine salinity, CO_2 solubility significantly reduces and salinity itself showed an adverse effect on the solubility of CO_2. Finally, the method was proved to be reproducible, fast, and accurate in comparison with the previous conventional methods.

On the other hand, application of ultrasound waves is an environmental friendly technology, which is known to improve chemical reactions in multiphase systems (Mason 1990). The use of such technique has successfully been implemented in a number of fields of science and engineering, such as food, chemical, oil and gas industries. In oil and gas industry, hydrocarbon recovery from oil reservoirs using vibration methods dates back to 1950s when an increase in oil production was observed as a result of cultural noise and earthquakes. During penetration of nonlinear low-frequency elastic waves through porous media, wave shapes might be distorted due to energy loss and attenuation, which makes them harmonic with higher frequencies. Then, one might expect the propagation of ultrasonic waves in the subsurface environment while sending low-frequency elastic waves from the surface (seismic) or due to earthquake nonlinear plane waves (Naderi and Babadagli 2010). A large number of studies have been conducted on the effects of ultrasound on enhanced oil recovery processes, emulsification of oil and brine, oil viscosity, oil mobilization in porous media, phase behaviour of a surfactant–brine–oil system, liquid–liquid interaction, and chemical treatment of horizontal wells (Hamidi et al. 2012, 2013, 2015a, b; Hamida and Babadagli 2006, 2007; Naderi and Babadagli 2010; Abramov et al. 2013, 2015, 2016). All of these studies have shown promising results after applying ultrasonic waves for testing processes.

Samenov et al. (2010) investigated the kinetics of dissolution of a single bubble of CO_2 in a glass column of water under ultrasound at low-pressure and temperature conditions. They studied the dissolution of CO_2 bubbles in water during free rise to the surface, in the presence and absence of ultrasound. Based on their studies, it was found that the dissolution of CO_2 bubbles was enhanced as a result of increase in the rate of mass transfer of CO_2 into water phase through imposed acoustic vibrations. This might be due to the fact that ultrasonic waves excited the molecular structure of water, where CO_2 molecules can easily get trapped between them.

Since there is very limited reported data on the effects of ultrasound on the solubility of CO_2, in this study, we investigated the effects of ultrasonic waves on the solubility of CO_2 at different brine salinity, temperature, and pressure conditions. We used the potentiometric titration technique to conduct our experiments. One of the main advantages of titration method for measuring the solubility of CO_2 is the conservation of samples, which inhibits CO_2 loss as a result of degassing during depressurization, while it is accurate and easy to use. Moreover, unlike previous studies, there is no need to estimate parameters, such as fugacity and density.

2 Materials and methods

2.1 Materials

Deionized and distilled water (Milli-Q) with 18.2 Ω resistivity was used to prepare brine with different salinities. In our experiments, we used NaCl with the purity of 0.995 (mass fraction) provided by Systerm, and CO_2 with a purity of 99.99% was provided by the SIG.

2.2 Methods

In this study, a series of experiments were performed to investigate the effects of ultrasound on the solubility of CO_2 in brine at high-pressure and temperature conditions. For this purpose, a customized experimental set-up was developed as shown in Fig. 1. It consists of a Genesis ™ XG-500-6 ultrasonic wave generator, CO_2 cylinder, ISCO pump made by Teledyne (model 100 DX), 100-mL autoclave reactor equipped with ultrasound transducers, and a magnetic stirrer. The reactor was enclosed in an electric heater. The ultrasonic generator emits ultrasonic waves at a frequency of 40 kHz with a power of 500 W. A thin ($d = 3.2$ mm) dipping tube was attached to a floating-piston-type sampler fabricated locally, which was operated by a KD Scientific syringe pump. The dip tube was used for the sampling from the solution of CO_2-saturated brine.

To regulate the pressure in our experiments, the ISCO pump was used. The autoclave reactor was placed in an electric oven with adjustable temperature with a precision of 0.1 °C. Even with a thick reactor base of 0.85 cm, sufficient coupling between the magnetic stirrer and the bead was established that in turn leads to a thoroughly homogeneous solution. The reactor was heated up to the desired temperatures after the brine was introduced to it. CO_2 was injected into the reactor and pressurized to the desired level where 2/3 of the reactor (~ 70 mL) was filled with an aqueous phase (brine or distilled water). In a closed system, the solution was stirred for 3 h until equilibrium was achieved. Required time to reach the equilibrium has been reported to be between 10 min and 24 h in previous studies (El-Maghraby et al. 2012; Peper and Dohrn 2012; Portier and Rochelle. 2005; Yan et al. 2011).

A sample of CO_2-laden brine was taken from the reactor using the dipping tube, which was equipped with a floating piston. Through the sampling chamber, it reacts with the base solution (0.5 M NaOH) that filled half of the chamber (~ 3 mL). This process helps in preserving of any sort of dissolved carbon types in the solution.

The other half of the sampling chamber was filled with distilled water, which was slowly withdrawn to allow CO_2-saturated brine to react with the basic solution of NaOH. It should be noted that a time period of 10 min was set for the

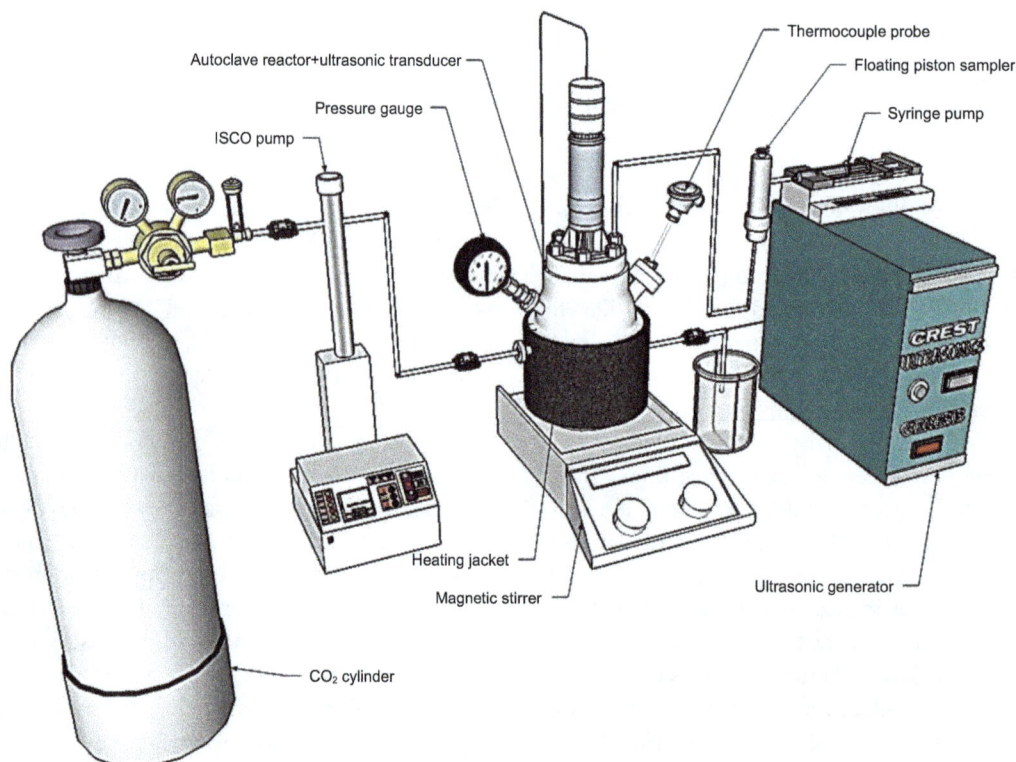

Fig. 1 Experimental set-up

reaction of solution with NaOH prior to its removal from the chamber.

During the withdrawal, no gas bubbles were generated as all types of carbon species (including H_2CO_3) were dissolved in the NaOH and converted to basic solution (NaOH) (Portier and Rochelle 2005; Duan and Sun 2003).

The potentiometric titration method was then used to analyse the samples of the aqueous NaOH solution. The equivalence point of titration can be identified via various methods, such as potentiometric (used in this research), indicators, and conductivity methods. Consumed volume of the reactant was measured and used to determine the concentration of analyte through the following correlation:

$$C_a = \frac{C_t \times V_t \times N}{M_a}$$

where C_a is the concentration of the analyte, which is equal to CO_2 solubility in brine, in molal units (mol kg^{-1}); C_t is the concentration of titrant (HCl), typically in mol L^{-1}; V_t is the consumed volume of titrant, in mL; N is the mole ratio of the analyte and reactant from the balanced chemical equation; and M_a is the mass of sample that is titrated in grams. The advantage of using mass of the solution (M_a), instead of its volume V_a, which has been commonly used in previous studies, is that the mass of solution is not a function of pressure or temperature. Therefore, the level of uncertainty in solubility measurements is lowered. For acidity analysis, 0.5 M HCl used with a 5 mL sample as a titrant. The pH of the sample was closely monitored with added volume of titrant, and the titration continued until a pH less than 2 was achieved. The equivalence points were obtained using the derivative curves of titrant volume versus pH. All these experiments were also conducted for the cases in which ultrasonic waves were applied.

3 Results and discussion

In the first series of experiments, effects of ultrasound (US) on the solubility of CO_2 in distilled water at different pressure (1–210 atm) and temperature (60, 80, and 100 °C) conditions were investigated, and the results were compared with the cases where no source of ultrasound (NUS) was used (Fig. 2). In the second series of experiments, effects of ultrasound on the solubility of CO_2 in brine solutions with different salinities (1000 and 10,000 ppm) were investigated. These tests are designed to ensure that results are comparable to the conditions of saline aquifers (Figs. 3, 4). To analyse the experimental results, data analysis software (SPSS 18) was used, where it can eliminate any abnormally low or high data points. A couple of experiments were repeated three times (namely the solubility of carbon dioxide at 60, 80, and 100 °C in distilled

water, for both 1000 and 10,000 ppm brine salinity in the absence of ultrasound), and tolerance of the observed data was in the order of ±5%.

Figure 2 shows the solubility of CO_2 in distilled water at 60, 80, and 100 °C. The error bars represent 5% deviation from the base case (60 °C). It is apparent from Fig. 2 that an increase in pressure improved the solubility of CO_2 in brine solution at all tested temperatures in this study. In addition, the solubility of CO_2 at high-pressure conditions became independent from pressure at all temperatures. Therefore, the solubility curves showed a plateau behaviour at pressures higher than 150 atm. On the other side, temperature showed an inverse relationship with the solubility of CO_2. It was observed that an increase in temperature caused a reduction in the solubility of CO_2 in brine. For instance, as shown in Fig. 2, the solubility at 80 atm and 60 °C was 0.929 mol kg^{-1}, whereas under the same pressure and at 80 and 100 °C, the solubility was 0.771 and 0.679 mol kg^{-1}, respectively. The solubility of CO_2 in brine solutions (NaCl) with concentration of 1000 and 10,000 ppm is presented in Figs. 3 and 4. The error bars in these three figures illustrate 5% deviation from the base case (60 °C). Brine solutions showed similar responses to temperature and pressure changes as distilled water. For example, in 10,000 ppm brine (Fig. 4) as the temperature was raised from the initial value of 60–100 °C, a solubility reduction, depending on the pressure of between 5% and 24%, was observed. It can be concluded from the results presented in Figs. 2, 3 and 4 that regardless of the salinity and temperature, an increase in pressure resulted in a higher CO_2 dissolution in brine. This is due to the fact that at higher pressure, CO_2 molecules are closer to each other and their vibration energies are decreased; therefore, they can be more easily dissolved in the aqueous phase and make bonds with water molecules. Overall, it is apparent that the solubility of CO_2 is more sensitive to the changes of pressure and temperature where CO_2 is in subcritical (liquid or gas) conditions, rather than the supercritical state.

Thermodynamically, a rise in temperature leads to an increase in the kinetic energy of the molecules, more rapid motions of the molecules, and therefore easier breakage of intermolecular bonds. This process enables the molecules to escape from liquid phase into the gas phase (Zumdahl 2002). Tests at different salinities of brine showed that the solubility of CO_2 decreased with an increase in the salinity of brine at different pressure and temperature conditions. This can be explained by the fact that Na^+ and Cl^- ions from the salt attract water molecules to "solvate"; thus, the number of free water molecules available to be attracted by CO_2 molecules is decreased. Therefore, salinity decreases the weak affinity of CO_2 molecules to water and drives the dissolved CO_2 away from the polar water molecules. It can be concluded

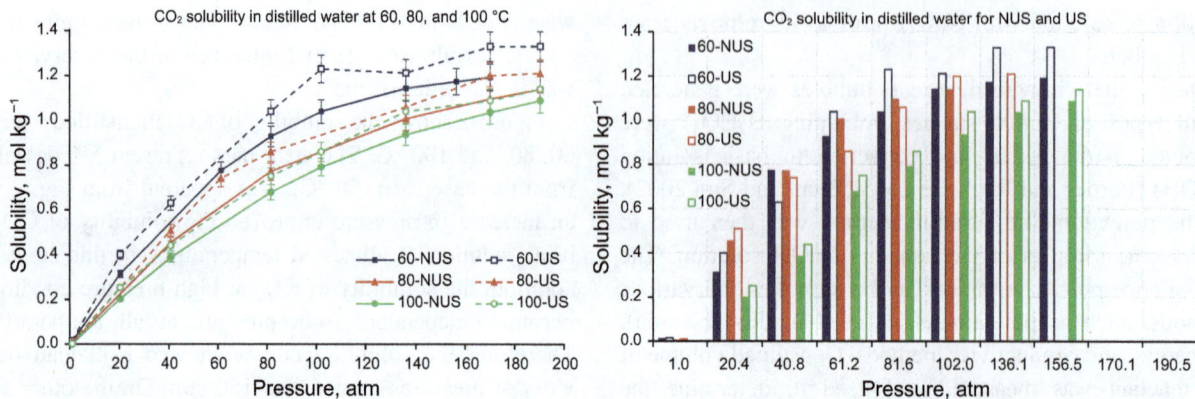

Fig. 2 The effect of ultrasound on solubility of CO_2 in distilled water at 60, 80, and 100 °C (*US* with ultrasound, *NUS* no ultrasound)

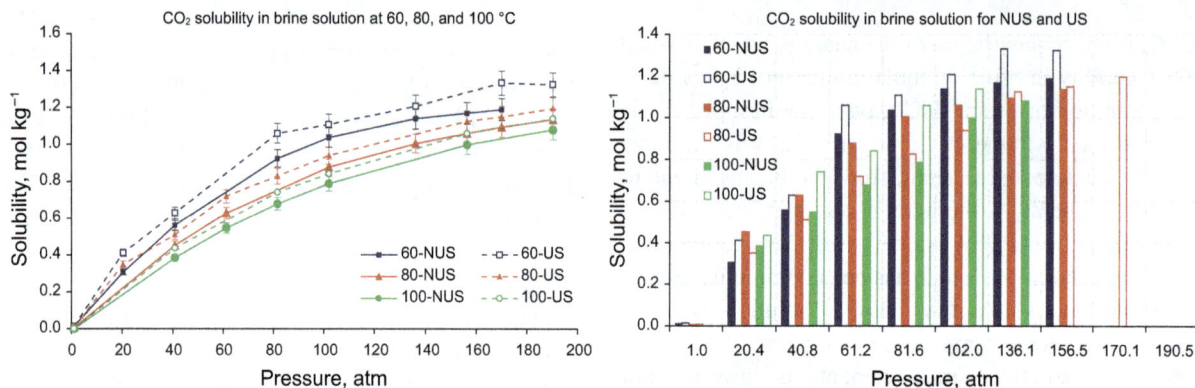

Fig. 3 The effect of ultrasound on solubility of CO_2 in brine solution (1000 ppm) at 60, 80, and 100 °C (*US* with ultrasound, *NUS* no ultrasound)

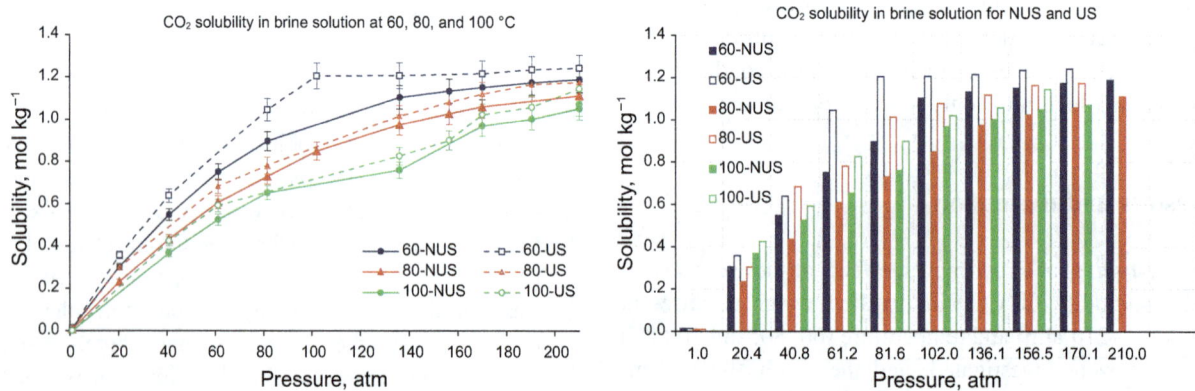

Fig. 4 The effect of ultrasound on solubility of CO_2 in brine solution (10,000 ppm) at 60, 80, and 100 °C

that the solubility mechanism becomes less effective in aquifers with a high level of salinity.

The effects of ultrasound on the solubility of CO_2 in distilled water at different temperature and pressure conditions are shown in Fig. 2. In addition, the effects of ultrasound (US) on the solubility of CO_2 in brine solutions (1000 and 10,000 ppm NaCl) are shown in Figs. 3 and 4, and the results were compared with the cases where no

ultrasound (NUS) was used. It is apparent from the figures that in all the cases, the solubility of CO_2 was improved by applying ultrasound. For instance, in the experiments using brine solution (1000 ppm), the solubility at 80 atm and 60 °C under ultrasound effects was 1.06 mol kg^{-1}, whereas under the same pressure and temperature conditions, without applying ultrasound, the solubility was 0.924 mol kg^{-1}. Thus, around 13% increase in the solubility of CO_2 with the application of ultrasound was observed. The increase in the solubility of CO_2 might be attributed to cavitation generated by ultrasound where cavitation could be defined as development, growth, and implosive collapse of bubbles in a liquid. Based on previous studies, the collapse of these bubbles could be considered as an adiabatic process, which leads to an enormous accumulation of energy inside the bubbles (Suslick 1989). This mechanism causes extremely high-temperature and pressure conditions in a microscopic region under the influence of sound waves. Therefore, we might conclude that as the solubility of CO_2 in water is high, more nuclei would be available for cavitation to take place, and therefore, bubble growth become easier. And as the cavitation continues, the induced high pressure pushes CO_2 molecules into the aqueous phase. Therefore, pressure surges that occur in the system as a result of cavitation could be the main reason behind higher solubility of CO_2 in the presence of ultrasound.

A vast range of outcomes could result from the cavitation such as increased chemical activity in the solution due to the formation of primary and secondary radical reactions. Therefore, this might lead to an improvement of mass and heat transfer processes between the gas and liquid phases. However, the detailed analysis of these mechanisms during CO_2 dissolution is not the focus of this study (Samenov et al. 2010; Santos et al. 2011; Laugier et al. 2008; Chen 2012).

It should also be noted that application of ultrasonic waves is associated with an increase in the temperature (Hamidi et al. 2013); therefore, one may argue that the increase in the temperature as a result of cavitation supposedly should result in a lower solubility of CO_2 in brine (Mohammadian et al. 2015; Duan and Sun 2003). This might be explained as the dominance pressure increases due to cavitation over the temperature effects. The later discussion could be the subject of further micro-scale studies to clarify the mechanisms involved with the solubility of CO_2 in brine in the presence of cavities.

In addition, higher external pressure reduces the vapour pressure of brine, which means higher intensity is needed to induce the cavitation by CO_2 (Vajnhandl and Megharaj 2005). The higher pressure may enhance the cavitation and collapse of CO_2 bubbles process; thus, it increases the dissolution of CO_2 in brine. Also incremental temperature induced by the ultrasonic equipment results in an increase in sonochemical reaction rates. The increase in cavitation intensity is caused by lowering vapour pressure and thus reducing the amount of vapour diffused into the bubbles to cushion the cavitational collapse (Adewuyi 2001). There are other mechanisms that involve molecular dynamics and phase behaviour of CO_2 and water which are beyond the scope of this study and need more investigation to understand the dissolution process.

4 Conclusion

Increasing pressure can enhance the dissolution of CO_2 over a range of pressure; afterwards, a plateau for the solubility of CO_2 can be observed. On the other hand, temperature and salinity inversely affect the dissolution of CO_2. There are different mechanisms including molecular dynamics, vibration energy and phase behaviour responsible for such processes. Increasing pressure can reduce the vibration energy of CO_2 molecules, and this can help in making bonds between CO_2 and water molecules. On the other hand, higher temperature can increase the vibration energy of CO_2 molecules, and higher salinity can decrease the chance of bonding of CO_2 with water molecules as Na^+ and Cl^- free ions occupy some of the places for intramolecular bonding with water molecules. As a result, both of these phenomena reduce the solubility of CO_2.

In the ultrasound-assisted dissolution experiments, it was found that ultrasonic waves can improve the solubility of CO_2 through complex mechanisms. These mechanisms might involve processes such as cavitation, bubble formation and sonochemical effects associated with ultrasonic waves among others. As a result, ultrasonic waves can improve the solubility of CO_2 in brine. It was found that a specific range of ultrasonic wave frequencies could be applied to improve the solubility of CO_2, beyond which the application of ultrasound can have a destructive impact on the dissolution process.

References

Abramov VO, Abramova AV, Bayazitov VM, et al. Sonochemical approaches to enhanced oil recovery. Ultrason Sonochem. 2015;25:76–81. doi:10.1016/j.ultsonch.2014.08.014.

Abramov VO, Mullakaev MS, Abramova AV, et al. Ultrasonic technology for enhanced oil recovery from failing oil wells and the equipment for its implementation. Ultrason Sonochem. 2013;20:1289–95. doi:10.1016/j.ultsonch.2013.03.004.

Abramov VO, Abramova AV, Bayazitov VM, et al. Selective ultrasonic treatment of perforation zones in horizontal oil wells for water cut reduction. Appl Acoust. 2016;103:214–20. doi:10.1016/j.apacoust.2015.06.017.

Adewuyi YG. Sanochemistry: environmental science and engineering applications. Eng Chem Res. 2001;40:4681–715.

Azdarpour A, Asadullah M, Mohammadian E, et al. A review on carbon mineral carbonation through pH-swing process. Chem Eng J. 2015;279:615–30. doi:10.1016/j.cej.2015.05.064.

Bachu S. Sequestration of CO_2 in geological media: criteria and approach for site selection in response to climate change. Energy

Convers Manag. 2000;41(9):953–70. doi:10.1016/S0196-8904(99)00149-1.

Bamberger A, Sieder G, Maurer G. High-pressure (vapor + liquid) equilibrium in binary mixtures of (carbon dioxide + water or acetic acid) at temperatures from 313 to 353 K. J Supercrit Fluids. 2000;17:97–110. doi:10.1016/S0896-8446(99)00054-6.

Brooks CEP. Climatic fluctuations and the circulation of the atmosphere. Weather. 1950;5(3):113–9. doi:10.1002/j.1477-8696.1950.tb01161.x.

Bryant S. Geologic CO_2 storage—Can the oil and gas industry help save the planet? J Pet Technol. 2007;59(9):98–105. doi:10.2118/103474-JPT.

Carroll JJ, Slupsky JD, Mather AE. The solubility of carbon dioxide in water at low pressure. J Phys Chem Ref Data. 1991;20:1201–9. doi:10.1063/1.555900.

Chen D. Applications of ultrasound in water and wastewater treatment. Handbook on application of ultrasonic: sonochemistry for sustainability. Boca Raton: CRC Press, Taylor and Francis Group; 2012.

Duan ZH, Sun R. An improved model calculating CO_2 solubility in pure water and aqueous NaCl solutions from 273 to 533 K and from 0 to 2000 bar. Chem Geol. 2003;193:257–71. doi:10.1016/S0009-2541(02)00263-2.

El-Maghraby RM, Pentland CH, Iglauer S, et al. A fast method to equilibrate carbon dioxide with brine at high pressure and elevated temperature including solubility measurements. J Supercrit Fluids. 2012;62:55–9. doi:10.1016/j.supflu.2011.11.002.

Gibson-Poole CM, Edwards S, Langford RP, et al. Review of geological storage opportunities for carbon capture and storage (CCS) in Victoria-summary report. Cooperative Research Centre for Greenhouse Gas Technologies, ICTPL-RPT07-0526, 2007.

Hamida T, Babadagli T. Investigations on capillary and viscous displacement under ultrasonic waves. J Can Pet Technol. 2006;45(2):16–9. doi:10.2118/06-02-TN2.

Hamida T, Babadagli T. Fluid–fluid interaction during miscible and immiscible displacement under ultrasonic waves. Eur Phys J B. 2007;60:447–62. doi:10.1140/epjb/e2008-00005-5.

Hamidi H, Rafati R, Junin RB, et al. A role of ultrasonic frequency and power on oil mobilization in underground petroleum reservoirs. J Pet Explor Prod Technol. 2012;2(1):29–36. doi:10.1007/s13202-012-0018-x.

Hamidi H, Rafati R, Junin R, et al. A technique for evaluating the oil/heavy-oil viscosity changes under ultrasound in a simulated porous medium. Ultrasonics. 2013;54(2):655–62. doi:10.1016/j.ultras.2013.09.006.

Hamidi H, Mohammadian E, Asadullah M, et al. Effect of ultrasound radiation duration on emulsification and demulsification of paraffin oil and surfactant solution/brine using Hele-Shaw models. Ultrason Sonochem. 2015a;26:428–36. doi:10.1016/j.ultsonch.2015.01.009.

Hamidi H, Mohammadian E, Rafati R, et al. Effect of ultrasonic waves on the phase behavior of a surfactant–brine–oil system. Colloids Surf A. 2015b;482:27–33. doi:10.1016/j.colsurfa.2015.04.009.

King MB, Mubarak A, Kim JD, et al. The mutual solubilities of water with supercritical and liquid carbon dioxides. J Supercrit Fluids. 1992;5(4):296–302. doi:10.1016/0896-8446(92)90021-B.

Koschel D, Coxam JY, Rodier L, et al. Enthalpy and solubility data of CO_2 in water and NaCl (aq) at conditions of interest for geological sequestration. Fluid Phase Equilib. 2006;247:107–20. doi:10.1016/j.fluid.2006.06.006.

Paul EE, Naylor M, Stuart H, Curtis A. CO_2/brine surface dissolution and injection: CO_2 storage enhancement," SPE offshore Europe paper presentation, Aberdeen, UK, 2010;6(1):41–53. doi:10.2118/124711PA.

Laugier F, Andriantsiferana C, Wilhelm AM, et al. Ultrasound in gas-liquid systems: effects on solubility and mass transfer. Ultrason Sonochem. 2008;15:965–72. doi:10.1016/j.ultsonch.2008.03.003.

Li ZW, Dong MZ, Li SL, et al. Densities and solubilities for binary systems of carbon dioxide plus water and carbon dioxide plus brine at 59 °C and pressures to 29 MPa. J Chem Eng Data. 2004;49(4):1026–31. doi:10.1021/je049945c.

Mao S, Zhang D, Li Y, et al. An improved model for calculating CO_2 solubility in aqueous NaCl solutions and the application to CO_2 − H_2O − NaCl fluid inclusions. Chem Geol. 2013;347:43–58. doi:10.1016/j.chemgeo.2013.03.010.

Mason TJ. Chemistry with ultrasound, critical report on applied chemistry, published for SCI by Elsevier Science Publishers, vol. 28, London, 1990.

Metz B, Davidson O, Coninck H, et al. Carbon dioxide capture and storage. IPCC Special Report prepared by Working Group III of the Intergovernmental Panel on Climate Change, Cambridge, UK, 2005.

Mohammadian E, Hamidi H, Assadullah M, et al. Measurement of CO_2 solubility in NaCl brine solutions at different temperatures and pressures using the potentiometric titration method. J Chem Eng Data. 2015;60(7):2042–9. doi:10.1021/je501172d.

Naderi K, Babadagli T. Influence of intensity and frequency of ultrasonic waves on capillary interaction and oil recovery from different rock types. Ultrason Sonochem. 2010;17:500–8. doi:10.1016/j.ultsonch.2009.10.022.

Nguyen DN. Carbon dioxide geological sequestration: technical and economic reviews. In: SPE/EPA/DOE Exploration and Production Environmental Conference, San Antonio, Texas, USA, 2003. doi:10.2118/81199MS.

Peper S, Dohrn R. Sampling from fluid mixtures under high pressure: review, case study and evaluation. J Supercrit Fluids. 2012;66:2–15. doi:10.1016/j.supflu.2011.09.021.

Portier S, Rochelle C. Modelling CO_2 solubility in pure water and NaCl-type waters from 0 to 300 °C and from 1 to 300 bar: application to the Utsira Formation at Sleipner. Chem Geol. 2005;217(3):187–99. doi:10.1016/j.chemgeo.2004.12.007.

Reichle D, Houghton J, Kane B, et al. Carbon sequestration research and development. US Department of Energy Report DOE/SC/FE-1, Washington DC, USA, 1999.

Rumpf B, Nicolaisen H, Ocal C, et al. Solubility of carbon dioxide in aqueous solutions of sodium chloride: experimental results and correlation. J Solution Chem. 1994;23:431–48. doi:10.1007/BF00973113.

Samenov IA, Ulyanov BA, Kulov NN. Effect of ultrasound on the dissolution of carbon dioxide in water. Theor Found Chem Eng. 2010;45(1):21–5. doi:10.1134/S0040579511010106.

Santos R, Ceulemans P, Francois D, et al. Ultrasound-enhanced mineral carbonation. In: The 3rd European Process Intensification Conference. ICheme Symposium Series no. 157. 2011.

Sengul M. CO_2 sequestration: a safe transition technology. Paper SPE 98617 presented at SPE International Conference on Health, Safety, and Environment in Oil & Gas Exploration and Production, Abu Dhabi, UAE, 2006. doi:10.2118/98617MS.

Schrag DP. Preparing to capture carbon. Science. 2007;315(5813):812–3. doi:10.1126/science.1137632.

Shedid AS, Adel MS. Experimental investigations of CO_2 solubility and variations in petrophysical properties due to CO_2 storage in carbonate reservoir rocks. In: SPE North Africa Technical Conference and Exhibition, Cairo, Egypt, 2013. doi:10.2118/164632-MS.

Suslick KS. The chemical effects of ultrasound. USA: Scientific American; 1989. p. 62–8.

Tong DM, Vega-Maza D, Trusler M. Solubility of CO_2 in aqueous solutions of $CaCl_2$ or $MgCl_2$ and in a synthetic formation brine at temperatures up to 423 K and pressures up to 40 MPa. J Chem

Eng Data. 2013;58(7):2116–24. doi:10.1021/je400396s@proofing.

Vajnhandl S, Megharaj AML. Ultrasound in textile dyeing and the decolourization/mineralization of textile dyes. Dyes Pigments. 2005;65:89–101. doi:10.1016/j.dyepig.2004.06.012.

Wiebe R, Gaddy VL. The solubility in water of carbon dioxide at 50, 75 and 100 °C at pressures to 700 atm. J Am Chem Soc. 1939; 61:315–8. doi:10.1021/ja01871a025.

Yan W, Huang S, Stenby EH. Measurement and modelling of CO_2 solubility in NaCl brine and CO_2—saturated NaCl brine density.

Int J Greenh Gas Control. 2011;5(6):1460–77. doi:10.1016/j.ijggc.2011.08.004.

Zheng DQ, Guo TM, Knapp H. Experimental and modeling studies on the solubility of CO_2, $CHClF_2$, CHF_3, $C_2H_2F_4$ and $C_2H_4F_2$ in water and aqueous NaCl solutions under low pressures. Fluid Phase Equilib. 1997;129:197–209. doi:10.1016/S0378-3812(96)03177-9.

Zumdahl SS. Chemical principles. 4th ed. Boston: Houghton Mifflin Company; 2002.

10

Rheology of rock salt for salt tectonics modeling

Shi-Yuan Li[1] · Janos L. Urai[2]

Abstract Numerical modeling of salt tectonics is a rapidly evolving field; however, the constitutive equations to model long-term rock salt rheology in nature still remain controversial. Firstly, we built a database about the strain rate versus the differential stress through collecting the data from salt creep experiments at a range of temperatures (20–200 °C) in laboratories. The aim is to collect data about salt deformation in nature, and the flow properties can be extracted from the data in laboratory experiments. Moreover, as an important preparation for salt tectonics modeling, a numerical model based on creep experiments of rock salt was developed in order to verify the specific model using the Abaqus package. Finally, under the condition of low differential stresses, the deformation mechanism would be extrapolated and discussed according to microstructure research. Since the studies of salt deformation in nature are the reliable extrapolation of laboratory data, we simplified the rock salt rheology to dislocation creep corresponding to power law creep ($n = 5$) with the appropriate material parameters in the salt tectonic modeling.

Keywords Rock salt rheology · Power law creep · Dislocation creep · Modeling

✉ Shi-Yuan Li
15330011512@163.com

[1] School of Petroleum Engineering, China University of Petroleum, Beijing 102249, China

[2] Endogene Dynamik, Faculty of Geo-Resources and Materials Technology, RWTH Aachen University, Lochnerstrasse 4-20, 52056 Aachen, Germany

Edited by Yan-Hua Sun

1 Introduction

In recent years, a number of numerical studies about the deformation of salt structures have been published (Schultz-Ela and Jackson 1993; Poliakov et al. 1993; van Keken et al. 1993; Koyi 1996, 1998; Chemia et al. 2008; Chemia and Koyi 2008; Chemia and Schmeling 2009; Li et al. 2009; Ings and Beaumont 2010; Li et al. 2012a). It can be an effective method to study constitutive equations for salt flow based on natural deformation and will deliver reliable reference in salt tectonics and also salt mining engineering. In order to prepare for salt tectonics modeling, the primary problem is what deformation mechanisms of salt can be suitable in the model. In this paper, through the introduction of deformation mechanisms of rock salt and reading previous references about salt creep in laboratories, we have summarized the experimental results and built a database about the strain rate and differential stress relationships. However, direct measurements of rheology at these low rates are not possible, and the rheology of rock salt during long-term deformation in nature is controversial (Li et al. 2012b).

Evaporite rock is mainly composed of carbonates, sulfates, and chlorides. Sodium chlorite (NaCl) is the main chloride, and its mineral form is halite which is also called rock salt. Rock salt influences the salt tectonics and basin evolution because of its specific characteristics including low porosity and permeability, low creep strength and low density (Urai et al. 2008).

The mechanical behavior of rock salt has been investigated in numerous studies during recent decades. The purpose of these studies is to design safe salt mines and develop the design and performance assessment for nuclear waste disposal and oil and gas storage (Wawersik and Zeuch 1986; Cristescu 1998; Hunsche and Hampel 1999).

On a geological time scale, rock salt is a ductile material and behaves like a Newtonian fluid (Schultz-Ela and Jackson 1993; van Keken et al. 1993; Koyi 2001; Gemmer et al. 2004). In recent years, salt mechanics including the theory, experiment and modeling of creep behavior has made progress in China. The research findings have been widely used in layered salt mechanics and engineering (Yang et al. 2009), salt-gypsum layer drilling engineering (Deng 1997; Tang et al. 2004; Yang et al. 2007) and CO_2 geological storage (Liang and Zhao 2007). However, the deformation mechanism of rock salt is relatively less researched in China. The research on the deformation mechanism can lead to a deep understanding of rock salt creep, and a constitutive relation can be built according to microphysics. The creep descriptions based on deformation mechanisms can be used to forecast the geological tectonics and rock engineering stability. It is also relevant to petroleum engineering. For instance, hydraulic fracturing mechanics is an important area of study in petroleum engineering (Zhang et al. 2008; Zhang and Chen 2009, 2010), and the creep properties of rock will also have impacts on fracturing, crack propagation and closure.

2 Deformation mechanisms of rock salt

Two main deformation mechanisms have been investigated in different ways such as laboratory experiments, microstructural investigations and analysis of displacement in actively deforming salt (Fig. 1). One is the dislocation creep mechanism which means crystal defects lead to a dislocation effect. The creep controlled by dislocation mechanisms is widely investigated in laboratory experiments (Carter and Hansen 1983; Carter et al. 1993; Rutter 1983; Urai et al. 1986; Wawersik and Zeuch 1986; Heard and Ryerson 1986; Senseny et al. 1992; van Keken et al. 1993; Franssen 1994; Peach and Spiers 1996; Weidinger et al. 1997; De Meer et al. 2002; Hampel et al. 1998; Brouard and Bérest 1998; Bérest et al. 2005; Ter Heege et al. 2005a, b). The creep equations mainly used in the salt mining industry are based on dislocation creep processes quantified in laboratory experiments (Ottosen 1986; Haupt and Schweiger 1989; Aubertin et al. 1991; Cristescu 1993; Munson 1979, 1997; Jin and Cristescu 1998; Hampel et al. 1998; Hampel and Schulze 2007; Hunsche and Hampel 1999; Peach et al. 2001; Fossum and Fredrich 2002). The steady-state strain rate is related to the flow stress σ using a power law creep (non-Newtonian) equation:

$$\dot{\varepsilon}_{DC} = A(\Delta\sigma)^n = A_0 \exp\left(-\frac{Q}{RT}\right)(\sigma_1 - \sigma_3)^n \qquad (1)$$

where $\dot{\varepsilon}_{DC}$ is the strain rate; $\Delta\sigma = \sigma_1 - \sigma_3$ is the differential stress; $A = A_0 \exp(-Q/RT)$ is the viscosity of the salt; A_0 is a material dependent parameter; Q is the specific activation energy, while R is the gas constant ($R = 8.314$ J mol^{-1} K^{-1}); and T is the temperature.

The other mechanism is solution-precipitation creep (also called diffusion creep), which means a crystal grain slides along the crystal boundary. It includes pressure solution and dynamic recrystallization processes, and it has a strong relation with water content and chemical reaction. This process is also an important deformation mechanism in rock salt as is shown in experiments and microstructure research (Urai et al. 1987; Spiers et al. 1990; Spiers and Carter 1998; Martin et al. 1999; Schenk and Urai 2004; Schenk et al. 2006; Ter Heege et al. 2005a, b). The solution-precipitation creep is described as following the Newtonian flow law:

$$\dot{\varepsilon}_{PS} = B(\Delta\sigma) = B_0 \exp\left(-\frac{Q}{RT}\right)\left(\frac{\sigma_1 - \sigma_3}{TD^m}\right) \qquad (2)$$

and the strain rate $\dot{\varepsilon}_{PS}$ is dependent on the grain size D; $B = B_0\exp(-Q/TD^m)$ is the viscosity of the salt; and B_0 is a material parameter. The order m influences the strain rate which is dependent on the grain size. Finally, if the dislocation creep and pressure solution creep act simultaneously, the total strain rate is the sum of the two strain rates:

$$\dot{\varepsilon} = \dot{\varepsilon}_{DC} + \dot{\varepsilon}_{PS} \qquad (3)$$

Halite rheology is mainly dominated by two main models of the deformation mechanisms in laboratory experiments (Eqs. 1 and 2). These are dislocation creep and solution-precipitation creep. One (pressure solution creep) corresponds to Newtonian viscous rheology with a viscosity dependent on grain size and temperature, the other (dislocation creep) corresponds to the power law creep with a high stress exponent, here creep is mainly temperature-dependent. Based on the deformation mechanisms of rock salt, modelers can choose an effective viscosity which is an implementation of power law creep, making a combination of viscosity associated with both dislocation and solution-precipitation creep (a similar approach is used in many other papers by the salt tectonics modeling community such as van Keken et al. 1993; Koyi 2001; Ings and Beaumont 2010).

Two important differences between Eqs. (1) and (2) are the power order relevant to the influence of stress on strain rate ($n = 1$ for solution-precipitation creep or pressure solution while $n > 1$ for dislocation creep) and secondly the dependence of strain rate on grain size. What needs to be pointed out is that the steady-state creep process is mainly discussed because it is dominant during salt deformation under the long-term conditions.

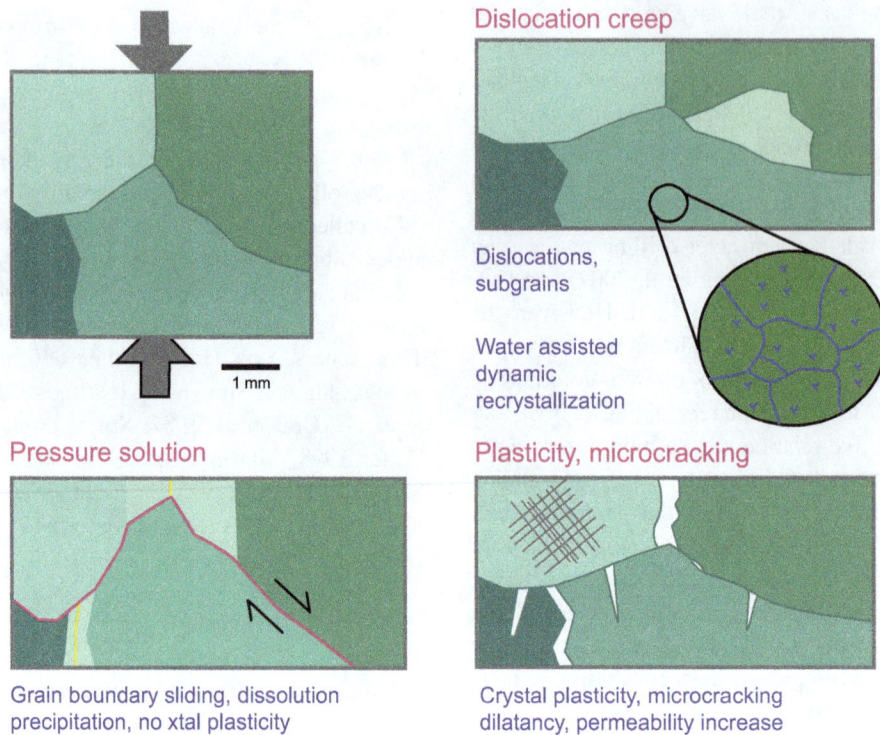

Fig. 1 Deformation mechanisms of rock salt at 20–200 °C. The crystals with different orientations are represented by different *green shades* (Urai and Spiers 2007)

3 Creep mechanisms at high stresses

3.1 Creep experiments of rock salt at 20–200 °C

Salt flow or halokinesis often occurs at temperatures ranging from 20 to 200 °C. Salt deformation at those temperatures has been widely investigated in laboratories (Heard 1972, 1986; Hansen 1979, Wawersik and Hannum 1980; Wawersik and Zeuch 1984, 1986; Hansen and Carter 1984; Wawersik 1985; Senseny 1988; DeVries 1988; Spiers et al. 1990; Wawersik and Zimmerer 1994; Weidinger et al.1997; Yang et al. 1999; Hunsche and Hampel 1999; Peach and Spiers 2001; Hunsche et al. 2003, Ter Heege et al. 2005a, b; Schoenherr et al. 2007). The purpose of collecting the experimental data is to provide a basis for deformation modeling and to discuss the influence of different physical parameters on creep properties. The data in Figs. 2 and 3 show the relationship between the strain rate and differential stress during steady-state creep of rock salt from different areas such as Asse, Avery Island, Gorleben, South Oman, and synthetic samples. Two main relationships can be observed, one is between the strain rate and the differential stress (Fig. 2) follows the power law equation which is relevant to dislocation creep, while the other relation (Fig. 3) follows the Newtonian equation which is relevant to pressure solution creep.

3.2 The analysis of the database of laboratory results

It is clearly demonstrated in Fig. 2 that the relation between the strain rate and the differential stress in these experiments is controlled by the power law equation with the power order 5 which is relevant to dislocation creep when the salt deforms at 20–200 °C (Heard 1972; Wawersik and Hannum 1980; Hansen and Carter 1984; DeVries 1988; Weidinger et al. 1997; Hunsche and Hampel 1999; Hunsche et al. 2003). We can see that salt rheology is strongly dependent on temperature, and higher temperature leads to higher strain rate. For every 50 °C increase in temperature, the range of the strain rate varies by around 1.5–2 orders of magnitude for each differential stress. One important thing is that, due to the limited time length in laboratory experiments and strong dependence of grain size and loss of water content, solution-precipitation creep is not often observed in experimental results. The power law related to dislocation creep is considered as engineering creep (Cristescu and Hunsche 1988; Hunsche and Hampel 1999; Fossum and Fredrich 2002).

Some experiments with synthetic samples, for instance, fine grained wet halite also show that solution-precipitation creep controls salt rheology (Urai et al. 1986; Spiers et al. 1990; Spiers and Brzesowsky 1993; Renard et al.

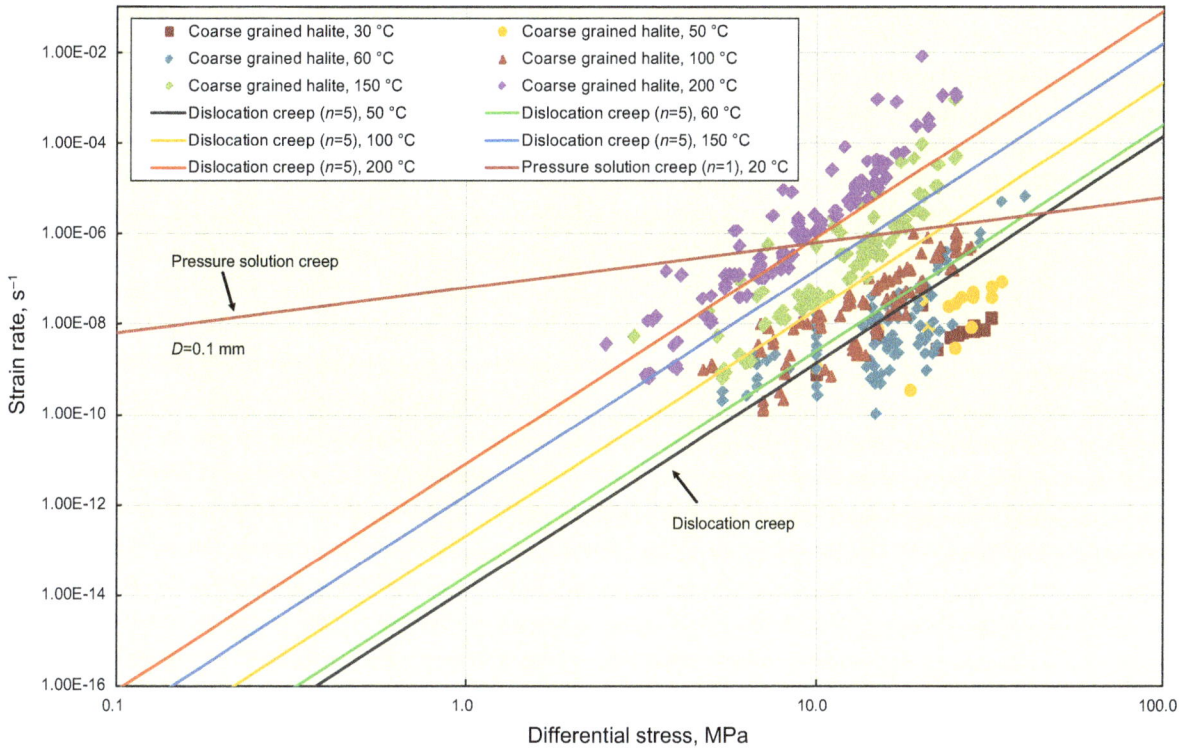

Fig. 2 Strain rate versus differential stress of coarse grain halite at 30–200 °C

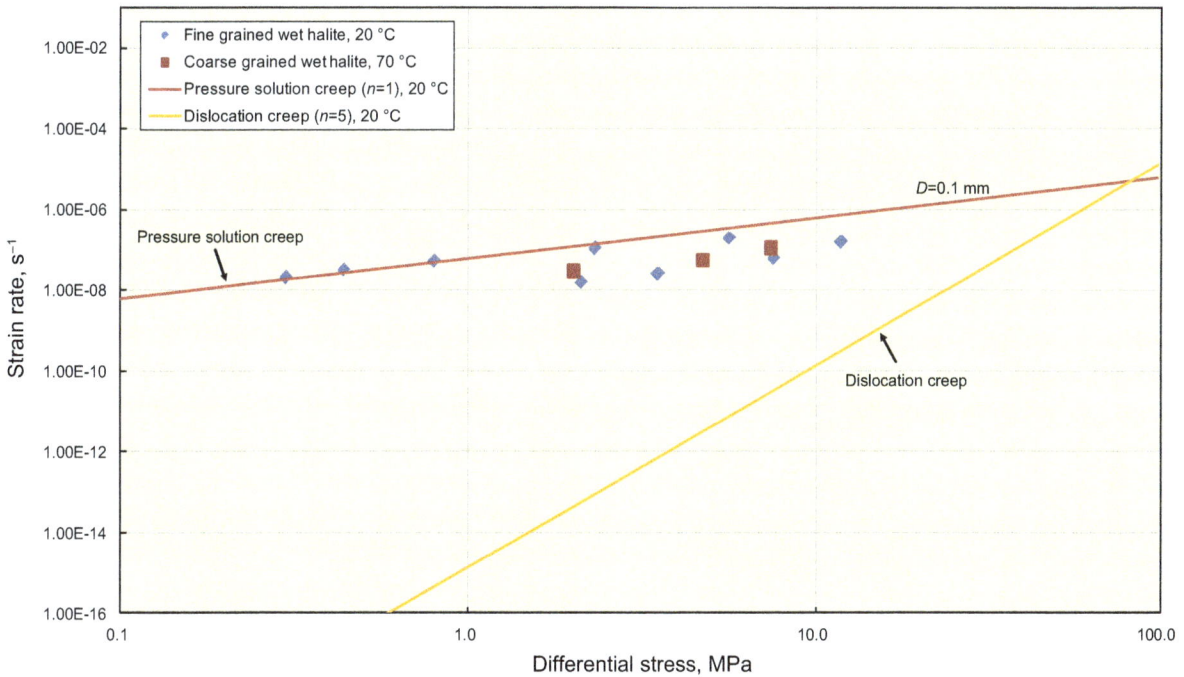

Fig. 3 Strain rate versus differential stress of fine and coarse grain wet halite at 20 and 70 °C

2002, 2004). The data in Fig. 3 show that the salt rheology is dependent on three important factors for wet halite deformation. These are grain size, water content and temperature. The creep strain rate of fine grained wet halite at

20 °C is above 10^{-8} s^{-1} and similar to the one of coarse grained wet halite at 70 °C.

Moreover, solution-precipitation creep is observed in some experiments (where the period of the experiment lasts more than one year) on coarse grained natural rock salt at low stresses (Bérest et al. 2005). For natural rock salt, solution-precipitation creep has an important influence on the strain rate, and it cannot be neglected. Previous research shows that solution-precipitation creep and dislocation creep occur at the same time and both of them control salt creep. Due to time scale constraints in experiments, it is rather difficult to measure the deformation process with the strain rate below 10^{-9} s^{-1}, in Figs. 2 and 3 we put two creep equations not only for comparison but also as an extrapolation of the data which were observed in experiments. The two creep equations and the rheological parameters A_0 and B_0 are based on the experiments of Spiers et al. (1990) and Wawersik and Zeuch (1986).

The rheological parameters for dislocation creep (Eq. 1) and pressure solution creep (Eq. 2) are listed in Table 1.

4 Numerical simulation of triaxial experiments

In the laboratory, the principal experimental approach is to apply uniaxial or triaxial tensile and compressive loading on a rock salt sample which is drilled from core from salt structures in order to investigate mechanical behavior (for example, the relation between stress and strain or strain rate). In order to prepare for salt tectonics modeling, we develop a numerical model of a triaxial creep experiment of rock salt to verify the specific model using the Abaqus package on salt deformation and evaluate the error due to the boundary conditions in real situations.

The height of the cylindrical sample is 20 cm, and the radius of the cross section is 5 cm (Figs. 4, 5). Table 1 shows the rheological parameters. As a simplified way, we use an axially symmetric model in Abaqus. The loading on

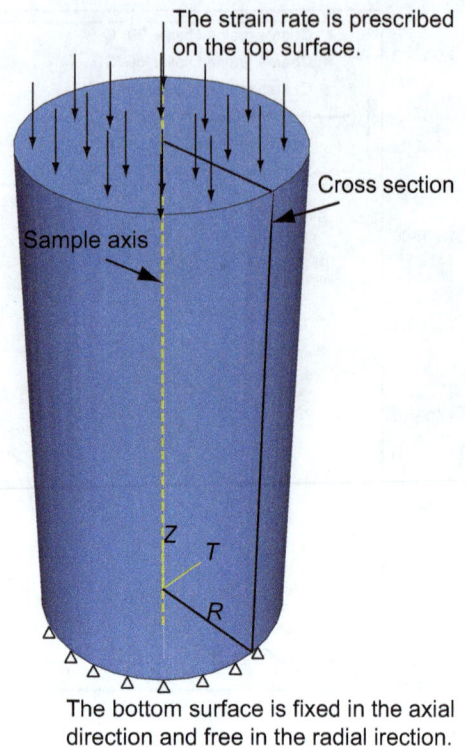

Fig. 4 Cylindrical sample with free boundary conditions in the radial direction

the sample is differential stress $\sigma_1 - \sigma_3$ (σ_1 and σ_3 are two principal stresses in the triaxial experiment). Considering the influence of boundary friction, we simulate two extreme cases, one is radial-free boundary conditions (frictionless) and the other is radial fixed boundary conditions (maximum friction). Two assumptions are taken into consideration, one is the volume which is not changed

Table 1 Material and geometry of rock salt samples in experiments

	Salt A: pressure solution creep (Newtonian flow)	Salt B: dislocation creep (non-Newtonian flow)
A_0, MPa^{-5} s^{-1}	–	7.26×10^{-6}
B_0, MPa^{-1} s^{-1}	4.70×10^{-4}	–
Q, J/mol	24,530	53,920
R, J/mol	8.314	8.314
T, K	323	293
n	1	5
m	3	–
Density ρ, kg/m^3	2200	2200
Young's modulus E, GPa	10	10
Poisson's ratio v	0.4	0.4
Sample height, cm	20	20
Sample radius, cm	5	5
Reference	Spiers et al. (1990)	Wawersik and Zeuch (1986)

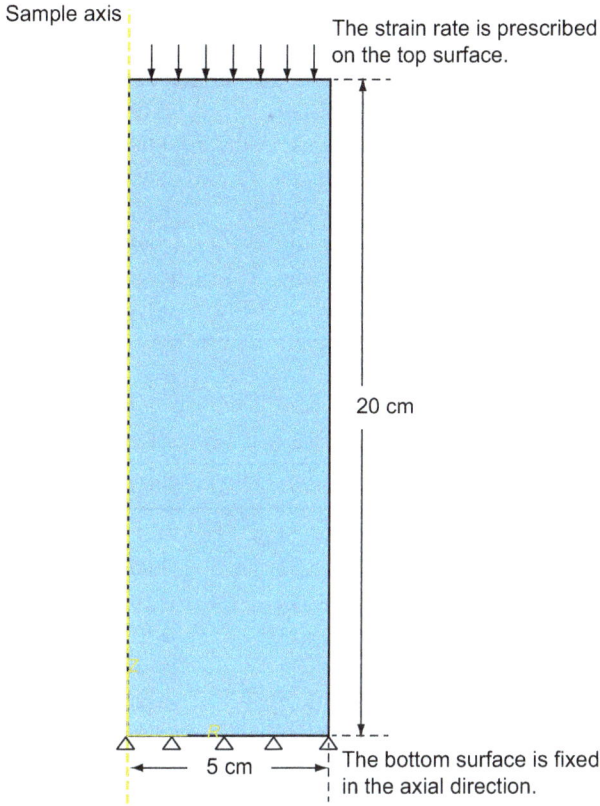

Fig. 5 Half cross section from the cylindrical sample, the *yellow line* is the sample axis, the *bottom surface* is fixed in the axial direction and free in the radial direction, the *top surface* is prescribed by the strain rate in the axial direction and free in the radius direction

during the compression, (incompressibility) and the second one is homogeneous material.

4.1 Simulation of the triaxial creep experiment on a sample with free boundary conditions in the radial direction

In order to obtain the value of differential stress, we calculate the total force $\sum F$ on the top boundary, and then, the total force is divided by the area of the cross section which is due to the radial expansion of the sample as a function of time. We keep the vertical displacement rate on the top surface constant and choose free radial movement on the boundary. On the bottom surface, it is fixed perpendicular to the boundary on the bottom, i.e., no radial movement at one corner and no axial movement at the bottom. For the free radial-movement boundary condition, the homogeneous sample is applied to a uniform displacement rate on the top boundary so that the deformation is also homogeneous. The area of the cross section is:

$$A(t) = \pi[R(t)]^2 = \pi[R_i + u_x(t)]^2 \tag{4}$$

where $A(t)$ is the cross-sectional area; $R(t)$ is the radius of the cylindrical sample; R_i is the initial radius of the sample;

u_x is the displacement in the radial direction. The differential stress $\sigma_1 - \sigma_3$ in the numerical simulation can be confirmed:

$$\sigma(t) = \frac{\sum F}{A(t)} \tag{5}$$

where $\sigma(t)$ is the differential stress; $\sum F$ is the total force acting on the cross section.

4.2 Simulation of the triaxial creep experiment on the sample with fixed boundary conditions in the radial direction

Boundary conditions are fixed perpendicular to the boundary on the bottom and the sides, i.e., no radial movement on the top and bottom edges and no axial movement at the bottom. On the top surface, the vertical displacement rate is applied in the axial direction, and the horizontal displacement in the radial direction is fixed. For the fixed radial-movement boundary condition, heterogeneous deformation occurs due to the fixed boundary condition. The average area of the cross section which is equal to the volume over the length:

$$A(t) = \frac{V}{l(t)} = \frac{V}{l_o \left(1 - \frac{u(t)\Delta t}{l_o}\right)} = \frac{A_o}{1 - \varepsilon_{yy}} \tag{6}$$

where $l(t)$ is the length of the cylindrical sample; V is the volume of the sample; A_o is the original area of the cross section, l_o is the original length of the cylindrical sample; and $u(t)$ is the displacement rate; ε_{yy} is the strain in the longitudinal direction. The stress is calculated through the total force $\sum F$ over the average area of the cross section $A(t)$, as described in Eq. (5).

4.3 Results of numerical modeling

The parameters and results in the models of free boundary conditions, where the power law and Newtonian law creep are used, respectively, are shown in Tables 2 and 3. Meanwhile, the parameters and results for validations for dislocation creep and solution-precipitation creep in the models of the fixed boundary condition are shown in Tables 4 and 5, respectively.

As we know, the stages of creep include the initial stage called primary creep and the second stage called steady-state creep. In the primary creep stage, the strain rate gradually decreases from a relatively high value. In the steady-state creep stage, the strain rate reaches an almost constant value. The creep strain rate means the rate in the steady-state creep stage. The relation between the differential stress and the creep strain rate can be described as

Table 2 Test parameters and the results of the differential stress (dislocation creep and free radial boundary condition)

	Vertical displacement on the top surface, cm	Total deformation time, s	Strain rate, s^{-1}	Stress $\sigma_1 - \sigma_3$, MPa
Strain rate 1	2.48	1.0E+19	1.41E−20	0.10
Strain rate 2	2.48	1.0E+14	1.41E−15	1.00
Strain rate 3	2.42	1.0E+12	1.37E−13	2.50
Strain rate 4	2.37	1.0E+10	1.34E−11	6.25
Strain rate 5	2.30	1.0E+08	1.30E−09	15.60
Strain rate 6	2.26	1.0E+06	1.27E−07	39.00
Strain rate 7	2.22	1.0E+04	1.25E−05	97.60

Table 3 Test parameters and the results of the differential stress (solution-precipitation creep and free radial boundary condition)

	Vertical displacement on the top surface, cm	Total deformation time, s	Strain rate, s^{-1}	Stress $\sigma_1 - \sigma_3$, MPa
Strain rate 1	2.54	1.2E+13	1.21E−14	0.10
Strain rate 2	2.54	1.2E+12	1.21E−13	1.00
Strain rate 3	2.16	4.0E+11	3.02E−13	2.50
Strain rate 4	2.40	1.8E+11	7.56E−13	6.25
Strain rate 5	2.34	7.0E+10	1.89E−12	15.60
Strain rate 6	2.49	3.0E+10	4.72E−12	39.00
Strain rate 7	2.13	1.0E+10	1.18E−11	97.60

Table 4 Test parameters and the results of the differential stress (dislocation creep and fixed radial boundary condition)

	Vertical displacement on the top surface, cm	Total deformation time, s	Strain rate, s^{-1}	Stress $\sigma_1 - \sigma_3$, MPa
Strain rate 1	2.48	1.0E+19	1.41E−20	0.10
Strain rate 2	2.48	1.0E+14	1.41E−15	1.03
Strain rate 3	2.42	1.0E+12	1.37E−13	2.58
Strain rate 4	2.37	1.0E+10	1.34E−11	6.44
Strain rate 5	2.30	1.0E+08	1.30E−09	16.06
Strain rate 6	2.26	1.0E+06	1.27E−07	40.12
Strain rate 7	2.22	1.0E+04	1.25E−05	100.13

Table 5 Test parameters and the results of differential stress (solution-precipitation creep and fixed radial boundary condition)

	Vertical displacement on the top surface, m	Total deformation time, s	Strain rate, s^{-1}	Stress $\sigma_1 - \sigma_3$, MPa
Strain rate 1	2.54	1.2E+13	1.21E−14	0.11
Strain rate 2	2.54	1.2E+12	1.21E−13	1.09
Strain rate 3	2.16	4.0E+11	3.02E−13	2.72
Strain rate 4	2.40	1.8E+11	7.56E−13	6.78
Strain rate 5	2.34	7.0E+10	1.89E−12	16.90
Strain rate 6	2.49	3.0E+10	4.72E−12	42.12
Strain rate 7	2.13	1.0E+10	1.18E−11	105.28

Eqs. (1) or (2). In the numerical simulation, we apply the elastic law and power law equations (the steady-state creep equations) on the rock salt sample.

We can obtain the relationship between the differential stress and the strain when dislocation creep dominates ($n = 5$) and evaluate the friction boundary effect in the experiment. We can obtain the relation between the differential stress and the strain rate in the steady-state creep stage and make a comparison between the numerical and experimental results. In Fig. 6, we can see that the thick dots represent the differential stress with the fixed boundary condition. The thin dots represent the differential stress with the free boundary condition. The homogeneous deformation results in a constant stress in the steady state. When the strain is the same, the differential stress with the radial fixed boundary condition is higher than the one with the radial-free boundary condition.

From the stress–strain curve in Fig. 7, we can see the mechanical behavior of rock salt includes elastic and steady-state creep processes. However, it is obvious there is a 'pseudo'-primary creep between the elastic and steady-state creep stages. In the numerical model, we do not apply primary creep in the material property. It is because creep which leads to stress relaxation can occur even when the elastic strength is not reached after the loading is applied for a short period time. In Fig. 7, we can see the relation between the differential stress and the strain when solution-precipitation creep dominates ($n = 1$). The thick dots represent the differential stress with the fixed boundary condition. The thin dots represent the differential stress with the free boundary condition. The differential stress increases slowly because in reality the top surface displacement velocity is constant and the strain rate increases slowly. The change in stress is very relevant to the slow increase in the strain rate when the Newtonian creep law ($n = 1$) dominates.

Figure 8 shows the diagram of the differential stress and strain rate, and it summarizes low-temperature laboratory data and numerical simulation results. The broken line is extrapolation of the dislocation creep equation, taking $n = 5$. The blue and red dots show the results of the relation between the differential stress and the strain rate for $n = 5$ and $n = 1$, respectively, from the numerical simulation with free x-movement boundary (the frictionless boundary condition) of the cylindrical sample. We can see that the numerical solutions fit the experimental results quite well. Although from the numerical results the relation between stress and strain has a small difference from the theoretical situation, the relation between the strain rate and the differential stress with the free boundary condition matches the theoretical situation very well. This shows that for the steady-state creep the numerical solution is in a good agreement with the experimental result despite a 'pseudo-primary creep' between the elastic and steady-state creep stages. Therefore, the numerical result of the parameter A and the power order n (Eq. 1) achieved from the free radial-movement boundary condition can be thought of as the correct rheology.

5 Creep mechanisms at low stresses

Some experiments with synthetic samples show that solution-precipitation creep can play a role at low stresses (Spiers et al. 1990). However, as an important extrapolation of laboratory data, it is necessary to study the rock salt

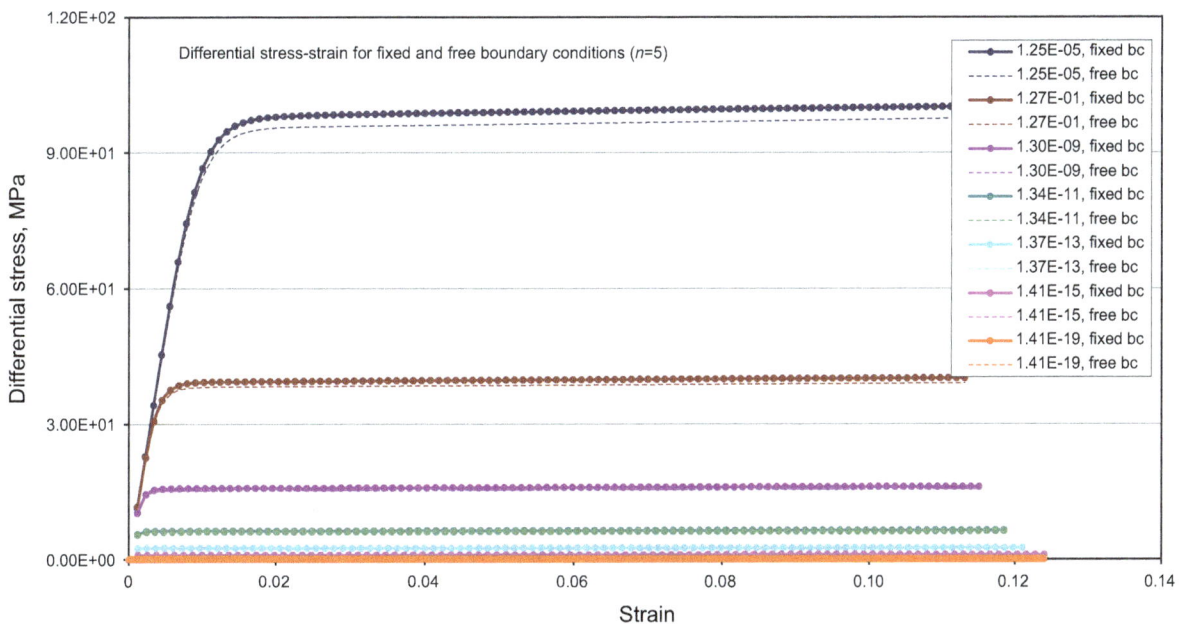

Fig. 6 Differential stress and strain in the y direction for fixed and free boundary conditions ($n=5$)

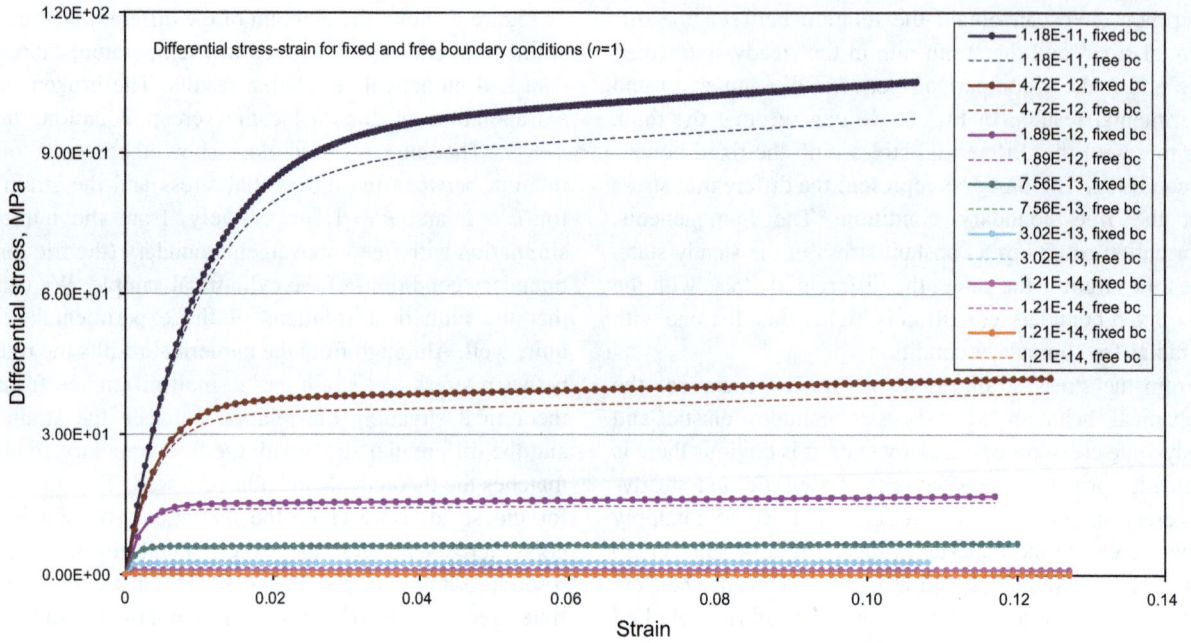

Fig. 7 Differential stress and strain in the *y* direction for fixed and free boundary conditions (*n*=1)

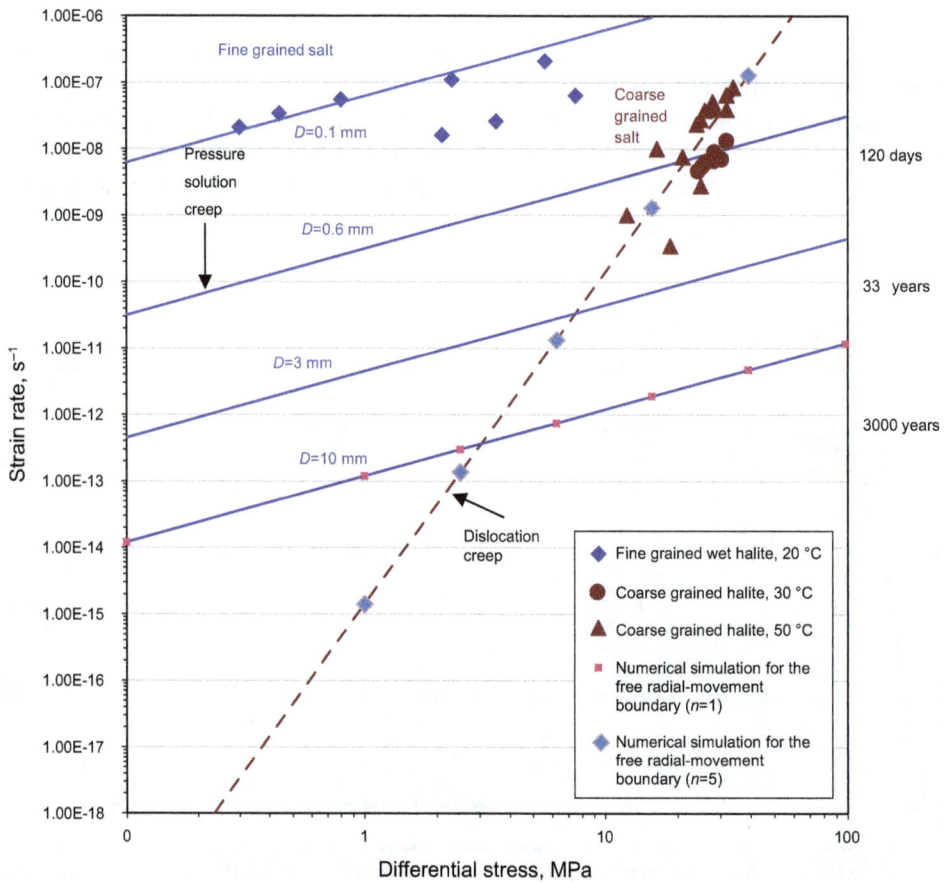

Fig. 8 Differential stress versus strain rate diagram. *Solid lines* are the pressure solution creep for fine and coarse grain sizes at room temperature. The *broken line* is the extrapolation of the dislocation creep. The *red* and *blue dots* represent numerical results with free radial-movement boundary condition (modified from Urai and Spiers 2007)

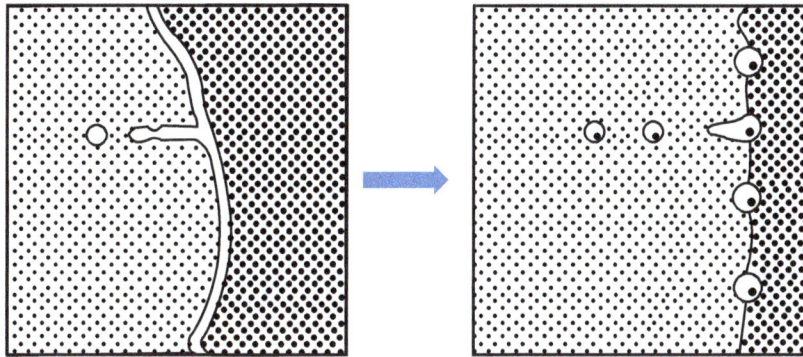

Fig. 9 Rock salt grain boundaries will tend to heal with a decrease in differential stress (Drury and Urai 1990)

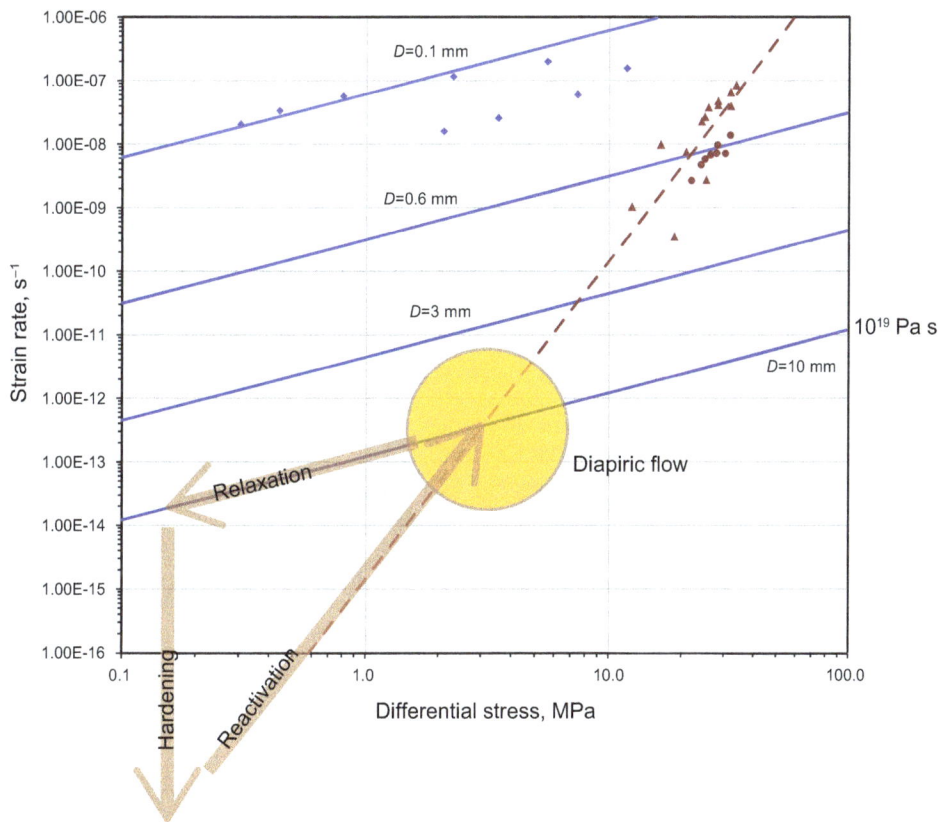

Fig. 10 Two deformation mechanisms at differential stresses around and below 2 MPa (Urai et al. 2008)

deformation in nature in order to observe detailed information for deformation mechanisms (Schléder and Urai 2005; Schléder et al. 2008). For creep of salt under high differential stresses, for example, 2.0 MPa, which is in agreement with many studies which used subgrain size piezometry on salt deformed by a combination of dislocation creep and pressure solution creep. The effective viscosity is a permissible model, because the dynamic recrystallization tends to bring the contributions of dislocation creep and pressure solution creep into balance and grain boundaries are mobile (De Bresser et al. 2001). However, the observation of rock salt deformation in the

natural laboratory through microstructural research provides the important information for the differential stress smaller than 2.0 MPa. Under this differential stress, the dynamic recrystallization changes the grain size and it contributes to the balance between the dislocation and pressure solution creep controlling rock salt deformation. If the differential stress is much lower, grain boundaries tend to heal or neck driven by interfacial energy, then the grain boundaries immobile (Fig. 9). Pressure solution creep is switched off, and deformation can only proceed by dislocation creep, at dramatically lower strain rates. In other words, the effective viscosity under these low differential stress values is many orders of magnitude higher. A more detailed recent review of these issues is Urai et al. (2008). In the initial stage of salt tectonics, the differential stresses in salt are very low, and recent studies suggest that under these conditions pressure solution creep in the salt should be inactive and salt deforms under the control of dislocation creep ($n = 5$). A related issue is raised by the observation that anhydrite–dolomite stringers do not sink through salt over long geologic periods (van Gent et al. 2011; Burchardt et al. 2011). In conclusion, the strain rates are similar at a differential stress around 2 MPa during salt tectonics under the controls of both dislocation creep and pressure solution creep. However, the pressure solution creep is inactive and switched off at lower stress below 2 MPa after salt tectonics stops. During and after salt tectonics, dislocation creep is always active and dominant for the rock salt deformation (Fig. 10).

6 Conclusions

The creep equations controlled by microstructure deformation mechanisms are the foundation of modeling rock salt creep properties under long-term conditions, and they are useful for salt tectonics modeling and salt mining engineering. In this paper, we summarized the research findings on creep controlled by two deformation mechanisms (dislocation creep and solution-precipitation creep) and built a database for rock salt rheology at 20–200 °C. Various rock salt has different water contents in the pore and grain boundaries, which strongly affects the rheological behavior. The temperature also plays an important role in rock salt deformation. Moreover, we modeled creep behavior of rock samples at various strain rates under the control of two deformation mechanisms. The numerical model in the Abaqus package can simulate the real triaxial experiment and can be used for salt tectonics modeling. The numerical simulation can also evaluate the effect of boundary conditions on the creep property. Finally, both deformation mechanisms lead to a similar strain rate during salt tectonics, and the pressure solution creep is inactive at a low stress after salt tectonics. Combined with experimental results in laboratories and microstructure research as the reliable exploration of experimental results, we conclude that rock salt rheology can be simplified to dislocation creep corresponding to power law creep ($n = 5$) with the appropriate material parameters in the salt tectonic modeling.

Acknowledgments We would like to thank Steffen Abe, Lars Reuning and Frank Strozyk for their contributions to the work. We also thank RWTH Aachen University and China University of Petroleum for the support of the work and Prof. Zhang Guangqing and the Department of Engineering Mechanics. The research is funded by the startup project of China University of Petroleum, Beijing (No. 2462014YJRC041) and supported by Science Foundation of China University of Petroleum, Beijing (No. C201601).

References

Aubertin M, Gill DE, Ladanyi B. An internal variable model for the creep of rocksalt. Rock Mech Rock Eng. 1991;24(2):81–97. doi:10.1007/BF01032500.

Bérest P, Blum PA, Charpentier JP, et al. Very slow creep tests on rock samples. Int J Rock Mech Min Sci. 2005;42(4):569–76.

Brouard B, Bérest P. A tentative classification of salts according to their creep properties. In: Proceedings of SMRI spring meeting, New Orleans; 1998. pp. 18–38.

Burchardt S, Koyi H, Schmeling H. Strain pattern within and around denser blocks sinking within Newtonian salt. J Struct Geol. 2011;33(2):145–53. doi:10.1016/j.jsg.2010.11.007.

Carter NL, Hansen FD. Creep of rocksalt. Tectonophysics. 1983;92(4):275–333. doi:10.1016/0040-1951(83)90200-7.

Carter NL, Horseman ST, Russell JE, Handin J. Rheology of rocksalt. J Struct Geol. 1993;15(9–10):1257–71. doi:10.1016/0191-8141(93)90168-A.

Chemia Z, Koyi H, Schmeling H. Numerical modeling of rise and fall of a dense layer in salt diapirs. Geophys J Int. 2008;172:798–816. doi:10.1111/j.1365-246X.2007.03661.x.

Chemia Z, Koyi H. The control of salt supply on entrainment of an anhydrite layer within a salt diapir. J Struct Geol. 2008;30(9):1192–200. doi:10.1016/j.jsg.2008.06.004.

Chemia Z, Schmeling H. The effect of the salt viscosity on future evolution of the Gorleben salt diapir, Germany. Tectonophysics. 2009;473(3–4):446–56. doi:10.1016/j.tecto.2009.03.027.

Cristescu ND, Hunsche U. Time effects in rock mechanics. New York: Wiley; 1988. p. 342.

Cristescu ND. A general constitutive equation for transient and stationary creep of rocksalt. Int J Rock Mech Min Sci Geomech Abstr. 1993;30(2):125–40.

Cristescu ND. Evolution of damage in rocksalt. In: Mechanical behavior of salt. Trans Tech Publications, Series on Rock and Soil Mechanics, Clausthal-Zellerfeld. 1998; 22:131–42.

De Bresser JHP, Ter Heege JH, Spiers CJ. Grain size reduction by dynamic recrystallization: can it result in major rheological weakening? Int J Earth Sci. 2001;90:28–45. doi:10.1007/s005310000149.

De Meer S, Spiers CJ, Peach CJ, et al. Diffusive properties of fluid-filled grain boundaries measured electrically during active pressure solution. Earth Planet Sci Lett. 2002;200(1–2):147–57. doi:10.1016/S0012-821X(02)00585-X.

Deng JG. Calculation method of mud density to control borehole closure. Chin J Rock Mech Eng. 1997;16(6):522–8 (**in Chinese**).

DeVries KL. Viscoplastic laws for Avery Island salt. Report for Stone Webster, RSI-0333, RE/SPEC, Inc., Rapid City, S.D. 1988.

Drury M, Urai J. Deformation-related recrystallization processes. Tectonophysics. 1990;172:235–53. doi:10.1016/0040-1951(90)90033-5.

Franssen RCMW. The rheology of synthetic rocksalt in uniaxial compression. Tectonophysics. 1994;233(1–2):1–40. doi:10.1016/0040-1951(94)90218-6.

Fossum AF, Fredrich JT. Salt mechanics primer for near-salt and sub-salt deepwater Gulf of Mexico Field Developments. Sandia report, SAND2002-2063. (2002).

Gemmer L, Ings SJ, Medvedev S, et al. Salt tectonics driven by differential sediment loading: stability analysis and finite-element experiments. Basin Res. 2004;16(2):199–218. doi:10.1111/j.1365-2117.2004.00229.x.

Hampel A, Hunsche U, Weidinger P, et al. Description of the creep of rock salt with the composite model—II. steady state creep. In:

Aubertin M, Hardy H Jr., editors. The mechanical behavior of salt IV. Trans Tech Publications, Clausthal. 1998. pp. 287–99.

Hampel A, Schulze O. The composite dilatancy model: a constitutive model for the mechanical behavior of rock salt. In: Proceeding of 6th conference of the mechanical behavior of salt understanding of THMC processes in salt; 2007. pp. 99–107.

Hansen FD. Creep behavior of bedded salt from Southeastern New Mexico at elevated temperatures. Report SAND 9-7030. Sandia National Laboratories. 1979.

Hansen FD, Carter NL. Creep of Avery Island rocksalt. In: Proceedings of the first conference on mechanical behavior of salt, Clausthal-Zellerfeld, Germany. Trans Tech Publications; 1984. pp. 53–69.

Haupt M, Schweiger HF. Development of a constitutive model for rock salt based on creep and relaxation tests. Comput Geotech. 1989;7(4):346.

Heard HC. Steady-state flow in polycrystalline halite at pressure of 2 kilobars. In: Heard HC, Borg IY, Carter NL, Raleigh CB, editors. Flow and fracture of rocks. AGU Geophysical Monograph Series. 1972;16:191–210. doi:10.1029/GM016p0191.

Heard HC, Ryerson FJ. Effect of cation impurities on steady-state flow of salt. In: Hobbs BE, Heard HC, editors. Mineral and rock deformation: laboratory studies. AGU Geophysical Monograph Series. 1986;36:99–115. doi:10.1029/GM036p0099.

Hunsche U, Hampel A. Rock salt—the mechanical properties of the host rock material for a radioactive waste repository. Eng Geol. 1999;52(3–4):271–91. doi:10.1016/S0013-7952(99)00011-3.

Hunsche U, Schulze O, Walter F, et al. Projekt Gorleben: Thermo-mechanisches Verhalten von Salzgestein. 2003. (in German).

Ings SJ, Beaumont C. Shortening viscous pressure ridges, a solution to the enigma of initiating salt 'withdrawal' minibasins. Geology. 2010;38(4):339–42. doi:10.1130/G30520.1.

Jin J, Cristescu ND. An elastic/viscoplastic model for transient creep of rock salt. Int J Plast. 1998;14(1–3):85–107. doi:10.1016/S0749-6419(97)00042-9.

Koyi H. Salt flow by aggrading and prograding overburdens. In: Alsop I, Blundell D, Davison I, editors. Salt tectonics. Geological Society, London, Special Publications. 1996;100(1):243–58.

Koyi H. The shaping of salt diapirs. J Struct Geol. 1998;20(4):321–38. doi:10.1016/S0191-8141(97)00092-8.

Koyi H. Modeling the influence of sinking anhydrite blocks on salt diapirs targeted for hazardous waste disposal. Geology. 2001;29(5):387–90. doi:10.1130/0091-7613(2001).

Li S, Abe S, Reuning L, Becker S, Urai JL, Kukla PA. Numerical modeling of the displacement and deformation of embedded rock bodies during salt tectonics—a case study from the South Oman Salt Basin. In: Alsop I, editors. Salt tectonics, sediments and prospectivity. Geological Society Special Publication. 2012a;363:503–20. doi:10.1144/SP363.24.

Li S, Strozyk F, Abe S, van Gent H, Kukla P, Urai JL. A method to evaluate long-term rheology of Zechstein Salt in the Tertiary. In: Berest P, Ghoreychi M, Hadj-Hassen F, Tijani M, editors. Mechanical behavior of salt VII. Taylor & Francis Group, London; 2012b. pp. 215–20.

Li SQ, Feng L, Tang PC, Rao G, Bao YH. Calculation of depth to detachment and its significance in the Kuqa Depression: a discussion. Pet Sci. 2009;6:17–20. doi:10.1007/s12182-009-0003-2.

Liang WG, Zhao YS. Investigation on carbon dioxide geological sequestration in salt caverns. Chin J Undergr Space Eng. 2007;z2:1545–50 (in Chinese).

Martin B, Röller K, Stoeckhert B. Low-stress pressure solution experiments on halite single-crystals. Tectonophysics. 1999;308(3):299–310. doi:10.1016/S0040-1951(99)00112-2.

Munson DE. Constitutive model for the low temperature creep of salt (with application to WIPP). Technical report, SAND79-1853. Sandia National Laboratories, Albuquerque, NM. 1979.

Munson DE. Constitutive model of creep in rock salt applied to underground room closure. Int J Rock Mech Min Sci. 1997;34(2):233–47. doi:10.1016/S0148-9062(96)00047-2.

Ottosen NS. Viscoelastic-viscoplastic formulas for analysis of cavities in rock salt. Int J Rock Mech Min Sci Geomech Abstr. 1986;23(3):201–12. doi:10.1016/0148-9062(86)90966-6.

Peach CJ, Spiers CJ. Influence of crystal plastic deformation on dilatancy and permeability development in synthetic salt rock. Tectonophysics. 1996;256(1–4):101–28. doi:10.1016/0040-1951(95)00170-0.

Peach CJ, Spiers CJ, Trimby PW. Effect of confining pressure on dilatation, recrystallization, and flow of rock salt at 150 °C. J Geophys Res. 2001;106:13315–28. doi:10.1029/2000JB900300.

Poliakov ANB, Podladchikov Y, Talbot C. Initiation of salt diapirs with frictional overburdens: numerical experiments. Tectonophysics. 1993;228:99–210. doi:10.1016/0040-1951(93)90341-G.

Renard F, Dysthe D, Feder J, Jamtveit B. Healing of fluid-filled microcracks. In: Proceedings of the second Biot conference on poromechanics. 2002; pp. 925–31.

Renard F, Bernard D, Thibault X, Boller E, et al. Synchrotron 3D microtomography of halite aggregates during experimental pressure solution creep and evolution of the permeability. Geophys Res Lett. 2004;31:L07607. doi:10.1029/2004GL019605.

Rutter EH. Pressure solution in nature, theory and experiment. J Geol Soc. 1983;140:725–40. doi:10.1144/gsjgs.140.5.0725.

Schenk O, Urai JL. Microstructural evolution and grain boundary structure during static recrystallization in synthetic polycrystals of sodium chloride containing saturated brine. Contrib Miner Pet. 2004;146:671–82. doi:10.1007/s00410-003-0522-6.

Schenk O, Urai JL, Piazolo S. Structure of grain boundaries in wet, synthetic polycrystalline, statically recrystallizing halite-evidence from cryo-SEM observations. Geofluids. 2006;6(1):93–104. doi:10.1111/j.1468-8123.2006.00134.x.

Schléder Z, Urai JL. Microstructural evolution of deformation-modified primary halite from Hengelo, The Netherlands. Int J Earth Sci. 2005;94(5–6):941–56. doi:10.1007/s00410-003-0522-6.

Schléder Z, Urai JL, Nollet S, Hilgers C. Solution-precipitation creep and fluid flow in halite: a case study on Zechstein (Z1) rocksalt from Neuhof salt mine (Germany). Int J Earth Sci. 2008;97(5):1045–56. doi:10.1007/s00531-007-0275-y.

Schoenherr J, Schléder Z, Urai JL, Fokker PA, Schulze O. Deformation mechanisms and rheology of Pre-Cambrian rocksalt from the South Oman Salt Basin. In: Proceedings of the 6th conference on the mechanical behavior of salt—understanding of THMC processes in salt. Hannover, Germany, 22–25 May 2007. pp. 167–73.

Schultz-Ela DD, Jackson MPA. Evolution of extensional fault systems linked with salt diapirism modeled with finite elements. AAPG Bull. 1993;77:179.

Senseny PE. Creep properties of four rock salts. In: Proceeding of the 2nd conference on mechanical behavior of salt. Trans. Tech. Pub; 1988. pp. 431–44.

Senseny PE, Hansen FD, Russell JE, Carter NL, Handin JW. Mechanical behaviour of rock salt: phenomenology and micromechanisms. Int J Rock Mech Min Sci Geomech Abstr. 1992;29(4):363–78.

Spiers CJ, Schutjens PMTM, Brzesowsky RH, Peach CJ, Liezenberg JL, Zwart HJ. Experimental determination of constitutive parameters governing creep of rocksalt by pressure solution. In: Knipe RJ, Rutter EH, editors. Deformation mechanisms, rheology and tectonics. Geological Society, London, Special

Publications. 1990;54(1):215–27. doi:10.1144/GSL.SP.1990.054.01.21.

Spiers CJ, Brzesowsky RH. Densification behaviour of wet granular salt: theory versus experiment. In: Kakihana H, Hardy HR, Jr. Hoshi T, Toyokura K, editors. Seventh symposium on salt. Elsevier Amsterdam, 1993;(1):83–92.

Spiers CJ, Carter NL. Microphysics of rocksalt flow in nature. In: Aubertin M, Hardy HR, editors. The mechanical behavior of salt proceedings of the 4th conference, Trans Tech. Publ. Series on Rock and Soil Mechanics, 1998;22:15–128.

Tang JP, Wang SQ, Chen M. Salt-gypsum drilling theory and application. Beijing: Petroleum Industry Press; 2004 (in Chinese).

Ter Heege JH, De Bresser JHP, Spiers CJ. Dynamic recrystallization of wet synthetic polycrystalline halite: dependence of grain size distribution on flow stress, temperature and strain. Tectonophysics. 2005a;396(1–2):35–57. doi:10.1016/j.tecto.2004.10.002.

Ter Heege JH, De Bresser JHP, Spiers CJ. Rheological behaviour of synthetic rocksalt: the interplay between water, dynamic recrystallization and deformation mechanisms. J Struct Geol. 2005b;27(6):948–63. doi:10.1016/j.jsg.2005.04.008.

Urai JL, Spiers CJ, Zwart HJ, Lister GS. Weakening of rock salt by water during long-term creep. Nature. 1986;324:554–7. doi:10.1038/324554a0.

Urai JL, Spiers CJ, Peach CJ, Franssen RCMW, Liezenberg JL. Deformation mechanisms operating in naturally deformed halite rocks as deduced from microstructural investigations. Geol Mijnbouw. 1987;66:165–76.

Urai JL, Spiers CJ. The effect of grain boundary water on deformation mechanisms and rheology of rocksalt during long-term deformation. In: Wallner M, Lux K, Minkley W, Hardy H Jr., editors. Proceeding of the 6th conference of mechanical behavior of salt understanding of THMC processes in salt; 2007. pp. 149–58.

Urai JL, Schléder Z, Spiers CJ, Kukla PA. Flow and transport properties of saltrocks. In: Littke R, Bayer U, Gajewski D, Nelskamp S, editors. Dynamics of complex intracontinental basins: The Central European Basin System. Berlin: Springer; 2008. p. 277–90.

van Gent H, Urai JL, De Keijzer M. The internal geometry of salt structures—a first look using 3D seismic data from the Zechstein of the Netherlands. J Struct Geol. 2011;33(2):292–311. doi:10.1016/j.jsg.2010.07.005.

van Keken PE, Spiers CJ, van den Berg AP, Muyzent EJ. The effective viscosity of rocksalt: implementation of steady-state creep laws in numerical models of salt diapirism. Tectono-physics. 1993;225(4):457–76. doi:10.1016/0040-1951(93)90310-G.

Wawersik WR. Determination of steady state creep rates and activation parameters for rock salt. In: Pineus HJ, Hoskins ER, editors. Measurement of rock properties at elevated pressures and temperatures, ASTM STP 869. Philadelphis: American Society for Testing and Materials; 1985. pp. 72–92.

Wawersik WR, Hannum DW. Mechanical behavior of New Mexico rock salt in triaxial compression up to 200 °C. J Geophys Res. 1980;85:891–900. doi:10.1029/JB085iB02p00891.

Wawersik WR, Zeuch DH. Creep and creep modeling of three domal salts—a comprehensive update. Technical report, SAND84-0568. Sandia National Laboratories, Albuquerque, NM (United States). 1984.

Wawersik WR, Zeuch DH. Modeling and mechanistic interpretation of creep of rocksalt below 200 °C. Tectonophysics. 1986;121(2–4):125–52. doi:10.1016/0040-1951(86)90040-5.

Wawersik WR, Zimmerer DJ. Triaxial creep measurements on rock salt from the Jennings Dome. Technical report. Louisiana, Borehole LA-1, Core #8*. Technical report, SAND94-1432. Sandia National Laboratories, Albuquerque, NM (United States). 1994.

Weidinger P, Hampel A, Blum W, Hunsche U. Creep behaviour of natural rock salt and its description with the composite model. Mater Sci Eng. 1997;234:646–8. doi:10.1016/S0921-5093(97)00316-X.

Yang CH, Daemen JJK, Yin JH. Experimental investigation of creep behavior of salt rock. Int J Rock Mech Min Sci. 1999;36:233–42. doi:10.1016/S0148-9062(98)00187-9.

Yang CH, Li YP, Chen F, Shi XL, Qu DA. Advances in researches of the mechanical behaviors of deep bedded salt rocks in China. In: Proceedings of the 43rd U.S. rock mechanics symposium and 4th U.S.—Canada rock mechanics symposium, Asheville. 2009.

Yang HL, Chen M, Zhang GQ. Deformation mechanisms and safe drilling fluids density in extremely thick, salt formations. Pet Sci. 2007;4(4):56–61. doi:10.1007/BF03187456.

Zhang GQ, Chen M, Zhao YB. Study on initiation and propagation mechanism of fractures in oriented perforation of new wells. Acta Pet Sin. 2008;29(1):116–9 (in Chinese).

Zhang GQ, Chen M. Complex fracture shapes in hydraulic fracturing with orientated perforations. Pet Explor Dev. 2009;36(1):103–7. doi:10.1016/S1876-3804(09)60113-0 (in Chinese).

Zhang GQ, Chen M. Dynamic fracture propagation in hydraulic re-fracturing. J Pet Sci Eng. 2010;70(3–4):266–72. doi:10.1016/j.petrol.2009.11.019.

11

Origin of dolomite in the Middle Ordovician peritidal platform carbonates in the northern Ordos Basin, western China

Xiao-Liang Bai[1,2] · Shao-Nan Zhang[1,2] · Qing-Yu Huang[3] · Xiao-Qi Ding[4] ·
Si-Yang Zhang[5]

Abstract The carbonates in the Middle Ordovician Ma$_5^5$ submember of the Majiagou Formation in the northern Ordos Basin are partially to completely dolomitized. Two types of replacive dolomite are distinguished: (1) type 1 dolomite, which is primarily characterized by microcrystalline (<30 μm), euhedral to subhedral dolomite crystals, and is generally laminated and associated with gypsum-bearing microcrystalline dolomite, and (2) type 2 dolomite, which is composed primarily of finely crystalline (30–100 μm), regular crystal plane, euhedral to subhedral dolomite. The type 2 dolomite crystals are truncated by stylolites, indicating that the type 2 dolomite most likely predated or developed simultaneously with the formation of the stylolites. Stratigraphic, petrographic, and geochemical data indicate that the type 1 dolomite formed from near-surface, low-temperature, and slightly evaporated seawater and that the dolomitizing fluids may have been driven by density differences and elevation-related hydraulic head. The absence of massive depositional evaporites in the dolomitized intervals indicates that dolomitization was driven by the reflux of slightly evaporated seawater. The $\delta^{18}O$ values (-7.5 to -6.1 ‰) of type 1 dolomite are slightly lower than those of seawater-derived dolomite, suggesting that the dolomite may be related to the recrystallization of dolomite at higher temperatures during burial. The type 2 dolomite has lower $\delta^{18}O$ values (-8.5 to -6.7 ‰) and Sr^{2+} concentration and slightly higher Na^+, Fe^{2+}, and Mn^{2+} concentrations and $^{87}Sr/^{86}Sr$ ratios (0.709188–0.709485) than type 1 dolomite, suggesting that the type 2 dolomite precipitated from modified seawater and dolomitic fluids in pore water and that it developed at slightly higher temperatures as a result of shallow burial.

Keywords Carbonate platform · Dolomitization · Dolomite · Middle Ordovician · Ma$_5^5$ submember · Dolomitizing fluids

✉ Shao-Nan Zhang
zsn@cdut.edu.cn

[1] State Key Laboratory of Oil and Gas Reservoir Geology and Exploitation, Southwest Petroleum University, Chengdu 610500, Sichuan, China

[2] School of Geoscience and Technology, Southwest Petroleum University, Chengdu 610500, Sichuan, China

[3] Research Institute of Petroleum Exploration & Development, PetroChina, Beijing 100083, China

[4] College of Energy, Chengdu University of Technology, Chengdu 610059, Sichuan, China

[5] Department of Geology, University of Regina, Regina, SK S4S 0A2, Canada

Edited by Jie Hao

1 Introduction

The origin of dolomitization has long been a subject of discussion (Warren 2000; Machel 2004; Gregg et al. 2015). Various models have been proposed to explain the origin of dolomite in carbonate platforms (Warren 2000; Machel 2004; Swart 2015), including regional subsurface flow models (sometimes referred to as burial-flow models) at elevated temperatures (Jones and Rostron 2000) and "early" synsedimentary models, such as those involving seepage reflux (Adams and Rhodes 1960). These models have been used to explain the massive dolomite in syndepositional evaporites (Jones and Rostron 2000; Qing et al. 2001) or in seawater with elevated salinity (Rott and Qing 2013; Read et al. 2012; Rivers et al. 2012). However, the development of small-scale, discrete dolomites in platform carbonates in association with elevated salinity seawater and little gypsum is less well documented. This study focuses on the early, pervasive dolomitization of shallow marine platform carbonates by penesaline seawater related to sea-level changes and the subsequent recrystallization of this early-formed dolomite during burial. This study then interprets the discrete, regional-scale distribution of dolomite that may have originated from modified seawater with high-frequency sequence cyclic changes and the precipitation of dolomitic fluids during burial.

The carbonates in the Middle Ordovician Ma_5^5 submember of the Majiagou Formation in the northern Ordos Basin are partially to completely dolomitized and are an attractive target for hydrocarbon exploration. In particular, the discovery of gas reservoirs in the dolomitic strata in the Jingbian Gas Field was an important finding involving the Ma_5^5 submember, which displays good natural gas exploration prospects among the old dolomite strata of the Ordos Basin. Tests in well Su-203 resulted in a large gas flow rate of over 104×10^4 m^3/d from the dolomite reservoirs (Yang and Bao 2011; Zhao et al. 2014), and horizontal well PG3 had a gas flow rate of over 10×10^4 m^3/d from the dolomite reservoir of the Ma_5^5 submember. Since then, exploration and development of the dolomite gas pools in the Ma_5^5 submember have been the key goal.

The dolomite in the Ma_5^5 submember is characterized by an uneven and discontinuous distribution of dolomitic carbonates and is classified as a low-seepage lithologic reservoir (Yang et al. 2006; Fang et al. 2009; Zhao et al. 2014), which is attributed to depositional facies and diagenetic alteration. Dolomitization can greatly influence porosity and permeability as limestone is replaced (Warren 2000; Zhang et al. 2010); however, the heterogeneity of the rock is closely related to the distribution and origin of the dolomite. Previous studies of the Ma_5^5 submember in the study area focused on its depositional environments and stratigraphic distribution (Wang et al. 2014). Additionally, many scholars have studied the origin of dolomite in the Ma_5 member in the Ordos Basin, and various models of its diagenesis have been proposed, including those involving mixed water zones and dolomitization (Zhao et al. 2005), evaporative reflux (Liu et al. 2011; He et al. 2014), burial dolomitization (Wang et al. 2009; Huang et al. 2010; Su et al. 2011), hydrothermal dolomitization (Huang et al. 2010; Wang et al. 2015), and microbial dolomitization (Fu et al. 2011). However, most of these studies have not provided an integrated stratigraphic, petrographic, or geochemical framework. This study focuses on the dolomitization of the carbonate deposits in the Ma_5^5 submember using petrography, stratigraphy, and geochemistry, and assesses the implications for hydrocarbon resource exploration.

2 Geologic setting

The Ordos Basin is a large-scale, multicycle, stable craton in northwestern China. The basement of the Ordos platform is composed of Precambrian crystalline schist, gneiss, and marble. In the Paleozoic Era, the Ordos Basin experienced steady uplift and subsidence with weak tectonic activity in the interior of the platform. During this period, particularly during the Ordovician, the depositional paleogeomorphology was generally higher in the north and middle and lower in the east, west, and south, and the Ordos Basin developed a gently inclined carbonate platform in a very shallow epicontinental sea (Wang et al. 2006). The Early Paleozoic Ordos platform developed shallow water carbonate deposits with thicknesses of 400–1600 m (Li and Zheng 2004). In the Late Ordovician, the oceanic crust near the southern and northern portions of the North China block began to be subducted, which led to the uplift of the North China block as a whole and a hiatus in deposition during the Silurian, Devonian, and Late Cretaceous that spanned more than 130 million years (Yang et al. 2006). The Daniudi area is located on a gentle monoclonal slope called the Yishan slope in the northern Ordos Basin (Fig. 1). This area is bounded by the Yimeng uplift to the north and the Jingxi fault-fold zone to the east and covers an area of approximately 2600 km^2. Its structure is a low uplift trending approximately NE-SW, with a structural crest located north of Yulin (Fig. 1).

A comprehensive summary of the stratigraphic units and their distribution in the carbonate platform is illustrated in Fig. 2a, and related information can also be found in it (Lei et al. 2010). The Ordovician carbonate rocks are divided into the following formations: Yeli, Liangjiashan, Majiagou, Pingliang, and Beiguoshan. The focus of this study is the Majiagou Formation, which is widely distributed and is subdivided into six members. The lithology in the Majiagou Formation consists of two parts: one is composed of gray muddy limestone, gypsum-bearing dolomite, and argillaceous dolomite (found in the Member 1, Member 3, and Member 5), and the other is composed of massive muddy limestone, moderately to thickly bedded muddy dolomite, and chert-band-bearing dolomite (found in the Member 2, Member 4, and Member 6). These changes in sedimentary lithology in the Majiagou Formation generally represent the carbonate rock formations of the Middle Ordovician sedimentary sequence, and the sequence can be divided into three secondary transgression–regression cycles. The Middle Ordovician carbonate rock in the Member 5 of the Majiagou Formation developed in a restricted, shallow, and hypersaline environment. The carbonates were deposited as a set of dolomite, banded

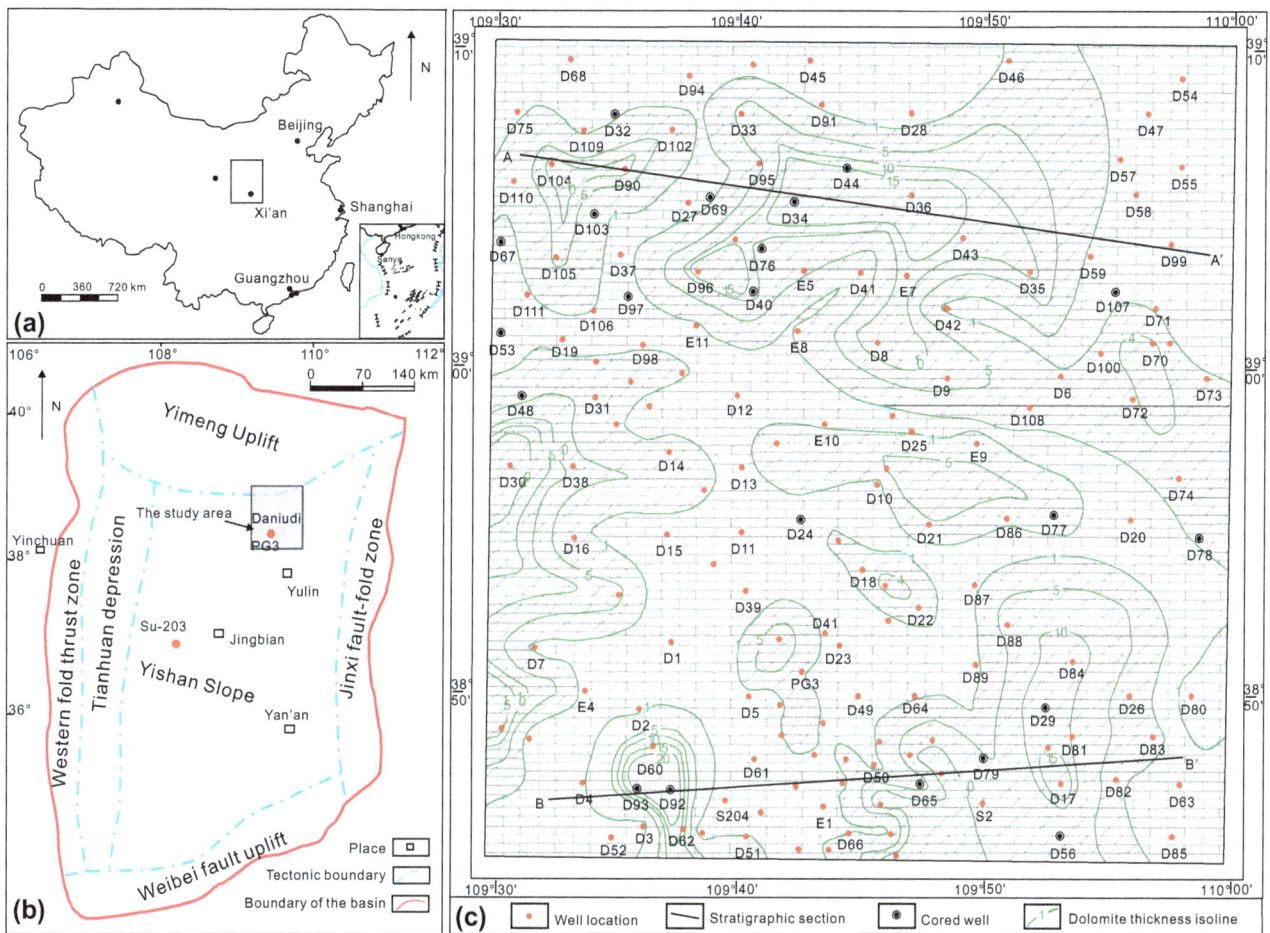

Fig. 1 **a** The Ordos Basin and study areas. **b** Sketch map of the Ordos Basin showing the location of the study area. **c** Detailed map showing the study area and related stratigraphic distribution, locations of wells, and dolomite distribution

limestone, and evaporated and represented the development of the regression cycles (Hu et al. 2014) (Fig. 2b).

The sedimentary facies has been studied by Feng and Bao (1999), who thought that the Ma_5^5 submember consists of restricted, shallow marine peritidal carbonate facies and that the depositional environment includes supratidal mud flats, intertidal flats, and shallow subtidal shoals (Zhou et al. 2011). The lithology of the Ma_5^5 submember is associated with a rapid transgression and a slow regression (Fig. 2b), and high-frequency meter-scale cycles are indicative of a restricted platform facies with subtidal and intertidal–supratidal sediment succession. The following lithofacies have been identified (Fig. 3): (1) columnar to hummocky stromatolites, peloidal mudstone, packstone, and wackestone; (2) thin-bedded to massive wackestone, rare grainstone/packstone, and laminated mudstone with fine anhydrite; and (3) bioturbated, thin-bedded wackestones/mudstones with scarce fossils and finely laminated mudstone with mottled anhydrite. Normal marine bioclasts and fossils are absent from most of the succession, but dasycladacean algae, microbial laminites, benthic

foraminifera, and gastropods, i.e., restricted-marine biota, occur in the Ma_5^5 submember limestones (Fig. 3). The restricted peritidal carbonate successions lack massive evaporitic minerals (such as gypsum/anhydrite), suggesting that the Ma_5^5 submember limestones developed in a restricted evaporitic environment with slightly increased salinity, where the marine water only rarely achieved gypsum saturation and did not reach gypsum precipitation.

3 Methods

Approximately 650 samples and 21 cores from the Ma_5^5 submember in the Majiagou Formation from the Daniudi area were taken for facies, stratigraphic, and diagenetic studies. The cores were from wells at intervals of approximately 620 m.

Approximately 140 thin sections were stained with Alizarin Red S and potassium ferricyanide to distinguish calcite and dolomite. The samples were examined using

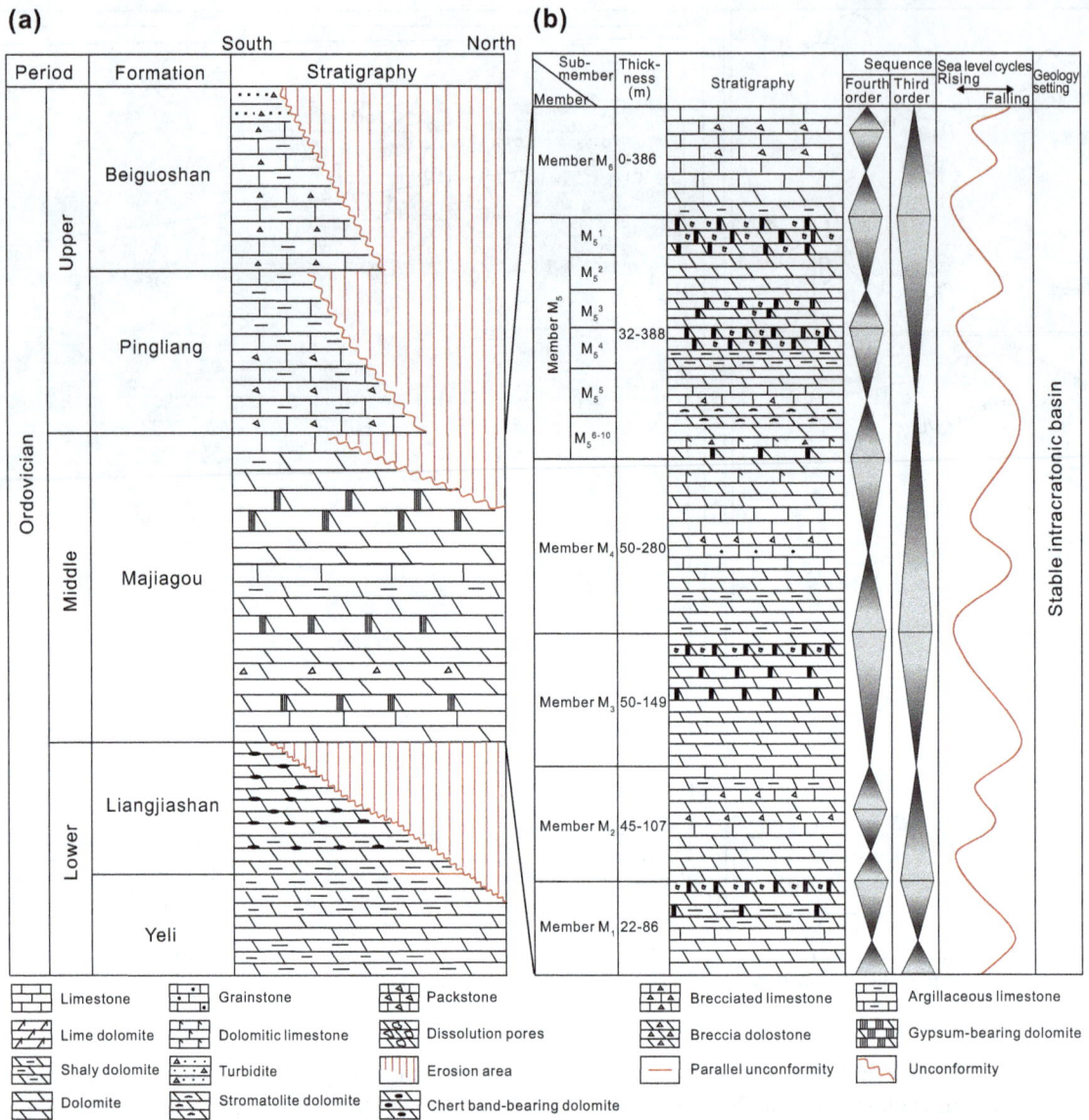

Fig. 2 Schematic diagram of the Ordovician stratigraphy and platform evolution in the Ordos Basin [Cited by Editorial Department of Chinese Stratigraphy of the Ordovician (1996)] showing the related sequence cycle changes and corresponding sea-level fluctuations in the study area

both normal and cathodoluminescent petrography with a Technosyn Cold Cathode Luminescence Model 8200 Mk II with a beam voltage of 17 kV and a current of 600 µA. Blue epoxy was used to make approximately 60 thin section casts to determine the porosity. More than 300 hand specimens were collected for further detailed studies.

Forty-seven samples were drilled out with a dental drill for stable C, O, and Sr isotopic analysis, which was performed at the Analytical Laboratory of Beijing Research Institute of Uranium Geology. The carbon and oxygen isotope values were measured using a Finnigan Kiel-III carbonate preparation device (Perkin-Elmer Inc., Wellesley, MA, USA) directly coupled to the inlet of a Finnigan MAT 253 isotope ratio mass spectrometer. The results are

reported in per mil notation relative to the Vienna Pee Dee Belemnite (VPDB) standard. The precision and calibration of the data were monitored through routine analysis of the National Bureau of Standards (NBS)-18 and NBS-19 carbonate standards. The precision was better than ± 0.1 ‰ internally for both the carbon and oxygen isotope values.

The $^{87}Sr/^{86}Sr$ isotope ratios of 18 samples were analyzed at the Chengdu University of Technology (CDUT, Chengdu). The Sr isotopic measurements were performed on a Finnigan MAT 261 instrument, and the errors associated with these analyses are reported as 2 sigma values. The maximum 2 sigma value for all of the matrix replacement Sr samples was ± 0.000064, with an average for all the samples of ± 0.000028. The NBS 987 standard,

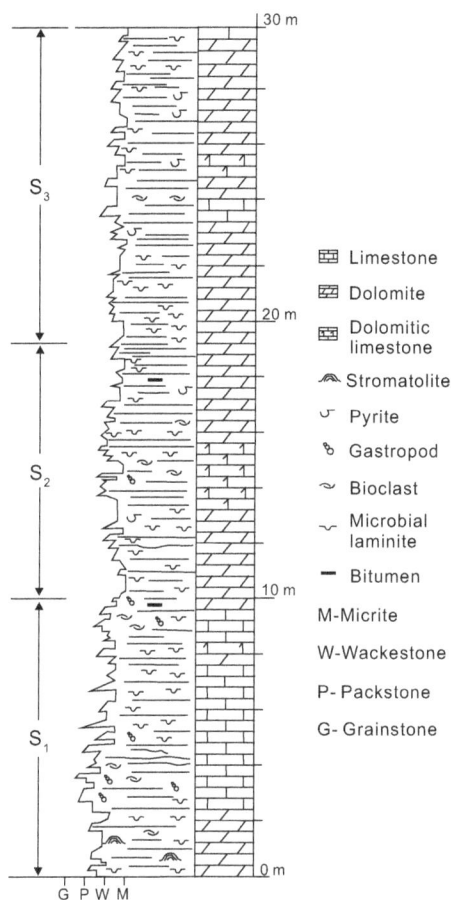

Fig. 3 Stratigraphic column of the platform carbonates in the study area. The dolomites are concentrated in the middle and upper Ma_5 strata and decrease volumetrically downward. S1–S3 are the depositional sequences

which has a value of 0.710250, yielded values between 0.710136 and 0.710312 (mean of 0.710253).

Major and trace element analyses were performed using a Perkin-Elmer Optima 3300 DVICP-Atomic Emission Spectrometer at the Chengdu University of Technology (CDUT, Chengdu). Multielemental high-purity solution standards were used for the calibration. An internal standard was used to correct the matrix differences. The analytical errors for Ca, Mg, Fe, Mn, Sr, and Na were ≤0.5 %, 0.8 %, 1.1 %, 1.1 %, 1.4 %, and 0.9 %, respectively.

4 Dolomite petrography and distribution

4.1 Dolomite petrography

The microcrystalline type 1 dolomite is characterized by euhedral to subhedral inclusion-rich dolomite (Fig. 5a). In the core samples, type 1 dolomite is generally present in algal laminae and displays preserved sedimentary fabric

(Fig. 6a, c). In thin section, type 1 dolomite is primarily characterized by microcrystalline (<10 μm), euhedral dolomite crystals associated with pyrite and anhydrite. Anhydrite was replaced by the precipitation of calcite (Fig. 5b), and some very finely crystalline (10–30 μm), subhedral dolomite crystals appear as bright rhombs and cut across the dolomite crystals (Fig. 5a). A few of the type 1 dolomites show algae laminae (Fig. 5c), such as algal stromatolite dolostone and algae laminae dolostone (Fig. 6a, c). The dolomite exhibits dull luminescence under CL examination.

The finely crystalline type 2 dolomite is the most abundant type of dolomite and consists of pervasively subhedral to euhedral dolomite crystals (Fig. 5d). In hand specimens, type 2 dolomite appears to have accumulated alongside stylolites and organic seams. The precursor sedimentary textures are partially to completely obliterated, and the type 2 dolomite displays visible light gray zones (Fig. 6b). In thin sections, the type 2 dolomite occurs primarily as finely crystalline (30–100 μm), euhedral to subhedral dolomite crystals that were truncated by stylolites, which indicates that the type 2 dolomite most likely predated or developed alongside the stylolites (Fig. 5d, e). This type of dolomite is commonly associated with abundant intercrystalline pores (Fig. 5d) and exhibits dark red to dull luminescence (Fig. 5f).

4.2 Dolomite distribution

The carbonate rock of the Ma_5^5 submember is extensively dolomitized in the study area. The interpretation of the logging data suggests that the dolomite displays extensive lateral and vertical variations (Fig. 4). The maximum dolomite thickness exceeds 20 m in the southern, western, and northern parts of the Daniudi area. In these locations, the microcrystalline and finely crystalline dolomites are well developed and thick (16–22 m). An examination of the cores indicates that there is a close spatial relationship between the abundances of the microcrystalline and the finely crystalline dolomites: the microcrystalline dolomite developed on top of finely crystalline dolomite, reaching substantial thicknesses, and the presence and thickness of type 2 dolomite is related to the abundance of type 1 dolomite. Type 1 dolomite is generally laminated and associated with gypsum-bearing microcrystalline dolomite that was deposited in a supratidal environment. Type 2 dolomite is primarily present as discrete, thickly bedded (6–22 m) bodies distributed unevenly in the middle–upper part of the Ma_5^5 submember in the carbonate platform (Fig. 4). In general, the abundance of dolomite decreases downward (Figs. 3, 4). Thin, interbedded micritic limestone/wackestone developed in the upper part and

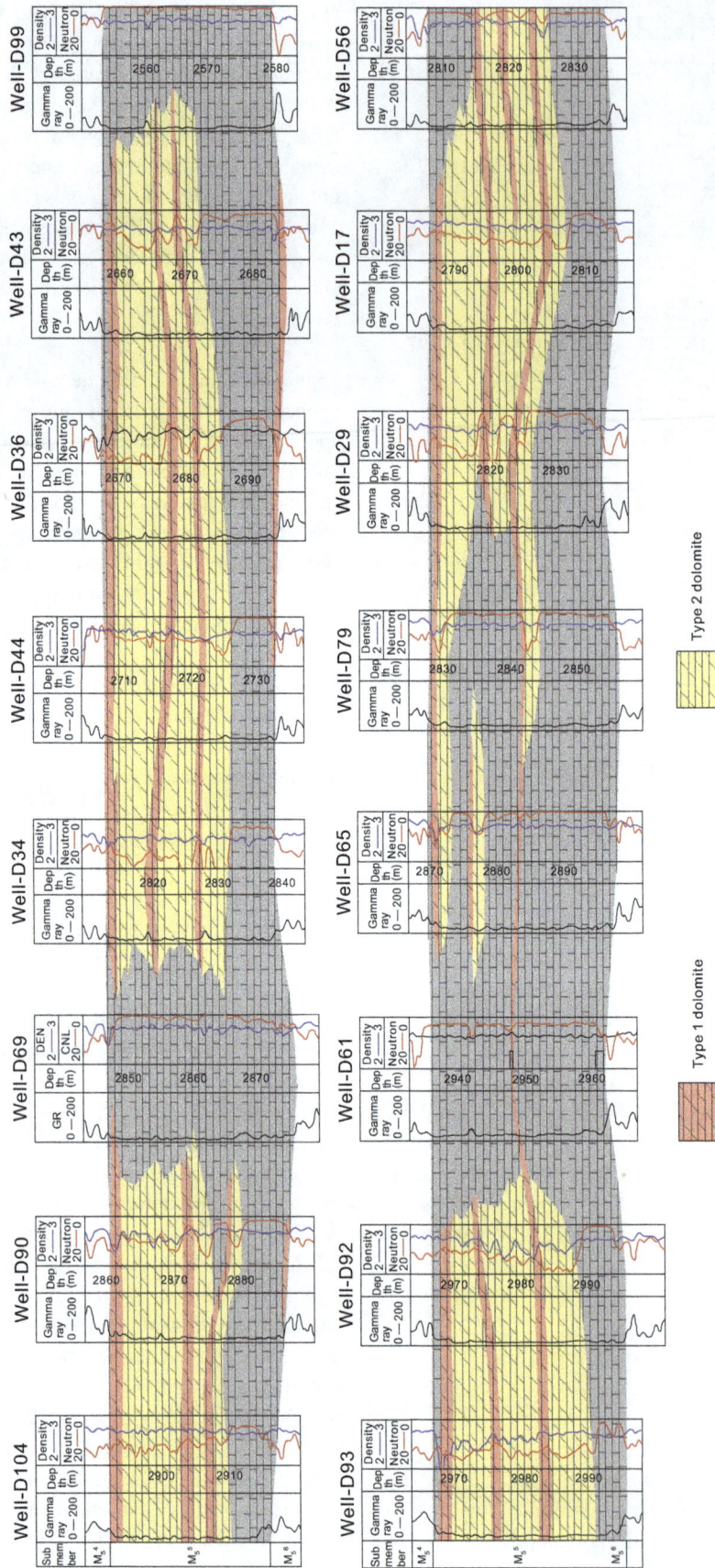

Fig. 4 Stratigraphic sections A–A′ and B–B′ showing the distribution of the dolomites. The type 1 dolomite is a set of thinly bedded, continuous, laminar cyclical sequences that overlie the type 2 dolomite which are discrete, unevenly distributed rocks in the middle–upper part of the Ma_5^5 submember that decrease downward. See Fig. 1 for the locations

packstone/wackestone developed in the lower part of the carbonate strata in the Ma_5^5 submember (Fig. 5g, h).

5 Geochemical results

5.1 Isotope characteristics

The stable isotope analysis results for oxygen and carbon are presented in Table 1 and Fig. 7, and the isotope analysis results for strontium are presented in Table 1 and Fig. 8. The $\delta^{18}O$ and $\delta^{13}C$ values of the Ma_5^5 submember limestone overlap with the estimated $\delta^{18}O$ and $\delta^{13}C$ ranges of the Middle Ordovician marine limestone (Qing and Veizer 1994; Veizer et al. 1999). Oxygen isotope difference between calcite and coevally precipitated dolomite is approximately 2.5 % (Swart and Melim 2000); thus, dolomites precipitated from normal Middle Ordovician seawater should have $\delta^{18}O$ values between -7.0 and -5.0 ‰ PDB (Fig. 7). The $^{87}Sr/^{86}Sr$ ratios of the Ma_5^5 submember limestone fall within the estimated $^{87}Sr/^{86}Sr$ range (0.7078–0.7093) of the Middle Ordovician marine carbonates (Qing et al. 1998; Veizer et al. 1999; McArthur et al. 2001; Davies and Smith 2006). These data indicate that the $^{87}Sr/^{86}Sr$ ratios and $\delta^{13}C$ and $\delta^{18}O$ values of the Ma_5^5 submember limestone can be used as a baseline for Middle Ordovician marine carbonate deposits.

The type 2 dolomite yields $\delta^{18}O$ values ranging from -8.5 to -6.7 ‰ (average -7.5 ‰), lower than those of the type 1 dolomite (Table 1). The $^{87}Sr/^{86}Sr$ ratios of the type 1 and type 2 dolomites are slightly higher than those of the limestone.

5.2 Major and trace elements

The results of the trace element concentration analyses are presented in Table 2. The average Sr, Na, Fe, and Mn concentrations in the Ma_5^5 submember limestone are 209, 274, 308, and 26 ppm, respectively (Table 2). The type 1 dolomite contains Ca^{2+} molar concentrations ranging from 54.3 % to 62.6 % (average 59.1 %), higher than those of the type 2 dolomite, which range from 50.8 % to 55.7 % (average 54.3 %). The type 2 dolomite has higher Fe and Mn concentrations and lower Sr and Na concentrations than the type 1 dolomite.

6 Discussion and interpretation

6.1 Petrographic implications

Type 1 dolomite is primarily characterized by microcrystalline (<30 µm), euhedral to subhedral dolomite crystals associated with anhydrite that has been replaced by the precipitation of calcite (Fig. 5b). It has been suggested that microcrystalline dolomite may form in evaporitic settings at relatively low temperatures (Gregg and Shelton 1990; Fu et al. 2006; Loyd and Corsetti 2010). The appearance of restricted-marine biota with laminar, distorted, green stromatolite algae indicates that the original structure of the limestone has been retained in the microcrystalline dolomite, which most likely originated from near-surface dolomitization in a restricted, shallow environment during early diagenesis (Fig. 5c). Type 1 dolomite pervasively replaces the matrix of all the facies, particularly those associated with the complete replacement of laminites. The textures of the type 1 dolomite and its close association with the restricted-marine deposits suggest that this type of dolomite most likely formed in a near-surface, low-temperature, saline environment with a high density of nucleation sites (Gregg and Shelton 1990). The very finely crystalline (10–30 µm), subhedral dolomite rhombs are interpreted as partial recrystallization during burial based on petrographic observations. The finely crystalline bright rhombs appear to cut across the subhedral dolomite crystals (Fig. 5a), suggesting that the recrystallization of very fine dolomite postdates the early penecontemporaneous dolomite.

Type 2 dolomite is present primarily as finely crystalline (30–100 µm), euhedral to subhedral dolomite crystals that preferentially replaced the limestone matrix. The type 2 dolomites are characterized by partial to complete replacement of muddy limestone/wackestone (Fig. 5g, h). This phenomenon suggests that the dolomitizing fluids were somewhat less supersaturated with respect to dolomite or insufficient for complete dolomitization. The dolomite commonly exhibits destruction of the fabric and indistinct original depositional features (Fig. 5d). The crystals in type 2 dolomite are truncated by the early stylolites, which indicates that the type 2 dolomite most likely predated or developed simultaneously with the formation of the stylolites (Fig. 5e). The early stylolites in the carbonate rocks may have developed at <300 m depth (Fabricius and Borre 2007), which indicates that the type 2 dolomite formed during a period of shallow burial.

6.2 Stratigraphic constraints

The type 1 dolomites in the Ma_5^5 submember have been shown to have formed in an evaporite-restricted condition based on their stratigraphic relationships. Type 1 dolomite is primarily represented by laminated, distorted green algae, and gypsum-bearing muddy dolomite originating in a supratidal environment (Figs. 5b, c, 6a), and its textural characteristics and geographic distribution may be correlated with the composite sea-level changes recorded in the

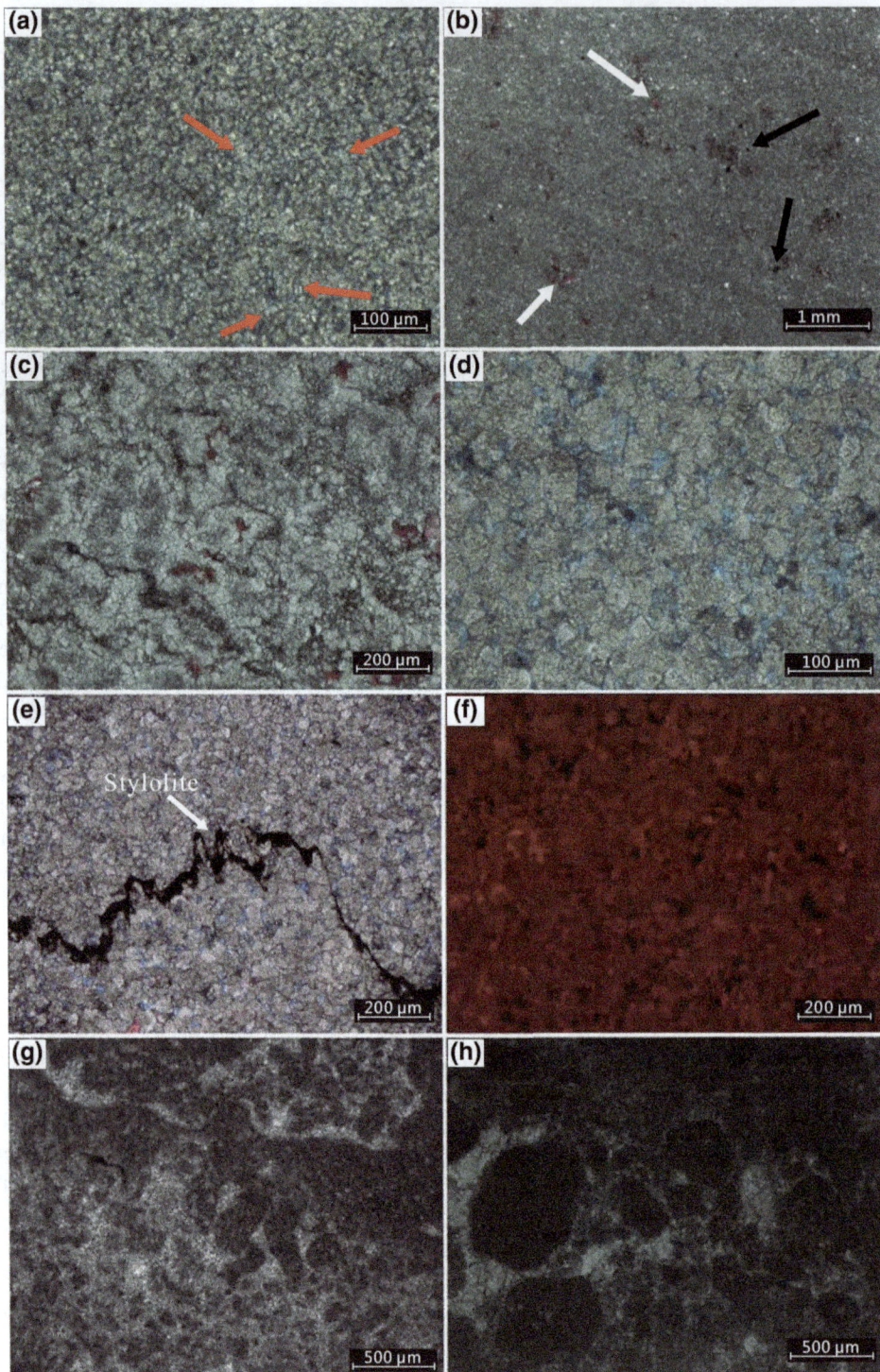

Fig. 5 Photomicrographs showing the petrographic characteristics of the dolomite. **a** Peloidal packstone–wackestone that developed in the lower part of the whole M_5^5 submember carbonate strata. **b** Peloidal grainstone that is cemented by fine equant calcite cement, which is related to early marine cementation. **c** Type 1 dolomite that is primarily characterized by microcrystalline (<30 μm), euhedral to subhedral dolomite crystals. **d** Layered type 1 dolomite that is associated with pyrite and anhydrite, and the anhydrite is replaced by the precipitation of calcite, though some retain the algal laminae structure. **e** Laminar algal dolomite that is characterized by crumpled textures and retains the original features and primary sedimentary structures. **f** Type 2 dolomite that is primarily present as finely crystalline (30–100 μm), regular crystal plane, euhedral to subhedral dolomite crystals with abundant intercrystalline pores; this type 2 dolomite commonly exhibits the destruction of the fabric and indistinct original depositional features as a result of diagenesis. **g** Type 2 dolomite that occurs primarily as finely crystalline, euhedral to subhedral dolomite crystals that were truncated by stylolites. This indicates that the Rd2 dolomite most likely predated or developed alongside the stylolites. **h** Type 2 dolomite that exhibits dull to dark red luminescence, with calcite cement pore fillings displaying orange luminescence

carbonate succession (Figs. 2b, 3). The Ma_5^5 submember deposits were largely replaced by type 1 dolomite on the shallow water carbonate platform, whereas the thinly bedded dolomite sequences (Fig. 4) and the lack of massive gypsum/anhydrite suggest that the type 1 dolomite formed from penesaline seawater in a near-surface environment, with magnesium ions provided by a combination of high- and low-frequency sea-level changes (Qing et al. 2001). The development of dolomite in a zone of mixed meteoric and marine water has been questioned and considered to be unlikely by many researchers (Hardie 1987; Melim et al. 2004). A mixed water process for dolomitization is not consistent with the type 1 dolomite, particularly because the dolomite strata in the Ma_5^5 submember are generally approximately horizontal rather than present as lenses (as in the mixing zone model) or related to unconformities.

The geometric configurations and distribution of the type 2 dolomite bodies are apparent constraints on the origin of the dolomite (Warren 2000). Type 2 dolomite is primarily present as discrete, unevenly distributed rocks in the middle–upper part of the Ma_5^5 submember (Fig. 4), and its distribution and abundance are closely related to those of the type 1 dolomite. Type 1 dolomite is present as a set of thinly bedded, continuous, laminar algal subtidal–intertidal cyclical sequences that overlie the type 2 dolomite and are associated with a downward decrease in the abundance of the type 2 dolomite (Figs. 3, 4). This association suggests that the dolomitizing fluids may have originated in the overlying strata. Another possibility is that type 2 dolomite developed as a result of the minor to extensive modification of type 1 dolomite. Studies suggest that early-formed dolomites in ancient carbonates commonly experience diagenetic modification (stabilization or recrystallization) (Rott and Qing 2013).

6.3 Implications of the geochemical data

The type 1 dolomite yields $\delta^{18}O$ values ranging from -7.5 to -6.1 ‰ (average -6.86 ‰), which are higher than the estimated values of the Middle Ordovician marine limestone -9.5 to -7.5 ‰) (Table 1; Fig. 7) and slightly lower than those of dolomite precipitated from Middle Ordovician seawater. Therefore, the dolomite may be related to dolomite recrystallization at higher temperatures during burial, but the replacement of the carbonate sediments by dolomitizing fluids took place in the early contemporaneous seawater. The $\delta^{13}C$ values and $^{87}Sr/^{86}Sr$ ratios in the type 1 dolomite are comparable with those in the corresponding limestone in the Ma_5^5 submember and were inherited from the precursor limestone (Fig. 7). The $\delta^{13}C$ values of carbonates are usually rock-buffered during diagenesis if no significant amount of organic CO_2 is involved

(Banner and Hanson 1990; Warren 2000). The appearance of algae laminae with rare gypsum pseudomorphs suggests that the type 1 dolomite formed in a near-surface, low-temperature, and penesaline environment.

The type 2 dolomite yields carbon isotopic values that overlap with the estimated values of the Middle Ordovician marine limestone in the Ma_5^5 submember, which is interpreted to reflect the carbon isotopic compositions of the precursor limestone and type 1 dolomite. However, the type 2 dolomite displays lower $\delta^{18}O$ values and higher $^{87}Sr/^{86}Sr$ ratios compared to the type 1 dolomite (Fig. 8). The more negative $\delta^{18}O$ values have two causes, namely, dolomitization at deeper burial depths and elevated temperatures or the participation of fresh water (Gregg and Shelton 1990; Durocher and Al-Aasm 1997). It is unlikely that meteoric water was responsible for the low $\delta^{18}O$ values in the type 2 dolomite based on its petrographic features and the lack of evidence of meteoric diagenesis. The shift to depleted $\delta^{18}O$ values in the type 2 dolomite was most likely due to recrystallization at elevated temperatures during burial. The higher $^{87}Sr/^{86}Sr$ ratios in the type 2 dolomite relative to the type 1 dolomite are most likely related to recrystallization in more radiogenic pore water with slightly elevated $^{87}Sr/^{86}Sr$ ratios.

The type 1 dolomites (54.3 %–62.6 %, average 59.1 % mol% $CaCO_3$) are typically nonstoichiometric, and the type 2 dolomites have more stoichiometric compositions (50.8 %–55.7 %, average 54.3 % mol% $CaCO_3$) than type 1 dolomites, indicating that these rocks underwent diagenetic modification (Table 2; Fig. 9). It is possible that the nearly stoichiometric compositions of the type 2 dolomites are due to the recrystallization of dolomites or dolomitization in pore water during burial. Studies have suggested that this stoichiometry is a function of both the chemistry and the duration of interaction with dolomitizing/stabilizing fluids (Sibley 1990). In addition, the Ca concentrations in syndepositional dolomites decrease with increasing crystal size (Warren 2000). These previous findings suggest that the initially Ca-rich type 1 dolomite formed near the surface during a period of rapid penecontemporaneous crystallization and that the type 2 dolomite formed via the recrystallization of dolomites or via dolomitization during burial. The Sr concentrations of the type 2 dolomites (69–124 ppm, average 92 ppm) are lower than those of the evaporitic marine type 1 dolomite (138–190 ppm, average 165 ppm). Recrystallization and dolomitization assist in stoichiometric enhancement and commonly lower the Sr concentrations in dolomite due to the low distribution coefficient of this element ($\ll 1$), a low Sr/Ca ratio in the solutions, and/or low reaction rates (Banner 1995) (Fig. 10).

The Na concentrations in the type 1 dolomite (387–559 ppm) are significantly higher than those in the

Table 1 Isotope data of the Ma_5^5 submember carbonate rocks in the Daniudi area

Well number	Lithology	Strata	Depth, m	$\delta^{18}O$, ‰ (PDB)	$\delta^{13}C$, ‰ (PDB)	Sr^{87}/Sr^{86} (±SE)
D92	Type 1 dolomite	Ma_5^5	2970.44	−6.7	−0.1	
D92	Type 1 dolomite	Ma_5^5	2985.45	−6.2	−0.1	
D56	Type 1 dolomite	Ma_5^5	2826.1	−6.6	−0.6	
D56	Type 1 dolomite	Ma_5^5	2825.4	−6.2	−0.3	
D92	Type 1 dolomite	Ma_5^5	2984.38	−6.4	−0.2	
D92	Type 1 dolomite	Ma_5^5	2983.97	−6.1	−0.1	
D65	Type 1 dolomite	Ma_5^5	2871.55	−7.2	−0.7	
D65	Type 1 dolomite	Ma_5^5	2871.7	−7.1	−0.8	
D56	Type 1 dolomite	Ma_5^5	2836.61	−6.8	−0.5	0.709242 ± 0.000064
D78	Type 1 dolomite	Ma_5^5	2637.96	−7.0	−0.4	0.709128 ± 0.000098
D92	Type 1 dolomite	Ma_5^5	2973.28	−7.5	−1.0	0.709326 ± 0.000052
D32	Type 1 dolomite	Ma_5^5	2881.82	−7.0	−0.6	
D32	Type 1 dolomite	Ma_5^5	2884.05	−7.1	−0.1	
D29	Type 1 dolomite	Ma_5^5	2834.55	−6.6	−0.8	0.709082 ± 0.000072
D53	Type 1 dolomite	Ma_5^5	3012.5	−7.2	−1.0	0.709224 ± 0.000065
D78	Type 1 dolomite	Ma_5^5	2638.56	−6.5	−0.6	0.709148 ± 0.000024
D78	Type 1 dolomite	Ma_5^5	2643.32	−6.6	−0.5	0.709168 ± 0.000012
D65	Type 2 dolomite	Ma_5^5	2873.44	−7.0	0.3	
D56	Type 2 dolomite	Ma_5^5	2842.67	−8.5	0.4	
D48	Type 2 dolomite	Ma_5^5	3009.53	−7.4	0.3	
D48	Type 2 dolomite	Ma_5^5	3013.65	−7.3	0	
D48	Type 2 dolomite	Ma_5^5	3017.32	−7.1	−0.6	
D48	Type 2 dolomite	Ma_5^5	3023.54	−8.1	0.1	
D44	Type 2 dolomite	Ma_5^5	2720.31	−7.7	0.1	0.709382 ± 0.000072
D44	Type 2 dolomite	Ma_5^5	2720.98	−7.5	0.1	0.709404 ± 0.000044
D44	Type 2 dolomite	Ma_5^5	2722.3	−7.9	−0.1	0.709485 ± 0.000054
D44	Type 2 dolomite	Ma_5^5	2723.07	−6.9	0.2	0.709188 ± 0.000072
D29	Type 2 dolomite	Ma_5^5	2819.74	−7.3	−0.6	0.709326 ± 0.000026
D48	Type 2 dolomite	Ma_5^5	3013.65	−7.2	−0.5	0.709254 ± 0.000090
D48	Type 2 dolomite	Ma_5^5	3014.88	−7.5	−0.9	
D78	Type 2 dolomite	Ma_5^5	2639.06	−7.9	−0.4	0.709308 ± 0.000034
D78	Type 2 dolomite	Ma_5^5	2639.44	−8	−0.3	
D78	Type 2 dolomite	Ma_5^5	2638.20	−7.7	−0.3	
D78	Type 2 dolomite	Ma_5^5	2638.68	−7.9	−0.2	
D65	Type 2 dolomite	Ma_5^5	2872.32	−7.5	−0.3	
D65	Type 2 dolomite	Ma_5^5	2872.97	−7.8	0.2	
D65	Type 2 dolomite	Ma_5^5	2873.45	−7.6	−0.6	
D65	Limestone	Ma_5^5	2873.82	−8.8	−1.3	
D53	Limestone	Ma_5^5	2986.14	−8.7	−0	
D53	Limestone	Ma_5^5	2987.97	−8.5	−1.0	
D53	Limestone	Ma_5^5	2994.5	−8.3	−0.4	
D53	Limestone	Ma_5^5	3003.7	−8.3	−0.1	
D29	Limestone	Ma_5^5	2831.5	−9.0	−1.5	0.708992 ± 0.000084
D29	Limestone	Ma_5^5	2821.95	−9.2	−0.9	0.709158 ± 0.000036
D56	Limestone	Ma_5^5	2831.61	−7.9	−0.1	0.708998 ± 0.000052
D101	Limestone	Ma_5^5	2777.00	−7.8	−1.4	0.709106 ± 0.000045
D92	Limestone	Ma_5^5	2990.55	−9.0	−2.1	

Table 2 Elemental concentrations of the Ma$_5^5$ submember carbonate rocks in the Daniudi area

Well number	Lithology	Ca, mol%	Ca, ppm	Mg, ppm	Na, ppm	Fe, ppm	Sr, ppm	Mn, ppm
D29	Limestone	99.2	434,600	2040	234	406	158	26
D29	Limestone	99.0	422,400	2640	292	420	152	18
D56	Limestone	99.0	353,900	2220	274	260	144	17
D101	Limestone	98.7	420,400	3300	297	145	171	42
D29	Type 1 dolomite	62.6	257,600	92,400	514	1565	138	50
D53	Type 1 dolomite	61.9	278,800	103,100	444	910	153	44
D56	Type 1 dolomite	56.5	245,800	113,400	387	1145	146	40
D78	Type 1 dolomite	58.9	245,900	102,800	455	876	152	42
D92	Type 1 dolomite	61.4	269,700	101,600	559	1685	169	48
D78	Type 1 dolomite	59.2	248,400	102,800	442	1325	190	44
D44	Type 1 dolomite	54.3	219,700	110,900	442	1270	188	56
D78	Type 1 dolomite	58.0	225,600	98,000	437	1837	181	53
D29	Type 2 dolomite	55.0	229,000	112,300	392	2237	94	52
D48	Type 2 dolomite	55.7	249,400	118,800	434	1948	69	79
D78	Type 2 dolomite	55.3	233,900	113,300	409	1973	104	54
D44	Type 2 dolomite	50.8	200,400	116,500	349	1682	92	62
D44	Type 2 dolomite	54.5	224,600	112,400	465	2042	124	70
D44	Type 2 dolomite	54.4	220,900	111,100	462	2491	69	80

Ma$_5^5$ submember limestone (average 274 ppm) and are similar to those in the type 2 dolomite (349–465 ppm), which is inferred to have formed from evaporitic or slightly modified seawater (Staudt et al. 1993). The Na concentrations in dolomite generally increase with the salinity of dolomitizing fluids and decrease with stoichiometric enhancement (Mazzullo 1992). However, the distribution coefficient of the Na concentrations in dolomite is not well known (Wheeler et al. 1989). The variable Na concentrations in dolomite may not reflect changes in the salinity of the dolomitizing fluids, and the Na concentrations during dolomitization are not useful indicators of the chemistry of the parent fluid (Wheeler et al. 1989; Fu et al. 2006).

The Fe and Mn concentrations in the type 2 dolomites are higher than those in the type 1 dolomites and the Ma$_5^5$ submember limestone, suggesting that the dolomitic fluids in the pore water were slightly reducing, incorporating Fe^{2+} and Mn^{2+} more easily into the crystal lattice. The high concentrations of Mn and Fe in the type 2 dolomites and the lower concentrations of Mn and Fe in the precursor limestone show that there were no significant sources of these elements, which suggests that the dolomitic fluids precipitated under slightly reducing conditions or that the dolomite formed in a late burial environment (Warren 2000; Franzolin et al. 2012) (Fig. 11).

6.4 Hydrology of the dolomitizing fluids

The hydrology of dolomitizing fluids is such that dolomitization requires both the active circulation of Mg-rich

fluids and favorable geochemical conditions to overcome kinetic barriers to dolomitization (McKenzie 1980). Dolomite associated with evaporites is commonly interpreted to be related to brine seepage reflux (Warren 2000; Fu et al. 2006). The type 1 dolomite is primarily represented by laminated, meter-scale, peritidal, cyclic, gypsum-bearing muddy dolomite deposited in a supratidal environment. It is suggested that the type 1 dolomite formed in marine conditions of slightly elevated salinity and that the dolomitization may have occurred in extensive platform-top environments associated with the repeated flooding and reflux of marine water with slightly higher salinity (Friedman and Sun 1995). The flux distribution varies by orders of magnitude along the flow path (Jones and Rostron 2000). The discontinuous (lengths of 40–50 km and widths of 20–30 km), meter-scale, stratiform dolomites were controlled by fourth- or fifth-order regressions associated with major storms during which seawater was transported tens of kilometers (Montañez and Read 1992). The stratigraphic distribution of the dolomite was also controlled by its position within third-order depositional sequences during the early deposition of the Ma$_5$ member (Fig. 2b). The regressive facies of the third-order sequences tends to be represented primarily by dolomite, whereas the transgressive facies retains abundant limestone (Montañez and Read 1992). The reflux model is based on the presence of transgressive facies in third-order depositional sequences with fourth- or fifth-order regressions (Fig. 2b). This regional-scale reflux is a viable mechanism for regional-scale dolomitization during shallow burial (<500 m)

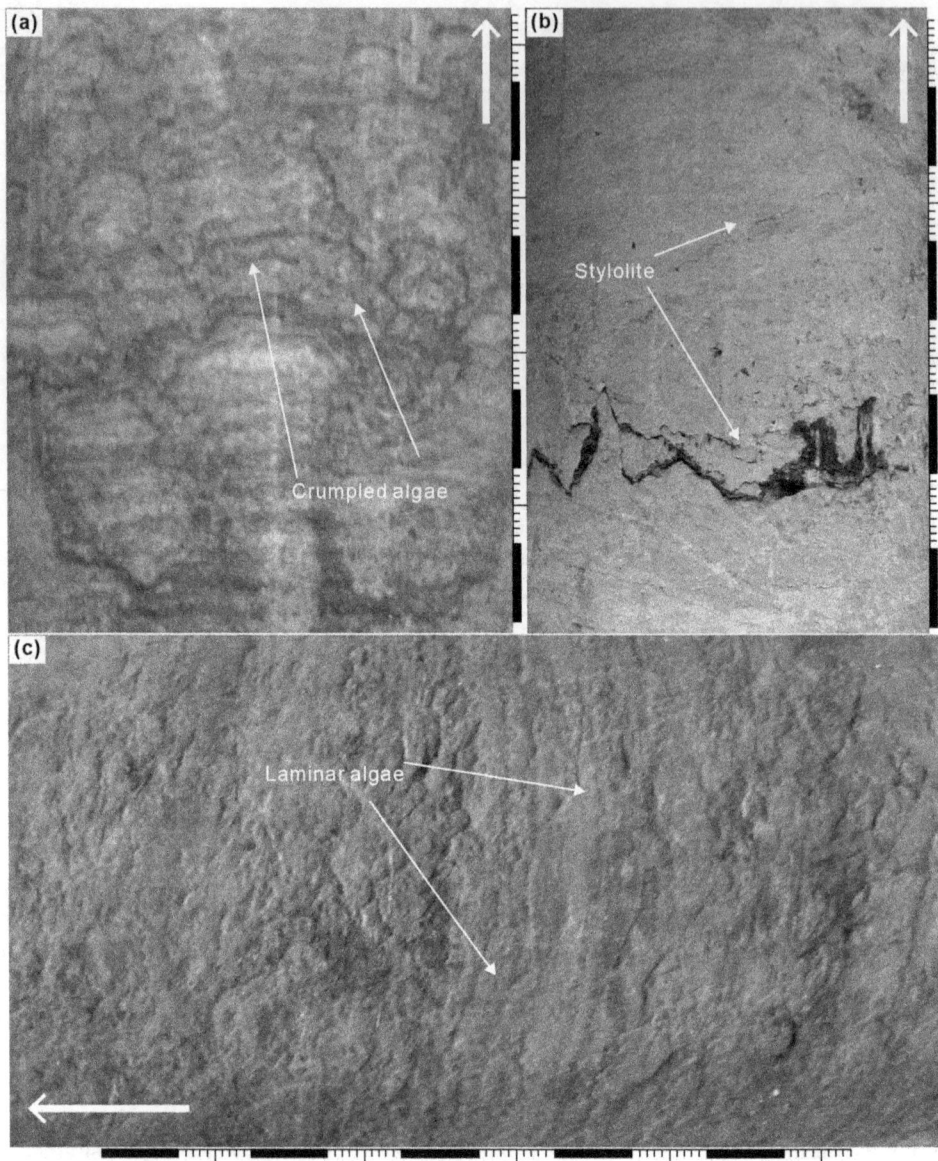

Fig. 6 Photomicrographs showing the depositional characteristics of the dolomite. **a** Laminar algal dolomite with stromatolites that are characterized by crumpled textures and retain the original features and primary sedimentary structures, displaying a gradual upward-deepening sequence. **b** Type 2 dolomite that appears to have accumulated alongside stylolites; the organic seams may be contemporaneous with the early stylolites. **c** Continuous, well-laminated dolomite that generally developed in the algal laminae in the supralittoral zone and may preserve the sedimentary fabric

(Haeri-Ardakani et al. 2013). Seawater with slightly elevated salinity can result in the generation of dense platform-top brines. These dense brines are potential dolomitizing fluids that can descend into underlying pore networks under the influence of gravity. However, the flow of diagenetic fluids critically depends on spatial variations in the permeability, generating complex dolomite bodies at a range of scales (Al-Helal et al. 2012). Intervals of peritidal cycles dominated by dolomite are abundant in the areas of regional highs and on the inner platform during regressive periods, whereas limestone-rich cyclic intervals

develop best near depocenters during transgression periods (Zhang et al. 2015). In the study area, this trend most likely resulted in the uneven distribution of the dolomite but was also related to sea-level fluctuations (Fig. 12). Of course, the dolomitizing fluids were easily partitioned by the regional highs, and the gently inclined, very shallow epicontinental carbonate platform also encourages vertical infiltration of the dolomitizing fluids while discouraging lateral migration, which may be the cause of the discontinuity in the dolomite distribution (as exemplified in well-D61 and well-D69). The bottom of the Ma_5^5 submember

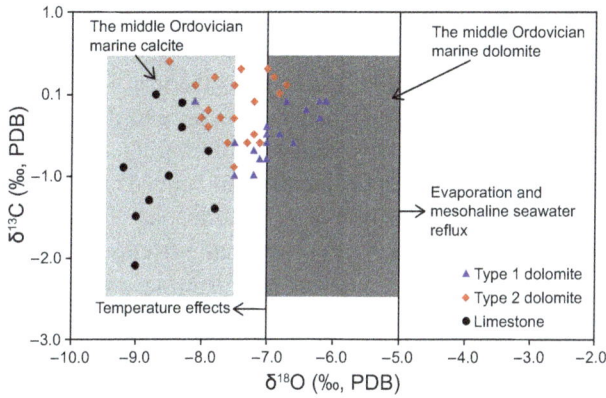

Fig. 7 $\delta^{13}C$ values versus $\delta^{18}O$ values for the dolomites

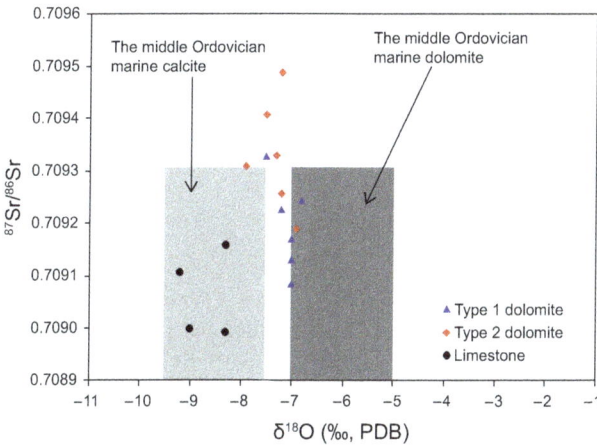

Fig. 8 $^{87}Sr/^{86}Sr$ ratios versus $\delta^{18}O$ values for the dolomites

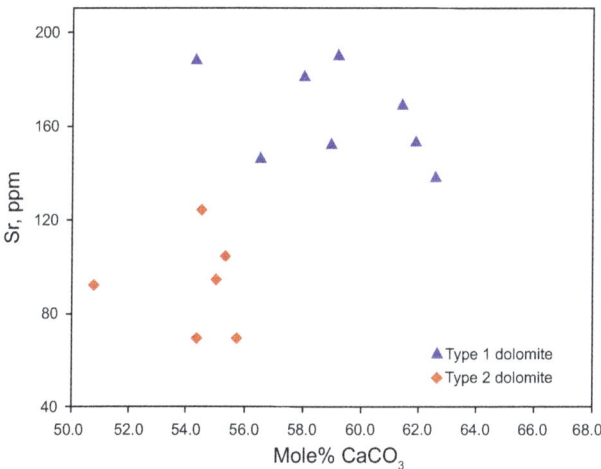

Fig. 9 Sr concentration versus mol% $CaCO_3$ for the dolomites

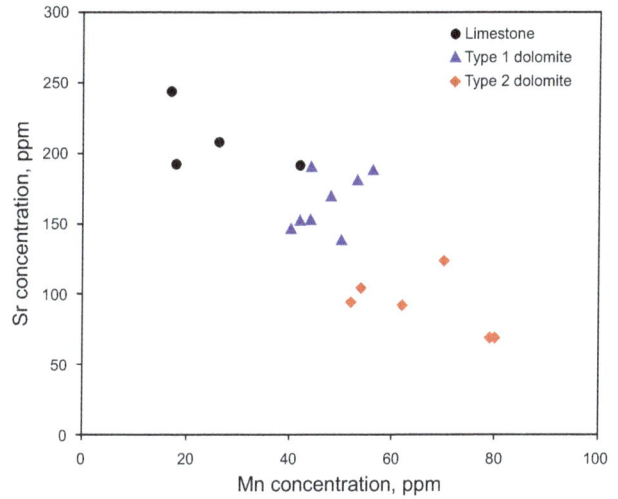

Fig. 10 Sr versus Mn concentrations in the dolomites

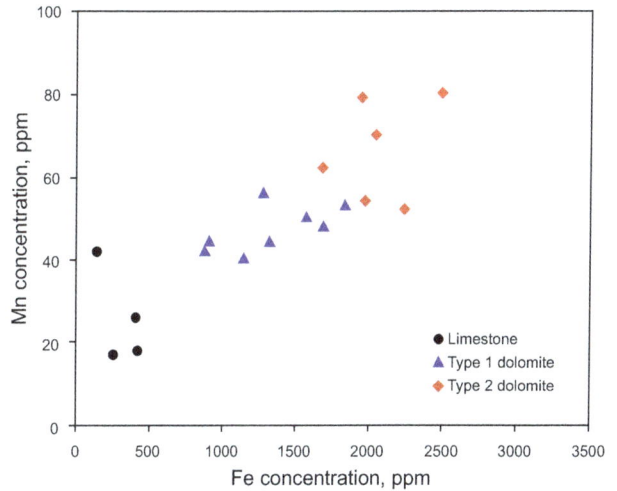

Fig. 11 Mn versus Fe concentrations in the dolomites

illustrated by the fact that a few of the transgressive depositional facies contain stacks of partially dolomitized rocks and limestone cycles (Figs. 3, 4). This pattern indicates that the distribution of the syndepositional dolomite is closely associated with sea-level changes.

6.5 Dolomitization mechanism

The Middle Ordovician Ordos platform was protected from the open sea to the west by the central uplift and developed gentle slope (<1°), restricted evaporative tidal-flat facies in the Ma_5 member of the Majiagou Formation (Zhou et al. 2011) (Fig. 12). The precipitation of anhydrite was conducive to elevated Mg^{2+}/Ca^{2+} molar ratios and lowered the SO_4^{2-} concentrations of the brines. Dolomitization was likely further promoted by low-sulfate concentrations due to sulfate reduction in the tidal-flat sediments (Shelton

contains a set of layered algal dolomites, which are an important indicator of the early genesis of the type 1 dolomite (Harwood and Sumner 2012). The relationship between the petrography and distribution of the dolomite is

et al. 2009). Abundant pyrite has been shown to be syn-depositional in tidal-flat sedimentary conditions and the meter-scale stratigraphy of the dolostone, which lacks massive gypsum/anhydrite, indicates that the synsedimentary deposits of the Ma_5^5 submember developed in a very shallow and slightly altered seawater environment. Elevated salinity seawater percolated downward through the underlying carbonates, resulting in the dolomitization.

The rate of brine flux is proportional to the concentration of the brine and critically depends on the magnitude and distribution of the permeability (Jones et al. 2004). However, synsedimentary unconsolidated deposits generally have higher permeability and porosity, and the dolomitization may have occurred in extensive platform-top environments associated with the repeated flooding and reflux of marine water with slightly higher salinity.

The type 2 dolomite primarily consists of finely crystalline, regular crystal plane, euhedral to subhedral dolomite crystals that replaced the limestone matrix. The type 2 dolomite is closely associated with the type 1 dolomite in the stratigraphic column and is most likely related to the downward flow of modified seawater at elevated temperatures during burial (Figs. 3, 4) based on the following evidence. First, the type 2 dolomite is characterized by finely crystalline, regular crystal plane, euhedral to subhedral dolomite crystals that most likely predated or were coeval with early stylolitization at intermediate burial depths. Second, the bodies of the type 2 dolomite are scattered and discontinuous, vary greatly in thickness and generally decrease in abundance downward. The type 2 dolomites are characterized by the partial to complete replacement of muddy limestone/wackestone but rarely replace packstone that experienced early marine cementation (Fig. 5h), which resulted in a decrease in permeability. Thus, the replacement by dolomite likely postdates the calcite cementation. Finally, the type 2 dolomite displays lower $\delta^{18}O$ values and Sr^+ concentrations and slightly higher Na^+, Fe^{2+}, and Mn^{2+} concentrations and $^{87}Sr/^{86}Sr$ ratios, suggesting that the type 2 dolomites precipitated from modified seawater and dolomitic fluids in the pore water and that their development was affected by elevated temperatures.

Thermal convection has been proposed by some researchers (Morrow 1998; Jones et al. 2004), but the type 2 dolomites are not likely related to thermal convection. The northern Ordos Basin may have experienced tectonism and large-scale heat flow anomalies during the Late Paleozoic (Wang et al. 2006), but thermal anomaly events are not consistent with the timing of the formation of the type 2 dolomites. Fault and fracture conduit systems are commonly important and efficient in delivering dolomitizing fluids to the overlying strata, thus causing extensive dolomitization (Qing and Mountjoy 1989). Faults are rare in the northern Ordos Basin, but fractures are well developed in the Ma_5^5 submember, and parts of this unit have been affected by calcite cementation. However, the fractures are mostly present in the limestone, and the underlying fractured limestone displays no dolomitization features. Furthermore, the type 2 dolomite did not form from upward-migrating basinal fluids in a burial environment because the volume of the dolomite decreases downward.

7 Conclusions

The Middle Ordovician Ma_5^5 submember carbonate unit in the northern Ordos Basin is partially to completely dolomitized. Two types of replacive dolomite are distinguished: (1) type 1 dolomite, which is primarily characterized by microcrystalline (<30 μm), euhedral to subhedral dolomite crystals, mimetically replaces the precursor limestone, and is generally laminated and associated with gypsum-bearing microcrystalline dolomite; and (2) type 2 dolomite, which is present primarily as finely crystalline (30–100 μm), regular crystal plane, euhedral to subhedral dolomite crystals that are truncated by stylolites, indicating that the type 2 dolomite most likely predated or developed simultaneously with the formation of the stylolites. Stratigraphic, petrographic, and geochemical data indicate that the type 1 dolomite formed in slightly evaporated Middle Ordovician seawater and that dolomitizing fluids may have been driven by density differences and elevation-related hydraulic head. The absence of

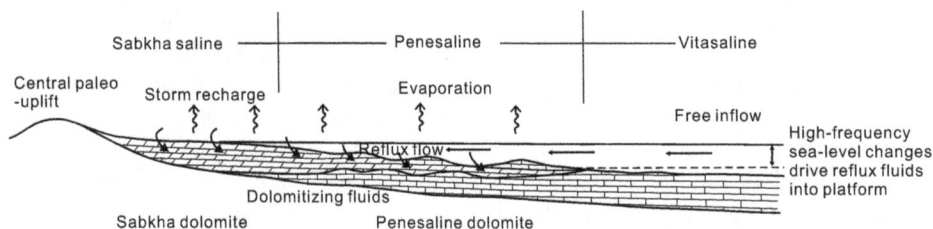

Fig. 12 Diagram indicating the dolomitization model of peritidal carbonates by slightly evaporated (penesaline) sea water in a restricted platform setting driven by high-frequency sea-level changes (after Adams and Rhodes 1960; Qing et al. 2001)

depositional evaporites in the dolomitized intervals suggests that dolomitization was driven by the reflux of slightly evaporated seawater. The $\delta^{18}O$ values of type 1 dolomite are slightly lower than those of the seawater-derived dolomite, suggesting that dolomite may have been recrystallized at higher temperatures during burial.

The type 2 dolomite has lower $\delta^{18}O$ values and Sr^+ concentrations and slightly higher Na^+, Fe^{2+}, and Mn^{2+} concentrations and $^{87}Sr/^{86}Sr$ ratios than the type 1 dolomite. These results suggest that the type 2 dolomites precipitated from modified brines and dolomitic fluids from pore water and that their development was affected by elevated temperatures. Their distribution and abundance are closely related to those of the type 1 dolomite, but their downward decrease in abundance and discontinuous lateral distribution suggest that the dolomitizing fluids were most likely related to the infiltration and diffusion of penesaline water that replaced the carbonate sediments in association with higher temperatures related to shallow burial. The relationship between the petrography and distribution of the dolomite is illustrated by the fact that a few of the transgressive depositional facies contain stacks of partially dolomitized rocks and limestone cycles, indicating that the distribution of the syndepositional dolomite is closely associated with cyclic sea-level changes, which may account for the development of the discrete dolomite bodies.

Acknowledgments This work was supported by the Major National Science and Technology Projects of China (Grant No. 2011ZX05045) and Sinopec (Grant No. 34550000-13-FW0403-0010). We thank Dr. Meng and Dr. Fu at the Isotope Laboratory of the State Key Laboratory of Oil and Gas Reservoir Geology and Exploitation, who checked the stable isotope analysis results.

References

Adams JE, Rhodes ML. Dolomitization by seepage refluxion. AAPG Bull. 1960;44(12):1912–20.

Al-Helal AB, Whitaker FF, Xiao Y. Reactive transport modeling of brine reflux: dolomitization, anhydrite precipitation, and porosity evolution. J Sediment Res. 2012;82:196–215.

Banner JL. Application of the trace element and isotope geochemistry of strontium to studies of carbonate diagenesis. Sedimentology. 1995;42(5):805–24.

Banner JL, Hanson GN. Calculation of simultaneous isotope and trace element variations during water-rock interaction with applications to carbonate diagenesis. Geochim Cosmochim Acta. 1990;54(11):3123–37.

Davies GR, Smith LB. Structurally controlled hydrothermal dolomite reservoir facies: an overview. AAPG Bull. 2006;90(11):1641–90.

Durocher S, Al-Aasm IS. Durocher S. Dolomitization and neomorphism of Mississippian (Visean) upper Debolt Formation, Blueberry field, northeastern British Columbia: geologic, petrologic, and chemical evidence. AAPG Bull. 1997;81(6):954–77.

Fabricius IL, Borre MK. Stylolites, porosity, depositional texture, and

silicates in chalk facies sediments. Ontong Java Plateau-Gorm and Tyra fields, North Sea. Sedimentology. 2007;54(1):183–205.

Fang SX, Hou FH, Yang X, et al. Reservoirs pore space types and evolution in M_5^5 to M_5^1 sub-members of Majiagou Formation of Middle Ordovician in central gasfield area of Ordos Basin. Acta Pet Sin. 2009;25:2425–41 (**in Chinese**).

Feng ZZ, Bao ZD. Lithofacies paleogeography of Majiagou age of Ordovician in Ordos Basin. Acta Sedimentol Sin. 1999;17(1):1–8 (**in Chinese**).

Franzolin E, Merlini M, Poli S, et al. The temperature and compositional dependence of disordering in Fe-bearing dolomites. Am Miner. 2012;97:1676–84.

Friedman GM, Sun SQ. A reappraisal of dolomite abundance and occurrence in the Phanerozoic: discussion and reply. J Sediment Res. 1995;65(1a):244–6.

Fu JH, Wang BQ, Sun LY, et al. Dolomitization of Ordovician Majiagou Formation in Sulige region, Ordos Basin. Pet Geol Exp. 2011;33:268–73 (**in Chinese**).

Fu QL, Qing HR, Bergman K. Dolomitization of the Middle Devonian Winnipegosis carbonates in south-central Saskatchewan, Canada. Sedimentology. 2006;53(4):825–48.

Gregg JM, Bish DL, Kaczmarek SE, et al. Mineralogy, nucleation and growth of dolomite in the laboratory and sedimentary environment: a review. Sedimentology. 2015;62(6):1749–69.

Gregg JM, Shelton KL. Dolomitization and dolomite neomorphism in the back reef facies of the Bonneterre and Davis Formations (Cambrian), southeast Missouri. J Sediment Res. 1990;60(4):549–62.

Haeri-Ardakani O, Al-Aasm I, Coniglio M, et al. Diagenetic evolution and associated mineralization in Middle Devonian carbonates, southwestern Ontario, Canada. Bull Can Pet Geol. 2013;61(1):41–58.

Hardie LA. Dolomitization: a critical view of some current views. J Sediment Res. 1987;57:166–83.

Harwood CL, Sumner DY. Origins of microbial microstructures in the Neoproterozoic Beck Spring Dolomite: variations in microbial community and timing of lithification. J Sediment Res. 2012;82(9):709–22.

He XY, Shou JF, Shen AJ, et al. Geochemical characteristics and origin of dolomite: a case study from the middle assemblage of Ordovician Majiagou Formation Member 5 of the west of Jingbian Gas Field, Ordos Basin, North China. Pet Explor Dev. 2014;41(3):417–27.

Hu B, Kong FJ, Zhang YS, et al. The homogenization temperature in the fluid inclusions of Ordovician halite and paleoclimatic implication. Acta Geol Sin. 2014;88(S1):10–11.

Huang QY, Zhang SN, Ding XQ, et al. Origin of dolomite of Ordovician Majiagou Formation, western and southern margin of the Ordos Basin. Pet Geol Exp. 2010;32(2):146–7 (**in Chinese**).

Jones GD, Rostron BJ. Analysis of fluid flow constraints in regional-scale reflux dolomitization: constant versus variable-flux hydrogeological models. Bull Can Pet Geol. 2000;48(3):230–45.

Jones GD, Whitaker FF, Smart PL, et al. Numerical analysis of seawater circulation in carbonate platforms: II. The dynamic interaction between geothermal and brine reflux circulation. Am J Sci. 2004;304(3):250–84.

Lei BJ, Fu JH, Sun FJ, et al. Sequence stratigraphy of the Majiagou Formation, Ordos Basin: sedimentation and early diagenesis related to eustatic sea-level changes. J Stratigr. 2010;34(2):145–53 (**in Chinese**).

Li ZH, Zheng CB. Evoluation process of palaeokarst and influence to reservoir—a case for Ordovician of Ordos Basin. Nat Gas Geosci. 2004;15:247–52 (**in Chinese**).

Liu Y, Fu JH, Li JM. Dolomite genetic analysis on Ordovician Majiagou Formation in eastern Ordos Basin. J Oil Gas Technol. 2011;33:46–50 (**in Chinese**).

Loyd SJ, Corsetti FA. The origin of the millimeter-scale lamination in the Neoproterozoic lower Beck Spring Dolomite: implications for widespread, fine-scale, layer-parallel diagenesis in Precambrian carbonates. J Sediment Res. 2010;80:678–87.

Machel HG. Concepts and models of dolomitization: a critical reappraisal. Geol Soc Lond Spec Publ. 2004;235(1):7–63.

Mazzullo SJ. Geochemical and neomorphic alteration of dolomite: a review. Carbonates Evaporites. 1992;7(1):21–37.

McArthur JM, Howarth RJ, Bailey TR. Strontium isotope stratigraphy: LOWESS version 3: Best fit to the marine Sr-isotope curve for 0-590 Ma and accompanying look-up table for deriving numerical age. J Geol. 2001;109:155–70.

McKenzie J. Movement of subsurface waters under the sabkha Abu Dhabi, UAE, and its relation to evaporative dolomite genesis. Soc Econ Palaeontol Mineral. 1980;28:11–30.

Melim LA, Swart PK, Eberli GP. Mixing-zone diagenesis in the subsurface of Florida and the Bahamas. J Sediment Res. 2004;74(6):904–13.

Montañez IP, Read JF. Fluid-rock interaction history during stabilization of early dolomites, Upper Knox Group (Lower Ordovician), U.S. Appalachians. J Sediment Res. 1992;62(5):753–78.

Morrow DW. Regional subsurface dolomitization: models and constraints. Geosci Can. 1998;25(2):57–70.

Qing HR, Barnes CR, Buhl D, et al. The strontium isotopic composition of Ordovician and Silurian brachiopods and conodonts: relationships to geological events and implications for coeval seawater. Geochim Cosmochim Acta. 1998;62(10):1721–33.

Qing HR, Bosence DWJ, Rose EPF. Dolomitization by penesaline sea water in early Jurassic peritidal platform carbonates, Gibraltar, western Mediterranean. Sedimentology. 2001;48(1):153–63.

Qing HR, Mountjoy EW. Multistage dolomitization in Rainbow buildups, Middle Devonian Keg River Formation, Alberta, Canada. J Sediment Res. 1989;59(1):114–26.

Qing HR, Veizer J. Oxygen and carbon isotopic composition of Ordovician brachiopods: implications for coeval seawater. Geochim Cosmochim Acta. 1994;58(20):4429–42.

Read JF, Cangialosi M, Husinec A, et al. Coarse, fabric destructive post-depositional dolomites, Late Jurassic-Early Cretaceous Adriatic platform, Croatia: origin by mesohaline reflux. 2012 Geological Society of America Annual Meeting in Charlotte. 2012, 44: 456.

Rivers JM, Kyser TK, James NP. Salinity reflux and dolomitization of southern Australian slope sediments: the importance of low carbonate saturation levels. Sedimentology. 2012;59(2):445–65.

Rott CM, Qing HR. Early dolomitization and recrystallization in shallow marine carbonates, Mississippian Alida Beds, Williston Basin (Canada): evidence from petrography and isotope geochemistry. J Sediment Res. 2013;83(11):928–41.

Shelton KL, Gregg JM, Johnson AW. Replacement dolomites and ore sulfides as recorders of multiple fluids and fluid sources in the southeast Missouri Mississippi valley-type district: halogen $^{87}Sr/^{86}Sr$ ^{18}O-^{34}S systematics in the Bonneterre dolomite. Econ Geol. 2009;104(5):733–48.

Sibley DF. Unstable to stable transformations during dolomitization. J Geol. 1990;98:739–48.

Staudt WJ, Oswald EJ, Schoonen MAA. Determination of sodium, chloride and sulfate in dolomites: a new technique to constrain the composition of dolomitizing fluids. Chem Geol. 1993;107(s1–2):97–109.

Su ZT, Chen HD, Xu FY. Geochemistry and dolomitization mechanism of Majiagou dolomites in Ordovician, Ordos, China. Acta Pet Sin. 2011;27(8):2230–8 (in Chinese).

Swart PK. The geochemistry of carbonate diagenesis: the past, present and future. Sedimentology. 2015;62(5):1233–304.

Swart PK, Melim LA. The origin of dolomites in Tertiary sediments from the margin of Great Bahama Bank. J Sediment Res. 2000;70(3):738–48.

Veizer J, Ala D, Azmy K, et al. $^{87}Sr/^{86}Sr$, $\delta^{13}C$, $\delta^{18}O$ evolution of Phanerozoic seawater. Chem Geol. 1999;161(1):1586.

Wang BQ, Qiang ZT, Zhang F, et al. Isotope characteristics of dolomite from the fifth member of the Ordovician Majiagou Formation, the Ordos Basin. Geochimica. 2009;38(5):472–9 (in Chinese).

Wang DX, Wang AG, Zhang BG, et al. Geochemical characteristics of rare earth element of dolostones from Ma5 member of Majiagou Formation in Jingbian Gasfield, Ordos Basin: implications for the origin of diagenetic fluids. Nat Gas Geosci. 2015;26(4):641–9 (in Chinese).

Wang QF, Deng J, Yang LQ, et al. Tectonic constraints on the transformation of Paleozoic framework of uplift and depression in the Ordos Area. Acta Geol Sin. 2006;80(6):944–53.

Wang Z, Zhang Y, Zheng M, et al. Sr isotope geochemistry and depositional setting of carbonate in Ordovician, Ordos Basin, China. Acta Geol Sin. 2014;S1:262–4.

Warren J. Dolomite: occurrence, evolution and economically important associations. Earth Sci Rev. 2000;52(s 1–3):1–81.

Wheeler CW, Aharon P, Ferrell RE. Successions of late Cenozoic platform dolomites distinguished by texture, geochemistry, and crystal chemistry: Niue, South Pacific. J Sediment Res. 1989;69:239–55.

Yang H, Bao H. Characteristics of hydrocarbon accumulation in the middle Ordovician assemblages and their significance for gas exploration in the Ordos Basin. Nat Gas Ind. 2011;31(12):11–20 (in Chinese).

Yang SZ, Jin WH, Li ZH. Multicycle superimposed basin form and evolution of Ordos Basin. Nat Gas Geosci. 2006;17:494–8 (in Chinese).

Zhang XF, Liu B, Cai ZX, et al. Dolomitization and carbonate reservoir formation. Geol Sci Technol Inf. 2010;29(3):79–85 (in Chinese).

Zhang YQ, Chen DZ, Zhou XQ, et al. Depositional facies and stratal cyclicity of dolomites in the lower Qiulitag Group (Upper Cambrian) in northwestern Tarim Basin, NW China. Facies. 2015;61(1):1–24 (in Chinese).

Zhao JX, Chen HD, Zhang JQ, et al. Genesis of dolomite in the fifth member of Majiagou Formation in the middle Ordos Basin. Acta Pet Sin. 2005;26:38–41 (in Chinese).

Zhao WZ, Shen AJ, Zheng JF, et al. The porosity origin of dolostone reservoirs in the Tarim, Sichuan and Ordos basins and its implication to reservoir prediction. Sci China Earth Sci. 2014;57(10):2498–511 (in Chinese).

Zhou JG, Zhang F, Guo QX, et al. Barrier-lagoon sedimentary model and reservoir distribution regularity of Lower-Ordovician Majiagou Formation in Ordos Basin. Acta Sedimentol Sin. 2011;29(1):64–71 (in Chinese).

12

Quantitative characterization of polyacrylamide–shale interaction under various saline conditions

Samyukta Koteeswaran[1] · Jack C. Pashin[2] · Josh D. Ramsey[1] · Peter E. Clark[1]

Abstract Interaction of polymer-containing injected fluids with shale is a widely studied phenomenon, but much is still unknown about the interaction of charged polyacrylamides such as anionic and cationic polyacrylamides with shale. The nature of interaction of charged polyacrylamides with shale is not well understood, especially from the perspective of assessing the potential for polyacrylamides to cause formation damage. Zeta potential and rheological measurements were made for Chattanooga and Pride Mountain shales suspended in polyacrylamide solutions with and without inorganic salts and tetramethyl ammonium chloride (TMAC). The change in zeta potential and viscosity with time was recorded. The magnitude of decrease in the absolute value of zeta potential with time is indicative of adsorption of polymer on the surface of shale and serves as a measure of the extent of polymer interaction with shale. The salts that were used in this study are potassium chloride (KCl), sodium chloride (NaCl). This study quantified the interaction of anionic and cationic polyacrylamide with different North American shales. From the experimental results, it was determined that the polyacrylamides can interact strongly with shale, particularly the cationic polyacrylamide. The objective of this study was to determine the extent of interaction of anionic and cationic polyacrylamide with each shale sample in the presence of additives such as salts.

Keywords Anionic and cationic polyacrylamides · Chattanooga shale · Pride Mountain shale · Zeta potential · Slurry rheology

1 Introduction

Interaction of injected fluids such as drilling, fracturing and completion fluids with shale has been a problem for many decades in the oil field, and shale constitutes 75% of all the formations drilled by the oil and gas industry (Khodja et al. 2010). Over the years, many studies have been conducted to quantify shale–fluid interaction and also to minimize this interaction. Interactions between shale and injected fluids are of concern for a variety of reasons. The interaction of injected fluids with shale leads to wellbore instability (Tan et al. 1996; Yu et al. 2003; Muniz et al. 2005), and the productivity of the wells decreases due to this instability, which also increases the drilling cost (Lal 1990; Mahto and Sharma 2004). Water-based mud (WBM) is the most commonly used type of drilling fluid, and shale is highly sensitive to the additives and the clays present in the WBM (Gomez and He 2012; He et al. 2014). The common additives used in WBM are friction reducers, acids, gellants, crosslinkers, clay controlling agents and other polymers (Harris 1988; Aften and Watson 2009). It is important to use all the necessary additives in injected fluids, but it is also equally important to use additives that do not potentially weaken the shale.

The way shale interacts with the injected fluid depends on shale properties, such as mineralogy, rock mechanical properties, porosity, clay composition and permeability, as well as the properties of injected fluids such as ionic strength and salt concentration (Gomez and He 2012; Lal 1990; Horsrud et al. 1998). Clay in shale has a great

✉ Peter E. Clark
peter.clark@okstate.edu

[1] School of Chemical Engineering, Oklahoma State University, Stillwater, OK 74078, USA

[2] Boone Pickens School of Geology, Oklahoma State University, Stillwater, OK 74078, USA

Edited by Xiu-Qin Zhu

influence on the chemical and mechanical properties of shale. Clay minerals have a tendency to absorb water and cause an increase in the swelling pressure—a phenomenon called hydration, and this is attributed to the hydrophilic surface of the clay (Lu 1988). The clay minerals present in shale are mostly classified into five categories: montmorillonite, illite, smectite, kaolinite and attapulgite (van Olphen 1977; Luckham and Rossi 1999). The presence of abundant clay minerals changes the interaction properties of the shale with injected fluids, and the composition of the clay affects reactivity, with montmorillonitic clay being highly prone to swelling and high crystalline illite being less prone to swelling.

Much research is being conducted to study the rock mechanics to understand the interaction of shale with fluids. Conventional techniques such as dispersion and swelling tests do not fully reveal the effects of polymer–shale interaction. Studies such as pressure transmission tests are carried out to measure the effect of anions, cations and salts present in injected fluid that affect shale–fluid interaction (van Oort et al. 1995; Ghassemi and Diek 2003). The presence of ions in injected fluid alters the membrane efficiency of shale, thereby influencing ion transport from the fluid to the shale that causes the shale to swell/disperse (Zhang, et al. 2006; Mody and Hale 1993; van Oort 2003; Al-Bazali 2005).

High molecular weight polyacrylamides are commonly used friction reducers in hydraulic fracturing of shale formations. The large volumes of friction reducers (liquid volumes can be as high as four million gallons for one well), especially synthetic polymers such as polyacrylamides, are difficult to break and are proven to form membranes over shales and are associated with causing fracture and formation damage (Carman and Cawiezel 2007). Formation damage caused by the adsorption of polyacrylamides on the shale surface alters the surface properties. In this work, the study of the shale–polyacrylamide interaction focused on the extent to which polyacrylamides adhering to the shale can potentially cause formation damage.

Some of the commonly used methods such as swelling and dispersion tests do not give a true representation of the shale–fluid interaction and are qualitative in nature. Other sophisticated methods such as the autonomous triaxial and high-pressure triaxial tests can give a good quantitative measure of shale–fluid interaction by measuring the axial load, sample deformation, cell and pore pressures, but they are tedious and intensive processes (Mody et al. 2002). Hence, a simple testing method was devised that can produce reproducible semiquantitative data, which will aid in better understanding the interaction of different fluids and their components with shale.

One such method that was devised to probe the polymer–shale interaction is by rheological measuring the interactions. The rheology of shale slurries suspended in the polymer was analyzed. The factors that affect the rheology of the particle suspension are concentration, particle shape, interactions among particles, and interaction between particles and the bulk fluid (Mueller et al. 2010). Characterizing the interaction between the particle and the bulk fluid is the key to the research. When shale particles interact strongly with the bulk fluid, viscosity increases with increasing polymer concentration. This is used as a measure of the interaction of bulk fluid with shale particles. Additionally, the polymer tends to adsorb on the surface of the shale. Rheological methods were used in this work to assess the interaction of anionic and cationic polyacrylamide with samples of North American shale, the Pride Mountain shale and the Devonian-age Chattanooga shale. The interaction of shale with anionic and cationic polyacrylamide was studied rheologically by a series of flow ramps.

The second method uses zeta potential measurements over time to quantify polymer–shale interaction. The zeta potential is an electric potential developed at the solid–liquid interface due to the relative movement of solid particles in water (Vane and Zang 1997). Zeta potential at the solid–liquid interface is an indirect measure of solid–liquid interactions (Menon and Wasan 1987b; Werner et al. 2001; Petersen and Saykally 2008). The electrokinetic measurements made at the solid–liquid interface are a relative measure of surface charge and adsorption (Delgado et al. 2007; Hunter 2013). Zeta potential measurements have long been used to measure the stability of colloidal systems (Heurtault et al. 2003; Jiang et al. 2003; Hunter 2013). The colloidal system in the present study is shale dispersed in polyacrylamide. By measuring the stability of the shale system as a function of zeta potential over time, we will be able to quantify polymer–shale interaction. A comparison is made between different salt–polymer solutions (also called as shale inhibitors) for study of their role in preventing polymer adsorption on shale. Salts such as KCl and NaCl are widely used for shale inhibition (Lee et al. 2001; Patel 2009), and in the past, amines were widely used for this purpose (He et al. 2014). In this work, TMAC is compared with KCl and NaCl as an additive to anionic and cationic polyacrylamide systems for shale inhibition.

In this work, the impact of anionic and cationic polyacrylamide in injected fluids on the alteration of the surface properties of shale is studied. Using zeta potential and rheological measurements to quantify shale–polymer interaction is a novel technique and is extensively researched and studied in this work.

2 Materials and methods

2.1 Polyacrylamides

Anionic polyacrylamide and cationic polyacrylamide with average molecular weight 10^7 g/gmol were obtained from Kemira Supplies. The polyacrylamides are highly water absorbent and form soft gels even at low concentration. The anionic and cationic polyacrylamide samples were measured by weight and added to deionized water slowly and mixed on a shaker table for 15 min at a speed of 200 RPM. The time and speed of mixing of the sample were chosen carefully so that shear damage in polyacrylamide samples was kept to a minimum before the experiments. The samples were left to hydrate for 24 h. All of the solutions were tested within 36 h of preparation.

2.2 Shale samples

Pride Mountain and Chattanooga shale samples were prepared using a mortar and pestle. They were ground using a Bel-Art mixer to obtain smaller particles, and the sample was sieved to obtain fairly homogenous particles, with particle size smaller than 75 μm. The particles were small enough to remain suspended in the polymer solution and big enough to make accurate rheological measurements of slurry. The shale was kept at a constant concentration of 0.5 lb/bbl (pounds/barrel) for all of the rheology and zeta potential experiments.

2.3 Sample information

The Chattanooga shale sample is from an exploratory well in southwestern Tuscaloosa County, Alabama and is typical of Devonian shale reservoir rock in the eastern USA. The Pride Mountain sample is from the Gorgas #1 borehole, which was drilled to explore the CO_2 storage potential at a large coal-fired power facility in the Black Warrior Basin, Walker County, Alabama. The Pride Mountain sample is more representative of a sealing formation and is rich in expandable mixed-layer clay—wellbore stability was a significant problem during the drilling of this zone.

2.4 Characterization of shale

The shale samples were analyzed for clay and non-clay content by X-ray diffraction (XRD) (Clark et al. 2012) (Table 1). Other parameters such as total organic carbon (TOC), pressure decay permeability, and effective porosity were determined for both shale samples (Table 2) (Clark et al. 2012).

The whole rock mineralogy is shown in Fig. 1 as a bar graph for better understanding of the difference in mineralogy between the two shales used in this study.

Table 1 Whole rock mineralogy of shale samples determined by XRD

Analysis	Chattanooga	Pride Mountain
Depth, ft	9167	2863
Clay content, wt%		
Smectite	0	1
Illite/smectite	5	16
Illite + mica	24	37
Kaolinite	0	12
Chlorite	0	4
Non-clay mineral content, wt%		
Quartz	41	21
K feldspar	16	3
Plagioclase	2	2
Calcite	0	1
Ankerite/Fe dolomite	0	1
Dolomite	5	0
Pyrite	5	1
Fluorapatite	0	0
Barite	1	1
Siderite	0	1
Magnetite	0	0

2.5 Equipment

A Discover DHR-3 stress controlled rheometer was used to make rheological measurements. Vane geometry was used for the polymer–shale samples. Vane geometry helps prevent wall slippage at higher shear rates, helps disrupt flow inhomogeneity while shearing, and also works well for samples with suspended solids (Goh et al. 2011). A cone-and-plate geometry was used for polymer solutions. Cone-and-plate is useful for solutions that have low viscosity and that do not have any dispersions with suspended solids larger than 64 μm. Cone-and-plate geometry (diameter: 60 mm and cone angle 2°) provides homogenous shear, shear rate, and stress in the geometry gap when used to measure the rheological properties of a solution. All the experiments were performed at a temperature of 25 °C ± 0.03 °C.

Since the cationic polyacrylamide can form agglomerates with shale, it is not possible to quantify the polymer–shale interaction rheologically. Due to agglomeration or in other words due to the flocculation of the shale particles in the solution, accurate rheological measurements cannot be made. The shale particles have to be suspended in the solution and have minimal settling velocity in order to perform rheological studies. In the cationic polyacrylamide medium, flocculation can result in excessive gravitational settling of the agglomerated shale particles. Hence, only

Table 2 TOC, effective porosity and pressure decay permeability

Parameters	Chattanooga	Pride Mountain
TOC, wt%	3.33	0.80
Effective porosity, % of bulk volume	2.32	12.30
Pressure decay permeability, mD	0.00032	0.00048

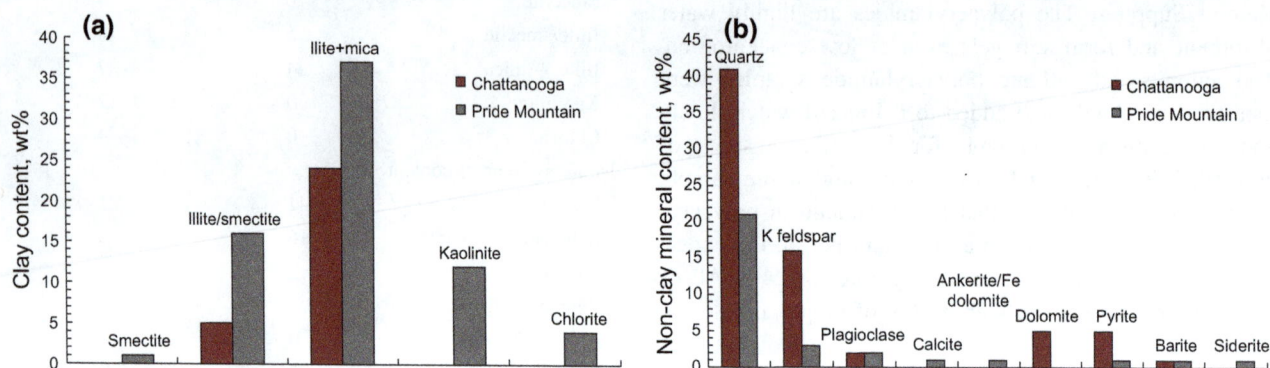

Fig. 1 **a** Percentage of clay content in Chattanooga and Pride Mountain shale, **b** Percentage of non-clay mineral content in Chattanooga and Pride Mountain shale

the anionic polyacrylamide was used to rheologically quantify polymer–shale interaction. However, both cationic and anionic polyacrylamides were used to quantify polymer–shale zeta potential. The anionic polyacrylamide concentration was 0.1–0.2 wt%, such that the concentration is well above C^* (critical overlap concentration) and below C^{**} (critical entanglement concentration). The concentration of shale was kept constant at 0.5 lb/bbl, and the concentration of anionic polyacrylamide was varied from 0.1 to 0.2 wt%.

Figure 2 shows the experimental setup, including the cone-and-plate and vane geometry.

2.6 Zeta potential analyzer

A phase analysis light scattering technique (PALS) is used to measure the zeta potential of polyacrylamide–shale interfaces. A Zeta PALS measurement system manufactured by Brookhaven Instruments Corporation (Holtsville, NY) was used. The experiments were conducted at 25 °C in triplicate. A platinum electrode and H-Ne laser light source were used to measure the electrophoretic mobility of colloidal suspensions. The polyacrylamide–shale sample was prepared by adding polyacrylamide to deionized (DI) water, and it was kept on a shaker table at a speed of 200 RPM for 15 min. The shale sample was weighed and added to DI water. Both samples were left to hydrate at room temperature for 24 h. The shale particles were filtered using a 1-μm syringe filter and added to the polyacrylamide sample. The solution was shaken and added to the cuvette using a pipette. The size of the shale particles is in the

colloidal range (1×10^{-9} m), in which physiochemical forces such as van der Waals attractive forces and double layer repulsive forces are stronger than gravitational forces (Kaya et al. 2003). Figure 3 shows the particle size distribution of shale particles before filtering it to get particle sizes lesser than 1 μm.

A 1-cm^3 sample was used for all the measurements, and the tip of the cuvette was immersed in the sample to prevent formation of air bubbles. The Pt electrode was then placed in the cuvette, and the zeta potential measurements were recorded. In order to study the influence of salt on polymer–shale interaction, salt solutions of KCl, NaCl and TMAC were used. To study the increase in average particle size with time, dynamic light scattering using the Zeta PALS was used. A 0.45-μm syringe filter was used to filter dust from the samples before loading the sample to the Zeta PALS. A zeta potential measurement was recorded every 20 min and for each data point, ten readings were taken, and the average effective diameter and the associated standard error were plotted vs. time. The compositions of the various suspensions used are given in Table 3. The composition of the suspension was chosen such that the salt concentration met the Zeta PALS instrument specification, and the polyacrylamide concentration, which was just enough to keep the shale suspended in the polyacrylamide, was chosen.

2.7 Analysis with the Carreau model

In order to determine the zero shear rate viscosity of the fluid, the Carreau model was used. This model describes a

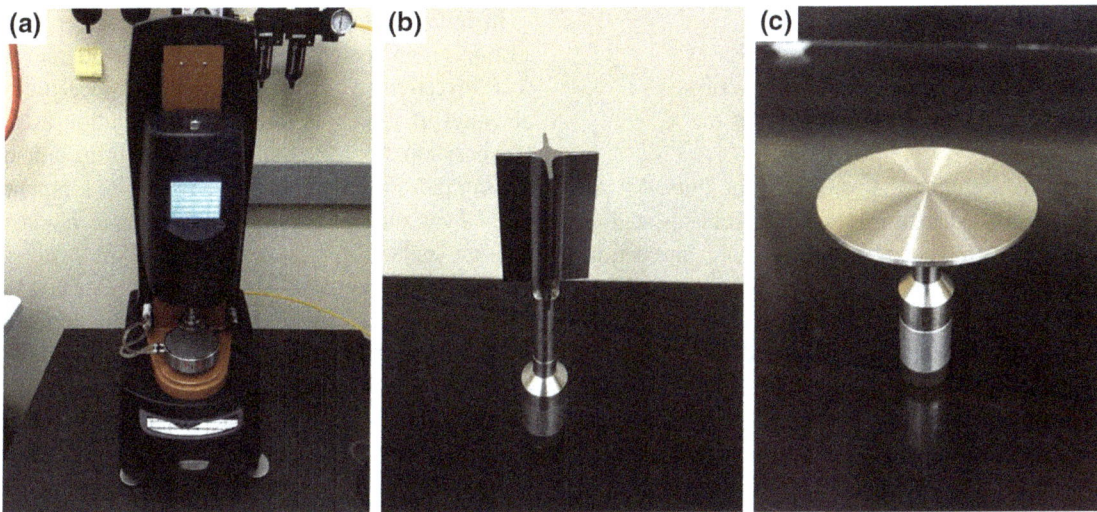

Fig. 2 a DHR-3 Rheometer. b Vane geometry. c Cone-and-plate geometry

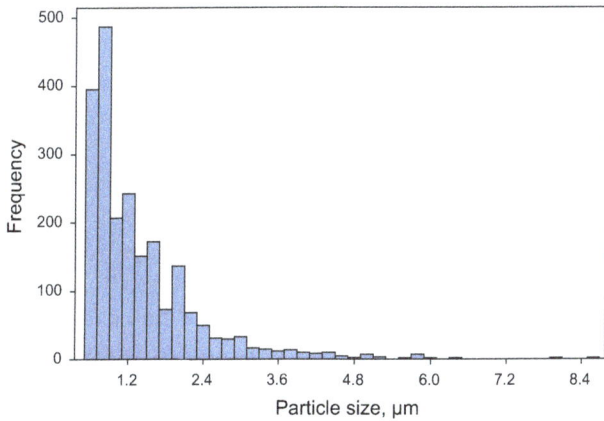

Fig. 3 Histogram of particle size distribution of shale particles used in the study

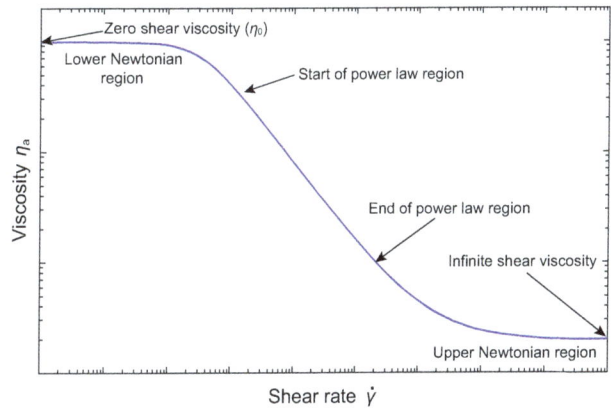

Fig. 4 Carreau model logarithmic fit for viscosity versus shear rate

Table 3 Compositions of different suspension media used in the study

Sample	Medium
1	0.05 wt% anionic polyacrylamide
2	0.05 wt% cationic polyacrylamide
3	0.05 wt% anionic polyacrylamide + 0.05 wt% KCl
4	0.05 wt% cationic polyacrylamide + 0.05 wt% KCl
5	0.05 wt% anionic polyacrylamide + 0.05 wt% NaCl
6	0.05 wt% cationic polyacrylamide + 0.05 wt% NaCl
7	0.05 wt% anionic polyacrylamide + 0.05 wt% TMAC
8	0.05 wt% cationic polyacrylamide + 0.05 wt% TMAC

wide range of non-Newtonian behavior by curve fitting within the Newtonian and the shear thinning non-Newtonian regions (Rao 2014). This model can be applied over a wide range of shear rates. The Carreau model is a variant of the Cross model and is used for logarithmic data sets. This viscosity model allows data to be fitted to the following model,

$$\frac{\eta - \eta_\infty}{\eta_0 - \eta_\infty} = \frac{1}{\left(1 + \left(\lambda \dot{\gamma}\right)^2\right)^{n/2}}$$

where η_0 the Newtonian viscosity, η_∞ the infinite viscosity, $\dot{\gamma}$ the shear rate, λ the relaxation time, and n the power law index.

Figure 4 shows the plot of apparent viscosity vs. shear rate for a shear thinning Carreau fluid identifying three separate regions. The zero shear viscosity represents the lower Newtonian region at lower shear rates; the infinite shear viscosity captures the higher shear rate, which is the upper Newtonian region; the power law region is characterized by the power law index and the relaxation time which gives the time estimate, at which the lower Newtonian region ends.

3 Results and discussion

3.1 Characterizing polymer–shale interaction through zeta potential measurements

Zeta potential measurements were made for Pride Mountain and Chattanooga shale samples in different suspending media to quantify the polyacrylamide–shale interaction. The measured zeta potential is a function of the surface charge of the suspended particle, any adsorbed layer at the particle–liquid interface, and the nature and composition of the surrounding medium (Jia and Williams 1990). For the same experimental conditions, the change in zeta potential over time is indicative of polymer adsorption on shale. The higher the absolute values of negative zeta potential, the bigger the double layer thickness of the shale particle. Higher negative zeta potential value is also indicative of swelling and dispersion of clay (Zhong et al. 2011). The zeta potential values measured for the shale samples were ~ -24 mV for both Chattanooga and Pride Mountain shale.

The zeta potential of cationic and anionic polyacrylamide with Chattanooga and Pride Mountain shale was measured immediately after adding the shale sample to the polyacrylamide sample. In cationic polyacrylamide (with no salts), there was not a significant difference in zeta potential values for the shales, whereas in the presence of salts (KCl and NaCl) or TMAC (Fig. 5), Chattanooga shale had higher zeta potential values which is indicative of higher polyacrylamide adsorption density than Pride Mountain shale. Similarly in anionic polyacrylamide, Pride Mountain shale had higher absolute zeta potential values in the presence of KCl and TMAC, indicative of higher polyacrylamide adsorption density (Fig. 6). The change in zeta potential with time for the same system will be discussed in the following sections.

In order to determine the influence factor for polymer adsorption on shale, the change in zeta potential with time was investigated. It is important to measure the zeta potential of the shale-free polymer solution as a control. Polyacrylamides were stable for 48 h from preparation of the sample. Figures 7 and 8 show the zeta potential measured over time for anionic and cationic polyacrylamide with no shale.

The zeta potential remained almost constant over time (Figs. 7, 8). This proved that the polyacrylamide remained stable during the time of experiment and the change in zeta potential after adding shale to the polymer was solely because of the changes in the surface properties of shale when in contact with polyacrylamide.

Figures 9 and 10 show the change in zeta potential with time for Chattanooga shale incubated in different media containing anionic and cationic polyacrylamide, respectively. A dotted line is drawn at 20 mV (Figs. 10, 12) to show the point below which the colloidal system is unstable due to flocculation.

Figures 11 and 12 show the change in zeta potential with time for Pride Mountain shale in different media containing anionic and cationic polyacrylamide, respectively.

In the absence of salt, the overall magnitude (i.e., absolute value) of the zeta potential increased for both shale samples in anionic polyacrylamide (Figs. 9, 11) and decreased in cationic polyacrylamide (Figs. 10, 12). In cationic polyacrylamide, the decrease in zeta potential of shale with polymer adsorption is due either to a decrease in charge density or a shift in the shear plane. The zeta potential also decreases more rapidly when the double layer is compressed at high ionic strength (Brooks and Seaman 1973; Vane and Zang 1997). The hydrophilic ends of the cationic polyacrylamide attached themselves to the positively charged edges of clay particles and cause bridging of clay particles. This created clusters of large particles to resist flow and lead to decrease in mobility and zeta potential (Yalçın et al. 2002). Addition of salts

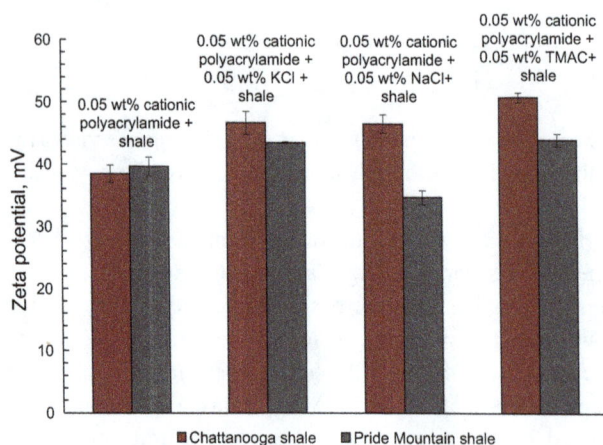

Fig. 5 Change of zeta potential of shales in cationic polyacrylamide with salts or TMAC

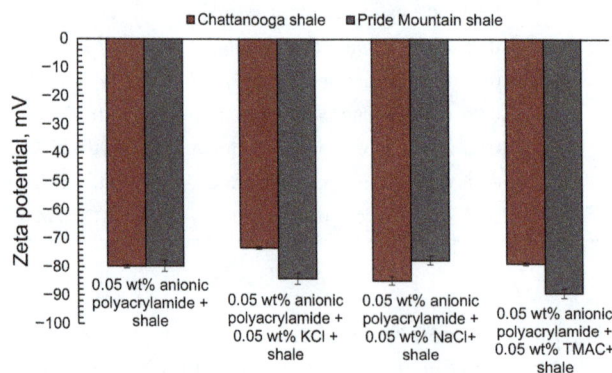

Fig. 6 Change of zeta potential of shales in anionic polyacrylamide with salts or TMAC

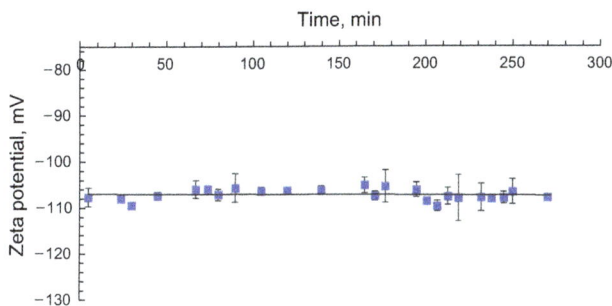

Fig. 7 Zeta potential versus time for anionic polyacrylamide

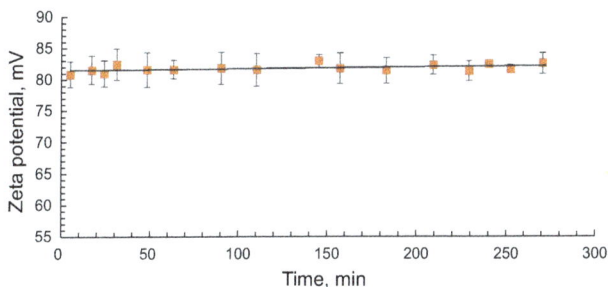

Fig. 8 Zeta potential versus time for cationic polyacrylamide

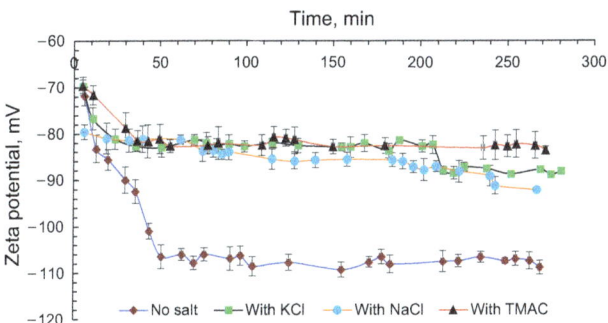

Fig. 9 Zeta potential of Chattanooga shale incubated with anionic polyacrylamide under various conditions

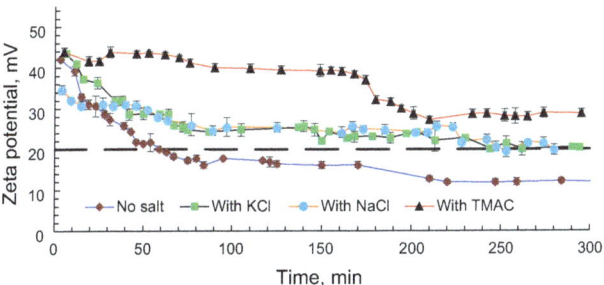

Fig. 10 Zeta potential of Chattanooga shale incubated with cationic polyacrylamide under various conditions

increased the net positive charge of the medium, leading to the increase in zeta potential. Zeta potential values between −20 mV and 20 mV had an effective charge low enough for flocculation to occur (Johnson et al. 2010).

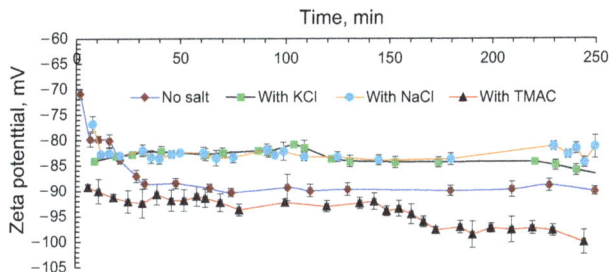

Fig. 11 Zeta potential of Pride Mountain shale incubated with anionic polyacrylamide under various conditions

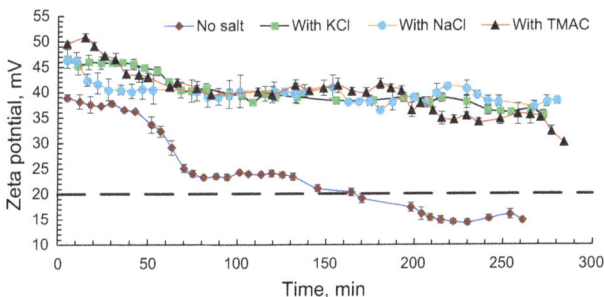

Fig. 12 Zeta potential of Pride Mountain shale incubated with cationic polyacrylamide under various conditions

Colloidal particles in suspension either flocculate or deflocculate depending on which force predominates, the van der Waals attractive force or the double layer repulsive force (Street and Wang 1966). In the absence of salt, cationic polyacrylamide caused flocculation of shale particles with time because the attractive forces predominated. Since there was rapid flocculation as the shale came in contact with cationic polyacrylamide, it was difficult to determine whether adsorption density increased with time. Additional studies will need to be performed with cationic polyacrylamide and shale to determine the effect of adsorption on zeta potential. The observed flocculation was a sign of strong interaction of polymer with shale, and the addition of salt inhibited flocculation.

In the anionic polyacrylamide system for Chattanooga and Pride Mountain shale, there is an increase in the absolute value of the zeta potential with time. This is indicative of the increase in the double layer thickness, which, in turn, is due to increasing adsorption density of polyacrylamide. In the Chattanooga shale, the absence of salt causes the absolute value of the zeta potential to increase to a point and then level off. Salt helps to decrease the ionic nature of clay and thus leaves fewer sites remaining for the polymer to adsorb (Menon and Wasan 1987a; Kulshrestha et al. 2004). In previous work, it has been shown that salts such as KCl minimize clay hydration and swelling, thereby minimizing the interaction of shale with fluid (van Oort 1994; Patel et al. 2001; van Oort 2003;

Patel 2009; Anderson et al. 2010; Lane and Aderibigbe 2013). However, with Pride Mountain shale, the zeta potential values are higher in the presence of TMAC. This is attributed to both the shale and the polyacrylamide having predominantly negative surface charge, which leads to an overall increase in charge of the system and also the Pride Mountain shale (being rich in smectites) has more exchangeable sodium ions. Ammonium ions from TMAC exchange with smaller sodium ions; ammonium with its larger hydration radius increases the swelling leading to an increase in zeta potential values.

The zeta potential of Chattanooga and Pride Mountain shales was measured in different saline media before adding the anionic and cationic polyacrylamide. Figure 13 shows the increase in the absolute value of the zeta potential after adding the anionic polyacrylamide to the shale–salt solution (i.e., the difference in the value of zeta potential before and after adding anionic polyacrylamide). In anionic polyacrylamide, KCl was the most effective shale inhibitor followed by TMAC and NaCl for Chattanooga shale. For Pride Mountain shale, KCl also was the most effective shale inhibitor, but NaCl was slightly more effective than TMAC. The reason for KCl providing better inhibition was because potassium ions have a smaller hydration radius and can easily exchange with the more swellable sodium ions on shale surface and due to their small hydration radius, they reduce swelling and provide better shale inhibition.

In order to observe flocculation of shale with cationic polyacrylamide, particle size measurements were made with time for the Pride Mountain shale–cationic polyacrylamide system. Figure 14 shows the increase in effective diameter of the shale particles with time. The system became unstable after 120 min because of flocculation and particle settling.

As shown in Fig. 14, the effective diameter increased with time indicative of flocculation.

The results are in agreement with previous work on the effect of adsorption density on zeta potential. As adsorption density increases, the zeta potential of the shale polyacrylamide complex increases and then levels off when the adsorption density approaches capacity (Menon and Wasan 1987a). In the presence of cationic polyacrylamide, by contrast, the absolute value of zeta potential decreases due to flocculation. In summary, salt tends to decrease the adsorption density of polymer on clay surfaces and leaves fewer active sites on the clay surfaces for the polyacrylamides to interact.

3.2 Rheological study of polymer–shale interaction

In this section, we discuss the rheology of the anionic and cationic polyacrylamide before and after adding the ground shale particles. The interaction of anionic polyacrylamide with the different shale samples was plotted as a function of anionic polyacrylamide concentration. The concentration of shale was kept constant at 0.5 lb/bbl, and the concentration of anionic polyacrylamide was varied from 0.1 to 0.2 wt%. The change in zero shear rate viscosity for the change in anionic polyacrylamide concentration is shown in Fig. 15.

Figure 15 demonstrates that each shale interacts differently with anionic polyacrylamide. The Chattanooga shale sample has the highest viscosity in a given polyacrylamide concentration, and the Pride Mountain sample has the lowest viscosity. Usually the sample with highest viscosity is considered to have strong interaction of the bulk fluid with the shale particles, but in this case, the viscosity decreases after adding the shale to the polyacrylamide, i.e., polyacrylamides without shale have higher viscosity values at a given polyacrylamide concentration. This was indicative that the polyacrylamides adsorbing onto the shale and leaving the solution are causing the decrease in the viscosity. Hence, Pride Mountain shale has stronger interactions with anionic polyacrylamide.

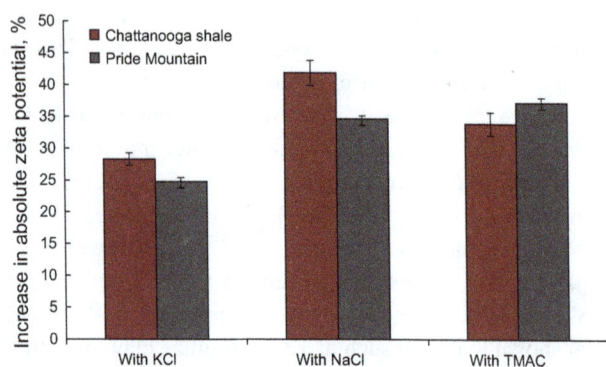

Fig. 13 Increase in zeta potential for shale in anionic polyacrylamide with salt and TMAC

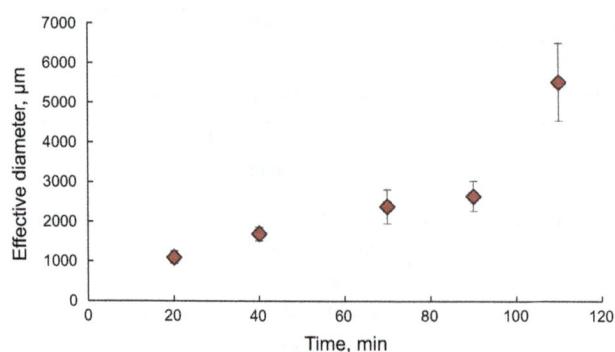

Fig. 14 Flocculation of Pride Mountain shale in the presence of cationic polyacrylamide as measured by dynamic light scattering

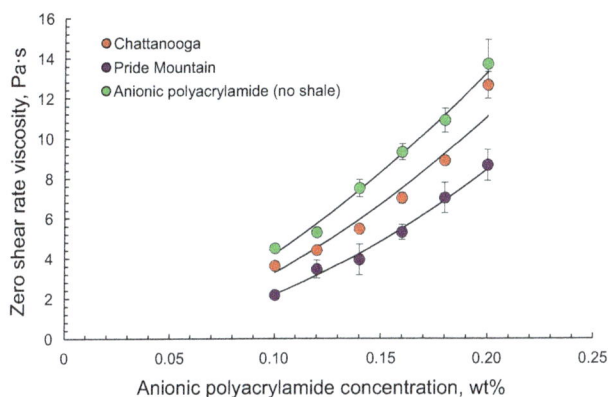

Fig. 15 Change in zero shear rate viscosity with increasing anionic polyacrylamide concentration for Chattanooga and Pride Mountain shale

Interestingly, the viscosity of the anionic polyacrylamide remains unchanged for the same experimental conditions, which proves that the anionic polyacrylamide remains stable over course of the experiment. In comparison with the zeta potential tests, rheological studies were easier to perform and the results are easier to interpret. Simple rheological methods like this can be used to assess shale–fluid interaction qualitatively.

4 Conclusion

A method was developed to characterize polyacrylamide–shale interaction. Zeta potential and rheological measurements were made to semi-quantify these interactions. Based on the studies, cationic polyacrylamide interacts with both the shales strongly even in the presence of salt and TMAC, whereas anionic polyacrylamide interacts less with the shales. Each type of shale analyzed interacts differently with polyacrylamide. All samples interact strongly with cationic polyacrylamide because of the negative surface charge on clay platelets. It is recommended to use anionic polyacrylamide because of its minimal interaction and also compatibility with other fluid additives. Due to the cationic polyacrylamides interacting strongly with shale, it can potentially cause formation damage. Both the rheological studies and the zeta potential tests gave the same results. Rheological methods are easier to perform and require less time compared to zeta potential experiments and can be used for qualitative understanding of shale–fluid interaction while zeta potential tests can be used for semiquantitative understanding alterations to shale surface when in contact with different fluids. It is imperative to understand fluid–rock interaction extensively, and this is especially true for polyacrylamide. Additives that are widely used as good shale inhibitors for one formation need not necessarily work well for another formation. For instance, in this study TMAC was an effective inhibitor for Chattanooga shale but increased swelling in Pride Mountain shale. This study reiterates the importance of testing shale for additives that can cause wellbore instability before injecting the fluids. Further studies are being performed to model the polymer–shale interaction and to identify additives that would facilitate effective friction reduction while minimizing these interactions.

In order to determine the change in viscosity of the shale–polymer samples with time, flow ramp tests were conducted on the samples for 5 days at equal intervals. The concentration of the anionic polyacrylamide and shale was kept constant at 0.16 wt% and 0.5 lb/bbl, respectively. After taking the first reading, the sample was left undisturbed in the geometry for few hours before the next reading. The sample was manually stirred in order to suspend the shale particles in the anionic polyacrylamide sample before starting the experiment. Figure 16 shows the change in viscosity of the shale–polymer sample with time.

The viscosity curve (Fig. 16) follows the same trend for both Pride Mountain and Chattanooga shales. After 2000 min, the viscosity remains constant. This signifies the point at which the clay particles have reached saturation in the anionic polyacrylamide solution. The percentage of reduction of viscosity was approximately same for both the shales at the end of Day 5, which is ∼34%. The polyacrylamide was adsorbed onto the surface of shale particles, which leads to decreasing viscosity with time.

References

Aften C, Watson WP. Improved friction reducer for hydraulic fracturing. SPE Hydraul Fract Technol Conf, The Woodlands, Texas, Soc Pet Eng. 2009. doi:10.2118/118747-MS.

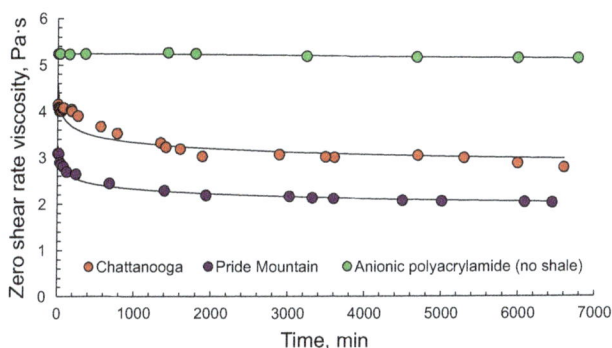

Fig. 16 Change in zero shear rate viscosity with time of Chattanooga and Pride Mountain shale in anionic polyacrylamide

Al-Bazali TM. Experimental study of the membrane behavior of shale during interaction with water-based and oil-based muds. Dissertation, University of Texas; 2005.

Anderson RL, Ratcliffe I, Greenwell HC, et al. Clay swelling—a challenge in the oilfield. Earth Sci Rev. 2010;98(3):201–16. doi:10.1016/j.earscirev.2009.11.003.

Brooks DE, Seaman GVF. The effect of neutral polymers on the electrokinetic potential of cells and other charged particles: I. Models for the zeta potential increase. J Colloid Interface Sci. 1973;43(3):670–86. doi:10.1016/0021-9797(73)90413-X.

Carman PS, Cawiezel K. Successful breaker optimization for polyacrylamide friction reducers used in slickwater fracturing. SPE Hydraul Fract Technol Conference, College Station, Texas, Soc Pet Eng. 2007. doi:10.2118/106162-MS.

Clark P, Pashin J, Carlson E, et al. Site Characterization for CO_2 Storage from Coal-fired Power Facilities in the Black Warrior Basin of Alabama, University Of Alabama. DOE report. 2012.

Delgado ÁV, González-Caballero F, Hunter RJ, et al. Measurement and interpretation of electrokinetic phenomena. J Colloid Interface Sci. 2007;309(2):194–224. doi:10.1016/j.jcis.2006.12.075.

Ghassemi A, Diek A. Linear chemo-poroelasticity for swelling shales: theory and application. J Pet Sci Eng. 2003;38(3):199–212. doi:10.1016/S0920-4105(03)00033-0.

Goh R, Leong Y-K, Lehane B. Bentonite slurries—zeta potential, yield stress, adsorbed additive and time-dependent behaviour. Rheol Acta. 2011;50(1):29–38. doi:10.1007/s00397-010-0498-x.

Gomez SL, He W. Fighting wellbore instability: customizing drilling fluids based on laboratory studies of shale-fluid interactions. IADC/SPE Asia Pac Drill Technol Conf Exhib Tianjin, China, Soc Pet Eng. 2012. doi:10.2118/155536-MS.

Harris PC. Fracturing-fluid additives. J Pet Technol. 1988;40(10):1277–279. doi:10.2118/17112-PA.

He W, Gomez SL, Leonard RS, et al. Shale-fluid interactions and drilling fluid designs. IPTC 2014: International Petroleum Technology Conference. 2014. 19–22 Jan, Doha, Qatar.

Heurtault B, Saulnier P, Pech B, et al. Physico-chemical stability of colloidal lipid particles. Biomaterials. 2003;24(23):4283–300. doi:10.1016/S0142-9612(03)00331-4.

Horsrud P, Bostrom B, Sonstebo EF, et al. Interaction between shale and water-based drilling fluids: laboratory exposure tests give new insight into mechanisms and field consequences of KCl contents. SPE Annu Tech Conf Exhib Soc Pet Eng. 1998. doi:10.2118/48986-MS.

Hunter RJ. Zeta potential in colloid science: Principles and applications. Cambridge: Academic Press; 2013.

Jia X, Williams RA. Particle deposition at a charged solid/liquid interface. Chem Eng Commun. 1990;91(1):127–98. doi:10.1080/00986449008940706.

Jiang L, Gao L, Sun J. Production of aqueous colloidal dispersions of carbon nanotubes. J Colloid Interface Sci. 2003;260(1):89–94. doi:10.1016/S0021-9797(02)00176-5.

Johnson JD, Schoppa D, Garza JL, et al. Enhancing Gas and Oil Production With Zeta Potential Altering System. SPE Int Symp Exhib Form Damage Control, Lafayette, Louisiana, Soc Pet Eng. 2010. doi:10.2118/128048-MS.

Kaya A, Oren A H, Yukselen Y. Settling behavior and zeta potential of kaolinite in aqueous media. The Thirteenth International Offshore and Polar Engineering Conference, Honolulu, Hawaii, International Society of Offshore and Polar Engineers. 2003.

Khodja M, Canselier JP, Bergaya F, et al. Shale problems and water-based drilling fluid optimisation in the Hassi Messaoud Algerian oil field. Appl Clay Sci. 2010;49(4):383–93. doi:10.1016/j.clay.2010.06.008.

Kulshrestha P, Giese RF, Aga DS. Investigating the molecular

interactions of oxytetracycline in clay and organic matter: insights on factors affecting its mobility in soil. Environ Sci Technol. 2004;38(15):4097–105. doi:10.1021/es034856q.

Lal M. Shale stability: drilling fluid interaction and shale strength. SPE Asia Pac Oil and Gas Conf Exhib Jakarta, Indonesia Soc Pet Eng. 1990. doi:10.2118/54356-MS.

Lane RH, Aderibigbe AA. Rock/fluid chemistry impacts on shale fracture behavior. SPE Int Symp Oilfield Chem Soc Pet Eng. 2013. doi:10.2118/164102-MS.

Lee L, Patel A D, Stamatakis E. Glycol based drilling fluid. 2001; US Patent No 6291405.

Lu CF. A new technique for the evaluation of shale stability in the presence of polymeric drilling fluid. SPE Prod Eng. 1988;3(03):366–74. doi:10.2118/14249-PA.

Luckham PF, Rossi S. The colloidal and rheological properties of bentonite suspensions. Adv Colliod Interface Sci. 1999;82(1):43–92. doi:10.1016/S0001-8686(99)00005-6.

Mahto V, Sharma VP. Rheological study of a water based oil well drilling fluid. J Pet Sci Eng. 2004;45(1–2):123–8. doi:10.1016/j.petrol.2004.03.008.

Menon VB, Wasan DT. Adsorption of maltenes on sodium montmorillonite. Colloids Surf. 1987a;25(2–4):387–92. doi:10.1016/0166-6622(87)80316-5.

Menon VB, Wasan DT. Particle—fluid interactions with applications to solid-stabilized emulsions Part III. Asphaltene adsorption in the presence of quinaldine and 1,2-dimethylindole. Colloids Surf. 1987b;23(4):353–62. doi:10.1016/0166-6622(87)80276-7.

Mody FK, Hale AH. Borehole-stability model to couple the mechanics and chemistry of drilling-fluid/shale interactions. J Pet Technol. 1993;45(11):1093–101. doi:10.2118/25728-PA.

Mody FK, Tare UA, Tan CP, et al. Development of novel membrane efficient water-based drilling fluids through fundamental understanding of osmotic membrane generation in shales. SPE Annu Tech Conf Exhib San Antonio, Texas, Soc Pet Eng. 2002. doi:10.2118/77447-MS.

Mueller S, Llewellin EW, Mader HM. The rheology of suspensions of solid particles proceedings: mathematical. Phys Eng Sci. 2010;466(2116):1201–28. doi:10.1098/rspa.2009.0445.

Muniz ES, Fontoura SAB, Lomba RFT. Rock-drilling fluid interaction studies on the diffusion cell. SPE Latin American and Caribbean Pet Eng Conf, Rio de Janeiro, Brazil, Soc Pet Eng. 2005. doi:10.2118/94768-MS.

Patel AD. Design and development of quaternary amine compounds: shale inhibition with improved environmental profile. SPE Int Symp Oilfield Chem, Soc Pet Eng. 2009. doi:10.2118/121737-MS.

Patel A D, Stamatakis E, Davis E. Shale hydration inhibition agent and method of use; US Patent 6247543. (2001).

Petersen PB, Saykally RJ. Is the liquid water surface basic or acidic? macroscopic vs. molecular-scale investigations. Chem Phys Lett. 2008;458(4):255–61. doi:10.1016/j.cplett.2008.04.010.

Rao MA. Flow and functional models for rheological properties of fluid foods. Rheology of fluid, semisolid, and solid foods. Berlin: Springer; 2014. p. 27–61.

Street N, Wang F D.(1966) Surface potentials and rock strength. 1st ISRM Congress, Lisbon, Portugal, International Society for Rock Mechanics.

Tan CP, Richards BG, Rahman SS. Managing physico-chemical wellbore instability in shales with the chemical potential mechanism. SPE Asia Pac Oil and Gas Conf, Adelaide, Australia, Soc Pet Eng. 1996. doi:10.2118/36971-MS.

van Olphen H. An introduction to clay colloid chemistry. 2nd ed. Hoboken: Wiley; 1977.

van Oort E. A novel technique for the investigation of drilling fluid induced borehole instability in shales. Rock Mech Pet Eng, Delft, Netherlands, Soc Pet Eng. 1994. doi:10.2118/28064-MS.

van Oort E. On the physical and chemical stability of shales. J Pet Sci Eng. 2003;38(3):213–35. doi:10.2118/28064-MS.

van Oort E, Hale AH, Mody FK. Manipulation of coupled osmotic flows for stabilisation of shales exposed to water-based drilling fluids. SPE Annu Tech Conf Exhib, Dallas, Texas, Soc Pet Eng. 1995. doi:10.2118/30499-MS.

Vane LM, Zang GM. Effect of aqueous phase properties on clay particle zeta potential and electro-osmotic permeability: implications for electro-kinetic soil remediation processes. J Hazard Mater. 1997;55(1–3):1–22. doi:10.1016/S0304-3894(97)00010-1.

Werner C, Zimmermann R, Kratzmüller T. Streaming potential and streaming current measurements at planar solid/liquid interfaces for simultaneous determination of zeta potential and surface conductivity. Colloids Surf A. 2001;192(1):205–13. doi:10.1016/S0927-7757(01)00725-7.

Yalçın T, Alemdar A, Ece ÖI, et al. The viscosity and zeta potential of bentonite dispersions in presence of anionic surfactants. Mater Lett. 2002;57(2):420–4. doi:10.1016/S0167-577X(02)00803-0.

Yu M, Chenevert ME, Sharma MM. Chemical–mechanical wellbore instability model for shales: accounting for solute diffusion. J Pet Sci Eng. 2003;38(3):131–43. doi:10.1016/S0920-4105(03)00027-5.

Zhang J, Al-Bazali TM, Chenevert ME, et al. Sharma factors controlling the membrane efficiency of shales when interacting with water-based and oil-based muds. Int Oil and Gas Conf Exhib China, Beijing, China, Soc Pet Eng. 2006. doi:10.2118/100735-MS.

Zhong H, Qiu Z, Huang W, et al. Shale inhibitive properties of polyether diamine in water-based drilling fluid. J Pet Sci Eng. 2011;78(2):510–5. doi:10.1016/j.petrol.2011.06.003.

13

Influence of friction on buckling of a drill string in the circular channel of a bore hole

Valery Gulyayev[1] · Natalya Shlyun[1]

Abstract Enhancement of technology and techniques for drilling deep directed oil and gas bore hole is one of the most important problems of the current petroleum industry. Not infrequently, the drilling of these bore holes is attended by occurrence of extraordinary situations associated with technical accidents. Among these is the Eulerian loss of stability of a drill string in the channel of a curvilinear bore hole. Methods of computer simulation should play a dominant role in prediction of these states. In this paper, a new statement of the problem of critical buckling of the drill strings in 3D curvilinear bore holes is proposed. It is based on combined use of the theory of curvilinear elastic rods, Eulerian theory of stability, theory of channel surfaces, and methods of classical mechanics of systems with nonlinear constraints. It is noted that the stated problem is singularly perturbed and its solutions have the shapes of localized harmonic wavelets. The calculation results showed that the friction effects lead to essential redistribution of internal axial forces, as well as changing the eigenmode shapes and sites of their localization. These features make the buckling phenomena less predictable and raise the role of computer simulation of these effects.

Keywords Directed bore hole · Drill string · Critical states · Singular perturbation · Friction forces · Harmonic wavelet

✉ Valery Gulyayev
valery@gulyayev.com.ua

[1] Department of Mathematics, National Transport University, Kiev, Ukraine

Edited by Yan-Hua Sun

1 Introduction

Not long ago, rather shallow wells with simple outlines were drilled in oil and gas fields. However, at the present time, deeper and more complicated trajectories of bore holes are designed in connection with exhaustion of easily accessible hydrocarbon sources. In 2015, the record 13.5 km horizontal bore hole was drilled in the Sakhalin region, Russia. According to experts' opinions, most of the substantial achievements in the power engineering of the current century are associated with this technical direction. Particularly, some are related to the pioneering investigation of industrial extraction of shale oil and gas, whose deposits in the world essentially exceed conventional reserves. However, as a rule, drilling of such bore holes is attended by extraordinary phenomena bringing emergency situations. One of them is unstable bending buckling of a drill string (DS) in the channel of a curvilinear bore hole (Brett et al. 1989; Dawson and Paslay 1984; Kyllingstad 1995; Sawaryn et al. 2006; Gulyayev et al. 2009; Gao and Liu 2013; Huang and Gao 2014; Gao and Huang 2015). This effect is associated with deterioration of conditions of contact interaction between the DS and the bore hole wall, enlargement of friction forces, impossibility of transferring the required axial force to the bit, and the DS lockup situation. To predict these effects and exclude them in practice, computer simulation should be employed.

In parallel with the static phenomena of buckling of the DSs, there also can occur very complicated nonlinear dynamic processes accompanied by extraordinary stable and unstable changes of the DS rotation. Among them there are axial, torsional, bending, and whirl vibrations, inevitably linked with deterioration of the drilling efficiency (Liu et al. 2013, 2014a, b).

The effects of loss of equilibrium stability in these mechanical systems are manifestations of one of the most general law of nature—the law of quantitative changes transferring to qualitative ones. In different spheres of reality, these changes are realized by different ways. In mechanics, they are studied on the basis of bifurcation theory.

The problem of mechanical instability and bifurcational buckling acquires crucial urgency in the technology of long curvilinear bore hole drilling because it is specified by essential complications but is not yet understood. Current experience testifies that no well is drilled without problems. They are connected with the complexity of mechanical phenomena accompanying the drilling process and the absence of dependable methods of computer modeling providing the possibility to predict emergency situations and to exclude them in advance. In a vertical bore hole, the DS stability loss occurs at its lower part following the spiral buckling mode typical for a rod stretched, compressed, and twisted simultaneously (Lubinski et al. 1962; Gulyayev et al. 2009).

However, the problem of theoretic simulation of DS buckling in the channel of a curvilinear bore hole acquires supplementary difficulties associated with the necessity of integrating differential equations with variable coefficients in the full range of the large length of the DS. Besides, the problem possesses essential complications stemming from appearance of additional constraints, imposed on the DS by the well wall surface, and application of contact and friction distributed forces, as well as change of orientation of gravity forces compressing the DS to the well wall (Dawson and Paslay 1984; Wang and Yuan 2012; Gulyayev et al. 2014).

Comprehensive reviews of results achieved in this direction are presented in the literature (Cunha 2004; Mitchell 2008; Gao and Huang 2015). It stems from these analyses that, as a rule, the approaches used in bifurcational analysis are based on the eigenmode approximations by regular sinusoids and spirals, while the critical values of loads and buckling shapes are rather guessed. As Cunha notes apparently this circumstance is the reason of conclusions that solutions of the problem on stability loss of the DSs in curvilinear bore holes gained by different authors are in contradiction with each other and reality (Cunha 2004).

Mitchell emphasizes "that there are still challenging problems to solve and difficult questions to answer" in the domain of the DS buckling (Mitchell 2008). Among them, the fundamental unresolved questions remain:

- What is the critical buckling load in curved, 3D bore holes?
- What effect does friction play in DS buckling?

The last achievements in this domain are discussed by Gao and Huang (2015).

Gulyayev et al. (2014, 2015) elaborated a new mathematic model of a DS stability loss in smooth curvilinear channels. Without taking into consideration friction effects, they showed that the stated problem was singularly perturbed and so, typically, the modes of buckling were represented by boundary and localized effects in the shapes of wave packages or wavelets.

In this problem, fundamental unresolved questions remain: What role is played by the friction factor in stability loss phenomena and how to include it in the analytical or numerical model. Mitchell remarks: "Perhaps the most important force, and the force least studied in the analysis of buckling, is friction" (Mitchell 2008). The magnitude of the friction force is usually not that difficult to determine. The difficulty is determining the direction of the friction vector. We agree that this vector cannot be determined for stationary elastic systems with friction contacts because this problem is statically indeterminable (Mitchell and Samuel 2009). But if one of the contacting bodies slides on the surface of another, then the vectors of sliding velocity and friction force are collinear and the last one can be easily determined. This peculiarity facilitates the problem on the analysis of friction effects on critical buckling of the DS.

In the first place, it is necessary to point out that every balanced stationary state of the DS is preceded by its steady sliding motion associated with tripping in or out operation and drilling is accompanied by kinematic friction. As a rule, the coefficients of kinematic friction exceed their magnitudes established in static equilibrium. Besides, usually, the indicated technological procedures happen with certain ultimate longitudinal velocities, while the DS buckling occurs with very small velocities. In this case, the lateral buckling velocities (and the appropriate lateral friction forces) can be assumed to equal zero. Then, all the friction forces are axial and are oriented in one direction. Then, the moving DS experiences action of the more intensive friction forces which have deleterious effects upon the stability of its quasi-static equilibrium. Therefore, at this state, the bifurcational buckling of the DS should first of all be analyzed.

Secondly, the friction forces acting on the DS and all the functions of its total stress–strain state can be specified by rather simple calculation means.

Thirdly, with the use of these functions, the constitutive linearized homogeneous equations of critical equilibrium of the DS can be constructed. Their eigenvalues and eigenmodes determine critical loads on the DS in the bore hole channel and shapes of its bifurcational buckling.

To realize this approach, the nonlinear theory of elastic curvilinear rods is used. Its foundations are stated in monographs (Antman 2005; Gulyayev et al. 1992). A three-dimensional statement of this theory is expounded by Gulyayev and Tolbatov (2004). Its application to analysis of the DSs buckling in inclined rectilinear bore holes is described by Gulyayev et al. (2014). Below, it is formulated in a concomitant reference frame moving on the bore hole surface constraining the DS transformation. Owing to this, the total order of the differential equation is reduced to four. A two-step algorithm is proposed. At the first step, the stress–strain state of the moving DS under action of gravity and friction forces is determined; at the second step, the eigenvalue problem for linearized equations is solved. It is shown, that the buckling modes have the shapes of harmonic wavelets with localization segments depending on the friction forces.

2 Basic assumptions concerning drill string bending in a curvilinear bore hole

The problem about nonlinear elastic bending of a DS relative to the immovable coordinate system $OXYZ$ inside a channel cavity of a curvilinear bore hole is considered and shown in Fig. 1. In the considered case, the DS is lowering along its axial line but the surfaces of the DS and the bore hole are in contact throughout the DS length. The DS does not rotate; hence, the distributed friction torques equal zero. Assuming that the DS movement is quasi-static, then, the inertia forces are small and can be disregarded, so only the distributed gravitational ($\mathbf{f}^{gr}$), contact ($\mathbf{f}^{cont}$), and

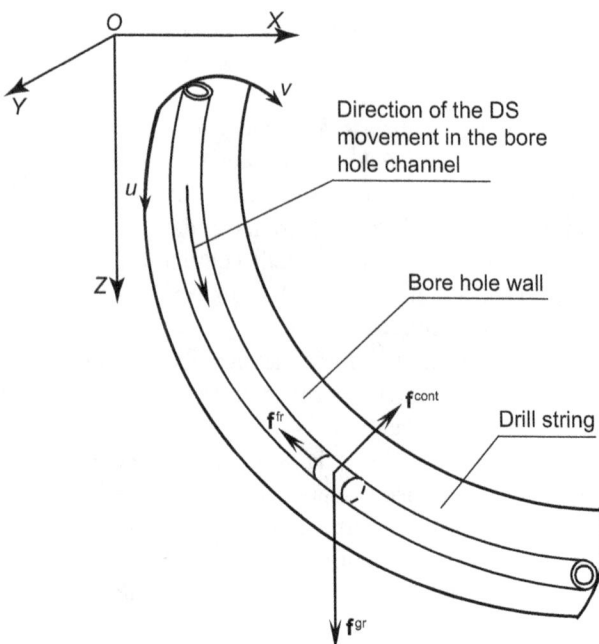

Fig. 1 Scheme of a drill string in a bore hole channel

frictional ($\mathbf{f}^{fr}$) forces are acting on every element of the DS. As shown in Fig. 1, the $\mathbf{f}^{gr}$ force is vertical, $\mathbf{f}^{cont}$ is applied normally to the axis line at the contact point, and $\mathbf{f}^{fr}$ force is opposite to the axial velocity of the descending element.

It is believed, also, that the DS element displacements can be comparable with the bore hole cross-sectional dimensions, but the curvature radii of its axis line L are so large that the DS strains are small and its stress–strain states are elastic. They are specified by the principal vectors of internal forces $\mathbf{F}(s)$ and internal moments $\mathbf{M}(s)$, where s is the natural parameter defined by the length of the DS axis line L measured from some initial point to the considered one.

The external and internal force factors have to satisfy the following differential equations of equilibrium (Gulyayev et al. 1992, 2014):

$$\frac{d\mathbf{F}}{ds} = -\mathbf{f}^{gr} - \mathbf{f}^{cont} - \mathbf{f}^{fr}, \quad \frac{d\mathbf{M}}{ds} = -\mathbf{t} \times \mathbf{F}, \tag{1}$$

where $\mathbf{t}$ is the unit vector directed along the tangent to the axis line L.

If Eq. (1) is projected on the immovable coordinate system $OXYZ$, it will be possible to receive six scalar equations of the element equilibrium. However, in a general case, it is more convenient to express them in axes of some movable trihedron. Usually, if elastic bending of the unconstrained curvilinear rod is studied, the Frenet trihedron with unit vectors of normal $\mathbf{n}$, binormal $\mathbf{b}$, and tangent $\mathbf{t}$ are used. They are calculated by the formulae:

$$\mathbf{t} = \frac{d\mathbf{R}}{ds}, \quad \mathbf{n} = r\frac{d\mathbf{t}}{ds}, \quad \mathbf{b} = \mathbf{t} \times \mathbf{n} \tag{2}$$

where $\mathbf{R}(s)$ is the radius vector of the DS element in the $OXYZ$ coordinate system; $r(s)$ is the curvature radius of the line L.

Yet, if the DS is in contact with the well wall surface, then, the axis line L lies and slides in the channel surface Σ of radius a (Fig. 2) which is equal to half-difference

$$a = (d_1 - d_2)/2,$$

where d_1 and d_2 are the diameters of the bore hole and DS cross sections, respectively.

This surface can be parameterized by parameter u, determining the axis line T of the bore hole

$$X_T = X_T(u), \quad Y_T = Y_T(u), \quad Z_T = Z_T(u) \tag{3}$$

and parameter v, prescribing position of a point in the generating circle (Fig. 2). Here, u, v are curvilinear coordinates in the surface Σ; Σ is the channel surface of the bore hole wall; X_T, Y_T, Z_T are the X, Y, Z coordinates of the line T.

With their use, the DS bending is preset in the 2D space of the Σ surface by equalities

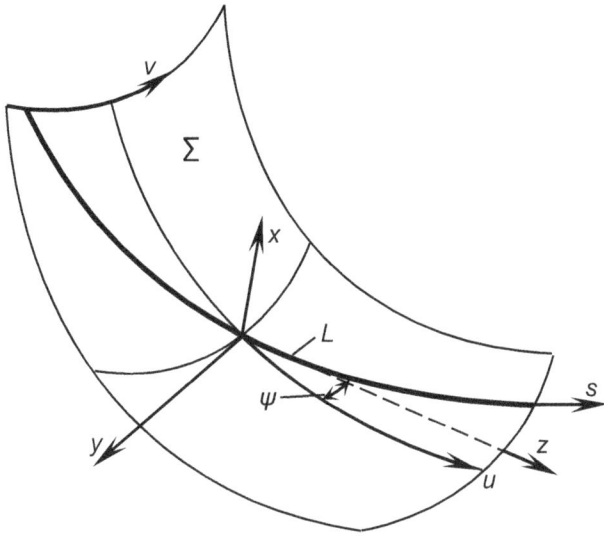

Fig. 2 Sliding of axial line L in reference surface Σ

$$u = u(s), \quad v = v(s) \tag{4}$$

Then, to describe the line L transforming, it is convenient to introduce additional right-hand reference frame $oxyz$ with unit vectors $\mathbf{i}, \mathbf{j}, \mathbf{k}$, moving on the constraining surface Σ along the line L. In doing so, the unit vector $\mathbf{i}$ is the internal normal to the surface Σ and the vector $\mathbf{k}$ is tangent to the curve L.

Now, it becomes possible to introduce an analogue

$$\boldsymbol{\omega} = k_x \mathbf{i} + k_y \mathbf{j} + k_z \mathbf{k} \tag{5}$$

of the Darboux vector (Dubrovin et al. 1992; Gulyayev et al. 1992).

In this equality, k_x and k_y are the appropriate components of the vector $\boldsymbol{\Omega} = \mathbf{b}/r$ of the curvature of the line L along the ox and oy axes; k_z is the value

$$k_z = \lim_{\Delta s \to 0} \Delta \psi / \Delta s \tag{6}$$

determining rotation of the $oxyz$ system around the vector $\mathbf{k}$ when this reference frame moves along the line L from the point s to the point $s + \Delta s$.

Here, $\Delta \psi$ is the elementary angle of vector $\mathbf{i}$ rotation.

Generally, if displacements of a rod are not constrained, its curvature $1/r$ can be calculated by the formula:

$$\frac{1}{r} = \sqrt{\frac{d^2 X}{ds^2} + \left(\frac{d^2 Y}{ds^2}\right)^2 + \frac{d^2 Z}{ds^2}} \tag{7}$$

and, subsequently, its components k_x and k_y can be determined. However, in the considered case, the rod axis L lies in the surface Σ with prescribed geometry and so it is more convenient to specify functions $k_x(s)$ and $k_y(s)$ in the terms of its geometrical parameters.

To do so, it is necessary to consider internal and external geometries of the surface Σ. Its internal geometry is defined by the first quadratic form:

$$\Phi_1(u, v) = a_{11}du^2 + 2a_{12}dudv + a_{22}dv^2, \tag{8}$$

where a_{11}, a_{12}, and a_{22} are parameters of the quadratic form. With their use, the geometrical objects, lying in the surface Σ, are described and calculated.

In the considered case, the surface Σ is a channel and then the coordinate lines $u = $ const and $v = $ const are orthogonal and $a_{12}(u, v) = 0$. Owing to this, the curvature k_x, coinciding with geodesic curvature k^{geod} of the curve L, is expressed with the use of the formula (Dubrovin et al. 1992)

$$k_x = k^{\text{geod}} = \sqrt{a_{11}a_{22}} \left[a_{11}(u')^2 + a_{22}(v')^2 \right]^{-3/2}$$
$$(u''v' - v''u' + Av' - Bu'). \tag{9}$$

Here, coefficients A and B are represented through the Cristoffel symbols Γ_{jj}^i by the equalities

$$A = \Gamma_{11}^1(u')^2 + \Gamma_{22}^1(v')^2, \quad B = \Gamma_{11}^2(u')^2 + \Gamma_{22}^2(v')^2 \tag{10}$$

The surface Σ shape, curvatures, and external geometry are determined by parameters b_{11}, b_{12}, b_{22} of the second quadratic form:

$$\Phi_2(u, v) = b_{11}du^2 + 2b_{12}dudv + b_{22}dv^2. \tag{11}$$

If the surface Σ is a channel, the correlation $b_{12} = 0$ is valid for the chosen coordinate u and v. Then, on the basis of the Euler theorem (Dubrovin et al. 1992), the curvature k_y of the line L can be equalized to appropriate normal curvature k^{norm} of the surface Σ in the direction of the curve L. In its turn, the curvature k^{norm} is expressed through principal curvatures k_1, k_2 of the surface Σ

$$k^{\text{norm}} = k_y = k_1 \cos^2 \theta + k_2 \sin^2 \theta. \tag{12}$$

Here, θ is the angle between the directions of curve L and coordinate line u; k_1 and k_2 are the normal curvatures of lines $v = $ const, $u = $ const, respectively. They can be represented as follows:

$$k_1 = b_{11}/a_{11}, \quad k_2 = b_{22}/a_{11} \tag{13}$$

The gained relations (6), (9), (12) permit one to study elastic bending of the DS in a movable reference frame (5).

3 Nonlinear constitutive equations of the drill string bending in a curvilinear channel

To deduce constitutive equations of bending, the drill string sliding along its axial line in a curvilinear bore hole Eq. (1) is represented in a moving reference frame $oxyz$ with unit vectors $\mathbf{i}$, $\mathbf{j}$, $\mathbf{k}$. Then, the absolute derivatives $d\mathbf{F}/ds$ and $d\mathbf{M}/ds$ in Eq. (1) can be expressed as follows:

$$dF/ds = \tilde{d}F/ds + \boldsymbol{\omega} \times \mathbf{F}, \quad dM/ds = \tilde{d}M/ds + \boldsymbol{\omega} \times \mathbf{M}, \tag{14}$$

Here, $\tilde{d}.../ds$ is the symbol of local derivative in the $oxyz$ system.

Vectors $\mathbf{F}$, $\mathbf{M}$, $\mathbf{f}^{gr}$, $\mathbf{f}^{cont}$, $\mathbf{f}^{fr}$ are resolved into their components in the $\mathbf{i}$, $\mathbf{j}$, $\mathbf{k}$ trihedron:

$$\begin{aligned} \mathbf{F} &= F_x\mathbf{i} + F_y\mathbf{j} + F_z\mathbf{k}, \quad \mathbf{M} = M_x\mathbf{i} + M_y\mathbf{j} + M_z\mathbf{k}, \\ \mathbf{f}^{gr} &= f_x^{gr}\mathbf{i} + f_y^{gr}\mathbf{j} + f_z^{gr}\mathbf{k}, \quad \mathbf{f}^{cont} = f_x^{cont}\mathbf{i}, \; \mathbf{f}^{fr} = f_z^{fr}\mathbf{k}. \end{aligned} \tag{15}$$

In these correlations, the advantages of the chosen approach and reference frame used are obvious. Indeed, the $\mathbf{f}^{cont}$ force is normal to the surface Σ, it is collinear with the vector $\mathbf{j}$, and has only one component. The problem on specification of friction forces is considerably harder. Assuming that the friction interaction between the DS tube and the bore hole wall obeys Coulomb's law (Berger 2002; Mitchell and Samuel 2009; Samuel 2010)

$$\mathbf{f}^{fr} \leq \mu\mathbf{f}^{cont} \tag{16}$$

where μ is the friction coefficient.

Following Eq. (16), the fraction can be subdivided into static friction ("stiction") between non-moving surfaces and kinetic friction generated between sliding ones (Fig. 3).

The static regimes occur under conditions when the motive forces cannot overcome the resistance of cohesion forces which have some ultimate value $\mathbf{f}^{ult} = \mu\mathbf{f}^{cont}$. Once the ultimate value has been achieved, the contacting bodies begin to move relative to each other with realization of kinetic friction which does not depend on the velocity $\mathbf{w}$ magnitude and is equal to $\mathbf{f}^{ult}$. Besides, the vector $\mathbf{f}^{fr}$ of this force is collinear with the vector $\mathbf{w}$ and the equality

$$\mathbf{f}^{fr} = -\mu|\mathbf{f}^{cont}|\frac{\mathbf{w}}{|\mathbf{w}|} \tag{17}$$

becomes valid.

This correlation permits one to investigate the processes of the DS buckling during its axial motion. Indeed, assuming that the DS is under conditions of tripping operations and internal axial force $F_z(s)$ does not achieve the critical value. Then, the DS moves along its axial line with the axial velocity w_z without buckling, the friction forces $f_z^{fr}(s)$ are directed along it, and lateral components of

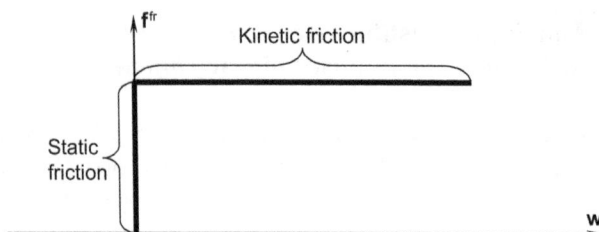

Fig. 3 Diagram of the Coulomb friction

these forces equal zero. Next, when the critical axial force $F_z^{cr}(s)$ is achieved and slightly exceeded, the DS begins to buckle with the small lateral velocity $w_y(s)$, generating small lateral friction forces

$$f_y^{fr}(s) = f_z^{fr}(s) \cdot w_y(s)/w_z \tag{18}$$

which impede the DS buckling with the induced velocity w_y.

Assuming that the generated friction forces (Eq. (18)) stopped the buckling process, then $w_y(s) = 0$ and the lateral forces $f_y^{fr}(s) = 0$. However, the critical axial forces are exceeded (though slightly), the DS remains unstable, and again it begins to buckle, but this time, it selects very small velocity $w_y(s)$ in Eq. (18), inducing very small lateral forces $f_y^{fr}(s)$, which cannot stop the buckling process. Therefore, it is considered that if in the DS, sliding along its axial line in the bore hole channel, the internal axial forces $F_z(s)$ exceed critical value F_z^{cr}, the influence of induced lateral friction forces $f_y^{fr}(s)$ on the buckling process can be neglected.

Therefore, the friction force is collinear with the vector $\mathbf{k}$ and its value constitutes (Mitchell and Samuel 2009) $f_z^{fr} = \pm\mu|f_x^{cont}|$, where signs "+", "−" are selected depending on the direction of DS movement.

Then, the system of vector correlations (1), (5), (14), and (15) can be reduced to the system of three scalar equations, describing equilibrium of internal and external forces:

$$\begin{aligned} dF_x/ds &= -k_yF_z + k_zF_y - f_x^{gr} - f_x^{cont}, \\ dF_y/ds &= -k_zF_x + k_xF_z - f_y^{gr}, \\ dF_z/ds &= -k_xF_y + k_yF_x - f_z^{gr} - f_z^{fr}, \end{aligned} \tag{19}$$

and three equations of internal moments equilibrium

$$\begin{aligned} dM_x/ds &= -k_yM_z + k_zM_y + F_y, \\ dM_y/ds &= -k_zM_x + k_xM_z - F_x, \\ dM_z/ds &= -k_xM_y + k_yM_x. \end{aligned} \tag{20}$$

Take into consideration that, according to the rod theory, bending moments M_x, M_y are determined by equalities

$$M_x = EIk_x, \quad M_y = EIk_y, \tag{21}$$

where E is Young's modulus; I is the moment of inertia of the DS cross-sectional area. Then, substituting Eq. (21) into the third equation of system (20), one gains

$$dM_z/ds = 0. \tag{22}$$

Hence, $M_z = const$ and its value can be calculated with the help of appropriate boundary conditions.

Now, with the introduction of the proposed movable reference frame $oxyz$ and vector (5), representing an analogue of the Darboux vector, it became possible to rewrite the second equation of system (20) in the form:

$$F_x = -EIk'_y - EIk_xk_z + M_zk_x. \tag{23}$$

Thereafter, contact force f_x^{cont} is found, using the first equation of system (19),

$$f_x^{\text{cont}} = EI\left(k''_y + k'_xk_z + k_xk'_z\right) + M_zk'_x - k_yF_z + k_zF_y - f_x^{\text{gr}} \tag{24}$$

In the results, systems (19) and (20) are reduced to three equilibrium equations

$$\begin{aligned} \frac{dF_y}{ds} &= EIk_xk_z^2 - M_zk_xk_z + EIk'_yk_z + F_zk_x - f_y^{\text{gr}}, \\ \frac{dF_z}{ds} &= -EIk_xk'_x - EIk_yk'_y - f_z^{\text{gr}} - f_z^{\text{fr}}, \\ \frac{dk_x}{ds} &= -\frac{M_z}{EI}k_y + k_yk_z + \frac{1}{EI}F_y. \end{aligned} \tag{25}$$

This system should be supplemented by the equations of the surface Σ, constraining displacements of the DS. They are formulated on the basis of channel surface properties with the use of Eq. (3). Generally, if the line T has a 3D geometry, the surface Σ can be represented as follows:

$$\begin{aligned} X &= X(a, X_T, Y_T, Z_T, u, v), \quad Y = Y(a, X_T, Y_T, Z_T, u, v), \\ Z &= Z(a, X_T, Y_T, Z_T, u, v) \end{aligned} \tag{26}$$

However, the most overwhelming obstacle associated with this problem consists of the necessity to simulate friction forces accompanying DS deformation. These forces are statically indeterminate for elastic systems. So, it is expedient to analyze particular cases of the DS movement inside curvilinear channels of simple trajectories and to study their stability. Of particular interest in this avenue of inquiry is incipient buckling of a DS in different segments of a plane circular channel because it adequately depicts the most general regularities of friction forces impact on the buckling phenomena.

4 Bifurcational equations of DS equilibrium in a circular bore hole

Let a DS be lowering in a plane circular bore hole. Then, it slides along the bore hole bottom line and its axis line is a circle of a radius $\rho + a$, where ρ is the radius of the bore hole axis T and a is the system clearance. In sliding, the DS is subjected to action of gravity (f^{gr}), contact (f^{cont}), and friction (f^{fr}) forces. In consequence of these forces, the DS can be compressed in some segments of its length where it can begin to buckle without losing its

contact with the well wall. It is necessary to predict the critical states of the DS and to construct the modes of its stability loss.

In this case, the surface Σ is a torus described by Eq. (26) in the form:

$$\begin{aligned} X &= a \sin v, \quad Y = \rho(1 - \cos u), \\ Z &= \rho \sin u + a \sin u \cos v \end{aligned} \tag{27}$$

By their application, the geometric parameters used in Eqs. (8), (11), and (13) are determined:

$$\begin{aligned} a_{11} &= (\rho + \cos v)^2, & a_{12} &= 0, & a_{22} &= a^2, \\ b_{11} &= (\rho + a \cos v)\cos v, & b_{12} &= 0, & b_{22} &= a, \\ k_1 &= \cos / (\rho + a \cos v), & k_2 &= 1/a \end{aligned} \tag{28}$$

The geodesic curvature $k^{\text{geod}} = k_x$ is calculated from Eq. (9)

$$k^{\text{geod}} = k_x = -a(\rho + a \cos v)\left[u''v' - v''u' - \sin vu' \frac{1 + a^2(v')^2}{a(\rho + a \cos v)}\right]. \tag{29}$$

The normal curvature $k^{\text{norm}} = k_y$ is determined by the equality

$$k^{\text{norm}} = k^y = \cos v \frac{1 - a^2(v')^2}{\rho + a \cos v} + a(v')^2. \tag{30}$$

Angle θ between directions of tangents to lines L and u at the considered point is introduced. Then, $\sin \theta = av'$, $\cos \theta = (\rho + a \cos v)u'$. Here, value u' is prescribed by the formula:

$$u' = du/ds = \pm \frac{\sqrt{(1 - auv')^2}}{\rho + a \cos v}. \tag{31}$$

With the use of these formulae, the appropriate components of the gravity force are constructed:

$$\begin{aligned} f_x^{\text{gr}} &= \mathbf{f}^{\text{gr}}\mathbf{i} = -f^{\text{gr}}\sin u \cos v, \\ f_y^{\text{gr}} &= \mathbf{f}^{\text{gr}}\mathbf{j} = f^{\text{gr}}(\sin u \sin v \cos \theta + \cos u \sin \theta), \\ f_z^{\text{gr}} &= \mathbf{f}^{\text{gr}}\mathbf{k} = f^{\text{gr}}(-\sin u \sin v \sin \theta + \cos u \cos \theta). \end{aligned} \tag{32}$$

Here, the distributed gravity force f^{gr} is defined by the formula

$$f^{\text{gr}} = g(\gamma_{\text{st}} - \gamma_{\text{mud}})\pi(d_1^2 - d_2^2)/4$$

where γ_{st} and γ_{mud} are the densities of steel and mud; d_1 and d_2 are the external and internal diameters of the DS tube.

Now, it became possible to formulate constitutive equations of the stated problem relative to unknown variables F_y, F_z, k_x, v, and u. They are deduced on the basis of Eqs. (25), (29), (31), and (32):

$$\begin{cases} \dfrac{dF_y}{ds} = EI k_x k_z^2 - M_z k_x k_z + EI k_y' k_z + k_x F_z - f^{\mathrm{gr}}(\sin u \sin v \cos\theta + \cos u \sin\theta), \\[2mm] \dfrac{dF_z}{ds} = -EI k_x k_x' - EI k_y k_y' - f^{\mathrm{gr}}(-\sin u \sin v \sin\theta + \cos u \sin\theta), \\[2mm] \dfrac{dk_x}{ds} = -\dfrac{M_z}{EI} k_y + k_y k_z + \dfrac{1}{EI} F_y, \\[2mm] \dfrac{dv}{ds} = v', \\[2mm] \dfrac{d(v')}{ds} = \dfrac{1}{a(\rho + a\cos v)u'} k_x - \sin v \dfrac{1 + a^2(v')^2}{a(\rho + a\cos v)} + \dfrac{u''v'}{u'}, \\[2mm] \dfrac{du}{ds} = \dfrac{\sqrt{1 - a^2(v')^2}}{(\rho + a\cos v)}. \end{cases}$$

$$(33)$$

This system with appropriate boundary conditions, determining external axial forces and torques, can be used for modeling nonlinear elastic bending of a DS in the circular channel cavity of a bore hole. Assuming that in the general case the DS can bend and take new deformed shapes remaining in contact with the bore hole wall throughout its length. If during this shape transformation small elastic displacements of the DS correspond to small increments of the external force perturbation, then the considered equilibrium state is stable. In the vicinity of this state, the linear differential equations deduced from the nonlinear system (33) with the use of linearization procedure are not degenerate and have only one solution. However, if the DS is loaded further, the coefficients of the linearized equations continue to evolve and the state can be reached when these equations become degenerate and acquire an additional (bifurcating) solution along with the initial one. Because of this, the state reached is critical (unstable) and the bifurcating solution represents the buckling mode.

The peculiarity of the nonlinear problem stated for the buckling of a DS in the circular channel is that the tube does not change its shape in the subcritical states and so the coefficients of the linearized equations of the equilibrium change owing to the step-by-step enlargement of the axial force $F_z(s)$ with the external load increase. In that event, the stated problem is analogous to the problem of Eulerian stability of a rectilinear rod because it is also associated with the eigenvalue search and eigenmode construction.

To identify the critical equilibrium and the stability loss of the DS moving along the bore hole bottom, system (33) should be linearized in the vicinity of the considered state and its eigenvalues and eigenmodes should be found. During the prescribed movement, the static and kinematic conditions $F_z = F_z(S)$, $u = u_0 + s/(\rho + a)$, $u' = 1/(\rho + a)$, $u'' = 0$, $v = 0$, $v' = 0$, $v'' = 0$, $k_x = 0$, $k_y = 1/(\rho + a)$, $k_z = 0$ are satisfied. As a consequence of bifurcational deformation, the system parameters assume small

variations δv, δk_x, δF_y. They are calculated with the help of linear homogeneous equations

$$\begin{cases} \dfrac{d}{ds}\delta F_y = F_z \delta k_x - f^{\mathrm{gr}}\sin\left(u_0 + \dfrac{s}{\rho + a}\right)\delta v + f^{\mathrm{gr}} a\cos\left(u_0 + \dfrac{s}{\rho + a}\right)\delta v', \\[2mm] \dfrac{d}{ds}\delta v = \delta(v'), \\[2mm] \dfrac{d}{ds}\delta(v') = -\dfrac{1}{a(\rho + a)}\delta v + \dfrac{1}{a}\delta k_x, \\[2mm] \dfrac{d}{ds}\delta k_x = \dfrac{1}{EI}\delta F_y, \end{cases}$$

$$(34)$$

arising from system (33) after taking into account that $\delta u = 0$, $\delta k_y = 0$, $\delta F_z = 0$.

Coefficient $F_z(s)$ in the first equation is determined with the help of the second equation of system (33). It results in

$$\frac{dF_z}{ds} = -f_z^{\mathrm{gr}} - f_z^{\mathrm{fr}} = -f^{\mathrm{gr}} \pm \mu f^{\mathrm{cont}}. \qquad (35)$$

Here, signs $\pm$ are selected for the operations lowering and hoisting of the DS, contact force f^{cont} is established through the use of the first equation of system (19) in the form:

$$f^{\mathrm{cont}} = -\frac{1}{\rho + a}F_z - f_x^{\mathrm{gr}} = -\frac{1}{\rho + a}F_z + f^{\mathrm{gr}}\sin u \qquad (36)$$

Substituting Eq. (36) into Eq. (35) gives the linear differential equation of the first order:

$$\frac{dF_z}{du} \pm \mu F_z = -(\rho + a)f^{\mathrm{gr}}\cos u \pm \mu(\rho + a)f^{\mathrm{gr}}\sin u \qquad (37)$$

Let $u = U$, $s = S$, then

$$F_z(u) = \frac{(\rho + a)f^{\mathrm{gr}}}{1 + \mu^2}\left[\pm 2\mu(\cos u - \cos U) - (1 - \mu^2)(\sin u - \sin U)\right] + F_z(U),$$

$$(38)$$

where $F_z(U)$ is the compressive force applied to the DS at its lower end $u = U$; U is the u coordinate value at $s = S$.

Four first-order differential Eq. (34) are equivalent to one homogeneous fourth order equation

$$\delta v^{\mathrm{IV}} + \left[\frac{1}{a(\rho + a)} - \frac{F_z}{EI}\right]\delta v'' - \frac{f^{\mathrm{gr}}}{EI}\cos\left(u_0 + \frac{s}{\rho + a}\right)\delta v' + \left[\frac{f^{\mathrm{gr}}}{aEI}\sin\left(u_0 + \frac{s}{\rho + a}\right) - \frac{F_z}{aEI(\rho + a)}\right]\delta v = 0. \qquad (39)$$

It is derived from the assumption that in buckling the friction force f^{fr} is directed along the DS axis line. To find states of bifurcational buckling of the DS, the Sturm–Liouville problem (eigenvalue problem) should be formulated for Eq. (39). Its statement is based on the finite difference method and the construction of the corresponding matrix of algebraic equation coefficients. The matrix elements depend on multiplier $F_z(s)$ before the δv value in Eq. (39) which in its turn is determined by the boundary force $F_z(U)$ and the distributed friction force in Eq. (38). Then, the critical (bifurcational or eigen) value of the external force $F_z(U)$ applied at the end $u = U$ (i.e., $s = S$) is found by its varying through the trial-and-error method application.

It is notable that the torque M_z is not present in Eq. (39). This means that critical states of DSs in circular bore holes do not depend on M_z values (as well as in rectilinear ones (Gulyayev et al. 2014)).

5 Critical buckling of a DS moving inside a channel of a circular bore hole

The primary objective of this paper lies in investigation of the influence of friction forces on the stability of a DS moving inside a curvilinear bore hole. Notwithstanding the fact that the simplest trajectory in the shape of a circular arc is chosen for the bore hole configuration, this example permits us to trace the principal peculiarities of the buckling processes proceeding in curve channels. It consists in the possibility to generate localized buckling wavelets in the most unexpected places of the hole length. Additional uncertainty is contributed to this situation by axial friction forces generated during lowering or hoisting the DS. It is considered that the motion is slow, and inertia forces can be disregarded. Then, the DS can be partially compressed and partially stretched by distributed variable gravity forces $f^{gr}(s)$, variable frictional forces $f^{fr}(s)$, and axial force $F_z(U)$ applied at the DS end $s = S$ with an angular coordinate $u = U$. So, Eq. (39) cannot be solved by analytical methods.

To find bifurcational states of the DS, Eq. (39) was algebraized by the finite difference method for different values of $F_z(U)$ force, and the states when the matrix of linear algebraic equations became degenerate were assumed to be critical. At this state, the eigenfunction $F_z^{cr}(s)$ and eigenmode $\delta v(s)$, representing the shape of the buckling DS, were constructed.

In numerical analyses, the DS section $0 \leq s \leq S$ was divided into 500 finite difference pieces. The calculation results were tested with a doubled number of pieces. The checking confirmed the adequate precision of the computations.

In order to trace the influence of friction forces on DS buckling, every calculation example was examined with the use of frictionless and frictional statements. The analytic results are compared.

In Fig. 4, the geometric scheme of the DS, lying inside the lower quarter of a circular channel, is presented. At its lower end, the DS axis is tangent to the horizontal, its spanning angle $\varphi = U - u_0$ and position of the top end

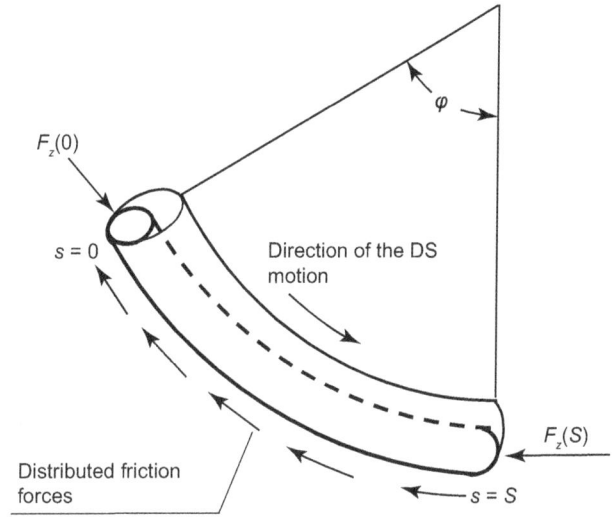

Fig. 4 Schematic of the circular DS segment

$s = 0$ were varied. The DS is pinned at both its ends. Influences of the DS length S, radius ρ, clearance a, and friction force f^{fr} on critical values $F_z^{cr}(U)$ were examined for the next values of the system parameters: $E = 2.1 \times 10^{11}$ Pa, $\gamma_{st} = 7.8 \times 10^3$ kg/m^3, $\gamma_{mud} = 1.3 \times 10^3$ kg/m^3, $d_1 = 0.1683$ m, $d_2 = 0.1483$ m, $\mu = 0.2$. Every example was studied for the cases $f^{fr} = 0$ and $f^{fr} \neq 0$ and clearance values $a = 0.5, 0.1, 0.05, 0.03$ m. Although the first value of a is not practicable, it is included into analysis to reveal the trend of critical states evolving with clearance change.

The findings of the calculations evidence that if the DS is rather short, the clearance a is not small, and the radius ρ is large, the DS buckles similarly to the Eulerian beam under critical axial forces $P^{cr} = \pi^2 EI/S^2$ and they are unaffected by the hole friction. However, the situation varies radically with the length S and angle φ enlargement. As it becomes longer, the system begins to exhibit properties of singularly perturbed structures and to localize unpredictably its short buckling waves in boundary layers (see Fig. 5a for frictionless case) or in inner zones with the larger values of compressive axial force (see Fig. 5b for the case of frictional interaction). In the theory of waves, such modes are termed the wave packages (Crawford 2011), in the applied mathematics they have come to known as harmonic wavelets.

In Table 1, the calculation results for the 1200-m DS inserted into the circular bore hole with a radius ρ of 1146 m are given. The spanning angle for this example is $\varphi = 60°$. It can be seen that the DS buckling character depends on the character of the $F_z(s)$ function distribution and its external value locations. Thus, if $a = 0.5$ m and $f^{fr} = 0$, the DS is stretched at its top end $s = 0$ and compressed at its lower one $s = S$. The critical value $F_z^{cr}(S) = -98.33$ kN of the external compressive axial

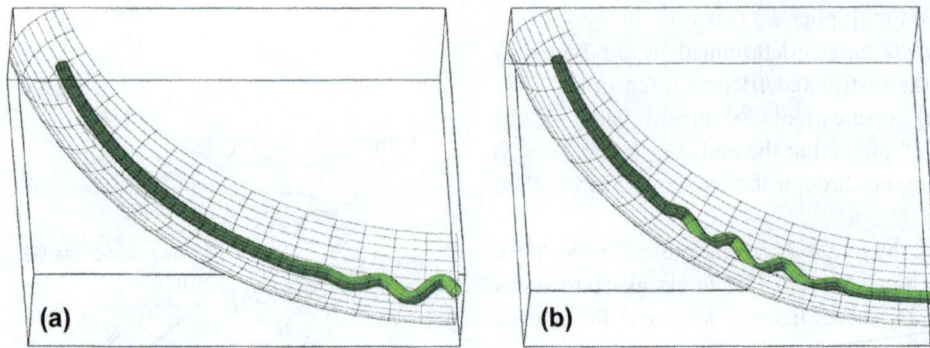

Fig. 5 Buckling of the DS in an inclined circular channel of a bore hole. **a** Frictionless model of the immovable DS. **b** Frictional model of the moving DS

force $F_z(s)$ applied at the pinned end $s = S$ is maximal throughout its length. So, the buckling wavelet is localized in the boundary zone, justifying the properties of singularly perturbed systems (Chang and Howes 1984; Elishakoff et al. 2001; Gulyayev et al. 2014).

The situations change if $f^{fr} \neq 0$ (see position 1 in Table 1). Then, the maximal compressive value $F_z^{fr} = -93.97$ kN of the axial force shifts to the bore hole interiority and the system becomes singularly perturbed inside its length. Then, the buckling wavelet also displaces inside the DS segment.

If $a \leq 0.1$ m (positions 2–4 in Table 1), the DSs buckle under the action of greater forces and the smaller a is, the more complicated is the mode of stability loss. It becomes also harder to predict the zone of the buckling localization in the presence of friction forces. Besides, the buckled zone propagates through a larger region of the DS length and at $a = 0.03$ m the boundary effect drifts from the right end of the diagram to its left end. As this takes place, the buckling mode pitch (λ—pitch of the eigenmode wavelet) becomes smaller and smaller. Thus, the wavelet pitch $\lambda \approx 26$ m for $a = 0.5$ m and it is 10.5 m for $a = 0.03$ m.

Now, consider the instance when the bore hole channel arc is symmetrically disposed relative to the vertical (Fig. 6). The computations are fulfilled for the values $S = 2400$ m, $\rho = 1146$ m, $\varphi = 120°$. If the friction effects are not taken into account, then, the representative functions, determining the DS buckling, can be obtained by simple symmetric prolongations relative to the vertical of the appropriate functions represented in Table 1 for the asymmetric arc (compare Figs. 5a, 6a). As this takes place, the critical values of boundary forces $F_z^{cr}(s)$ given in Table 1 for a frictionless asymmetric hole become equal to critical values $F_z^{cr}(S/2)$ for the corresponding DSs in Table 2.

The presence of friction effects leads to shifting the buckling wavelet positions in both cases (see Figs. 5b, 6b)

but the maximal values of the appropriate critical functions $F_z^{cr}(S)$ (located in the wavelets zones) and wavelet pitches remain approximately equal, though shapes of their buckling modes acquire some distinctions.

It is of interest that if the DS length is larger than the wavelet extension, then the critical value of $F_z^{cr}(s)$ does not depend on the DS size and with the enlargement of the spanning angle φ the noted peculiarities come into particular prominence. These conclusions are verified by Table 3 charted for $\varphi = 180°$, $\rho = 1146$ m, and $S = 3600$ m.

So then, juxtaposition of these results with the data displayed in Table 3 for $\varphi = 180°$ enables us to infer that in the case of the absence of friction the buckling wavelets are localized in the central zone of the DS length, they are identical for every value of clearance a, and are realized under similar values of axial force F_z^{cr} achieved at the middle point $s = S/2$. It is evident that in this case, the buckling effect does not depend on the boundary conditions at the ends $s = 0$ and $s = S$. As this takes place, the pitches λ of the eigenmode half-harmonics equal the distances between two adjacent zeros do not demonstrate essential differences. Their values at $S = 2400$ m and 3600 m ($\varphi = 120°$ and $180°$) are listed in Table 4. They are seen to be invariant if the bifurcation buckling zone is small and they diminish with a decrease in a. In the cases of frictionless contact of the DS with the bore hole wall, the λ values are identical for both length S at clearance values $a = 0.5$ m and 0.1 m, but their distinction becomes conspicuous at $a = 0.05$ m and $a = 0.03$ m.

The influence of the friction forces on buckling modes is more appreciable. They cause not only the diversification of the buckling zone positioning but lead to these zones widening as well (see Table 3). Besides, in the every zone limits, the pitches of conventional harmonics become also variable (see Table 4 for $f^{fr} \neq 0$).

Table 1 Functions of the critical axial force $F_z^{cr}(s)$ and buckling modes $\delta v(s)$ for the case $S = 1200$ m, $\rho = 1146$ m, $\varphi = 60°$

No.	a, m	Friction force value	Function of the critical axial force $F_z^{cr}(s)$, kN	Mode $\delta v(s)$ of stability loss
1	0.5	$f^{fr} = 0$	$F_z^{cr}(s) = -98.33$	
		$f^{fr} \neq 0$	-93.97 $F_z^{cr}(s) = -66.02$	
2	0.1	$f^{fr} = 0$	$F_z^{cr}(s) = -234.0$	
		$f^{fr} \neq 0$	-224.9 $F_z^{cr}(s) = -196.9$	
3	0.05	$f^{fr} = 0$	$F_z^{cr}(s) = -347.1$	
		$f^{fr} \neq 0$	-331.6 $F_z^{cr}(s) = -303.6$	
4	0.03	$f^{fr} = 0$	$F_z^{cr}(s) = -469.2$	
		$f^{fr} \neq 0$	-445.2 $F_z^{cr}(s) = -417.26$	

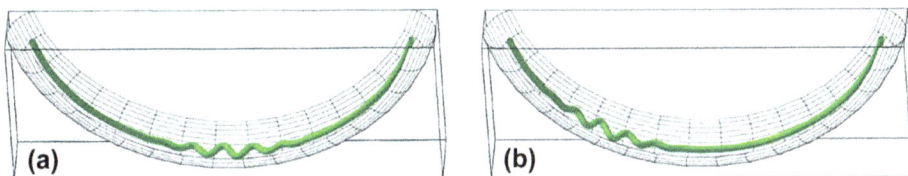

Fig. 6 Buckling of DS in a symmetric circular channel of a directed bore hole. **a** Frictionless model of the stationary DS. **b** Frictional model of the moving DS

Table 2 Functions of the critical axial force $F_z^{cr}(s)$ and buckling modes $\delta v(s)$ for the case $S = 2400$ m, $\rho = 1146$ m, $\varphi = 120°$

No.	a, m	Friction force value	Function of the critical axial force $F_z^{cr}(s)$, kN	Mode $\delta v(s)$ of stability loss
1	0.5	$f^{fr} = 0$	-98.29 $F_z^{cr}(s) = 83.40$	
		$f^{fr} \neq 0$	-94.11 $F_z^{cr}(s) = 222.65$	
2	0.1	$f^{fr} = 0$	-233.9 $F_z^{cr}(s) = -522$	
		$f^{fr} \neq 0$	-222.3 $F_z^{cr}(s) = 94.46$	
3	0.05	$f^{fr} = 0$	-347.0 $F_z^{cr}(s) = -165.3$	
		$f^{fr} \neq 0$	-332.8 $F_z^{cr}(s) = -16.09$	
4	0.03	$f^{fr} = 0$	-468.9 $F_z^{cr}(s) = -287.3$	
		$f^{fr} \neq 0$	-446.5 $F_z^{cr}(s) = -129.8$	

It is intriguing to compare the obtained results with the analytic solution deduced for the limiting case when the curvature radius ρ tends to infinity in the absence of friction effects. Then, it is valid to assume that the rectilinear bore hole is horizontal and the infinitely long DS is prestressed by the axial force $F_z(s)$ remaining unchanged throughout its length. In this event, the critical values of force $F_z(s)$ and pitch λ are determined by equalities (Cunha 2004; Mitchell 2008; Gulyayev et al. 2014):

$$F_z^{cr} = 2\sqrt{EIf^{gr}/a}, \quad \lambda^{cr} = \pi\sqrt[4]{EIa/f^{gr}}.$$

Their bracketed values for the corresponding states are tabulated in Table 4. It can be seen that the calculated results are closely related for large values of clearance a and the difference between them grows with a reduction in a.

Table 3 Functions of the critical axial force $F_z^{cr}(s)$ and buckling modes $\delta v(s)$ for the case $S = 3600$ m, $\rho = 1146$ m, $\phi = 180°$

No.	a, m	Friction force value	Function of the critical axial force $F_z^{cr}(s)$, kN	Mode $\delta v(s)$ of stability loss
1	0.5	$f^{fr} = 0$	-98.18; $F_z^{cr}(s) = 265.2$	
		$f^{fr} \neq 0$	-96.95; $F_z^{\sigma}(s) = 409.2$	
2	0.1	$f^{fr} = 0$	-233.7; $F_z^{cr}(s) = 129.7$	
		$f^{fr} \neq 0$	-225.9; $F_z^{\sigma}(s) = 277.2$	
3	0.05	$f^{fr} = 0$	-346.5; $F_z^{cr}(s) = 16.85$	
		$f^{fr} \neq 0$	-335.9; $F_z^{cr}(s) = 167.2$	
4	0.03	$f^{fr} = 0$	-473.5; $F_z^{cr}(s) = -110.7$	
		$f^{fr} \neq 0$	456.1; $F_z^{cr}(s) = 47.05$	

Summarizing obtained results, one can recognize that the found regularities of the realization of critical states and critical modes evolving are associated, in a large extent, with the circular geometry of a bore hole and invariability of its curvature radius ρ. One might expect that the considered phenomena will be far more intricate for the wells with a variable curvature both in the cases of frictionless contacts and when the frictional interactions occur.

Noteworthy also is the remark in relation to the influence of the bore hole geometry imperfections on the buckling process. The geometry imperfections entail enlargement of distributed friction forces and axial force $F_z(s)$, on the other hand, the geometry distortions result in a change of the curvature radius ρ. Both these factors imply the essential effect on the buckling process and should be specially studied.

Table 4 Extreme values of axial force $F_z^{cr}(s)$ and wavelet pitches λ at critical states of the DSs

No.	a, m	Friction force value	$\phi = 120°$, $S = 2400$ m		$\phi = 180°$, $S = 3600$ m	
			F_z^{ext}, kN	λ, m	F_z^{ext}, kN	λ, m
1	0.5	$f^{fr} = 0$	−98.29 (−91.26)	26 (26.65)	−98.18 (−91.26)	26 (26.65)
		$f^{fr} \neq 0$	−94.11	26	−96.95	23.5
2	0.1	$f^{fr} = 0$	−233.9 (−204.1)	15.5 (17.8)	−233.7 (−204.1)	15.5 (17.8)
		$f^{fr} \neq 0$	−222.3	15–15.5	−225.9	13.6–14
3	0.05	$f^{fr} = 0$	−347.0 (−288.6)	13 (14.9)	−346.5 (−288.6)	12 (14.9)
		$f^{fr} \neq 0$	−332.8	11–12.5	−335.9	9.5–10.5
4	0.03	$f^{fr} = 0$	−468.9 (−372.6)	10.5 (13.1)	−469.2 (−372.6)	9.3 (13.1)
		$f^{fr} \neq 0$	−446.5	10–10.5	−456.1	8.6–9

6 Conclusions

(1) On the basis of the theory of curvilinear elastic rods, a new statement of the problem of critical buckling of DSs in 3D curvilinear bore holes with allowance made for friction effects is suggested. It is assumed that a DS does not lose its stability in the state of its stationary equilibrium but it can buckle during its axial movement when friction forces and their directions can be easily determined.

(2) The fourth order system of linearized differential equations of the DS buckling in a curvilinear channel is deduced with the use of differential geometry methods, theory of channel surfaces, and classical mechanics of systems with nonlinear constraints. The method of numerical solutions of this system is elaborated.

(3) The problem is shown to belong to the singularly perturbed class, and for this reason, the buckling modes have the shapes of localized harmonic wavelets.

(4) As an example, the phenomena of DS stability in circular channels are studied. The cases of absence and presence of friction effects are considered. It is demonstrated that the friction forces stimulate redistribution of internal axial forces in the DSs and result in essential shifting of the buckling wavelet localizations, bringing the buckling phenomenon to the less predictable type.

(5) One might suppose that the considered phenomena will be far more complicated for the well with variable curvatures both in the cases of frictionless contacts and when the frictional interactions take place.

References

Antman SS. Nonlinear problems of elasticity. New York: Springer; 2005.

Berger EJ. Friction modeling for dynamic system simulation. Appl Mech Rev. 2002;55(6):535–76. doi:10.1115/1.1501080.

Brett JF, Beckett AD, Holt CA, et al. Uses and limitations of drill string tension and torque models for monitoring hole conditions. SPE Drill Eng. 1989;4:223–9. doi:10.2118/16664-PA.

Chang KW, Howes FA. Nonlinear singular perturbation phenomena. Berlin: Springer; 1984.

Crawford FS. Waves. Berkeley physics course, vol. 3. Noida: Mc Graw-Hill; 2011.

Cunha JC. Buckling of tubulars inside wellbores: a review on recent theoretical and experimental works. SPE Drill Complet. 2004;19(1):13–8. doi:10.2118/87895-PA.

Dawson R, Paslay RR. Drill pipe buckling in inclined holes. J Pet Technol. 1984;36(10):1734–8. doi:10.2118/11167-PA.

Dubrovin BA, Novikov SP, Fomenko AT. Modern geometry-methods and applications. Berlin: Springer; 1992.

Elishakoff I, Li Y, Starnes JH. Non-classical problems in the theory of elastic stability. Cambridge: Cambridge University Press; 2001.

Gao DL, Huang WJ. A review of down-hole tubular string buckling in well engineering. Pet Sci. 2015;12(3):443–57. doi:10.1007/s12182-015-0031-z.

Gao DL, Liu FW. The post-buckling behavior of a tubular string in an inclined wellbore. Comput Model Eng Sci. 2013;90(1):17–36. doi:10.3970/cmes.2013.090.017.

Gulyayev VI, Andrusenko EN, Shlyun NV. Theoretical modelling of post-buckling contact interaction of a drill string with inclined bore-hole surface. Struct Eng Mech. 2014;49(4):427–48. doi:10.12989/sem.2014.49.4.427.

Gulyayev VI, Gaidaichuk VV, Andrusenko EN, et al. Critical buckling of drill strings in curvilinear channels of directed bore-holes. J Pet Sci Eng. 2015;129:168–77. doi:10.1016/j.petrol.2015.03.004.

Gulyayev VI, Gaidaichuk VV, Koshkin VL. Elastic deforming, stability and vibrations of flexible curvilinear rods. Kiev: Naukova Dumka; 1992 (in Russian).

Gulyayev VI, Gaidaichuk VV, Solovjov IL, et al. The buckling of elongated rotating drill strings. J Petr Sci Eng. 2009;67:140–8. doi:10.1016/j.petrol.2009.05.011.

Gulyayev VI, Tolbatov EYU. Dynamics of spiral tubes containing internal moving masses of boiling liquid. J Sound Vib. 2004;274:233–48. doi:10.1016/j.jsv.2003.05.013.

Huang WJ, Gao DL. Helical buckling of a thin rod with connectors constrained in a cylinder. Int J Mech Sci. 2014;84:189–98. doi:10.1016/j.ijmecsci.2014.04.022.

Kyllingstad A. Buckling of tubular strings in curved wells. J Pet Sci Eng. 1995;12(3):209–18. doi:10.1016/0920-4105(94)00046-7.

Liu X, Vlajic N, Long X, et al. Nonlinear motions of a flexible rotor with a drill bit: stick-slip and delay effects. Nonlinear Dyn. 2013;72:61–77. doi:10.1007/s11071-012-0690-x.

Liu X, Vlajic N, Long X, et al. Multiple regenerative effects in cutting process and nonlinear oscillations. Int J Dyn Control. 2014a;2:86–101. doi:10.1007/s40435-014-0078-5.

Liu X, Vlajic N, Long X, et al. Coupled axial-torsional dynamics in rotary drilling state-dependent delay: stability and control. Nonlinear Dyn. 2014b;78(3):1891–906. doi:10.1007/s11071-014-1567-y.

Lubinski A, Althouse WS, Logan JL. Helical buckling of tubing sealed in packers. JPT. 1962;14(6):655–70. doi:10.2118/178-PA.

Mitchell RF. Tubing buckling—the state of the art. SPE Drill Complet. 2008. doi:10.2118/104267-PA.

Mitchell RF, Samuel R. How good is the torque/drag model? SPE Drill Complet. 2009;24(1):62–71. doi:10.2118/105068-PA.

Samuel R. Friction factors: what are they for torque, drag, vibration, bottom hole assembly, and transient surge/swab analysis. J Pet Sci Eng. 2010;73(3–4):258–66. doi:10.1016/j.petrol.2010.07.007.

Sawaryn SJ, Sanstrom B, McColpin G. The management of drilling-engineering and well-services software as safety-critical systems. SPE Drill Complet. 2006;21(2):141–7. doi:10.2118/73893-PA.

Wang X, Yuan Z. Investigation of frictional effects on the nonlinear buckling behavior of a circular rod laterally constrained in a horizontal rigid cylinder. J Pet Sci Eng. 2012;90–91:70–8. doi:10.1016/j.petrol.2012.04.011.

14

Effects of pH on rheological characteristics and stability of petroleum coke water slurry

Fu-Yan Gao[1] · Eric-J. Hu[2]

Abstract In this study, the effects of pH on slurrying properties of petroleum coke water slurry (PCWS) were investigated. The slurrying concentration, rheological characteristics and stability of PCWS were studied with four different types of additives at pH varying from 5 to 11. The results showed that the slurrying concentration, rheological characteristics and stability of PCWS all increased at first and then decreased with increasing pH from 5 to 11, and a pH of around 9 was found to be the most favorable acid–alkali environment to all these three slurrying properties. It was also indicated that only in a moderate alkaline environment can the additives be active enough to react with particle surfaces sufficiently to obtain good slurrying concentration and form a stable three-dimensional network structure, which can support strong pseudoplastic characteristics and good stability. An acid environment was a very unfavorable factor to the slurrying properties of PCWS.

Keywords Petroleum coke · Petroleum coke water slurry · pH · Slurrying concentration · Rheological characteristics · Stability

1 Introduction

Petroleum oil and its products are important fuels and chemical raw materials, which are widely used in almost all aspects of production and life. Along with the rapid development of the economy, the demand for petroleum oil keeps increasing (Hu 2014). More and more petroleum coke, as an end product of the petroleum refining process, is produced (Ren et al. 2012; Zhang et al. 2012). Petroleum coke, with its characteristics of high carbon content, high calorific value and low ash content, has become a popular fuel for power generation (Chen and Lu 2007; Milenkova et al. 2005; Anthony et al. 2001; Wang et al. 2004; Sheng et al. 2007) and has started to become a potential gasification fuel (Valero and Usón 2006; Fang et al. 2005).

With the development of coal water slurry (CWS) technology, increasing attention has been paid to petroleum coke water slurry (PCWS). CWS and PCWS are liquid fuels of low pollution and high efficiency and can be pumped like oil by pipeline and burned in power plants as an oil substitute (Zhan et al. 2010). They change the traditional combustion of solid fuels and show huge environmental protection and energy-saving advantages (Cen et al. 1997; Wu et al. 2015). Because of the strong hydrophobicity, PCWS generally possesses higher solid concentration than conventional CWS. Moreover, PCWS also can be a superior raw material for industrial gasification (Gao et al. 2012a, b; Zou et al. 2008). Hence, PCWS has become an important way to utilize petroleum coke efficiently and cleanly.

The slurrying properties are most important for industrial application of the slurry fuels. The solid concentration of the slurry fuels should be increased as much as possible to reach a high level of heat value and thus ensure efficient gasification and combustion, but the viscosity should be

✉ Fu-Yan Gao
gaofuyan@nit.zju.edu.cn

[1] Ningbo Institute of Technology, Zhejiang University, Ningbo 315100, China

[2] School of Mechanical Engineering, University of Adelaide, Adelaide, SA 5005, Australia

Edited by Xiu-Qin Zhu

low enough to facilitate preparation, pumping and atomization of the slurry. Many studies are aimed at influencing factors on PCWS's slurrying properties (Gao et al. 2012a, b; He et al. 2011; Xu et al. 2008; Vitolo et al. 1996; Wang et al. 2006). Yet, up to now, the effect of pH on slurrying properties of PCWS is rarely reported. Acid–alkali properties of the slurry can directly influence the interactions between the additives and the surface of petroleum coke particles and subsequently influence the slurrying properties of PCWS. In this work, the effects of pH on the slurrying properties of PCWS were investigated.

2 Materials and methods

2.1 Material

A petroleum coke from America was used in the experiments. The proximate and ultimate analysis results of the petroleum coke used in this work are shown in Table 1. The petroleum coke was ground in a ball mill to obtain the pulverized sample, and particles below 149 μm were selected by an electric sieve shaker to prepare PCWS. The granularity distribution of the selected petroleum coke particles was analyzed with a Mastersizer 2000 Granularity Meter (Malvern, UK), as shown in Fig. 1. The average particle diameter was approximately 27 μm.

Chemical additives are an important component of slurry fuel, for they can help particles to disperse stably in the slurry. Four kinds of anionic surfactants were used as additives in PCWS preparation in this work. These were sodium methylene naphthalene sulfonate-sodium styrene sulfonate-sodium maleate (NDF), methylene naphthalene sulfonate formaldehyde condensate (MF), lignin sulfonate (LS) and petroleum sulfonate (PS). The additive dosage

Table 1 Proximate and ultimate analysis results of petroleum coke

Issues	Results (air-dried basis)
Proximate analysis, wt%	
Moisture	1.24
Ash	0.85
Volatile	11.68
Fixed carbon	86.23
High heating value, kJ kg^{-1}	35,146
Ultimate analysis, wt%	
C	85.57
H	4.01
N	1.84
Total S	3.54
O	5.04

The oxygen datum was obtained by calculation

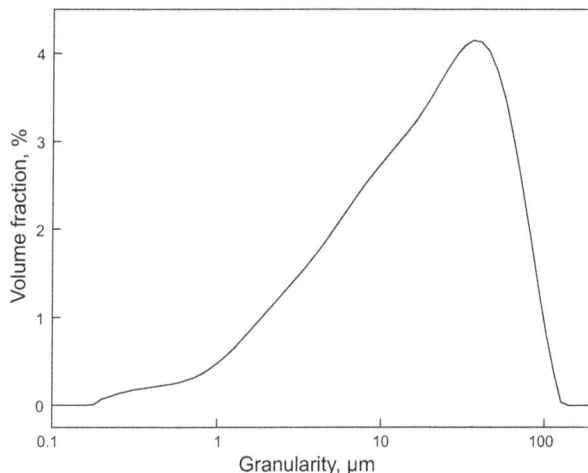

Fig. 1 Granularity distribution of petroleum coke particles

was fixed at 0.8 wt% based on dry petroleum coke (Gao et al. 2015).

2.2 Methods

The petroleum coke particles, deionized water, one additive and moderate HCl or NaOH were mixed with an electric mixer at 1000 r/min for 10 min to form a PCWS sample. With each additive, PCWSs were prepared at four different pH values (i.e., pH 5, 7, 9 and 11). The pH values were measured by using an E200 Portable pH Meter (Mont, China).

The apparent viscosity and rheological properties of PCWS were measured on a rotary viscometer (NXS-4C, China). A PCWS sample was first loaded into the viscometer, and then the shear rate was increased from 10 to 100 s^{-1}. The relationship of the shear stress and the shear rate can be revealed in this process. Keeping the shear rate at 100 s^{-1} for 5 min, the apparent viscosity data were recorded every 30 s during a 5-minute period. The average apparent viscosity at 100 s^{-1} was calculated from the ten apparent viscosity values recorded. During the entire process, temperature was controlled at 20 ± 1 °C.

The solid concentration of PCWS was determined by drying the slurry in an oven at 105 °C for 2 h and then weighing the dried residue.

Measurement of stability of PCWS was taken after the slurry was sealed in a container for 7 days. In order to ensure the reliability of the experimental results, the stability of PCWS was measured by both the rod-insertion method (Zhao 2009) and a visual method (Li et al. 2008). In the rod-insertion method, a steel rod was inserted vertically and freely from the slurry surface, and the first traveling length through the slurry was recorded. Then the steel rod was strongly pressed down to the bottom of the container, and the second traveling length through the slurry was recorded as well. The relative height of the hard sediment layer can be obtained by calculation, which is an index to evaluate slurry stability. A large relative height of hard sediment layer indicates poor stability of the PCWS. In the visual method, the changes in slurry properties, such as separated water, could be found through observation. The mass ratio of separated water to total slurry is used to evaluate the stability of slurry. A higher water-to-slurry ratio indicates a worse stability.

3 Results and discussion

3.1 Effects of pH on slurrying concentration of PCWS

Solid concentration at a specific viscosity of 1000 mPa s with the shear rate of 100 s^{-1} is used to evaluate the slurrying concentration of petroleum coke. The higher the solid concentration, the better the slurrying concentration of petroleum coke (Hu et al. 2009). Figure 2 shows the relationship of slurrying concentration of PCWSs (with different additives) with pH.

Figure 2 shows that the slurrying concentration increased first and then decreased with increasing pH and that an acid environment of pH 5 resulted in the worst slurrying concentration and an alkaline environment of pH 9 resulted in the best slurrying concentration. The reason is that the additives themselves possess moderate alkalinity,

and the activity of the additives can be restrained in acid or strong alkali conditions, leading to the dispersion of particles worsened and slurrying concentration decreased.

3.2 Effects of pH on rheological characteristics of PCWS

Rheological characteristics are very important to the industrial application of slurry fuels. These are not only related to the slurrying properties, but also directly affect the pumping, atomizing and combustion performances of the slurry fuels (Ma et al. 2012, 2013a, b; Li et al. 2010; Meikap et al. 2005; Chen et al. 2009). Usually PCWS is expected to be of high viscosity to promote stability during storage and low viscosity to ensure fluidity during transport; hence, "shear-thinning" pseudoplastic characteristics are generally required in industry.

Figure 3 shows the relationship of rheological characteristics of PCWS (with different additives) with pH at solid concentration of 69 wt%. It can be seen that the PCWSs with NDF and MF additives were dilatant fluids and had shear-thickening properties at pH from 5 to 11 and exhibited the feeblest shear-thickening at pH of 9, while the PCWSs with LS and PS additives possessed shear-thinning pseudoplastic characteristics at pH of around 9.

A three-parameter Herschel–Bulkley model (Ma et al. 2013a, b) expressed by Eq. (1) was used to fit the shear stress–shear rate data.

$$\begin{cases} \dot{\gamma} = 0 & \tau \leq \tau_y \\ \tau = \tau_y + k\dot{\gamma}^n & \tau > \tau_y \end{cases} \tag{1}$$

where $\dot{\gamma}$ is the shear rate, s^{-1}; τ is the shear stress, Pa; τ_y is the yield stress, Pa; k is the consistency coefficient, Pa s^n; n is the dimensionless flow characteristic exponent.

The parameter "n" can correctly reflect rheological characteristics of PCWS (Ma et al. 2013a, b). When $n > 1$, the PCWS is a dilatant fluid; otherwise, it is a pseudoplastic fluid. Furthermore, the smaller the value of n, the greater the pseudoplastic characteristics.

Figure 4 shows the relationship of flow characteristic exponents of the PCWSs with pH when the PCWSs were prepared with different additives at solid concentration of 69 wt%. It can be seen that the flow characteristic exponents all decreased first and then increased with increasing pH, and the smallest flow characteristic exponent with each additive appeared at pH of 9, indicating that pH of around 9 was the most favorable acid–alkali environment to strengthen pseudoplastic characteristics of PCWS.

This is because pH can affect interactions between the additive and particles and then affect the rheological characteristics of slurry. The alkaline additives can maintain their own activity and react sufficiently with surface of particles only in moderate alkaline environment, and make

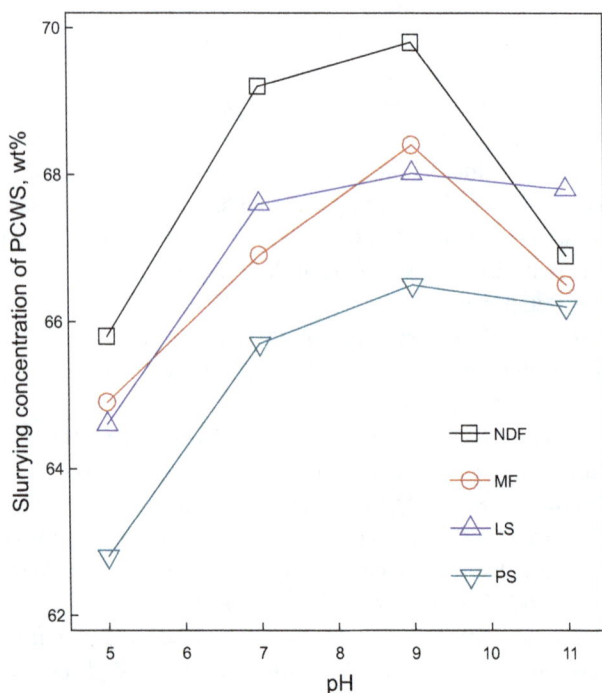

Fig. 2 Effects of pH on slurrying concentration of PCWS

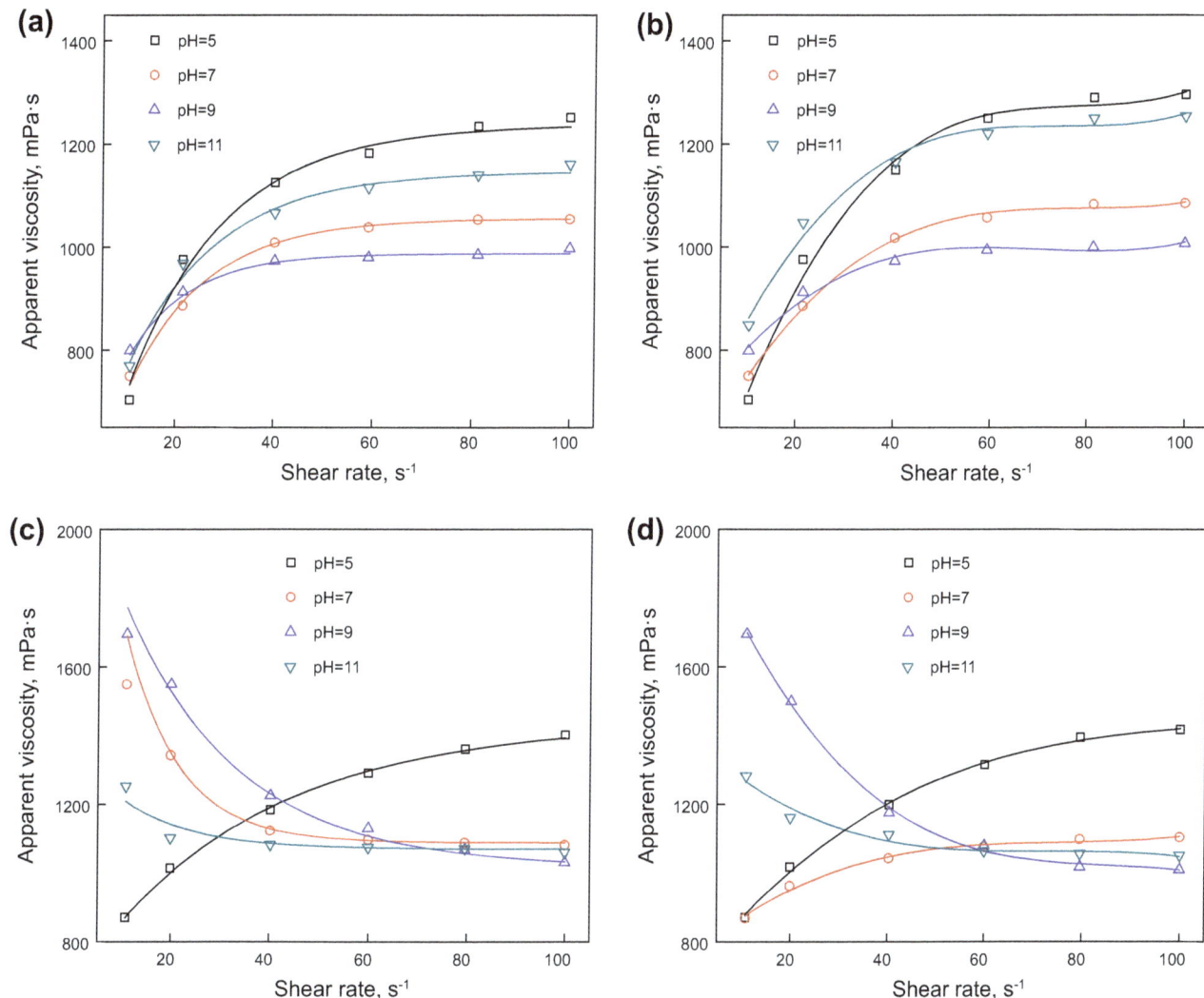

Fig. 3 Rheological characteristics of PCWS (with different additives) at different pH values. **a** PCWS with NDF, **b** PCWS with MF, **c** PCWS with LS, **d** PCWS with PS

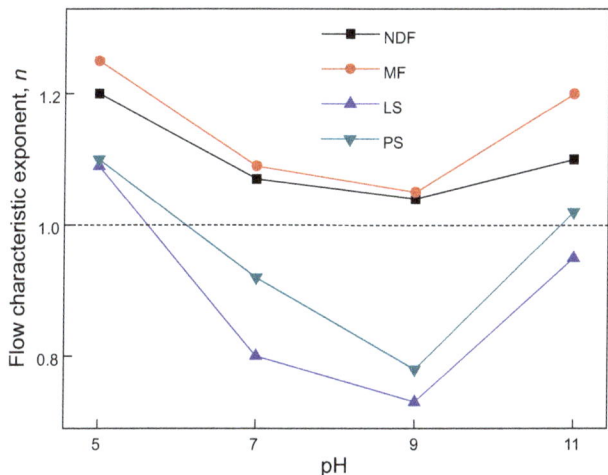

Fig. 4 Effects of pH on flow characteristic exponent

it possible to form a relatively stable three-dimensional network structure in slurry and present relatively strong pseudoplastic characteristics.

3.3 Effects of pH on stability of PCWS

Figure 5 shows the trends of stability indexes of the PCWSs with pH when the PCWSs were prepared with different additives at solid concentration of 67 wt%. It can be seen that both the relative height of hard sediment layer and the separated water rate decreased first and then increased with increasing pH, and the smallest relative height of hard sediment layer and the smallest separated water rate both appeared at a pH of 9. The PCWSs with LS or PS additives even did not produce hard sediment at pH of 9, as shown in Fig. 5a, indicating that a pH of around 9 was the most favorable acid–alkali environment to strengthen stability of PCWS.

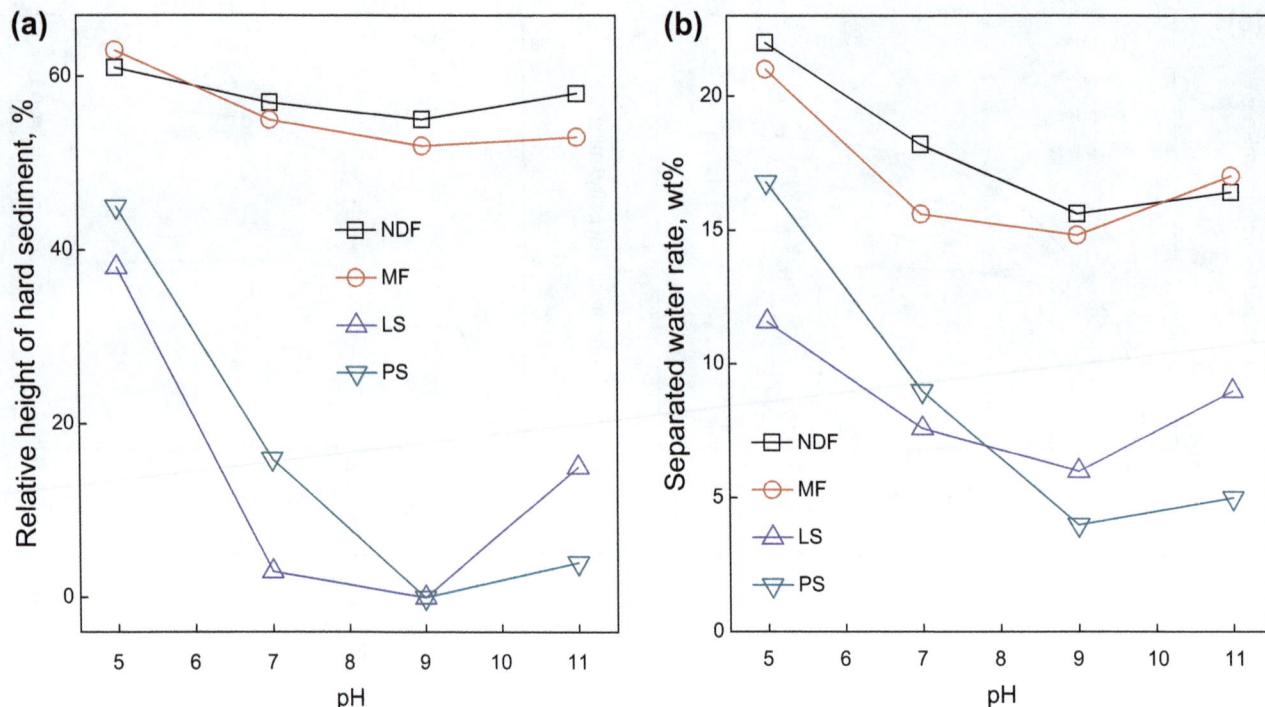

Fig. 5 Effects of pH on stability of PCWS by **a** rod-insertion method and **b** visual method

There is a positive correlation between stability and pseudoplastic characteristics of PCWS. The more stable the three-dimensional network structure of slurry is, the better the stability turns out to be. Therefore, a pH of around 9 can also be the best acid–alkali environment to obtain good stability of PCWS.

4 Conclusions

Through the study, the following conclusions can be drawn:

1. The slurrying concentration of the PCWS increased first and then decreased with increasing pH from 5 to 11. An acid environment was an unfavorable factor to the slurrying concentration. The optimal pH to obtain best slurrying concentration was 9. The additives used in this work themselves possess moderate alkalinity, and their activity can be restrained in acid or strong alkali conditions, leading to worse dispersion of particles and decreased slurrying concentration.

2. The pseudoplastic characteristics of the PCWS increased first and then decreased with increasing pH from 5 to 11, and a pH of around 9 was the most favorable acid–alkali environment to strengthen the pseudoplastic characteristics. The alkaline additives can maintain their own activity and react sufficiently with surface of particles only in a moderate alkaline environment, and make it possible to form a relatively stable three-dimensional network structure in slurry and present relatively strong pseudoplastic characteristics.

3. The stability of the PCWS increased first and then decreased with increasing pH from 5 to 11, and the best stability occurred when the pH was around 9. A pH of around 9 was the best acid–alkali environment to obtain good stability of PCWS. It shows a positive correlation between stability and pseudoplastic characteristics of PCWS.

Acknowledgments The authors would like to acknowledge the financial support from the National Natural Science Foundation of China (No. 51506185) and the Zhejiang Provincial Natural Science Foundation of China (No. LQ15E060002).

References

Anthony EJ, Iribarne AP, Iribarne JV, et al. Fouling in a 160 MWe FBC boiler firing coal and petroleum coke. Fuel. 2001;80:1009–14.

Cen KF, Yao Q, Cao XY, et al. theory and application of combustion, flow, heat transfer, gasification of coal slurry. Hangzhou: Zhejiang University Press; 1997. pp. 15–69 (**in Chinese**).

Chen JH, Lu XF. Progress of petroleum coke combusting in circulating fluidized bed boilers—a review and future perspectives. Resour Conserv Recycle. 2007;49:203–16.

Chen LY, Duan YF, Zhao CS, et al. Rheological behavior and wall slip of concentrated coal water slurry in pipe flows. Chem Eng Process. 2009;48:1241–8.

Fang YT, Wu JH, Li QF, et al. Industrial application of ash agglomerating fluidized bed coal gasification and gasification of petroleum coke for syngas and hydrogen production. In: 8th international conference on circulating fluidized beds. Hangzhou, China; 2005. p. 431–8.

Gao FY, Liu JZ, Wang CC, et al. Effects of the physical and chemical properties of petroleum coke on its slurrying concentration. Pet Sci. 2012a;2:251–6.

Gao FY, Liu JZ, Wang RK, et al. Surface properties of sludge-petroleum coke-slurry and its effect on the slurrying concentration. Proc CSEE. 2012b;32:37–43 (in Chinese).

Gao FY, Jiang L, Zhang XZ. Experimental research into optimum additive dosage for petroleum coke water slurries. Chem Eng (China). 2015;43(5):63–7 (in Chinese).

He QH, Wang R, Wang WW, et al. Effect of particle size distribution of petroleum coke on the properties of petroleum coke-oil slurry. Fuel. 2011;90:2896–901.

Hu B. Oil and gas cooperation between China and Central Asia in an environment of political and resource competition. Pet Sci. 2014;11(4):596–605.

Hu WB, He GF, Duan QB. Study of slurrying characteristics of low-volatile coal. Coal Process Compr Util. 2009;1:37–9 (in Chinese).

Li PW, Yang DJ, Lou HM. Study on the stability of coal water slurry using dispersion-stability analyzer. J Fuel Chem Technol. 2008;36:524–9.

Li WD, Li WF, Liu HF. Effects of sewage sludge on rheological characteristics of coal–water slurry. Fuel. 2010;89:2505–10.

Ma XY, Duan YF, Li HF. Wall slip and rheological behavior of petroleum-coke sludge slurries flowing. Powder Technol. 2012;230:127–33.

Ma XY, Duan YF, Liu M. Atomization of petroleum-coke sludge slurry using effervescent atomizer. Exp Therm Fluid Sci. 2013a;46:131–8.

Ma XY, Duan YF, Liu M. Effects of petrochemical sludge on the slurry-ability of coke water slurry. Exp Therm Fluid Sci. 2013b;48:238–44.

Meikap BC, Purohit NK, Mahadevan V. Effect of microwave pretreatment of coal for improvement of rheological character-istics of coal-water slurries. J Colloid Interface Sci. 2005;281:225–35.

Milenkova KS, Borrego AG, Alvarez D, et al. Coal blending with petroleum coke in a pulverized-fuel power plant. Energy Fuel. 2005;19:453–8.

Ren JD, Meng XH, Xu CM, et al. Analysis and calculation model of energy consumption and product yields of delayed coking units. Pet Sci. 2012;9(1):100–5.

Sheng GH, Li Q, Zhai JP, et al. Self-cementitious properties of fly ashes from CFBC boilers co-firing coal and high-sulphur petroleum coke. Cem Concr Res. 2007;37(6):871–6.

Valero A, Usón S. Oxy-co-gasification of coal and biomass in an integrated gasification combined cycle (IGCC) power plant. Energy. 2006;31:1643–55.

Vitolo S, Belli R, Mazzanti M, et al. Rheology of coal–water mixtures containing petroleum coke. Fuel. 1996;75:259–61.

Wang JS, Anthony EJ, Abanades JC. Clean and efficient use of petroleum coke for combustion and power generation. Fuel. 2004;83:1341–8.

Wang ZQ, Wang HF, Guo QJ. Effect of ultrasonic treatment on the properties of petroleum coke oil slurry. Energy Fuel. 2006;20:1959–64.

Wu JH, Liu JZ, Yu YJ, et al. Improving slurrying concentration, rheology, and stability of slurry fuel from blending petroleum coke with lignite. Pet Sci. 2015;12(1):157–69.

Xu RF, He QH, Cai J, et al. Effects of chemical and blending petroleum coke on the properties of low-rank Indonesian coal water mixtures. Fuel Process Technol. 2008;89:249–53.

Zhan XL, Zhou ZJ, Kang WZ, et al. Promoted slurrying concentration of petroleum coke-water slurry by using black liquor as an additive. Fuel Process Technol. 2010;91:1256–60.

Zhang YH, Lan XY, Gao JS. Modeling of gas-solid flow in a CFB riser based on computational particle fluid dynamics. Pet Sci. 2012;9(4):535–43.

Zhao WD. Micromechanism and combustion characteristics of low-rank coal water slurry upgraded by hot water treatments. Hangzhou: Zhejiang University; 2009. p. 59–60 (in Chinese).

Zou JH, Yang BL, Gong KF, et al. Research on slurrying concentration of petroleum coke. Chem Eng (China). 2008;36(3):22–5 (in Chinese).

15

Formation of fine crystalline dolomites in lacustrine carbonates of the Eocene Sikou Depression, Bohai Bay Basin, East China

Yong-Qiang Yang[1] · Long-Wei Qiu[1] · Jay Gregg[2] · Zheng Shi[1] · Kuan-Hong Yu[1]

Abstract The genesis of the fine crystalline dolomites that exhibit good to excellent reservoir properties in the upper fourth member of the Eocene Shahejie Formation (Es_4^s) around the Sikou Sag, Bohai Bay Basin, is uncertain. This paper investigates the formation mechanisms of this fine crystalline dolomite using XRD, SEM, thin section analysis and geochemical data. The stratigraphy of the Sikou lacustrine carbonate is dominated by the repetition of metre-scale, high-frequency deposition cycles, and the amount of dolomite within a cycle increases upward from the cycle bottom. These dolomite crystals are 2–30 μm in length, subhedral to anhedral in shape and typically replace both grains and matrix. They also occur as rim cement and have thin lamellae within ooid cortices. Textural relations indicate that the dolomite predates equant sparry calcite cement and coarse calcite cement. The Sr concentrations of dolomites range from 900 to 1200 ppm. Dolomite $\delta^{18}O$ values (−11.3 to −8.2 ‰ PDB) are depleted relative to calcite mudstone (−8.3 to −5.4 ‰ PDB) that precipitated from lake water, while $\delta^{13}C$ values (0.06–1.74 ‰ PDB) are within the normal range of calcite mudstone values (−2.13 to 1.99 ‰ PDB). High $^{87}Sr/^{86}Sr$ values (0.710210–0.710844) indicate that amounts of Ca^{2+} and Mg^{2+} have been derived from the chemical weathering of Palaeozoic carbonate bedrocks. The high strontium concentration indicates that hypersaline conditions were maintained during the formation of the dolomites and that the dolomites were formed by the replacement of precursor calcite or by direct precipitation.

Keywords Dolomite · Lacustrine carbonate · Eocene · Sikou Sag · Bohai Bay Basin

1 Introduction

The controversy surrounding the origin of sedimentary dolomites is known as the "Dolomite Problem" (Van Tula 1916; Land 1985). Very fine crystalline dolomites are widely formed in varied types of lacustrine systems (Meister et al. 2011; Mauger and Compton 2011; Last et al. 2012; Casado et al. 2014; Meng et al. 2014a, b; Köster and Gilg 2015; Lu et al. 2015). They may form by replacement (dolomitization) of micrite matrix or allochems, or alternatively, they may occur as intraparticle and interparticle cements (Rosen and Coshell 1992). They consist of euhedral to anhedral crystals from a submicron size to 7 mm and are normally nonstoichiometric and poorly ordered (Last et al. 2012). Although the formation of fine dolomites in both modern and ancient lake records has been widely studied (Wright 1999; Bustillo et al. 2002; Casado et al. 2014; Lu et al. 2015), in many studies of lacustrine dolomites, there is insufficient evidence to determine whether the fine dolomite is of a primary or secondary (replacement) origin.

Fine crystalline dolomites are widely distributed at the Eocene lacustrine carbonate interval, in the Bohai Bay Basin (Jiang 2011; Peng 2011). Many previous studies have investigated the reservoir characteristics of lacustrine dolostones in the Bohai Bay Basin and have reported porosity improvement by the dolomitization (Xu 2013).

✉ Yong-Qiang Yang
yangyq_520@163.com

[1] School of Geosciences, China University of Petroleum, Qingdao 266580, Shandong, China

[2] Department of Geology, Oklahoma State University, Stillwater, OK 74075, USA

Edited by Jie Hao

However, the origin of dolomites is not as well documented, and the source of Mg^{2+} ions remains poorly resolved during the Es_4^s in the Bohai Bay Basin (Yuan et al. 2006). Understanding dolomite formation is a first step towards improved prediction of the architecture and distribution of dolomite reservoirs at regional scales.

In this study, the mineralogy, texture and stable isotope composition of minerals recovered from dolomites of the Sikou Sag, Bohai Bay Basin, are examined to better understand their origin. Carbon isotopes are used to determine the source of the carbonate ions, and strontium isotopes are used to determine the source of Ca^{2+} and Mg^{2+} ions. A detailed study of Sikou Sag dolomites may benefit strategies for lacustrine carbonate hydrocarbon exploration across the Bohai Bay Basin. This paper provides an example of early, high-frequency cyclic dolomite in lacustrine carbonate by high-salinity lake water.

2 Geological background

The Bohai Bay Basin, an important hydrocarbon-producing basin in China, is located on the eastern coast of China and covers an area of approximately 200,000 km^2. It is a complex rifted basin formed in the late Jurassic period through the early Tertiary period on the basement of the North China platform. The tectonic evolution of the Basin consists of a synrift stage (65.0–24.6 Ma) and a postrift stage (24.6 Ma to the present). The synrift stage can be further subdivided into an initial stage, an expansion stage, an expansion and deep subsidence stage, and a contraction stage. The sediments deposited at the synrift stage were restricted to the grabens and half grabens and were deposited in lacustrine environments. The postrift stage occurred during the deposition of the Guantao, Minghuazhen and Pingyuan Formations (Guo et al. 2010).

The Sikou Sag is located in the mid-western part of the Zhanhua Depression of the Bohai Bay Basin (Fig. 1a), which is adjacent to the Yihezhuang Uplift and bounded by the Yidong Fault (Pan and Li 2004). The northern part of the Sag is connected to the Chengdong Uplift; the western part of the sag is connected to the Yihezhuang Uplift through the Yidong Fault zone; the southern part is adjacent to the Chenjiazhuang Uplift bounded by the Gunan Fault zone, and the eastern part of the sag is connected to the Gudao Uplift and is also separated from the Bonan Sag by an east–west trending fault zone (Fig. 1b). In cross section, the research area can be divided into three parts: the Shaojia gentle ramp zone, the sag zone and the Yidong fault zone from the south-east to the north-west (Fig. 1c).

The Sikou Sag is filled by Cenozoic to Quaternary strata up to 5000 m thick in the depocentre. In ascending order, these strata consist of the Kongdian (Ek), Shahejie (Es),

Fig. 1 **a** Location map showing subunits of the Bohai Bay Basin. **b** Locations of the AA′ transect and normal faults for the top of the Es_4^s interval in the study area. **c** The cross section AA′ showing different tectonostructural zones within the Sikou Sag

Dongying (Ed), Guantao (Ng), Minghuazhen (Nm) and Pingyuan (Qp) Formations. The Es formation can be further subdivided into four members, from the base to the top: Es_4, Es_3, Es_2 and Es_1 (Fig. 2). The Es_4 interval is the focus of this study. It ranges from 0 to 1400 m thick and consists of a lower "red bed" unit (Es_4^x: red interbedded sandstone and shale) and an upper grey mudstone interbedded with fine-grained carbonates and thin-bedded sandstones (Es_4^s). The Es_4 overlaps the Ek formation unconformably and is overlain in turn by the Es_3 Member, which is composed of 1000-m-thick brownish oil shales and dark-grey shales interbedded with thin sandstones and carbonates.

The study area began to subside during the deposition of the Es_4 Member resulting in a large topographic relief adjacent to the sag, which is steep in the north–south and gentle in the east–west directions. During this time interval, the climate was dry and warm, resulting in evaporative conditions. Sedimentary rocks in the study area consist of dark conglomerate, sandstone, shale, limestone, dolomite and gypsum beds ranging from 4.5 to 1735.5 m in thickness. Siliciclastic rocks dominate in the lower Es_4, whereas carbonate rock and gypsum are distributed in the upper Es_4. During the upper Es_4, there was mainly terrestrial detritus and carbonate deposition on the edge of the sag with gypsum, dark carbonate mudstone and shale deposited in the centre of the sag.

The Es_4^s Member of the Shahejie Formation in the Sikou Sag is interpreted as a complete marine third-order depositional sequence that formed during the rifting phase of the Bohai Bay Basin. Metre-scale, high-frequency depositional cycles are fundamental genetic units which make up the lacustrine carbonates in the Sikou Es_4^s interval. It reveals the extreme susceptibility of this system to high-frequency fluctuations in basin hydrology. The ideal high-frequency depositional cycles are characterized by a shallowing-upward trend of depositional facies and by a minor depositional unconformity at the top, which is related to subaerial exposure. A transgressive and regressive hemicycle can be observed within each genetic unit.

Three main facies associated with the research area are identified: (1) reef mounds are mainly developed along the downthrown block of the Yidong Fault, where algal bindstones are the main constituents. (2) Shoal facies developed on the gentle slope of the Shaojia area. These are composed of various grainstones and packstones. (3) The higher-salinity central-lake deposits of dark shale interbedded with gypsum and halite and thin dolomitized mudstone are mainly deposited at the centre of the lake. Deposition of these associations is interpreted to have taken place in a rift

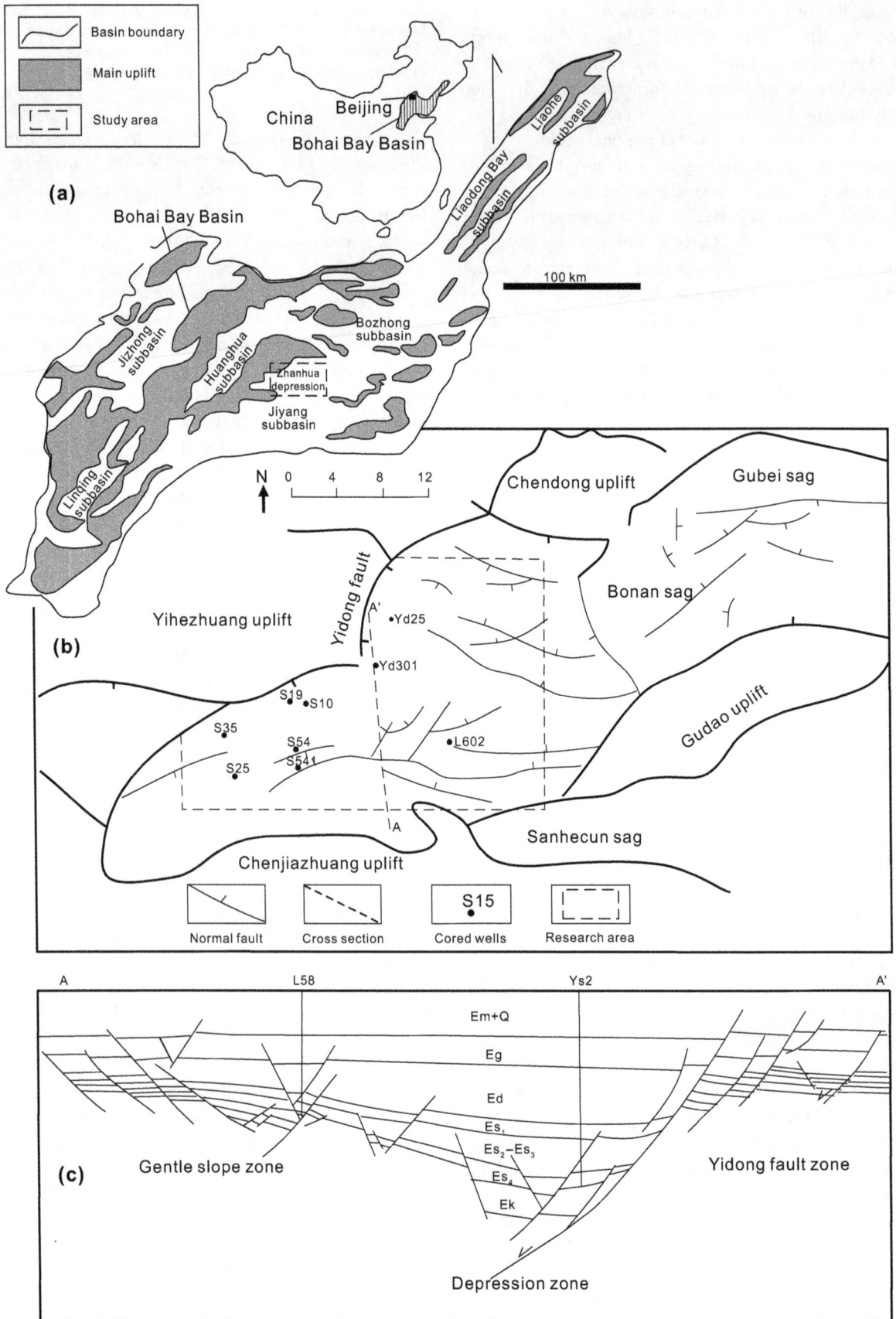

(a)

(b)

(c)

basin, where carbonate deposits changed to gypsum, halite and oil shale from the shore to the lake basin (Fig. 3). The sedimentation in the research area was dominantly controlled by wind-induced wave and wind action.

3 Data and methods

The petrographic characteristics of 52 polished thin sections were observed using a standard, transmitted light petrographic microscope. Staining with Alizarin red S and potassium ferricyanide differentiated between the ferroan and non-ferroan phases of calcite and dolomite. Twenty thin sections were studied with a CILT Mk5 cold-cathode luminoscope. The uncovered thin sections were exposed to electron radiation at 12 kV in a vacuum-sealed chamber that was attached to a Nikon Optiphot-pol petrographic microscope at Oklahoma State University. An FEI Quanta 600 F field emission scanning electron microscopy (SEM) at Oklahoma State University was used for SEM and energy-dispersive X-ray spectroscopy (EDS) of six dolomite samples. X-ray diffraction (XRD) analysis was carried out with powdered bulk rock materials on a Dmax 12 kW powder X-ray diffractometer. The samples were step-scanned at a $0°–70°$ 2θ interval.

Sample powders for chemical analysis were selected by the standard X-ray diffraction data and obtained from thin rock slabs using a dental drill. The strontium contents of dolomite were analysed by ICP-OES at Oklahoma State University. A total of 32 stable oxygen ($\delta^{18}O$) and carbon ($\delta^{13}C$) isotopes were analysed from powder dolomite and calcite samples using a Thermo Scientific Kiel IV automated carbonate device connected to a Thermo Scientific MAT253 dual-inlet isotope ratio mass spectrometer in the Analytical Laboratory for Paleoclimate Studies (ALPS) at the Jackson School of Geosciences, University of Texas at Austin. Dolomite and calcite powders were reacted with 100 % phosphoric acid at 90 °C using a Kiel III online carbonate preparation line connected to a ThermoFinnigan 252 mass spectrometer. Stable isotope ratios of carbonate samples are reported in per mil (‰) units VPDB. The long-term analytical precision based on >1500 stable isotopic determinations of a carbonate standard is ±0.12 (2σ) for $\delta^{18}O$ and ±0.06 (2σ) for $\delta^{13}C$. The Sr isotope ratio analysis of eight selected samples including dolomite and calcite was performed at Nanjing University using a VG Aldermaston Micromass 30 solid-source mass spectrometer.

4 Stratigraphic distribution

Fine crystalline dolomites are mainly found in various grainstones and packstones at all depths across the littoral to sublittoral zone, and a small proportion of very fine crystalline dolomite may also partially replace lime mudstone. The distribution and abundance of dolomites is closely related to the characteristics of metre-scale, high-frequency depositional cycles. Metre-scale, high-frequency depositional cycles are defined by integrating core-derived sedimentology data and well-log interpretations. They are the fundamental genetic units, which make up the lacustrine carbonates in the Es_4^s interval. A transgressive and regressive hemicycle can be observed within each genetic unit. The lower section of each transgressive hemicycle consists of low-energy lime mudstone (including shells), whereas the upper regressive hemicycle consists of higher-energy packstone and grainstone. The upper regressive facies are extensively dolomitized, whereas the lower transgressive facies are partially dolomitized or undolomitized (Fig. 4). The abundance of dolomite is proportional to the amount of precursor grain present in the rock.

5 Petrological characteristics

Two types of dolomite can be observed in the Sikou lacustrine carbonate: (1) fine crystalline dolomite (4–20 μm), which is dominantly grain-replacive, matrix-replacive and rim cement, represents approximately 95 % of all dolomite. (2) Medium crystalline dolomite occurring as cement in the interparticle porosity accounts for less than 5 %. The focus of this paper is on the first type of dolomite, which is positive for the reservoir quality of lacustrine carbonate.

In most grainy facies, including packstones and grainstones, fine crystalline dolomite has replaced both micrite and non-skeletal grains, which generally results in complete dolomitization (Fig. 5a), whereas the matrix of lime mudstone was partially dolomitized (Fig. 5b). The fossil fragments in the lime mudstone typically remained unaffected (Fig. 5c). Fine crystalline dolomite rhombs may also occur as isopachous cement filling intergranular pores (Figs. 5d, 7a). Most of the grains are composed of pure dolomite, but some ooids consist of alternating lamellae of calcite and dolomite (Fig. 5e). Variable proportions of slightly coarser dolomite rhombs "floating" in a very fine crystalline dolomite matrix can be observed in the research area (Fig. 5f).

All of the very fine crystalline dolomites in the matrix and grains exhibit dull red luminescence (Fig. 6a, b). The dolomite is associated with clay and anhydrite (Fig. 6c). The size of the fine dolomite crystals is 1–10 μm in length, and they are subhedral to anhedral in shape (Fig. 6d), and the dolomite formed a smooth interface due to two-dimensional nucleation growth (Fig. 6e, f). The very fine dolomite rhombs may partially or completely replace the intraclasts, algaes, ooids and peloids from outside to inside (Fig. 7).

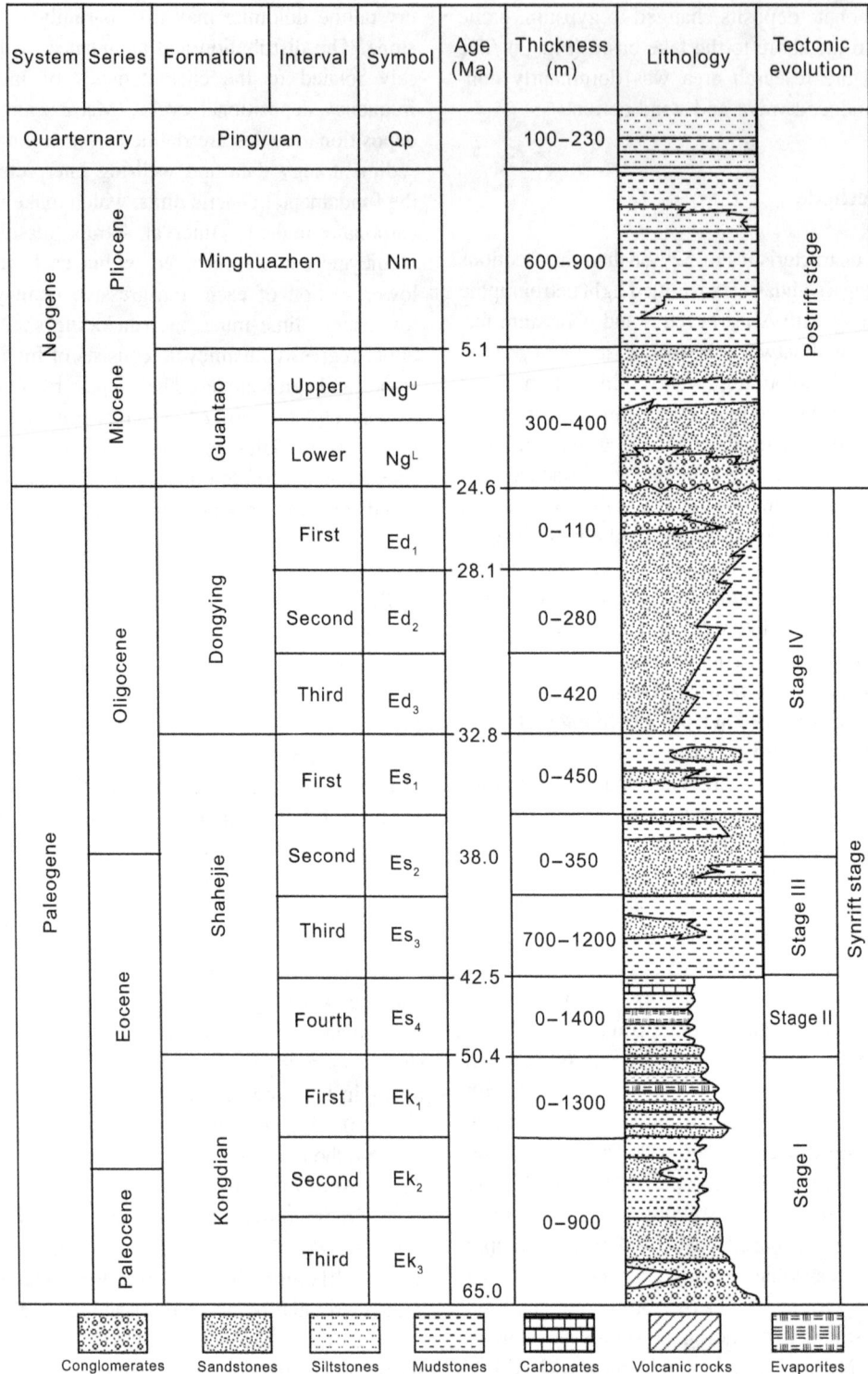

System	Series	Formation	Interval	Symbol	Age (Ma)	Thickness (m)	Lithology	Tectonic evolution
Quarternary		Pingyuan		Qp		100–230		
Neogene	Pliocene	Minghuazhen		Nm		600–900		Postrift stage
	Miocene	Guantao	Upper	Ng^U	5.1			
			Lower	Ng^L		300–400		
Paleogene	Oligocene	Dongying	First	Ed_1	24.6	0–110		Stage IV
			Second	Ed_2	28.1	0–280		
			Third	Ed_3		0–420		
		Shahejie	First	Es_1	32.8	0–450		
	Eocene		Second	Es_2	38.0	0–350		Stage III
			Third	Es_3		700–1200		
			Fourth	Es_4	42.5	0–1400		Stage II
	Paleocene	Kongdian	First	Ek_1	50.4	0–1300		Stage I
			Second	Ek_2		0–900		
			Third	Ek_3	65.0			

Legend: Conglomerates, Sandstones, Siltstones, Mudstones, Carbonates, Volcanic rocks, Evaporites

Fig. 2 Schematic Tertiary stratigraphy of the Sikou Sag in the Bohai Bay Basin (Guo et al. 2010)

6 Geochemistry

The MgCO$_3$ mol% of very fine dolomites ranges from 43 % to 50 %, with an average of 44.2 %. The strontium content in dolomites can generally be related to the texture or timing of formation. The strontium concentrations of fine dolomites range from 800 to 1200 ppm in the Sikou Sag.

Stable isotope analyses were carried out on samples of dolomites and calcites at different stratigraphic intervals

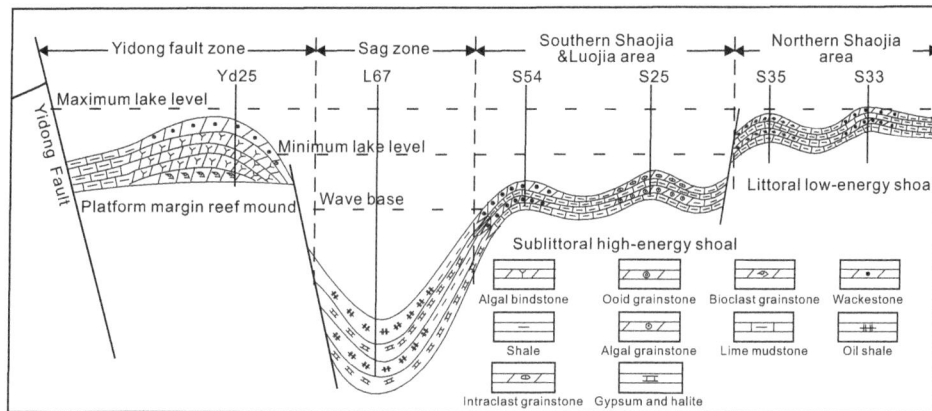

Fig. 3 Schematic model of the depositional distribution proposed for the lacustrine carbonate system of the upper Es$_4^s$ in Paleogene interval around the Sikou Sag. See Fig. 1b for well locations

and lithofacies, and the results are shown in Table 1. The $\delta^{13}C$ values of calcites from lime mudstone range from -2.13 to 1.99 ‰ PDB during this time interval. The carbon isotope composition of dolomite ranges from approximately 0.06 to 1.74 ‰ PDB, which is consistent with the carbonate isotope composition of Es$_4^s$ lacustrine water in the Sikou Sag (obtained from lime mudstone). The oxygen isotopic compositions of unaltered calcites from lime mudstone range from -8.3 to -5.4 ‰ PDB, while the oxygen isotopic compositions of fine crystalline dolomite range from -11.3 to -8.09 ‰ PDB.

The strontium isotope value of Eocene marine calcite was lower than 0.708000 (Veizer et al. 1999). The strontium isotope surrounding the Cambrian and Ordovician carbonate bedrock ranged from 0.7096 to 0.7140 (Liu et al. 2007). The strontium isotope signature of Sikou dolomites and calcite ranges from 0.710210 to 0.710844 and is significantly higher than the Eocene marine value, but this isotope signature is similar to the surrounding Cambrian and Ordovician carbonate bedrock.

7 Discussion and interpretation

7.1 Stratigraphic and petrographic considerations

Stratigraphic and petrographic observations indicated that very fine crystalline dolomite in the Sikou Sag is a product of penecontemporaneous or near-surface saline lake dolomitization. This interpretation is supported by the following lines of evidence:

1. In partially dolomitized metre-scale high-frequency cycles, the amount of dolomite within a metre-scale, shallowing-upward cycle increases upward from the cycle bottom. In the top grainy facies including packstones and grainstones, dolomite has replaced both micrite and non-skeletal grains generally resulting in complete dolomitization, whereas in the lower part, mudstones are dominated by calcite. These features illustrate that the dolomitization of the Es$_4^s$ occurred during the formation of each cycle, and the high porosity in packstones and grainstones favours the dolomitization fluids flowing through.

2. Textural relations between the dolomite and other phases indicated that the very fine crystalline dolomite formed relatively early in the diagenetic history. Very fine crystalline dolomites clearly predate equant sparry calcite cement and coarse calcite cement, which commonly filled the intergranular pores and is interpreted as a shallow to medium burial phase. The very fine rim cement is interpreted as an early cement formed at the lake floor. The dolomite has replaced the grain and matrix, generally resulting in complete dolomitization. This may reflect the occurrence of subaqueous evaporative conditions over longer periods, as opposed to a shallow tidal flat depositional system which may have experienced intermittent "flushing" by dolomitizing fluids (Rott and Qing 2013).

3. The occurrence of very fine crystalline dolomite as rim cement and lamellae within dolomitized ooids of oolitic grainstone suggests very early, syndepositional formation of the dolomite either by direct precipitation from saline lake water or by dolomitization of precursor calcite. The alternation of dolomitized and undolomitized layers of coated grains was likely produced before complete ooid formation.

4. The dolomite pervasively replaces the grains of all facies, whereas the matrix is normally only partially affected. Pervasive replacement of the calcite by very fine, poorly ordered crystalline structures (Fig. 6d) has been documented in numerous examples in the

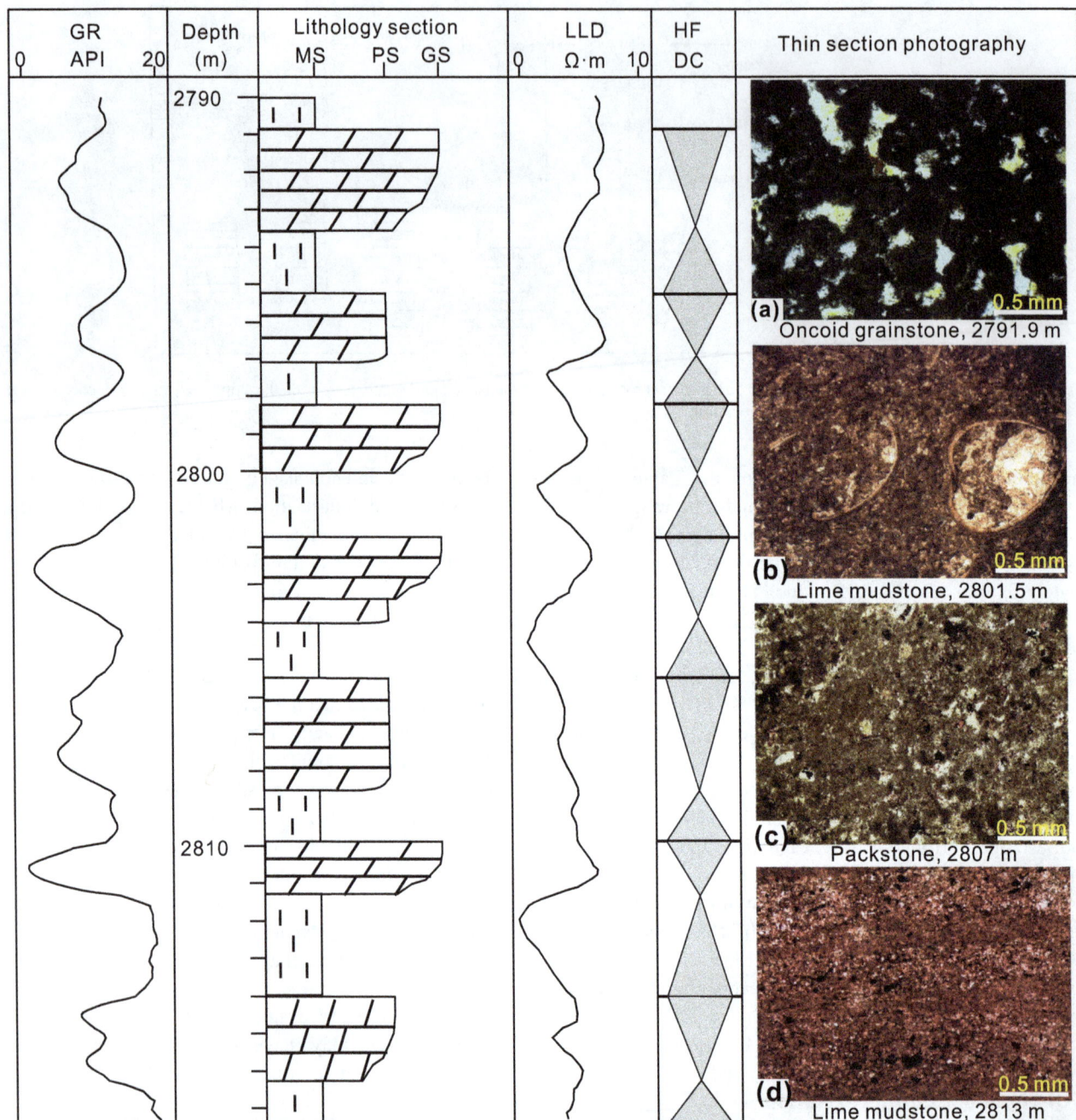

Fig. 4 High-frequency depositional cycles (HFDC) are recognized in the lacustrine strata, Well S54, the Sikou Sag. The amount of dolomites within a cycle increases upward from the cycle bottom. The *top part*, grainstone and packstone, of each cycle is completely dolomitized (*a, c*), whereas the *lower part* mudstones are dominated by calcite (*b, d*)

literature and is most often attributed to early diagenesis. In addition, some studies have shown that the dull red luminescence observed in dolomites (Fig. 6a, b) is commonly associated with early formed dolomites (Machel and Burton 1991). The crystal shape of dolomites is regulated only by the growth kinetics. At low temperatures, crystals grow by a layer-by-layer addition of atoms (Fig. 6e, f), resulting in faceted crystal sides (Sibley and Gregg 1987).

5. Based on petrographic observation, the euhedral dolomite rhombs are scattered throughout a very fine crystalline matrix and appear to overlap and cut across the finer crystals (Fig. 5f). They are interpreted as a product of recrystallization in its conventional sense, where dissolution–reprecipitation takes place on a micro-scale along crystal faces with concomitant increase in average crystal size (Malone et al. 1996; Rott and Qing 2013).

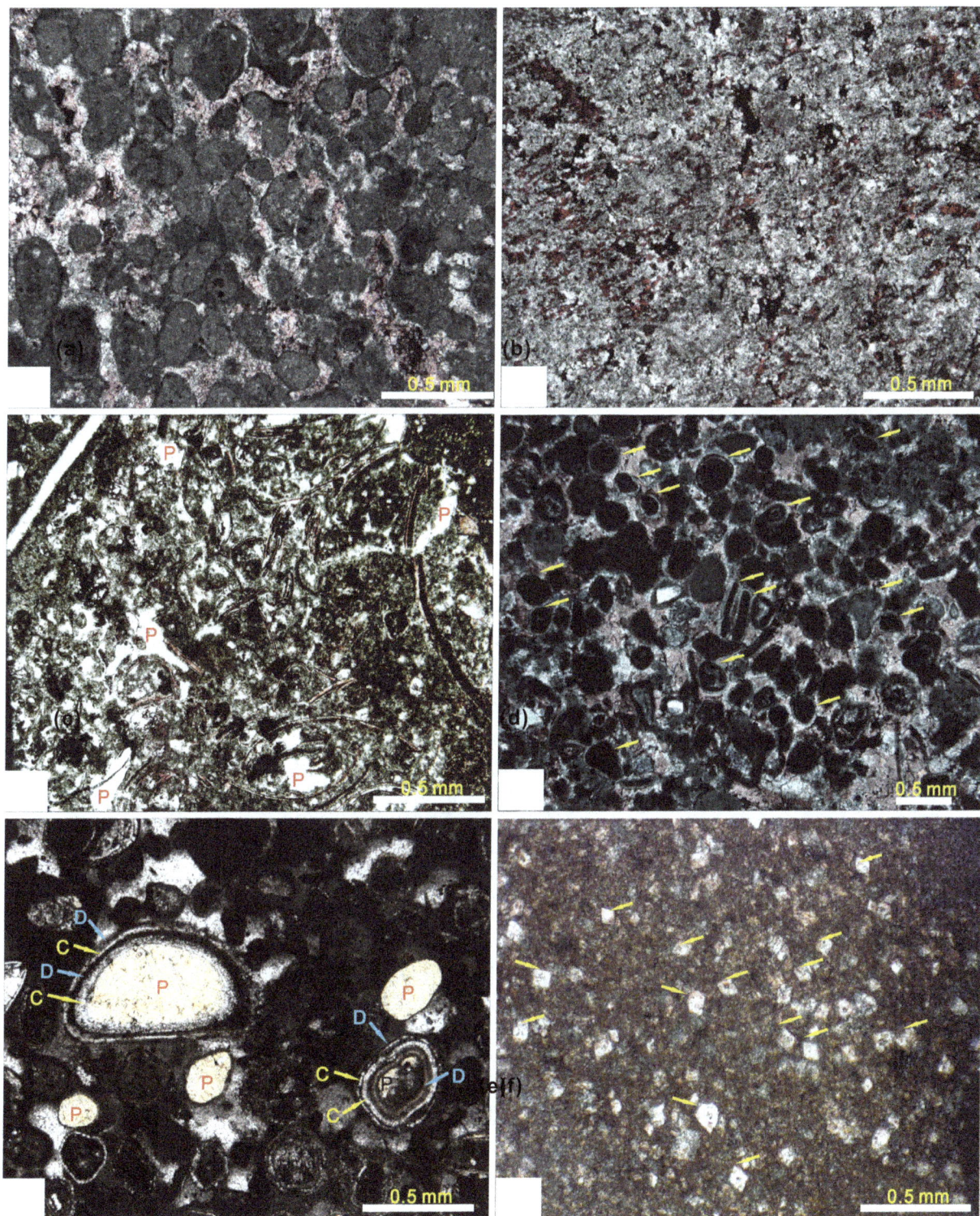

Fig. 5 Thin-section photomicrographs of fine crystalline dolomite. **a** Very fine dolomite in dolograinstone, the intergranular pores are filled by shallow-burial equant calcite cement, Well S541, depth 2681.6 m. **b** Partially dolomitized lime mudstone, Well Yd25, depth 3235.45 m. **c** Very fine dolomite in dolowackstone to packstone, undolomitized ostracoda shell fragments, Well S19, depth 2405.6 m. **d** Early isopachous dolomite cements, probably of lacustrine origin, in dolomitized grainstones, Well S54, depth 2798 m. **e** Pervasive dolomitized ooids with alternating lamellae of dolomite (*blue arrows*) and calcite (*yellow arrows*), parts of the core of ooids were dissolved. Well S25, depth 2318.7 m. **f** Euhedral, fine crystalline dolomite (*arrows*) scattered across very fine crystalline matrix in dolomitized limestone, Well S141, depth 2606.5 m. *Scale bars* are 0.5 mm

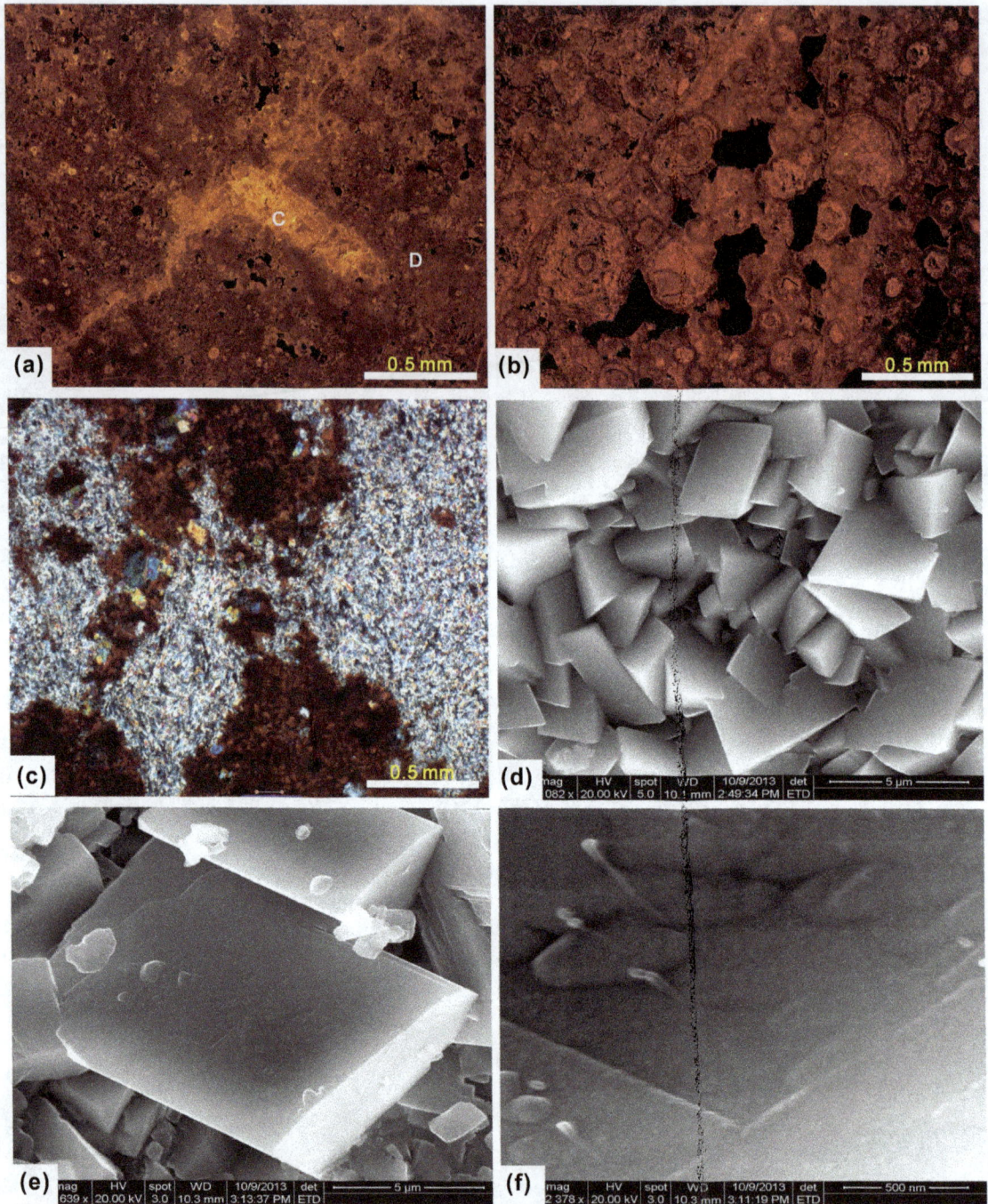

Fig. 6 a Cathodoluminescence photomicrograph of fine crystalline dolomite matrix (*dull red*) in packstone, Well Yd25, depth 3235.8 m. **b** Cathodoluminescence photomicrograph of fine crystalline dolomite grains (*dull red*) in oncoid grainstone, Well S54, depth 2796.52 m. **c** Nodular anhydrite cement in wackstone. Well Luo602, depth 2651.01 m. **d** SEM view of fine crystalline dolomite associated with clay in intraclastic grainstone. Well S541, depth 2181.6 m. **e, f** Planar dolomite forms a smooth interface due to two-dimensional nucleation growth. Well Yd25, depth 1890.4 m

7.2 Geochemical considerations

7.2.1 Source of Ca^{2+} and Mg^{2+} ions

Since the original studies of the upper fourth member of the Eocene Shahejie Formation in the Bohai Bay Basin, the source of Ca^{2+} and Mg^{2+} ions for the formation of lacustrine carbonate has been debated (Yuan et al. 2006)—is it marine or lacustrine in origin? There are four main processes cited in the literature that are beneficial to the lacustrine carbonate deposition in lakes. These include biogenic mediation, concentration through evaporation,

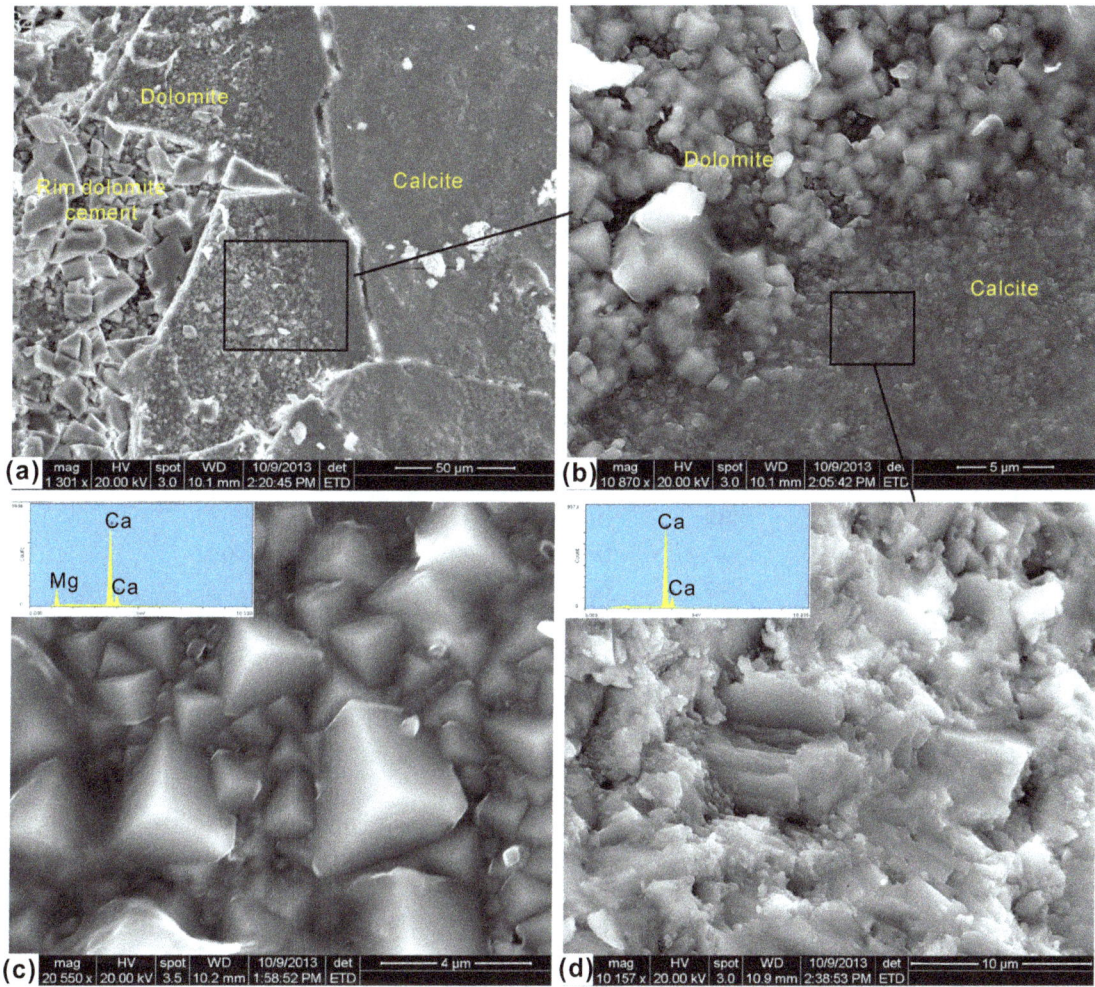

Fig. 7 SEM photomicrographs and EDX spectra of fine crystalline dolomites occurring as rim cements (**a**) and replacing the precursor grain from outside to inside (**b, c, d**). Well S25, depth 2318.7 m

waterborne clastic input and aeolian supply. The lake sediment type was controlled by the source rocks (Gebhardt et al. 2000). So the widespread carbonate rocks of the catchment area will allow for carbonate accumulations in the associated lake basin (Jones and Bowser 1978; Gierlowski-Kordesch 1998).

The carbon and oxygen stable isotopic composition of the sediments, suspended-load river input and the carbonate bedrock can be used to deduce the origins of lacustrine carbonate (Mauger and Compton 2011). The isotope values of bulk limestone samples indicate that the carbon and oxygen ratios of the Sikou carbonates differ significantly from the Eocene marine values, which range from δ^{13}C $\approx$ 1.5–2 ‰ VPDB and δ^{18}O $\approx$ −1 to 0 ‰ VPDB (Veizer et al. 1999).

The source of carbonate lacustrine sediments can be deduced from Sr isotopes. As strontium often substitutes for calcium, strontium isotopes can be used as a proxy to identify the source of the calcium; if the source of calcite is from marine water, then Sikou carbonate minerals would

be expected to have a sea water ^{87}Sr/^{86}Sr ratio, which was lower than 0.708000 during the Eocene (Veizer et al. 1999). The ^{87}Sr/^{86}Sr ratio of the lime mudstones and dolomitized mudstones ranges from 0.7107 to 0.7113 and is significantly higher than the Eocene marine value, but is similar to the surrounding Cambrian and Ordovician carbonate bedrock, which ranges from 0.7096 to 0.7140 (Fig. 8). The planar distribution of the lacustrine carbonates during the upper fourth member of the Eocene Shahejie Formation was also controlled by the surrounding Palaeozoic carbonate bedrock in the Jiyang Sub-basin (Jiang 2011). All of these features indicate that the source of the Ca^{2+} and Mg^{2+} ions may come from chemical weathering of the surrounding carbonate bedrock and that the Sikou carbonate is of lacustrine origin.

7.2.2 The formation environment of dolomites

In general, later-diagenetic, coarsely crystalline dolomites have lower strontium concentration compared to the early

Table 1 Result of geochemical analysis of lacustrine carbonate samples

Well	Depth, m	Lithofacies	Sr, ppm	δ^{13}C, ‰	δ^{18}O, ‰	^{87}Sr/^{86}Sr
S10	2663	Lime mudstone	–	0.73	−7.28	–
S10	2669	Dolograinstone	1083	0.35	−8.49	–
S10	2672	Lime mudstone	–	1.52	−6.24	–
S10	2676	Lime mudstone	–	0.77	−5.85	–
S10	2683	Dolomud-wackestone	1207	1.29	−10.3	–
S10	2695	Lime mudstone	–	−1.77	−6.65	0.70990
S10	2695.5	Lime mudstone	–	−1.48	−7.83	–
S10	2698	Lime mudstone	–	−0.09	−8.16	0.71014
S10	2699	Dolomud-wackestone	–	2.04	−8.54	–
S10	2700	Lime mudstone	–	−0.32	−5.81	–
S10	2704.8	Dolomud-wackestone	–	−2.10	−8.25	–
S10	2705	Lime mudstone	–	−2.1	−8.25	–
S13	2802	Dolomud-wackestone	971	1.74	−11.1	–
S13	2818	Dolomud-wackestone	939	0.57	−9.09	–
S141	2542.4	Dolomud-wackestone	–	−0.94	−8.42	–
S141	2607	Lime mudstone	–	−0.69	−8.10	–
S141	2609	Lime mudstone	–	−0.94	−8.11	–
S19	2403	Lime mudstone	–	−0.77	−8.19	0.71047
S19	2409	Lime mudstone	–	−2.13	−6.37	–
S25	2318.7	Dolograinstone	1102	0.92	−9.72	0.71042
S54	2428.3	Lime mudstone	–	−1.48	−7.83	0.71048
S54	2804.6	Dolograinstone	–	−1.75	−9.52	–
S541	2674	Dolograinstone	901	1.43	−8.76	0.70964
S541	2681.6	Dolograinstone	1081	0.06	−8.09	–
Yd25	3226.6	Dolograinstone	1205	1.14	−9.67	–
Yd25	3238	Lime mudstone	–	0.29	−7.38	–
Yd25	3235	Dolograinstone	1127	0.13	−11.02	0.70981
Yd25	3245	Lime mudstone	–	0.69	−6.63	–
Yd301	3590.3	Dolograinstone	–	2.06	−10.20	–
Yd301	3592.8	Dolograinstone	–	2.20	−9.88	–
Yd301	3593.3	Dolograinstone	1194	1.61	−10.62	0.71045
Yd301	3596	Lime mudstone	–	2.29	−6.65	–

fine crystalline dolomites (Hu et al. 2013). High strontium concentrations (more than 1000 ppm) occur in dolomites precipitated from intensively evaporated waters and from dolomitizing fluids with high Sr/Ca ratios (Morrow 1990). The strontium concentrations of fine crystalline dolomites ranged from 800 to 1200 ppm in the Sikou Sag, indicating that the fine dolomites formed in intensively evaporated waters (Table 1).

C–O isotopic compositions of dolomites are mainly affected by temperature and salinity in the medium. Carbon isotopic composition is influenced by different sources of carbon, while oxygen isotopic fractionation occurs during the evaporation of water as temperature rises (Yang et al. 2013). Overall, the δ^{18}O PDB and δ^{13}C PDB values become greater with increasing salinity (Lu et al. 2015). The dolomite in the Sikou Sag may have undergone alteration by diagenetic fluids, and as a result,

the oxygen isotopic signature was reset; however, the carbon was not affected (Fig. 9). As the isotope fractionation between calcites and coevally precipitated dolomite is approximately 2.5 % (Swart and Melim 2000; Wang et al. 2012), the δ^{18}O values of dolomite should be slightly higher than δ^{18}O values of calcite, if they formed from the same lacustrine water. However, the δ^{18}O values of dolomites from the Sikou Sag are significantly lower than the predicted values for dolomites derived from Es_4^s lacustrine water. The previous textural relations reveal that the initial calcite precursors were replaced relatively early by fine crystalline dolomite. Owing to the fluctuation of lake levels, the early formed dolomites were always affected by the meteoric water, and formed mouldic and vuggy pores (Figs. 5e, 6b), and caused a negative shift of the oxygen isotopic value of the early formed dolomites (Fig. 9).

7.3 Mechanism of dolomite formation

Owing to seasonal evaporation, carbonate mainly formed inshore, while gypsum and halite occurred at the centre of the lake in most of the modern and ancient saline lakes (Peng 2011; Mauger and Compton 2011). The existence of wet/dry cycles in surficial environments seems to be an essential requirement for the nucleation of dolomites (Deelman 2003). Based on the distribution of dolomites in high-frequency depositional cycles, the formation process of fine dolomite in the Sikou Sag was interpreted to undergo two stages of evolution (Fig. 10).

Early brackish lake stage: This stage represents the most humid interval, in which the lake level is high. The chemical weathering of Cambrian and Ordovician carbonate bedrock produces Ca^{2+}, Mg^{2+} and HCO^{3-} ions which are transported to the Sikou Sag by run-off or through flow. The sulphate reduction began at the centre of the sag, where it is likely anoxic and stratified (Mauger and Compton 2011). Early evaporation causes salinities to increase over time, and calcite is the least soluble of the major salts. As a result, it will precipitate in the shallow-water zone during evaporation, while the organic mudstone deposits occurred at the centre of the lake.

Late saline lake stage: This stage represents the aridity interval, and the greater evaporation rates resulted in the low lake level and the increase in salinity. The increase in the Mg/Ca ratio, following the removal of Ca^{2+} during gypsum precipitation, favours precipitation of Mg calcite. As the Mg/Ca ion ratio of the pore waters rises further, fine crystalline dolomite may replace existing calcite and high-Mg calcite.

The fine dolomite occurring in the grains and the matrix is interpreted to signify an early replacement origin, which is

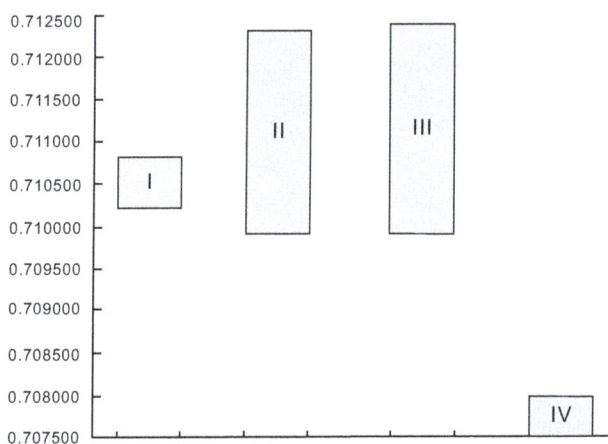

Fig. 9 $\delta^{13}C$ and $\delta^{18}O$ values of fine crystalline dolomites and calcites from Sikou lacustrine carbonates. Values are displayed in standard delta notation relative to the PDB standard. Expected $\delta^{13}C$ and $\delta^{18}O$ ranges of primary calcite precipitated from Eocene seawater (Veizer et al. 1999)

proven by the existence of the replacement (Fig. 7) and the coexistence of calcite and dolomite in the matrix (Fig. 5b, c). According to the early replacement origin, the fine dolomite may replace the tiny calcite crystals covering the grains during the increase in salinity in the lake water. It is likely that higher-magnesium calcite lamellae with higher microporosity are more prone to dolomitization than the lower-magnesium calcite lamellae, which formed the alternating lamellae of calcite and dolomite in the ooids (Rott and Qing 2013). However, replacement was not observed at the thin dolomite lamellae contained within some ooids.

The formation of lamellae and cement is considered in the published literature as primary dolomite in response to changing chemical conditions (Mitchell et al. 1987; Mazzullo et al. 1995). The precipitation of primary dolomites from modern and ancient lacustrine environments has been widely documented in the literature (De Deckker and Last 1988; Arenas et al. 1999). Some of the fine dolomite cement was also interpreted to be of primary origin in a shallow lacustrine environment (Mazzullo 2000). Two main models have been established for the primary very fine lacustrine dolomite. Firstly, the bacterial removal of sulphate and resultant increase in availability of Mg^{2+} ions made the very fine dolomite directly precipitate from the evaporating lake waters (Burn et al. 2000). Secondly, Meister et al. (2011) reported that fine-grained dolomite probably precipitates directly in the well-aerated water column from highly alkaline brine, whereas microbial alkalinity production and pH increases are negligible. So the precipitation of dolomites from saline lake water should be considered in light of the possible origin of very fine rim cement and dolomite lamellae within ooids in the Sikou Sag, and inferring a similar genesis for the origin of

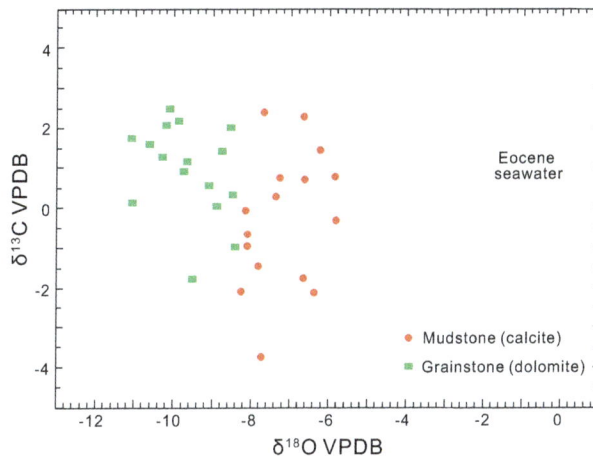

Fig. 8 $^{87}Sr/^{86}Sr$ isotope plot of different samples, *I* is from the samples in the Sikou Sag; *II* and *III* represent the surrounding Cambrian and Ordovician carbonate bedrock (Liu et al. 2007); *IV* represents the Eocene marine values (Veizer et al. 1999)

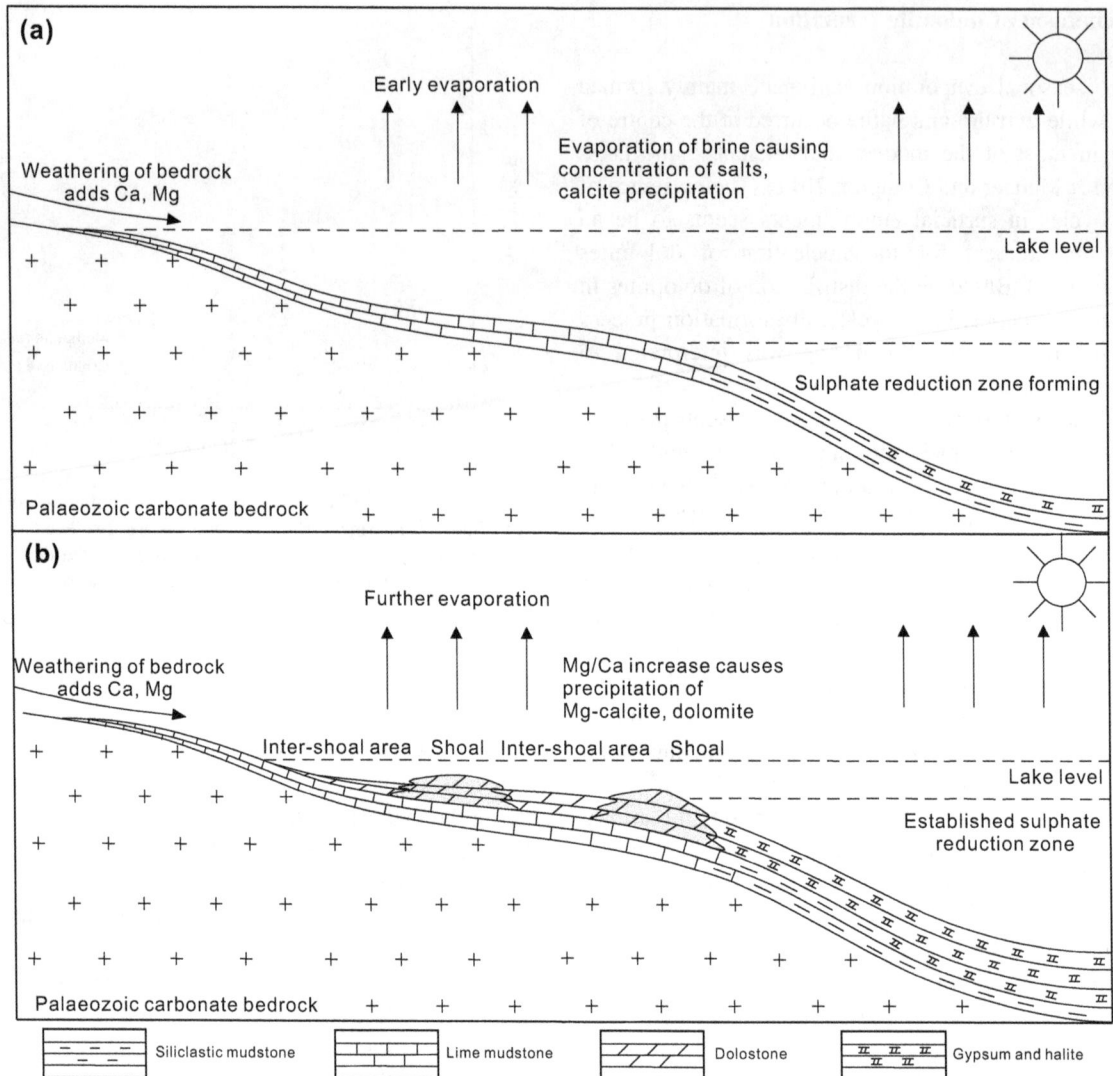

Fig. 10 Proposed mechanism of dolomitization for the upper fourth member of the Eocene Shahejie Formation (Es_4^s) around the Sikou Sag. **a** Saline lake with moderate salt and calcite precipitation during early stages of salt accumulation; **b** further evaporation elevated Mg/Ca ratios and sulphate reduced, which led to dolomite precipitation and replaced the precursor calcite

dolomite mud should also be explored in dolomitized mudstone.

7.4 Impact on development of hydrocarbon reservoir

Many previous studies have reported the porosity improvement by dolomitization (Moore 2001; Swei and Tucker 2012; Meng et al. 2014a, b). In this context, in most grainy facies and mudstones, fine crystalline dolomite has replaced both micrite and non-skeletal grains, generally resulting in a significant increase in reservoir quality, particularly permeability, compared with the precursor limestone (Xu 2013). With the development of intercrystalline pores, dolomites are more resistant to burial-related effects which therefore cause little reduction in porosity

relative to limestones (Rabbani 2004) and are more frequently affected by fracturing than limestones and sandstones (Sun 1995). In addition, the intercrystalline pores provided conduits for the influx of hydrocarbons and acidic fluids, enhancing porosity due to leaching.

8 Conclusions

The upper fourth member of the Eocene Shahejie Formation in the Sikou Sag consists of abundant grain- and matrix replacement very fine crystalline dolomite (2–30 μm). The dolomite is interpreted as a product of near-surface dolomitization of precursor calcite by lacustrine water of high salinity. Similar to the surrounding Cambrian and Ordovician bedrock strontium isotope values, the dolomite and calcite support a carbonate bedrock source for some of the Ca and Mg delivered to the sag. In particular, the chemical weathering of carbonate bedrock may raise the Mg/Ca ratios of the lacustrine waters to achieve the high Mg/Ca ratios required for dolomite precipitation. Most of the dolomite crystals typically replace precursor grains and matrix. However, the direct precipitation of dolomites from evaporating lake waters should be considered in light of the possible origin of very early rim dolomite cement and dolomite lamellae within ooids.

Acknowledgments This work was supported by the National Basic Research Program of China (Grant No. 2014CB239002) and Natural Science Foundation of Shandong Province, China (Grant No. ZR2014DQ016). The China University of Petroleum (East China) provided facilities for this research. The authors thank the Hekou Production Factory of the Shengli Oil Field Company for some data preparation.

References

Arenas C, Alonso Zarza AM, Pardo G. Dedolomitization and other diagenetic processes in Miocene lacustrine deposits, Ebro Basin (Spain). Sediment Geol. 1999;125:23–45.

Burn SJ, McKenzie JA, Vasconcelos C. Dolomite formation and biogeochemical cycles in the Phanerozoic. Sedimentology. 2000;47(s1):49–61.

Bustillo MA, Arribas ME, Bustillo M. Dolomitization and silicification in low-energy lacustrine carbonates (Paleogene, Madrid Basin, Spain). Sediment Geol. 2002;151:107–26.

Casado AI, Alonso-Zarza AM, La Iglesia A. Morphology and origin of dolomite in paleosols and lacustrine sequences. Examples from the Miocene of the Madrid Basin. Sediment Geol. 2014;312:50–62.

De Deckker P, Last WM. Modern dolomite deposition in continental, saline lakes, western Victoria, Australia. Geology. 1988;16:29–32.

Deelman JC. Low-temperature formation of dolomite and magnesite. Eindhoven: CD Publications; 2003. p. 504.

Gebhardt U, Merkel T, Szabados A. Karbonatsedimentation in siliziklastischen fluviatilen Abfolgen. Freib Forsch. 2000;490:133–68.

Gierlowski-Kordesch EH. Carbonate deposition in an ephemeral siliciclastic alluvial system: Jurassic Shuttle Meadow Formation, Newark Supergroup, Hartford Basin, U.S.A. Palaeogeogr Palaeoclimatol Palaeoecol. 1998;140:161–84.

Guo X, He S, Liu K, et al. Oil generation as the dominant overpressure mechanism in the Cenozoic Dongying Depression, Bohai Bay Basin, China. AAPG Bull. 2010;94:1859–81.

Hu ZW, Huang SJ, Li ZM, et al. Geochemical characteristics of the Permian Changxing Formation reef dolomites, northeastern Sichuan Basin, China. Pet Sci. 2013;10(1):38–49.

Jiang XL. Main controlling factors of lacustrine carbonate rock in Jiyang Depression. Pet Geol Recovery Effic. 2011;18:23–7 (**in Chinese**).

Jones BF, Bowser CJ. The mineralogy and related chemistry of lake sediments. In: Lerman A, editor. Lakes: chemistry, geology, physics. Springer: Berlin; 1978. p. 179–235.

Köster MH, Gilg HA. Pedogenic, palustrine and groundwater dolomite formation in non-marine bentonites (Bavaria, Germany). Clay Miner. 2015;50(2):163–83.

Land LS. The origin of massive dolomite. J Geol Educ. 1985;33:112–25.

Last FM, Last WM, Halden NM. Modern and late Holocene dolomite formation: Manito, Saskatchewan, Canada. Sediment Geol. 2012;281:222–37.

Liu SG, Shi HX, Wang GZ, et al. Formation mechanism of Lower Paleozoic carbonate reservoirs in Zhuanghai Buried Hill. Nat Gas Ind. 2007;27(10):1–5 (**in Chinese**).

Lu XC, Shi JA, Zhang SC, et al. The origin and formation model of Permian dolostones on the northwestern margin of Junggar Basin, China. J Asian Earth Sci. 2015;105:456–67.

Machel HG, Burton EA. Factors governing cathodoluminescence in calcite and dolomite and their implications for studies of carbonate diagenesis. In: Barker CE, Kopp OC, editors. Luminescence microscopy and spectroscopy: qualitative and quantitative applications: SPEM, short course; 1991. vol. 25, p. 37–57.

Malone MJ, Baker PA, Burns SJ. Recrystallization of dolomite: an experimental study from 50–200 °C. Geochim Cosmochim Acta. 1996;60:2189–207.

Mauger CL, Compton JS. Formation of modern dolomite in hypersaline pans of the Western Cape, South Africa. Sedimentology. 2011;58:1678–92.

Mazzullo SJ, Bischoff WD, Teal CS. Holocene subtidal dolomitization, north Belize. Geology. 1995;23:341–4.

Mazzullo SJ. Organogenic dolomitization in peritidal to deep-sea sediments. J Sediment Res. 2000;70:10–23.

Meister P, Reyes C, Beaumont W, et al. Calcium and magnesium-limited dolomite precipitation at Deep Springs Lake, California. Sedimentology. 2011;58:1810–30.

Meng WB, Wu HZ, Li GR, et al. Dolomitization mechanisms and influence on reservoir development in the Upper Permian Changxing Formation in Yuanba area, northern Sichuan Basin. Acta Pet Sin. 2014a;30(3):699–708.

Meng Y, Zhu HD, Li XN, et al. Thermodynamic analyses of dolomite dissolution and prediction of the zones of secondary porosity: a case study of the tight tuffaceous dolomite reservoir of the second member, Permian Lucaogou Formation, Santanghu Basin, NW China. Pet Explor Dev. 2014b;41(6):754–60 (**in Chinese**).

Mitchell JT, Land LS, Miser DE. Modern marine dolomite cement in a north Jamaican fringing reef. Geology. 1987;15:557–60.

Moore CH. Carbonate reservoirs. Developments in sedimentology, vol. 55. Amsterdam: Elsevier; 2001. p. 444.

Morrow DW. Dolomite, part 2: dolomitization models and ancient dolostones. In: McIlreath IA, Morrow DW, editors. Diagenesis. Geoscience Canada repint series 4; 1990. p. 125–39.

Pan YL, Li S. Large-scale continental rift basin sequence stratigraphy and subtle oil and gas reservoirs: a case study in the Jiyang Sag. Beijing: Petroleum Industry Press; 2004. p. 57 (**in Chinese**).

Peng CS. Distribution of favorable lacustrine carbonate reservoirs: A case from the Upper Es$_4$ of Zhanhua Sag, Bohai Bay Basin. Pet Explor Dev. 2011;38(4):435–43 (**in Chinese**).

Rabbani AR. Geochemical and petrographical study of dolomite facies in the Dalan and Kangan gas reservoirs in the south of Iran. Res Sci Eng Pet Bull. 2004;14(49):34–46.

Rosen MR, Coshell L. A new location of Holocene dolomite formation, Lake Hayward, Western Australia. Sedimentology. 1992;39:161–6.

Rott CM, Qing H. Early dolomitization and recrystallization in shallow marine carbonates, Mississippian Alida beds, Williston Basin (Canada): evidence from petrography and isotope geochemistry. J Sediment Res. 2013;83:928–41.

Sibley DF, Gregg JM. Classification of dolomite rock textures. J Sediment Res. 1987;57:967–75.

Sun SG. Dolomite reservoirs: porosity evolution and reservoir characteristic. AAPG Bull. 1995;79(21):186–204.

Swart PK, Melim L. The origin of dolomites in Tertiary sediments from the margin of Great Bahama Bank. J Sediment Res. 2000;70:738–48.

Swei GH, Tucker ME. Impact of diagenesis on reservoir quality in ramp carbonates: Gialo formation (Middle Eocene), Sirt Basin, Libya. J Pet Geol. 2012;35(1):25–48.

Van Tula FM. The present status of the dolomite problem. Science. 1916;44:688–90.

Veizer J, Ala D, Azmy K, et al. $^{87}Sr/^{86}Sr$, $\delta^{13}C$ and $\delta^{18}O$ evolution of Phanerozoic seawater. Chem Geol. 1999;161:59–88.

Wang SQ, Zhao L, Cheng XB, et al. Geochemical characteristics and genetic model of dolomite reservoirs in the eastern margin of the Pre-Caspian Basin. Pet Sci. 2012;2:161–9.

Wright DT. The role of sulphate reducing bacteria and cyanobacteria in dolomite formation in distal ephemeral lakes of the Coorong region, South Australia. Sediment Geol. 1999;126:147–57.

Xu L. Effects of dolomitization in carbonate rocks on reservoir porosity in the Dongying Depression. Bull Mineral Petrol Geochem. 2013;32:463–7 (in Chinese).

Yang ZH, Zhang N, Dong JX, et al. Carbon oxygen isotope analysis and its significance of carbonate in the Zhaogezhuang Section of Early Ordovician in Tangshan, North China. J Earth Sci. 2013;24(6):918–34.

Yuan WF, Chen SY, Zeng C. Study on marine transgression of Palaogene Shahejie Formation in Jiyang Depression. Acta Pet Sin. 2006;27:40–9 (in Chinese).

16

FCC riser quick separation system

Zhi Li[1] · Chun-Xi Lu[1]

Abstract The riser reactor is the key unit in the fluid catalytic cracking (FCC) process. As the FCC feedstocks become heavier, the product mixture of oil, gas and catalysts must be separated immediately at the outlet of the riser to avoid excessive coking. The quick separation system is the core equipment in the FCC unit. China University of Petroleum (Beijing) has developed many kinds of separation system including the fender-stripping cyclone and circulating-stripping cyclone systems, which can increase the separation efficiency and reduce the pressure drop remarkably. For the inner riser system, a vortex quick separation system has been developed. It contains a vortex quick separator and an isolated shell. In order to reduce the separation time, a new type of separator called the short residence time separator system was developed. It can further reduce the separation time to less than 1 s. In this paper, the corresponding design principles, structure and industrial application of these different kinds of separation systems are reviewed. A system that can simultaneously realize quick oil gas separation, quick oil gas extraction and quick pre-stripping of catalysts at the end of the riser is the trend in the future.

Keywords Fluidization · Quick separation · FCC · Post-riser system

✉ Chun-Xi Lu
  lcxing@cup.edu.cn; lcx725@sina.com

[1] State Key Laboratory of Heavy Oil Processing, China University of Petroleum, Beijing 102249, People's Republic of China

Edited by Xiu-Qin Zhu

1 Introduction

Fluid catalytic cracking (FCC), one of the most important conversion processes used in petroleum refineries, transforms the low-value heavy crude oil into a variety of high-value-added light oil products. Up to now, the total capacity of FCC units (FCCUs) in China has reached 150 Mt/a, of which the outputs roughly account for 70 %, 40 % and 30 % of gasoline, propylene and diesel pools, respectively. However, in recent years, the reaction temperature and cracking degree have increased to a large extent with the feedstocks becoming heavier and inferior. As a result, the incidental high coke formation rate may threaten the operation safety in some refineries. Two factors contribute to the coke formation: (1) long residence time of the mixture of oil, gas and catalysts in the separator and (2) the severe back-mixing of oil gas in the disengager (Li et al. 2011; Karthika et al. 2012; Xu 2014; Gao et al. 2013; Dong et al. 2013; Wang et al. 2016). To solve these problems, a series of efficient separation systems at the riser outlet have been developed. These separation systems concentrate on separating the production and catalysts as soon as possible. Besides, they should reduce the residence time of oil gas in the reaction systems (gas–solid separation unit and disengager) simultaneously. Technologies which can simultaneously realize quick mixture separation and reduce oil gas residence time in the reaction system have been studied. Results show that the separation systems have brought huge economic benefit for chemical industries all around the world.

2 Early technologies

The early industrial separation systems are shown in Fig. 1. The earliest versions are T type, inverted L type and clover type. In order to increase the separation

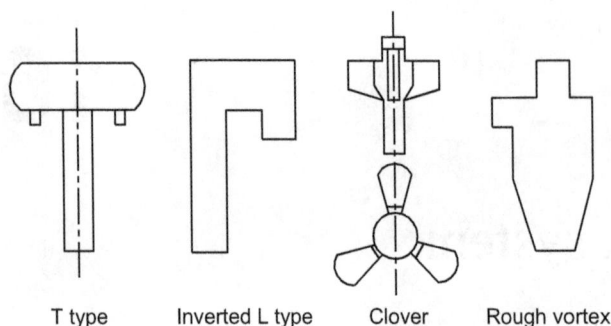

T type   Inverted L type   Clover   Rough vortex

Fig. 1 Schematic diagram of the quick separation systems used in the outlet of the early riser reactor

efficiency, the improved version called the rough vortex quick separation system was developed in 1990 (Amos and Avidant 1990). These systems with simple structures all use inertia force to separate the mixture. The separation efficiency has been improved to some degree with the optimization of the structures. However, problems like back-mixing and coke formation were not taken into consideration.

Table 1 shows the comparison of these four types of quick separation systems. It can be seen that T-type, inverted L-type and clover-type separation systems use inertia force generated by 180° rotation of gas and solid to separate the oil gas and catalysts, while the rough vortex quick separation system uses the centrifugal force resulted from the high spinning velocity. The T-type and inverted L-type separation systems usually use two-stage top cyclone separators in series to raise the separation efficiency. However, the pressure drop of the system is high and the residence time of oil gas is very long (about 20 s), which can result in severe coke formation problems. The improved rough vortex separation system uses only a one-stage top cyclone to separate the mixture. It shortens the residence time of gas oil and raises the separation efficiency to almost 98 %, but the high pressure drop is a threat to stable operation of the process. Though the rough vortex quick separation system has had many improvements recently, the separation efficiency of it still does not meet the requirements of the industry. Besides, the long contact time of mixture in the disengager can result in coking problems.

3 Foreign technologies

Foreign researchers have been working on the optimization and development of quick separation systems for a long time. Some technologies have already been applied to the industry production, and remarkable results have been achieved. These quick separation systems include the vortex disengager stripper (VDS) system and vortex separation system (VSS) from UOP Company (Gilbert 1995), the Rams horn system from the Stone & Webster Company and the closed cyclone technology from Mobil Company.

3.1 VDS system and VSS

For the first time, UOP Company (USA) developed a separation system with a closed cover outside the riser, namely VDS system and/or VSS. The VDS system has a stripping unit at the bottom of a rough vortex separation system and forms a two-stage stripping system. The rough cyclone standpipe is connected to the top cyclone entrance which forms a ring, and this ring makes the stripper gas flow into the VDS system easily. For the vortex separation system, a tube with a bent wall (split head) is located at the riser outlet to separate the mixture. Outside the split head is a sealed cover. The connection between the standpipe and top cyclone entrance is different from the VDS system (Gauthier et al. 2000; Chen and Brostem 2001). Figure 2 shows the basic structures of these two separation systems.

The common characteristic of VDS and VSS systems is that the oil gas and stripping steam flow through the quick separation system. Thus, the residence time is reduced dramatically with separation efficiency unchanged, and the coking problem is eased. It is reported that 98 % oil and gas flow out of the separation system with the residence time being decreased to around 6 s by using the VDS and VSS systems (UOP). In addition, compared with the traditional T-type separation system, VDS and VSS systems provide high gasoline yield, with decreased dry gas yield and coke formation (Tian and Zhang 2013).

3.2 Rams horn system

Stone & Webster Company (USA) developed a riser terminal device named the Rams horn separator (axial separator).

Table 1 Comparison among early technologies of quick separation systems

	T type; inverted L type; clover type	Rough vortex separation system
Principle	Inertia force	Centrifugal force
Pressure drop	Less than 1 kPa	About 2–4 kPa
Residence time	About 20 s	8–10 s, about 10 % oil gas back-mixing
System construct	Two-stage series of top cyclone	Single-stage of top cyclone
Problem	Severe coke formation	Severe coke formation

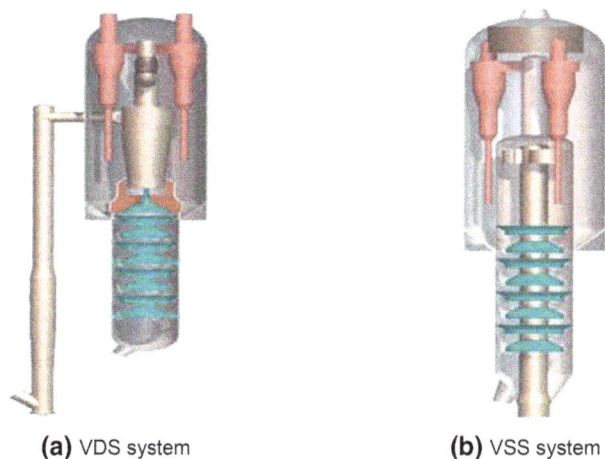

(a) VDS system   **(b)** VSS system

Fig. 2 Basic structures of the VDS and VSS systems

It can reduce the oil gas residence time in the disengager (from 15–51 to 4–9 s by recorded data), decrease the pressure drop and accelerate the gas solid separation (Gauthier et al. 2005). This system uses centrifugal force generated by 180° rotation of the oil/gas mixture and catalysts. So it can prevent catalyst particles from entering the fractionating tower and enables easy operation. As a result, the reaction temperature in the disengager can be lowered by 10 % with light oil yield increasing by about 2 %; meanwhile, the dry gas yield and coke formation can be decreased.

3.3 Closed cyclone technology

Mobil Company and Kellogg Company have developed a high-efficiency closed cyclone technology which can reduce secondary reactions and increase the desired products as well as the total liquid yield (Quinn and Silverman 1995). The main characteristic of this technology is that the catalysts and oil gas directly flow into the first cyclone separator located at the end of the riser. After the first stage of separation, the oil and gas mixture enters the second cyclone separator for further separation with a small amount of catalysts, which can effectively shorten the average contact time of oil with catalysts. The dry gas yield decreases, and the total liquid yield increases (Zhao et al. 2006). Using the tracer (helium) determination method, it is found that the closed cyclone technology can shorten the average residence time of oil and gas reaction process in the open riser from 20 s down to about 2 s with little back-mixing problem.

4 China's technologies

In China, the clover-type and rough vortex separation systems are widely used in FCC process. There are a few units still using the old T-type and inverted L-type separation systems. As mentioned above, these technologies are relatively behind the foreign technologies. In 1994, China University of Petroleum, Beijing, started to concentrate on developing new and efficient quick separation technologies. By now, many types of highly efficient separation systems have been applied in industry successfully, including the fender-stripping cyclone separation system (FSC), the vortex quick separation system (VQS) and the circulating-stripping cyclone separation system (CSC; Hu et al. 2008).

These three separation systems can quickly separate gas and solid, quickly pre-strip catalysts and quickly withdraw oil gas from reactor simultaneously. Besides, there are many other advantages such as no restrictions on unit start-up, good operating flexibility, high separation efficiency and good product distribution.

4.1 FSC system

Currently, the separation efficiency of rough cyclone separators is more than 98 %. Research has reduced the oil gas residence time by extending a rough vortex standpipe adjacent to the top cyclone entrance. The rough vortex dipleg operates with a positive pressure, which means part of the oil gas is also discharged through the dipleg with the pressure impact (the pressure in the dipleg is greater than that in the disengager) besides the dense-phase catalysts. The oil gas moves upward with a velocity of 0.1 m/s, through the 10 m of trip, and then enters the top cyclone.

Figure 3 shows the structural schematic diagram of the FSC system. This system mainly consists of four parts: riser, rough cyclone, pre-stripper and stripper burden surface. In the FSC system, the traditional rough cyclone dipleg is changed into a pre-stripper with a unique structured baffle inside. This baffle has a skirt border which can form a thin layer of catalysts on the baffle plate; thus, the pre-stripping efficiency can be improved. A socket connection structure located between the rough vortex standpipe and top cyclone outlet can achieve quick extraction of gas and oil (Lu and Shi 2007).

The characteristics of this system are shown as follows: The lower part of the pre-stripper is set downstream to absorb the oil and gas deposited on the catalysts easily. The pressure drop changes from positive to negative to extract the oil gas quickly. By this way, the separation efficiency can be improved to some extent. The connection between the rough cyclone and the top cyclone is open and flexible, which means the entrance is large, and it makes gas come out easily.

4.2 CSC system

Different from the baffles in the FSC and VQS systems, in the CSC system, a draft tube is set in the middle to increase

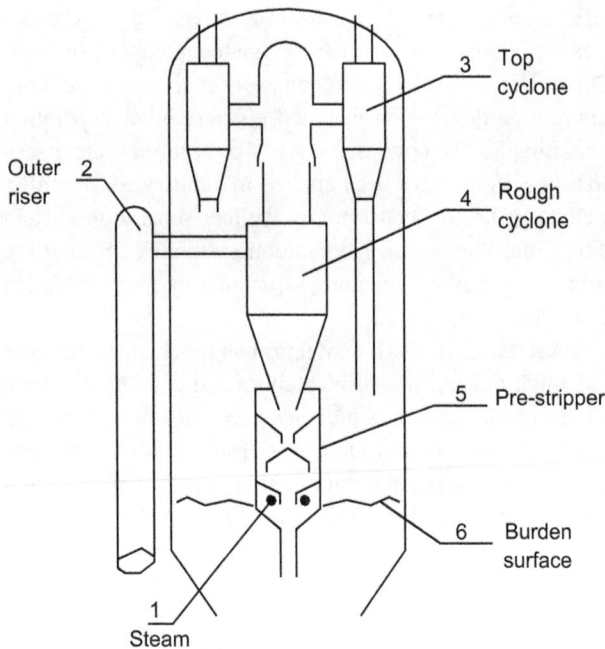

Fig. 3 Basic structure of the FSC separation system

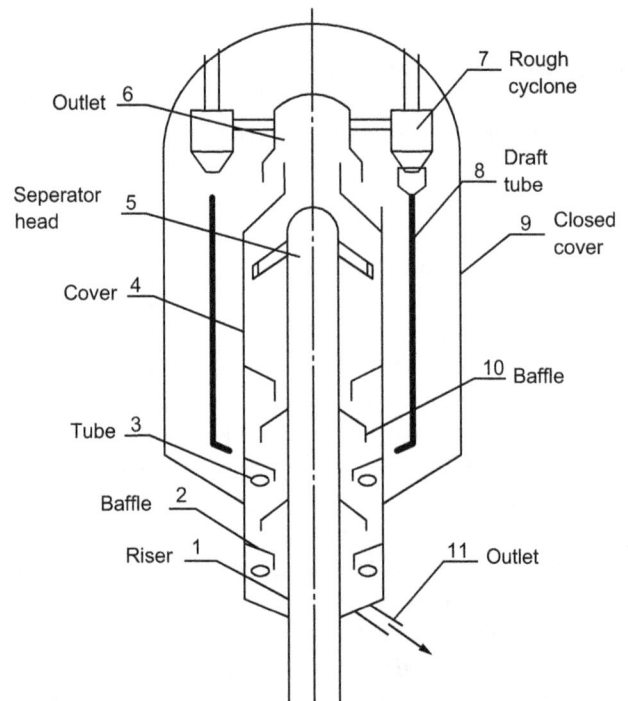

Fig. 4 Basic structure of the CSC separation system

the contact time of catalysts with steam. This kind of system has many advantages such as good stripping results, low coke formation, reduced dry gas yield and improved light oil yield (Rao et al. 2011).

Figure 4 shows the structure of the CSC system, which consists of three parts: rough cyclone with optimized structure size; dense-phase circulation pre-stripper with central feeding pipe; and socket-type (or open straight connected) fast oil gas withdrawal device. The oil gas residence time in the disengager is decreased; hence, this reduces secondary cracking and improves the product selectivity (Xia 2014; Pan 2011). Currently, 13 sets of the CSC system are being used industrially in China, and they can effectively reduce the coke formation and improve the light oil yield.

4.3 VQS

Due to its compact structure and excellent capability, the VQS also has the advantages of the FSC system. As shown in Fig. 5, a separator head is located at the riser outlet, and some baffles are settled under the sealed cover. In this system, the moving direction of oil gas and catalysts is changed into horizontal by the uniquely designed separator head located at the riser outlet. Therefore, the strong centrifugal force is generated by the rotation of catalyst particles, which results in a high separation efficiency and good product distribution.

This VQS is mainly used in heavy oil catalytic cracking units with large-scale inner risers. Researchers in China have tested the operation conditions of the VQS in a cold

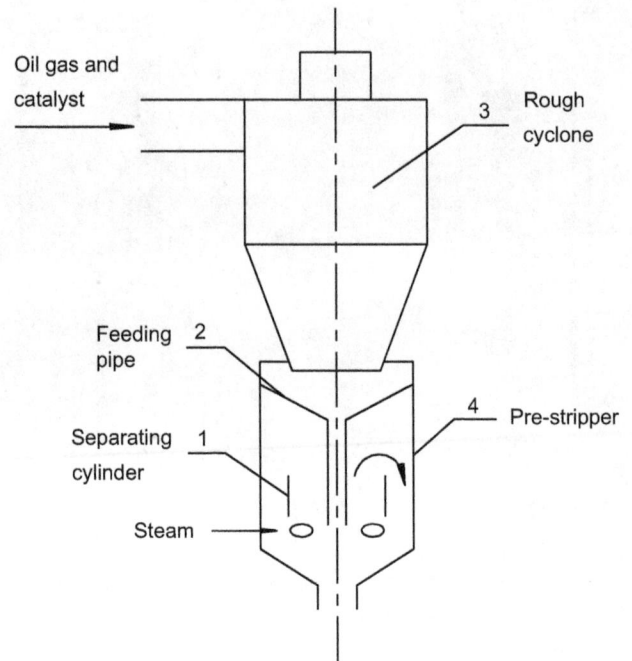

Fig. 5 Basic structure of the VQS

mode FCCU, and the results show that it has start-up flexibility, high gas–solid separation efficiency and improved product distribution (Sun et al. 2004; Lu et al. 2004; Hao and Cheng 2013; Mi 2014).

The VQS can simultaneously realize quick oil and gas separation, quick oil and gas extraction and quick pre-stripping of catalysts at the end of the riser. The pressure

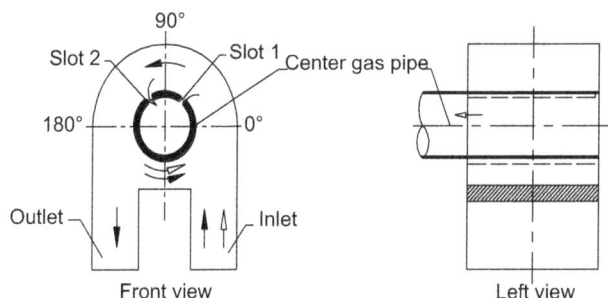

Fig. 6 Basic structure of the SRTS separation system

distribution is reasonable and pressure drop is low, which can prevent gas from gathering and jamming at the outlet. The solid catalyst particles are quickly and efficiently separated; hence, the gas residence time is shortened and the separation efficiency is enhanced. The total gas–solid separation efficiency at ambient temperature exceeds 98.5 %. The application of the VQS in the Sinopec Branch Jiujiang Company Refinery has achieved satisfying results. Their feedstocks have high viscosity and high carbon content. In September 1999, Sinopec Branch Jiujiang Company Refinery replaced the single rotation rough cyclone with a VQS to improve heavy oil processing capacity, reduce coking formation, prolong operating period and increase light liquid yield, and received great economic returns. From the data collected in the refinery, it is found that the products distribution is improved: The dry gas yield is reduced by 0.5 percentage points, and liquid yield increases by 1.2 percentage points.

4.4 SRTS system

The residence time of the oil gas and catalysts in the separation systems introduced above is still over 5 s, which means that coke formation is still a problem. In order to shorten the contact time between oil gas and catalyst particles in the separator, China University of Petroleum (Beijing) has developed a new type of quick separation system with an arch-shaped shell named SRTS (short

residence time separator, SRTS), and its structure is shown in Fig. 6. It has a central gas pipe, which is a guiding pipe with two slots along the circumferential direction. One end is in the separation chamber, while the other one extends out of the chamber and is connected with the subsequent separation system (Lu et al. 2008a, b). The gas oil and catalysts flow in the chamber through the inlet and enter in the center gas pipe. Under the effect of centrifugal force, the catalysts are thrown to the shell wall, while the oil gas is extracted through the slots of the gas pipe. The structure of SRTS system is similar to the 'goat head' separation system developed by Mobil Company (USA). The extraction time of oil gas shortens obviously after rotating a half circle in the separator. Although SRTS system has a lot of advantages compared with many other stable quick separation systems, but at present, the knowledge about this new type of separation system is limited. Meanwhile, there still is a lot of research work to be done before it is applied in industry.

5 Future

Our own separation technologies have wide applications. They can solve the unscheduled shut-down problem caused by coke formation in the disengager. Compared with the foreign separation systems, our separation systems have higher separation efficiency, larger operating flexibility and less cost, as shown in Table 2.

The focus of the above technologies is to reduce the residence time of oil gas and catalysts at the end of the reactor. However, for the quick separation unit, limited by its work mechanism, the gas–solid separation time is not reduced dramatically (the oil gas residence time is around 1–2 s). In that case, separation technology, which can realize ultrashort residence time, is required to further decrease the residence time of oil gas in the post-riser system.

Based on the knowledge of the gas–solid separation and experience in industrial application, China University of

Table 2 Comparison between China's new separation technologies and the foreign ones	Technology	China's system	Foreign VDS and VSS systems
	Pre-stripping method	Pre-stripper	Empty cylinder
	Unloading catalysts	Little negative	Positive
	Extracting	Socket	Straight
	Efficiency	99 %	95 %
	Residence time	<5 s	<5 s
	Flexibility	Large	Small
	Reforming cost (per unit)	<2 million RMB	>20 million RMB
	Application	57 U	5 U

Petroleum (Beijing) has been working on the development of this separation technology for more than 20 years (Liu et al. 2005, 2007; Lu et al. 2007, 2008a, b; Yan et al. 2007). This technology takes the advantages of both inertial and centrifugal separators. The inertial separator has short residence time but low separation efficiency, while the centrifugal separator has long residence time but high separation efficiency. The inertial separation is coupled with centrifugal separation, the separation efficiency is guaranteed by centrifugal separation, and the residence time is shortened by inertial separation. This technology features high separation efficiency, low pressure drop, compact structure and operation stability. Previous research results have shown that the residence time of oil gas in the separator is less than 0.5 s. At the same time, the separation efficiency is up to 99 %, while the pressure drop is only around 3 kPa, which is far less than the conventional rough vortex (6–8 kPa; Wang et al. 2016; Xia 2014).

References

Amos S, Avidant A. FCC close-cyclone system eliminates post-riser cracking. Oil Gas J. 1990;67(3):57–8.

Chen Y, Brostem J. Standpipe inlet enhancing particulate solids circulation for petrochemical and other processes. 2001; US patent 6228328.

Dong Q, Bai SL, Liu YX, et al. Research progress of regenerator in FCC unit. J Chem Ind Eng. 2013;34(2):1–4 (in Chinese).

Gao JS, Wang G, Lu CX, et al. Technical innovation of fluid catalytic cracking for heavy oil processing. J China Univ Pet. 2013;37(5):181–5 (in Chinese).

Gauthier T, Bayle T, Leroy P. FCC: fluidization phenomena and technologies. Oil Gas Sci Technol. 2000;55(2):187–207.

Gauthier T, Andreux R, Verstraete J, et al. Industrial development and operation of an efficient riser separation system for FCC Units. Int J Chem React Eng. 2005;. doi:10.2202/1542-6580.1247.

Gilbert T. Customized FCC revamps. In: European refining technology conference. Paris; 1995. p. 28.

Hao YJ, Cheng JL. Application and improvement of FCC VQS. Petrochem Ind Appl. 2013;32(2):82–6 (in Chinese).

Hu YH, Lu CX, Wei YD, et al. Study of the flow field in the annular space of a vortex quick separation system. Pet Process Petrochem. 2008;39(10):53–7 (in Chinese).

Karthika V, Brijet Z, Bharathi N. Design of optimal controller for fluid catalytic cracking unit. Procedia Eng. 2012;38:1150–60.

Li X, Li G, Xu Z, et al. A new downstream process design for a fluid catalytic cracking unit to raise propylene yield and decrease gasoline olefin content. Pet Sci Technol. 2011;29(24):2601–12.

Liu XC, Lu CX, Shi MX. Structural optimization of a novel gas-solid separator incorporating inertial and centrifugal separation. Chin J Process Eng. 2005;5(5):504–8 (in Chinese).

Liu XC, Lu CX, Shi MX. Post-riser regeneration technology in FCC unit. Pet Sci. 2007;4(2):91–6.

Lu CX, Shi MX. Novel catalytic cracking riser termination devices in China. Pet Technol Appl. 2007;25(2):142–6 (in Chinese).

Lu CX, Cai Z, Shi MX. Experimental study and industry application of a new vortex quick separation system at the FCCU riser outlet. Acta Petrolei Sinica (Pet Process Sect). 2004;3:24–9 (in Chinese).

Lu CX, Xu WQ, Wei YD, et al. Experimental studies of a novel compact FCC disengager. Acta Petrolei Sinica (Pet Process Sect). 2007;23(6):6–12 (in Chinese).

Lu CX, Li RX, Liu XC, et al. Gas–solid separation model of a novel FCC riser terminator device: super short quick separator (SSQS). J Chem Eng Chin Univ. 2008a;01:65–70 (in Chinese).

Lu CX, Li RX, Liu XC, et al. Gas–solid separation model of a novel FCC riser terminator device: super short quick separator (SSQS). J Chem Eng Chin Univ. 2008b;22(1):65–70 (in Chinese).

Mi YZ. Research on catalytic cracking unit optimization. Thesis for the Master Degree, Northeast Petroleum University; 2014 (in Chinese).

Pan QW. Measures to enhance adaptability of feedstock for RFCC Units and effectiveness. Sino-Glob Energy. 2011;16(7):76–80 (in Chinese).

Quinn GP, Silverman MA. FCC reactor product-catalyst separation: ten years of commercial experience with closed cyclones. NPRA Convention Center, NPRA AM-95-37, San Antonio, Texas; 1995.

Rao Z, Zhao SY, Pan QW. Optimization and effect of catalytic cracking reaction-regeneration system. Chem Eng. 2011;39(5):6–9 (in Chinese).

Sun FX, Lu CX, Shi MX. Experiment and numerical simulation of flow field in the multi-arm vortex quick separation system of FCC disengager. In: The second international symposium on multiphase, non-Newtonian and reacting flows. Hangzhou, China; 2004. pp. 146–150.

Tian WJ, Zhang L. Industrial application of UOP process technology in 3.5 MT/a RFCC unit. Technol Dev Chem Ind. 2013;42(227):62–6 (in Chinese).

Wang ZJ, Tang J, Lu CX. Fluidization characteristics of different sizes of quartz particles in the fluidized bed. Pet Sci. 2016;13(3):584–91.

Xia SH. The features and application of a circulating-stripping cyclone system on riser outlet. Guangzhou Chem Ind. 2014;42(5):109–11 (in Chinese).

Xu YH. Advances in fluid catalytic cracking (FCC) processes in China. Sci China. 2014;44(1):13–24 (in Chinese).

Yan CY, Lu CX, Liu XC, et al. Numerical simulation of the flow field in a novel gas–solids separator. J Chem Eng Chin Univ. 2007;21(3):392–7 (in Chinese).

Zhao HJ, Zhao F, Wang YL. The application of the VSS closed cyclone separation technique in FCC units. Chem Eng Oil Gas. 2006;35(3):211–3 (in Chinese).

17

Experiments on acoustic measurement of fractured rocks and application of acoustic logging data to evaluation of fractures

Bao-Zhi Pan[1] · Ming-Xin Yuan[1] · Chun-Hui Fang[1] · Wen-Bin Liu[1] · Yu-Hang Guo[1] · Li-Hua Zhang[1]

Abstract Fractures in oil and gas reservoirs have been the topic of many studies and have attracted reservoir research all over the world. Because of the complexities of the fractures, it is difficult to use fractured reservoir core samples to investigate true underground conditions. Due to the diversity of the fracture parameters, the simulation and evaluation of fractured rock in the laboratory setting is also difficult. Previous researchers have typically used a single material, such as resin, to simulate fractures. There has been a great deal of simplifying of the materials and conditions, which has led to disappointing results in application. In the present study, sandstone core samples were selected and sectioned to simulate fractures, and the changes of the compressional and shear waves were measured with the gradual increasing of the fracture width. The effects of the simulated fracture width on the acoustic wave velocity and amplitude were analyzed. Two variables were defined: H represents the amplitude attenuation ratio of the compressional and shear wave, and x represents the transit time difference value of the shear wave and compressional wave divided by the transit time of the compressional wave. The effect of fracture width on these two physical quantities was then analyzed. Finally, the methods of quantitative evaluation for fracture width with H and x were obtained. The experimental results showed that the rock fractures linearly reduced the velocity of the shear and compressional waves. The effect of twin fractures on the compressional velocity was almost equal to that of a single fracture which had the same fracture width as the sum of the twin fractures. At the same time, the existence of fractures led to acoustic wave amplitude attenuations, and the compressional wave attenuation was two times greater than that of the shear wave. In this paper, a method was proposed to calculate the fracture width with x and H, then this was applied to the array acoustic imaging logging data. The application examples showed that the calculated fracture width could be compared with fractures on the electric imaging logs. These rules were applied in the well logs to effectively evaluate the fractures, under the case of no image logs, which had significance to prospecting and development of oil and gas in fractured reservoirs.

Keywords Fractured rock · Acoustic wave velocity · Acoustic wave amplitude · Experimental measurement · Fracture width

1 Introduction

Rock fractures are important oil storages and transport channels. The physical properties of fractures, along with the fractures' development degree, are vital indices for evaluating reservoirs. Therefore, fractured rock's acoustic and electrical parameters have become a focus for geophysicists and reservoir engineers. Many types of methods have been used to evaluate fractures in seismic exploration (Li 1997; Zhao et al. 2014; Kong et al. 2012). Among these, the compressional wave anisotropy (Liu et al. 2012; Ass'ad et al. 1992) and shear wave splitting (Baird et al. 2015; de Figueiredo et al. 2013; Guéguen and Sarout 2012) are the most widely used methods at the present time to evaluate fractures. The anisotropy and fracture parameters

✉ Li-Hua Zhang
   zhanglh@jlu.edu.cn

[1] Faculty of Geo-exploration Science and Technology, Jilin University, Changchun 130012, Jilin, China

Edited by Jie Hao

were obtained by seismic prospecting. However, the relationships between anisotropy and fractures are complex and required further technical development. The qualitative identification of fractures through conventional logs also had certain developments (Sun et al. 2014; Deng et al. 2009; Wang 2013). However, using conventional logs affected by many factors, as well as the limitations in vertical resolution, has led to difficulties in obtaining accurate identification of fractures. With the development of computer technology, along with advancements in well logging technology, imaging logs, like Formation MicroScanner Image (FMI), had now become an accurate basis for fracture identification (Qiao et al. 2005). However, because of high costs and the large amount of data, it was difficult to use the imaging logs on an entire well and in all wells (Aldenize et al. 2015). In addition, laboratory fracture simulation experiments have been widely performed. Due to fractures, the collected core samples from boreholes broke easily, so that people cannot assess the actual situation underground, thus it is difficult to measure the fracture width of cores in the laboratory. Therefore, in the laboratory, the simulation method was used for the measurement of fractures in ultrasonic experiments (Faranak 2012). The physical simulation typically used a single material to simulate the rock, constructing through artificial means the pores and fractures to perform the measurement of the acoustic wave velocity, quality factor and other physical quantities with different fracture parameters (He et al. 2001; Li et al. 2016; Amalokwu et al. 2014; Wang et al. 2013). However, the single material simulation experiment had neglected the complexity of the mineral and pore distribution of the rock, so that the method was still faced with many problems in actual application. In addition, due to the fact that the acoustic wave attenuation was much more complicated than the change of the acoustic wave velocity, it was difficult to explain the principles of the acoustic wave attenuation using the physical model (Morris et al. 1964; Jose et al. 2013), thus most of the methods of acoustic wave amplitude were derived by means of numerical simulation, but the boundary conditions were too simple to match many problems in actual applications (Shi et al. 2004; Chen et al. 2012; Wang et al. 2015; Shragge et al. 2015).

It had been determined that the compressional and shear wave velocities will change when a large number of rock fractures exist (Quirein et al. 2015; Carcione et al. 2013). Fractures led the velocities to become abnormal. The relationship between the fracture width and the acoustic parameters obtained in the laboratory was the key to estimating fracture width. It has been found to be more accurate to obtain the compressional and shear wave velocities, as well as the acoustic wave amplitude, from the dipole shear wave logging (DSI) and array acoustic logging

data (Xu et al. 2014; Chen and Tang 2012). Using the inversion of DSI and array acoustic logging data to evaluate the fracture width, the application of this method for the identification of fractures had very broad prospects and feasibility (Wang et al. 2012).

This study was different from the physical experiments of simulating fractures with a single material (Wei and Di 2007; Cao et al. 2004). Actual core samples were used to simulate the fractured rocks. An ultrasonic experiment was used to examine the acoustic parameters of the fractured rocks. The effects of the pores of the rock itself on the acoustic waves were eliminated. This point was found to be more accurate than ignoring the rock porosity. The influences of different fracture width on the acoustic wave velocity and amplitude were studied. Two variables were defined: H represents the amplitude attenuation ratio of the compressional and shear wave, and x represents the transit time difference of shear wave and compressional wave divided by the transit time of the compressional wave. Transit time is the slowness of the acoustic wave. The effects of fracture width on the two physical quantities, H and x, were analyzed. The method of quantitative evaluation of fractures was achieved by using H and x. This theory provided a new method for the quantitative calculations of fracture width, as well as the evaluation of fractures in laboratory settings. This method has been applied to actual well logging data and has obtained good results. It also provided a basis for fractured reservoir evaluation, which can be of assistance in the exploration and development of oil fields in the future.

2 Experimental devices and measurement methods

2.1 Preparation of samples and fractures

In this study, sandstone samples were cut across to simulate fractured rocks. Table 1 displayed the parameters of the core samples. The No. 1 and No. 2 samples were cut to simulate fractured rocks.

Due to the evaporation of moisture, it was difficult to maintain a unified state during the measurement of the acoustic wave velocity in a fully saturated condition. The

Table 1 Parameters of the core samples

	Length, mm	Diameter, mm	Mass, g	Porosity, %
1	47.27	24.98	48.59	18.9
2	48.98	24.9	54.35	12.5
3	38.96	24.92	48.19	18.9

cores were kept dry during the measurement process, in order to facilitate the comparison of the velocity change.

PET film was used to simulate the width of the fracture. The PET film was formed into an annulus, with an outer diameter of 24 mm, inner diameter of 20 mm and thickness of 0.06 mm. The number of the PET film annuli was used to control the width of fracture.

2.2 Method of measurement

The instrument used to conduct the experiment was an HF-F Intelligent Ultrasonic Tester (Fig. 1a). A KDQS-II Full Diameter Acoustic Analyzer (Fig. 1b) was the core holder and used to measure the acoustic wave. The pass bandwidth of the instrument was set at 0.1 to 1000 kHz. The instrument launched the electrical signal, and the transmitter probe transformed the signal into vibration. Then the receiver probe converted the vibration into electrical voltage. The unit of amplitude was V, representing voltage. The compressional and shear wave voltage transmitted by the instrument was 250 V. Triggered by the computer, the recording time length was 812.5 µs, the sampling interval was 0.0625 µs, and the waveform length of each record was 13,000 points. The acoustic wavelength launched by the acoustic instruments was much less than the fracture width. The measurements were taken at normal temperature and pressure. Wave propagation was along the vertical axis of the rock. During the measurements, a good coupling between the transducer and the rock was always maintained, and the transmitting and receiving transducer were located at the ends of the center axis. There was a single

horizontal fracture in rock sample No. 1, and two parallel horizontal fractures in the rock sample No. 2 (Fig. 1c).

3 Repetitive experiment

The effects of accidental factors on acoustic propagation can be eliminated through repeated experiments. Due to the influence of different fracture width on the acoustic wave velocity and amplitude, the fracture width (W_f) was made at 0.18 mm with layers of PET film. The experiment was repeated three times, measuring the compressional and shear wave velocity (V_p, V_s) and amplitude (A_p, A_s) of the No. 1 samples. The test results are shown in Table 2.

As can be seen from Table 2, the four parameters of sample No.1 from three repeated measurements were very similar and the relative standard deviation values were relatively small. It was safe to conclude that the results were stable and repeatable, which provided a reliable basis for our subsequent data analysis.

4 Results and analysis

4.1 Velocity measurement and analysis of core with a single fracture

The influence of fracture width on the acoustic wave velocity was investigated. Figure 2 showed rock's variation of compressional wave velocity with the widths of the fracture of sample No. 1. The widths of the fractures were from 0 to 7 units, 0, 0.06, 0.12, 0.18, 0.24, 0.30, 0.36 and 0.42 mm. The thickness of each unit represented a layer of PET film.

When the fracture width was less than 0.42 mm, the variation of the V_p along with the fracture widths was obtained by fitting the measured data as follows.

$$V_p = -1509.8 \times W_f + 2797.6 \tag{1}$$

The variation of the V_s along with the fracture was as follows.

$$V_s = -399.48 \times W_f + 1597.3 \tag{2}$$

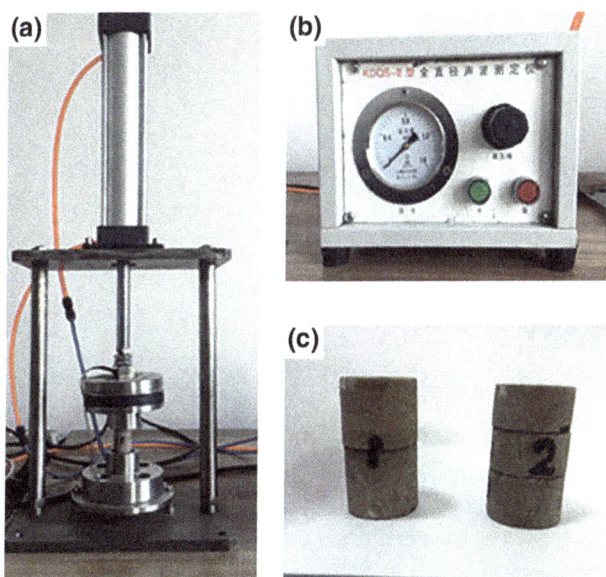

Fig. 1 Experimental instruments and rock samples. HF-F Intelligent Ultrasonic Tester (**a**), KDQS-II Full Diameter Acoustic Analyzer (**b**), two rock samples (**c**)

Table 2 Acoustic wave measurement of sample No. 1's repeatability of the $W_f = 0.18$ mm

	V_p, m/s	A_p, V	V_s, m/s	A_s, V
1	2525.8	1010.3	1525.4	846.9
2	2577.2	1007.5	1513.8	868.5
3	2581.8	985.7	1526.7	795.7
RSD (%)	2.79	3.09	1.07	1.1

RSD (%) = 100 × standard deviation/the arithmetic mean of the calculated results

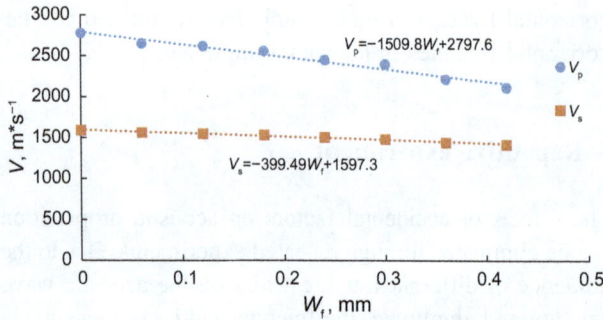

Fig. 2 Relationship between fracture widths and acoustic wave velocity of No. 1 rock. The *filled circles* and *squares*, respectively, represented the compressional and shear wave velocity of the core with the fracture. The fracture widths W_f in the core were 0; 0.06; 0.12; 0.18; 0.24; 0.3; 0.36 and 0.42 mm

The above results showed that there were obvious relationships between the acoustic wave velocity and the development degrees of fractures in the rock.

The existence of fractures led to velocity decrease for both the compressional and shear waves. With the increases of the fracture width, the velocity continued to decrease and displayed a linear relationship when the fracture width was less than 0.42 mm. From the slope of the fitting line, it could be seen that the changes of shear wave velocity were smaller than those of the compressional wave. Thus, the existence of fractures and fracture width were the influence factors of the compressional and shear wave velocity.

The measured values obtained in the acoustic logging were the acoustic transit time. Therefore, the relationships between the velocity of the acoustic wave and the width of the fractures could be converted into the relationships between the acoustic transit time and the fracture width. In this study, in order to eliminate the influence of the rock's porosity on the acoustic transit time and make the changes of the compressional and shear wave transit time only related to the fracture width, x was defined as follows:

$$x = (DT_s - DT_p)/DT_p \qquad (3)$$

where DT_s is the transit time of the shear wave, s/m, which was $1/V_s$; and DT_p is the transit time of the compressional wave, s/m, which was $1/V_p$.

The relationship between the fracture width and x of sample No. 1 is illustrated in Fig. 3.

In accordance with the data shown in Fig. 3, the formula for calculating the fracture width, W_f, mm, was determined as follows:

$$W_f = -1.6393x + 1.253 \qquad (4)$$

4.2 Velocity analysis of rocks with multiple fractures

In actual reservoirs, there are multiple fractures, rather than a single fracture. The impact of two parallel fractures on

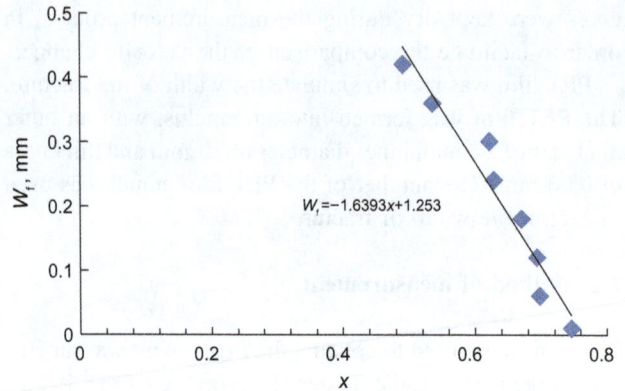

Fig. 3 Relationship between the fracture width and x

acoustic wave velocity was measured, and the acoustic wave velocities with the same width as a single fracture were compared. In dry conditions, between one and three PET films were placed in the two fractures of core sample No. 2, in order to simulate two parallel fractures with widths of 0.06, 0.12 and 0.18 mm. The relationships between the compressional wave velocity and the two parallel fracture widths were examined, and a comparison with the single fracture sample which had the same fracture width was performed. The results are shown in Fig. 4.

Figure 4 shows that multiple fractures caused the compressional wave velocity to be approximately linearly reduced. Furthermore, when the fracture width was less than 0.36 mm, two parallel fractures had almost the same influence on the compressional wave velocity as the single fracture.

4.3 Analysis of the relationship between the amplitude attenuation and fracture width

The fracture widths of core sample No. 1 were 0, 0.06, 0.12, 0.18, 0.24, 0.30 and 0.36 mm. In order to compare the

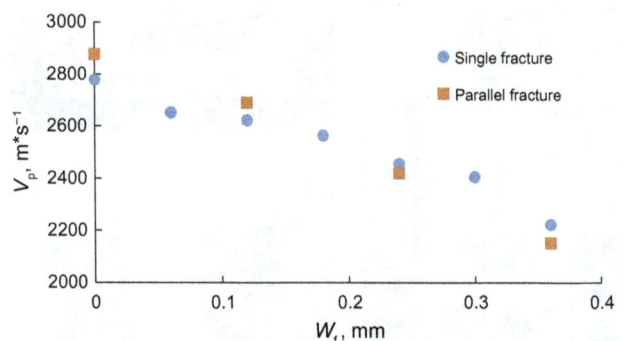

Fig. 4 Velocity contrasts of the single and two parallel fractures. The *filled circles* and *squares*, respectively, represent the single and twin fractures compressional wave velocities of the core with the fracture. The single fracture widths in the core were $W_f = 0$; 0.06; 0.12; 0.18; 0.24; 0.3 and 0.36 mm. The two fracture widths in the core were $W_f = 0$; 0.12; 0.24; 0.30 and 0.36 mm

attenuation of the acoustic wave amplitudes, the amplitude of the compressional and shear waves was corrected to the same gain value using Eq. (5):

$$A_2 = A_1 \times e^{0.1085 \times (y_2 - y_1)} \tag{5}$$

where y_1 was the gain before the correction; y_2 was the goal gain; A_1 was the amplitude before the correction, and A_2 was the amplitude after the correction. The amplitude after the correction was shown in Fig. 5.

The variation of the compressional wave amplitude, A_p, with the changes of the fracture width was obtained by fitting the measured data as follows:

$$A_p = 2.0663 \times e^{-3.5W_f} \tag{6}$$

The variation of shear wave amplitude, A_s, with the changes of the fracture widths was obtained by fitting the measured data as follows:

$$A_s = 1.1407 \times e^{-1.288W_f} \tag{7}$$

With the increases of the fracture width, the amplitudes of the shear and compressional wave were exponentially reduced. A_{pmax} was the maximum amplitude of the compressional wave without fractures; and A_{smax} was the maximum amplitude of the shear wave without fractures. $A_{max} - A$ was defined as the difference between acoustic wave amplitude without fracture and that with fractures. The relationship between $A_{max} - A$ and the fracture width is shown in Fig. 6.

From Fig. 6, it can be seen that, when there were fractures, the attenuation of the compressional wave was faster than that of the shear wave. The difference between $A_{pmax} - A_p$ and $A_{smax} - A_s$ increased gradually with the increase of fracture width. The ratio between the attenuation of compressional wave and the attenuation of shear wave was defined as H, in order to study the effect of fracture width.

$$H = (A_{pmax} - A_p) / (A_{smax} - A_s) \tag{8}$$

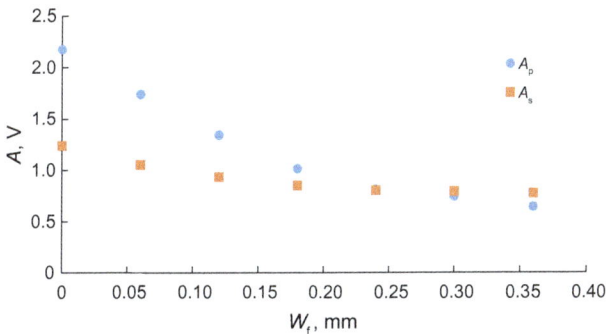

Fig. 5 Relationship between the fractures width and the amplitude of the acoustic wave. The *filled circle* and *square* represent the compressional and shear wave amplitudes of the core with the fracture. The fracture widths in the core were = 0; 0.06; 0.12; 0.18; 0.24; 0.3 and 0.36 mm

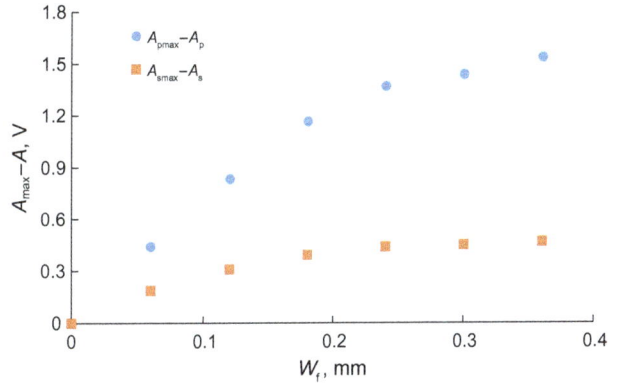

Fig. 6 Relationship between fracture width and acoustic wave amplitude attenuation. The *filled squares* and *circles* represent the $A_{pmax} - A_p$ and $A_{smax} - A_s$. The fracture widths in the core were $W_f = 0$; 0.06; 0.12; 0.18; 0.24; 0.3 and 0.36 mm

Figure 7 illustrates the relationship between H and the fracture width.

When the fracture width was less than 0.36 mm, the relationship between H and fracture width, W_f, was obtained by fitting the measured data as follows:

$$H = -12.49 + 14.43e^{-W_f} + 22.59W_f \times e^{-W_f} \tag{9}$$

Equation (9) shows that when the fracture width was less than 0.36 mm, the greater the fracture width was, the greater the H was, and the H tended to be stable. The relationship was summarized, and the empirical value of H was used to evaluate the fracture width. The experiment showed that when H was greater than 2, there was a fracture.

5 Application examples of identifying fractures based on acoustic logging data

In order to verify the accuracy of Eq. (4), the DSI data of a well were calculated to evaluate the fractures. The depths from 2780 m to 2810 m and from 3008 m to 3060 m were

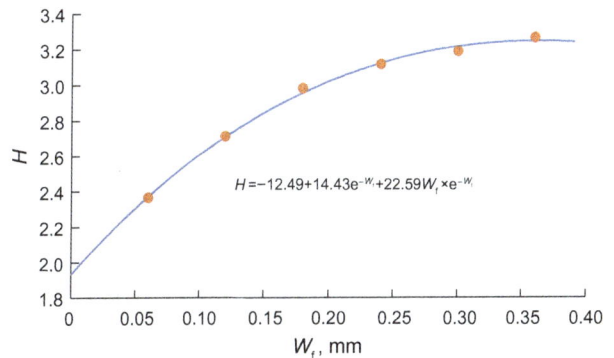

Fig. 7 Relationship between H and the fracture width

two sections in which fractures were developed (Fig. 8). Since electrical imaging logging was a good method for identifying fractures, imaging log data were used to validate the results.

In Fig. 8, the first track represents depth. The second track is the transit time of the compressional wave and shear wave, which were extracted from the DSI log data. The third track is the width of fractures by Eq. (4),

obtained by calculating x according to the transit time. The fourth is the electrical imaging of FMI. The following four tracks show the fracture characteristics calculated from FMI, which were used to compare the calculation results of the fracture width.

The fracture-developed sections were located at 2784 to 2799 m (I), 3010 to 3015 m (II), 3033 to 3037 m (III) and 3051 to 3054 m (IV). Figure 8 illustrated that the

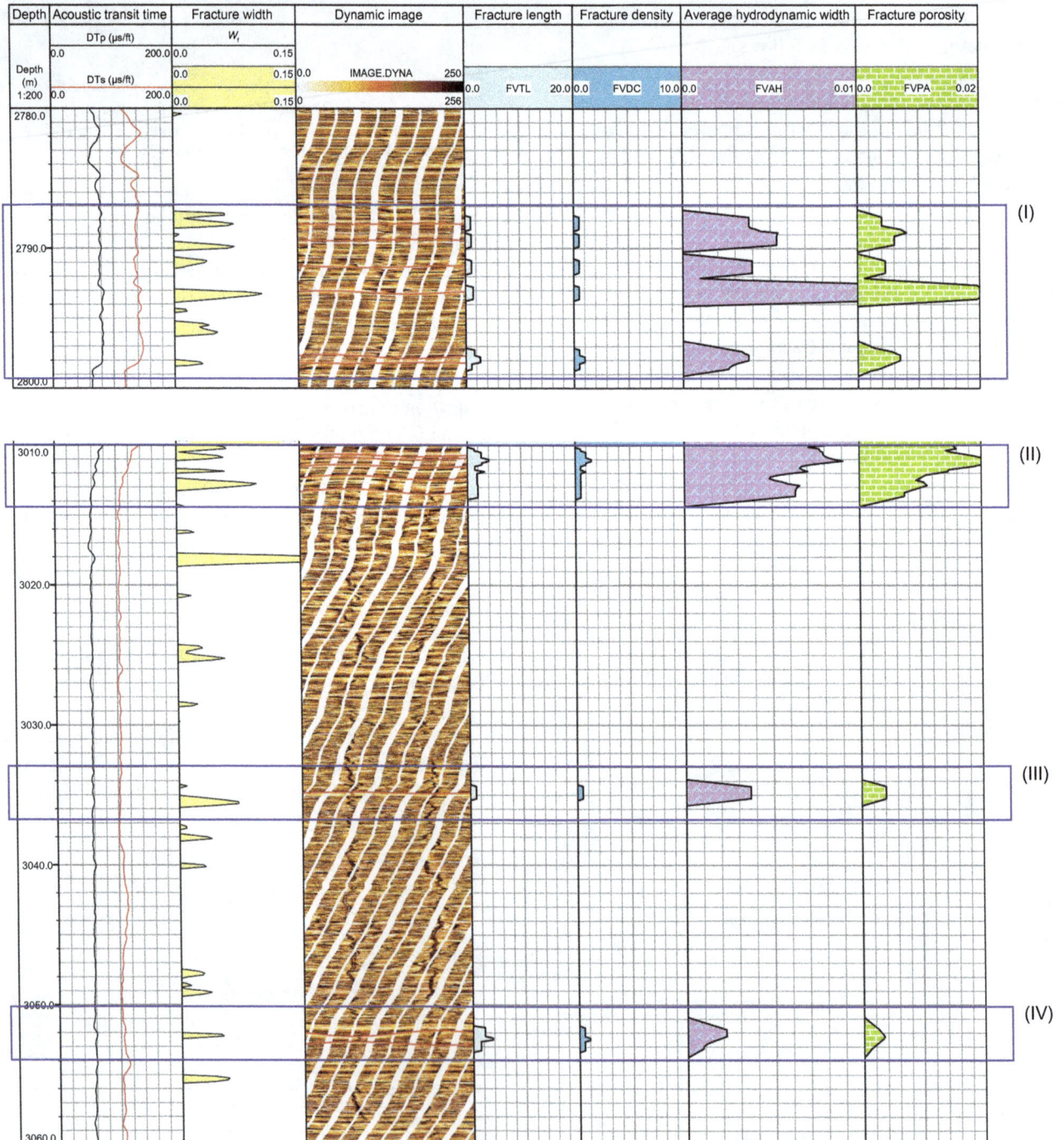

Fig. 8 Sections logging data at 2780 to 2800 m and 3010 to 3060 m in well A

calculated results through Eqs. (3) and (4) were coincident to the fractures of FMI.

Figure 9 illustrates the comparison of the H and FMI data of well B. The second track is the acoustic wave amplitude, and the third track is the acoustic wave attenuation. The fourth track represents H. The sixth is the static image and the fracture tracing. The seventh shows the tadpole plot of the fracture dip according to the imaging log, GR, and caliper curve. The last track is the dynamic image.

The pink area in Fig. 9 is the calculated H. It can be seen that the ratio was greater than 2 when the fracture developed. The results of the fracture evaluations were coincident to the fractures of FMI, which were in accordance with H. Due to the fact that the acoustic wave amplitude was sensitive to the response of the fracture, in the section of the fracture developed, the amplitude was attenuated. Therefore, when using the acoustic wave velocity to calculate the fracture, the resolution of the acoustic wave amplitude was higher. The exact calculation of fracture width from H remains to be studied.

6 Conclusions

1. The existence of fractures led to decrease of the acoustic wave velocity. The shear and compressional wave velocities were found to linearly decrease with the increases of the fracture width. The decrease of the compressional wave velocity was faster than the shear wave. The velocity of the acoustic wave was converted to the transit time. The ratios of the transit time of compressional wave and shear wave (x) were affected by fracture width, when the fracture width was less than 0.42 mm. With the increase of fracture width, x was becoming smaller and smaller. Fracture width can be calculated through the relationship between x and the fracture width.

2. The existence of fractures led to a decrease of the amplitudes of the compressional and shear wave. With the increase of the fracture width, the amplitudes of the compressional and shear wave decreased exponentially. The amplitude attenuation ratio (H) was defined, when there was a fracture, the H value was greater than

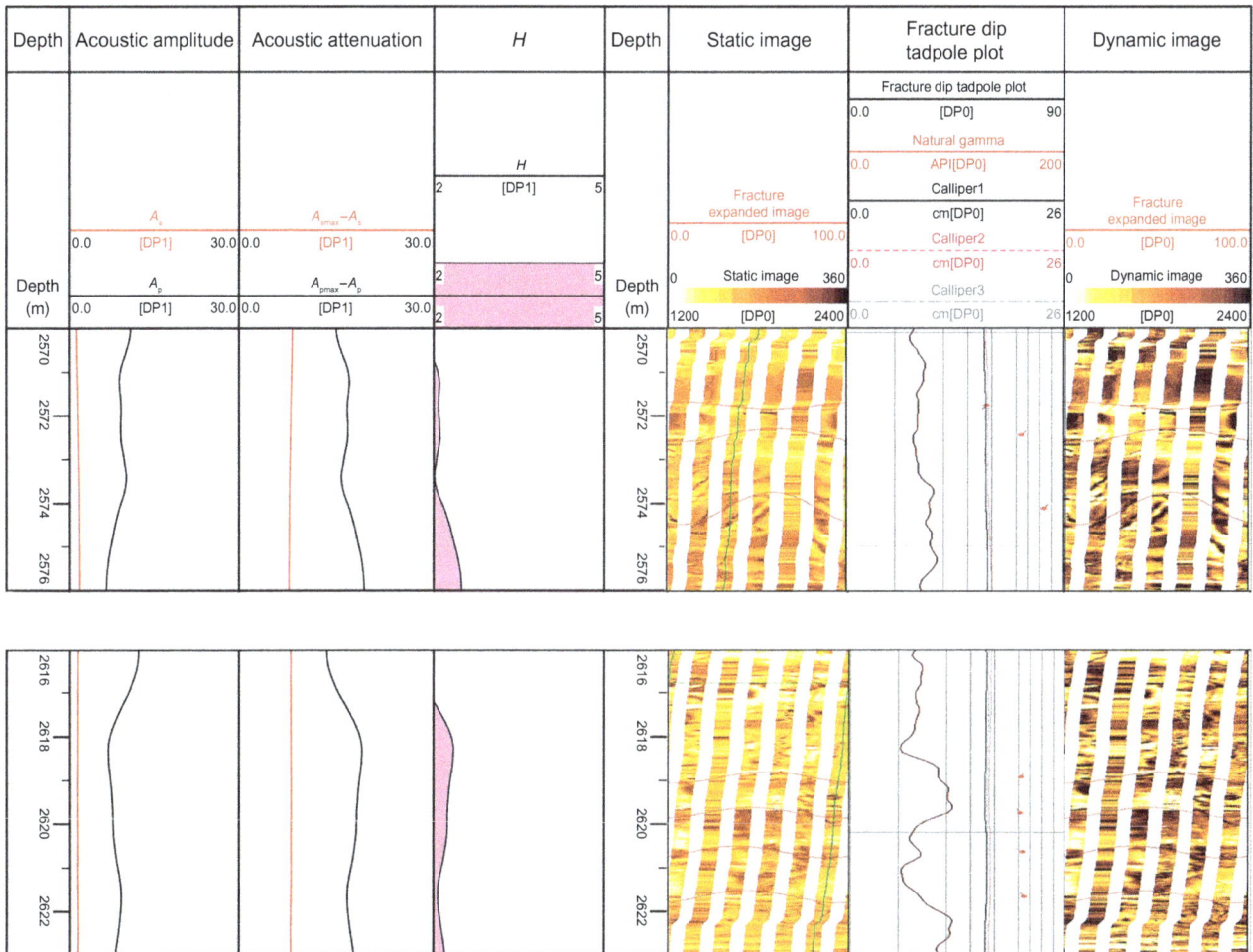

Fig. 9 Comparison of the amplitudes and imaging results at 2570 to 2576 m and 2616 to 2623 m

2. When the fracture width was less than 0.36 mm, the empirical value of H was gradually increased. According to this experience, the position of fractures in the reservoir can be located.

3. By measuring the effects of both the single and two parallel fractures on the compressional wave velocity, it was determined that when the width of the parallel fractures was equal to that of a single fracture, they had the same effect on the compressional wave velocity. This result was only applicable to the compressional wave, because the compressional wave velocity was less affected by the number of fractures.

4. Through experimental measurements, the method for evaluating fractures in the borehole by using x and H also was achieved. This study used wells A and B as application examples. The calculation results were compared with FMI data, with good agreement. The results verified that the rock acoustic parameters were affected by fractures in the borehole and provided a new method for the quantitative evaluation of fractures. In actual production, because of the influence of the fracture dip and fracture filler, this conclusion may exhibit a certain deviation, thus it will be studied by the authors in the future.

5. The next step will be to study the fracture of carbonate reservoirs and shale reservoirs.

Acknowledgements This work was supported in part by the National Natural Science Foundation of China (Grant No. 41174096) and was supported by the Graduate Innovation Fund of Jilin University (Project No. 2016103).

References

Aldenize X, Carlos EG, André A. Fracture analysis in borehole acoustic imaging using mathematical morphology. J Geophys Eng. 2015;3(12):492–501. doi:10.1088/1742-2132/12/3/492.

Amalokwu K, Best AI, Sothcott J, et al. Water saturation effects on elastic wave attenuation in porous rocks with aligned fractures. Geophys J Int. 2014;197:943–7. doi:10.1093/gji/ggu076.

Ass'ad JM, Tatham RH, McDonald JA. A physical model study of microcrack-induced anisotropy. Geophysics. 1992;57(12):1562–70. doi:10.1190/1.1443224.

Baird AF, Kendall JM, Sparks RSJ, et al. Transtensional deformation of Montserrat revealed by shear wave splitting. Earth Planet Sci Lett. 2015;425:179–86. doi:10.1016/j.epsl.2015.06.006.

Cao J, He ZH, Huang DJ, et al. Physical modeling and ultrasonic experiment of pore-crack in reservoirs. Prog Geophys. 2004;19(2):386–91 (in Chinese).

Carcione JM, Gurevich B, Santos JE. Angular and frequency-dependent wave velocity and attenuation in fractured porous media. Pure Appl Geophys. 2013;11(170):1673–83. doi:10.1007/s00024-012-0636-8.

Chen Q, Liu XJ, Liang LX, et al. Numerical simulation of the fractured model acoustic attenuation coefficient. Geophysics.

2012;55(6):2044–52. doi:10.6038/j.issn.0001-5733.2012.06.026 **(in Chinese)**.

Chen XL, Tang XM. Numerical study on the characteristics of acoustic logging response in the fluid-filled borehole embedded in crack-porous medium. Chin J Geophys. 2012;55(6):2129–40. doi:10.6038/j.issn.0001-5733.2012.06.035 **(in Chinese)**.

de Figueiredo JJS, Schleicher J, Stewart RR, et al. Shear wave anisotropy from aligned inclusions: ultrasonic frequency dependence of velocity and attenuation. Geophys J Int. 2013;193: 475–88. doi:10.1093/gji/ggs130.

Deng M, Qu GY, Cai ZX. Fracture identification for carbonate reservoir by conventional well logging. Geol J. 2009;33(1):75–8 **(in Chinese)**.

Faranak M. Anisotropy estimation for a simulated fracture medium using traveltime inversion: a physical modeling study. In: 2012 SEG annual meeting; 2012.

Guéguen Y, Sarout J. Characteristics of anisotropy and dispersion in cracked medium. Tectonophysics. 2012;503:165–72. doi:10.1016/j.tecto.2010.09.021.

He ZH, Li YL, Zhang F, et al. Different effects of vertically oriented fracture system on seismic velocities and wave amplitude. Comput Tech Geophys Geochem Explor. 2001;23(1):01–5 **(in Chinese)**.

Jose CM, Gurevich B, Santos JE. Angular and frequency-dependent wave velocity and attenuation in fractured porous media. Pure Appl. Geophys. 2013;11(170):1673–1683. doi:10.1007/s00024-012-0636-8.

Kong LY, Wang YB, Yang HZ. Fracture parameters analyses in fracture-induced HTI double-porosity medium. Geophysics. 2012;55(1):189–96. doi:10.6038/j.issn.0001-5733.2012.01.018 **(in Chinese)**.

Li TY, Wang RH, Wang ZZ. Experimental study on the effects of fractures on elastic wave propagation in synthetic layered rocks. Geophysics. 2016;81(4):441–51. doi:10.1190/geo2015-0661.1.

Li XY. Fractured reservoir delineation using multicomponent seismic data. Geophys Prospect. 1997;45:39–64. doi:10.1046/j.1365-2478.1997.3200262.x.

Liu ZF, Qu SL, Sun JG. Progress of seismic fracture characterization technology. Geophys Prospect Pet. 2012;51(2):191–8 **(in Chinese)**.

Morris RL, Grine DR, Arkfeld TE. Using compressional and shear acoustic amplitudes for the location of fractures. J Pet Technol. 1964;16(6):623–5.

Qiao DX, Li N, Wei ZL, et al. Calibrating fracture width using Circumferential Borehole Image Logging data from model wells. Pet Explor Dev. 2005;1:76–9 **(in Chinese)**.

Quirein J, Far M, Gu M, et al. Relationships between sonic compressional and shear logs in unconventional formations. In: SPWLA 56th annual logging symposium; 2015.

Shi G, He T, Wu YQ, et al. A study on the dual laterolog response to fractures using the forward numerical modeling. Chin J Geophys. 2004;47(2):359–63 **(in Chinese)**.

Shragge J, Blum TE, van Wijk K, et al. Full-wavefield modeling and reverse time migration of laser ultrasound data: a feasibility study. Geophysics. 2015;80(6):D553–63. doi:10.1190/geo2015-0020.1.

Sun W, Li YF, Fu JW. Review of fracture identification with well logs and seismic data. Pet Geol Exp. 2014;29(3):1231–42 **(in Chinese)**.

Wang RH, Wang ZZ, Shan X, et al. Factors influencing pore-pressure prediction in complex carbonates based on effective medium theory. Pet Sci. 2013;10:494–9. doi:10.1007/s12182-013-0300-7.

Wang RJ, Qiao WX, Ju XD. Numerical study of formation anisotropy evaluation using cross dipole acoustic LWD. Chin J Geophys.

2012;55(11):3870–82.   doi:10.6038/j.issn.0001-5733.2012.11. 035 **(in Chinese)**.

Wang RX. Summary of the convention logging to identify fractures. Shandong Ind Technol. 2013;7:128–9 **(in Chinese)**.

Wang ZZ, Wang RH, Li TY, et al. Pore-scale modeling of pore structure effects on P-wave scattering attenuation in dry rocks. PLoS ONE. 2015. doi:10.1371/journal.pone.0126941.

Wei JX, Di BR. Experimentally surveying influence of fractural density on P-wave propagating characters. Oil Geophys Prospect. 2007;42(5):554–9 **(in Chinese)**.

Xu S, Su YD, Chen XL, et al. Numerical study on the characteristics of multipole acoustic logging while drilling in cracked porous medium. Chin J Geophys. 2014;57(6):1992–2012. doi:10.6038/cjg20140630 **(in Chinese)**.

Zhao WH, Sun DS, Li AW, et al. Experimental study on the effect of fracture on seismic wave velocity. In: China earth sciences joint annual conference; 2014. p. 2896–99 **(in Chinese)**.

18

Comparative study of HFACS and the 24Model accident causation models

Gui Fu[1] · Jia-Lin Cao[1] · Lin Zhou[1] · Yuan-Chi Xiang[1]

Handling editor: Jian Shuai

Abstract A comparative study is conducted to compare the theory and application effect of two accident causation models, the human factors analysis and classification system (HFACS) and the accident causation "2-4" model (24Model), as well as to provide a reference for safety researchers and accident investigators to select an appropriate accident analysis method. The two models are compared in terms of their theoretical foundations, cause classifications, accident analysis processes, application ranges, and accident prevention strategies. A coal and gas outburst accident is then analyzed using both models, and the application results are compared. This study shows that both the 24Model and HFACS have strong theoretical foundations, and they can each be applied in various domains. In addition, the cause classification in HFACS is more practical, and its accident analysis process is more convenient. On the other hand, the 24Model includes external factors, which makes the cause analysis more systematic and comprehensive. Moreover, the 24Model puts forward more corresponding measures to prevent accidents.

Keywords HFACS · Accident causation "2-4" model · Comparative study · Unsafe acts · External causes · Coal and gas outburst accident

✉ Jia-Lin Cao
   cjl608@163.com

[1] School of Resources and Safety Engineering, China University of Mining and Technology (Beijing), Beijing 100083, China

Edited by Yan-Hua Sun

1 Introduction

Accidents are the main focus of research in safety science (Fu et al. 2004), and they are caused by a variety of reasons, including unsafe acts. To analyze the causes of a particular accident, it is necessary to determine which unsafe acts occur in relation to the event and also any latent failures that cause the unsafe acts (Reason 1990). Researchers and accident investigators apply various methods to analyze accidents, and these methods are crucial for the understanding of the underlying reasons accidents occur and how to improve system safety (Salmon et al. 2012).

Many accident analysis methods and accident models have been proposed. The accident causation theory consists of the accident causation chain, accident attribution theory, and accident triangle theory (Fu et al. 2013). The accident causation chain can be classified as the classical, modern, and contemporary accident causation chain according to the depth of analysis (Fu 2013). The classical accident causation chain, i.e. accident proneness theory (Greenwood and Woods 1919) and energy transfer theory (Sui et al. 2005), analyzes accidents from two viewpoints: the fault of people and physical reasons. The modern accident causation chain adds education and management factors as the root cause, such as Wigglesworth's education model (Fu 2013) and Bird's (1974) accident causation model. However, these models do not give specific explanations for the management factors, which make them difficult to use for practical accident analysis. As an advanced theory, the contemporary accident causation chain classifies management factors into several categories, which is helpful for practice application. The "Swiss cheese" model (Reason 1990), Stewart's accident cause model (Stewart 2011), the 24Model (Fu et al. 2016a), and the human factors analysis and classification system (HFACS) (Shappell and Wieg-mann 2000) all belong to the contemporary causation chains. HFACS, as one of the most famous models, was proposed by Shappell and Wiegmann in 2000 and estab-lished on the basis of the "Swiss cheese" model. The 24Model was proposed in 2005 (Fu et al. 2005b) and has become a common accident cause analysis method through continuous improvement. Compared with the famous HFACS, what advantages and disadvantages the 24Model actually has, as one of the latest accident cause chains, is what we need to focus on. However, until this time, few studies have researched the differences between the theo-ries and applications of these two models. Mi et al. (2014) analyzed a fire accident to illustrate the advantages and disadvantages of the "Swiss cheese" model, HFACS, and the 24Model. However, the 24Model used in this study was not the latest version, and the comparison was not comprehensive.

This paper discusses and analyzes the differences between the HFACS and the 24Model from various per-spectives: the theoretical foundations, cause classifications, accident analysis processes, application ranges, and acci-dent prevention strategies. Then, the two models are applied, respectively, to a rough analysis of a coal and gas outburst accident, with the aim of making a comparative analysis of the models' application results. The results of this study will provide a reference for safety researchers and accident investigators in their selection of an appro-priate accident analysis method.

2 Brief introduction to HFACS and the 24Model

2.1 Human factors analysis and classification system (HFACS)

Shappell and Wiegmann defined the latent failures and active failures in Reason's "Swiss cheese" model and described the "holes" of four level failures: unsafe acts, preconditions for unsafe acts, unsafe supervision, and organizational influences. They proposed the human fac-tors analysis and classification system (HFACS) (Wieg-mann and Shappell 2003) after analyzing thousands of aviation accidents caused by human factors. The frame-work of HFACS is shown in Fig. 1.

2.2 Accident causation "2-4" model (24Model)

The accident causation "2-4" model (24Model) is a behavior chain that was proposed based on Heinrich's (Heinrich et al. 1980), Bird's (Bird 1974), and Reason's (Reason 1990) accident causation chains and organiza-tional behavior theories (Tang 2015). This model

illustrates how organizational factors and individual fac-tors contribute to accidents (Fu 2015; Fu et al. 2016a). The 24Model divides the causes of accidents into two groups, the organization's internal factors and the orga-nization's external factors. The organization's internal factors are represented by two levels, the organizational level and the individual level. The organizational level is divided into two phases, guiding behavior and operating behavior, and the individual level is also divided into two phases, habitual behavior and one-off behavior and conditions. In this respect, guiding behavior refers to the defects of organizational safety culture. Operating behavior refers to the defects of the organizational safety management system. Habitual behavior includes the lack of safety knowledge, shortage of safety awareness, bad safety habits, and poor physiological status, and one-off behavior and conditions refer to unsafe acts and unsafe conditions. In the 24Model, each cause category has a clear definition. The accident causation "2-4" model is presented in Fig. 2.

3 Theoretical comparison of HFACS and the 24Model

With reference to relevant literature worldwide, the dif-ferences between HFACS and the 24Model can be ana-lyzed from multiple perspectives.

3.1 Theoretical foundations

HFACS was proposed based on Reason's ideas and theory. Reason's book, *Human Error*, was first published in 1990, and it continues to be one of the most widely cited and respected works in the field of behavior safety. HFACS provides a framework for applying Reason's ideas and theory (Wiegmann and Shappell 2003).

The 24Model, proposed by Fu, was based on Heinrich's, Bird's, and Reason's accident causation models and orga-nizational behavior theory as well. Unsafe acts and unsafe conditions, as the direct causes of Heinrich's theory, were retained. The management factors and the basic reason in Bird's theory are divided into several categories in the 24Model. Reason's ideas are considered as well in the 24Model. According to organizational behavior theory, the root cause of an accident lies in organization deficiencies. The 24Model established the relationship between organi-zational and individual factors to illustrate the mechanism of accidents (Fu 2013).

As described above, both models have strong theoretical foundations, and the 24Model is a scientific accident cause model.

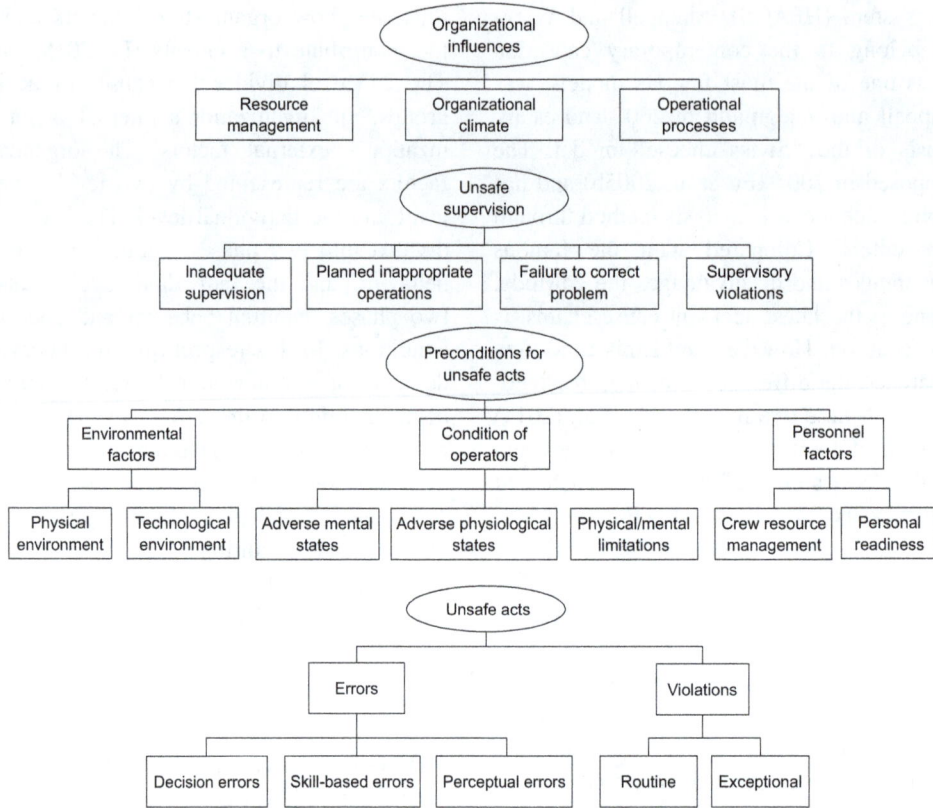

Fig. 1 Human factors analysis and classification system (HFACS) (Wiegmann and Shappell 2003)

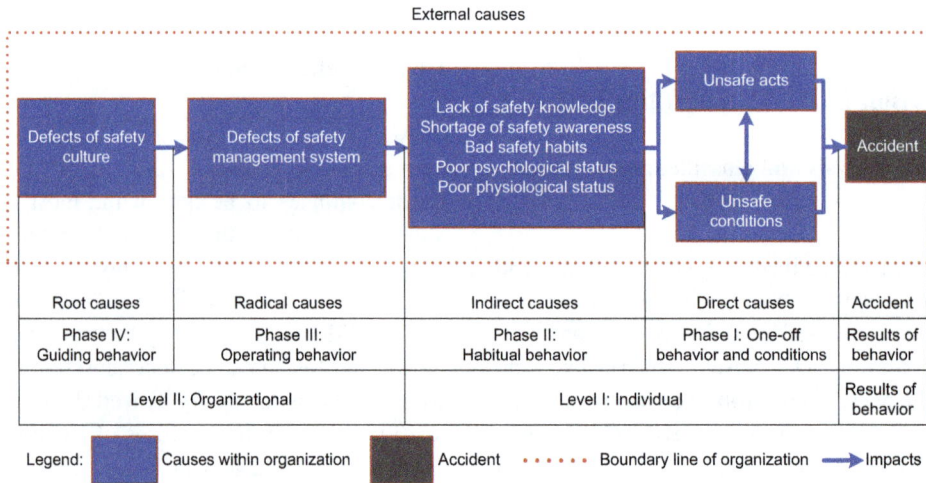

Fig. 2 Accident causation "2-4" model (24Model) (Fu et al. 2016a)

3.2 Accident cause classification

Comprehensive improvement and supplement to the classification of human factors is carried out by analyzing thousands of accidents. Therefore, the classification standard of causal factors in HFACS is practical (Shappell and Wiegmann 1997).

In the 24Model, the definition and classification of causal factors are obtained through abundant theoretical research and the analysis of many accidents (Fu et al. 2005a; Fu 2016; Ma 2016). However, the practicability of the classification results still needs to be verified through the model's application.

According to the definitions and categories of factors in the two models, the factors were compared. The results are shown in Table 1.

As shown in Table 1, all of the factors included in HFACS are correspondingly mentioned in the 24Model. In other words, the 24Model encompasses all of the factors included in HFACS. In addition, the 24Model takes external factors such as government regulation and social politics into consideration, which delivers a systematic and comprehensive accident analysis. Moreover, "unsafe acts" referred to in the 24Model also include operators' "unsafe acts" and supervisors' "unsafe supervision" in HFACS. In the 24Model, management relates solely to a manager's individual behavior control, which makes management behavior preventable and controllable and could thus effectively improve the current safety management status. Nevertheless, both models take into account organization and individual factors.

3.3 Accident analysis process

When applying the HFACS for accident analysis, we should first directly analyze the operators' unsafe acts leading to the accident, which can be classified into two categories: errors and violations, according to whether they are unintentional or intentional, respectively. Then, we can classify unsafe acts into specific error types (decision errors, skill-based errors, and perceptual errors) and violation types (routine violations and exceptional violations). Second, the reasons that the unsafe acts took place should be analyzed. Environmental factors, the condition of operators, and personnel factors can be identified and classified into various categories. Third, the factors of unsafe supervision affecting the condition of operators and

their environment should be identified and classified into inadequate supervision, planned inappropriate operations, failure to correct problems, and supervisory violations. Finally, the factors of organizational influences directly affecting supervisory practices, the conditions and actions of operators should be identified and classified into resource management, organizational climate, and operational processes.

However, the process of applying the 24Model through accident analysis is different. First, the organization where the accident occurred should be identified. Then, the causes within the organization are analyzed and are traced forward from the occurrence of the accident. Direct causes of the accident, related events and matters, are unsafe acts and unsafe conditions, while indirect causes are habitual behavior of members in the organization. Radical causes are defects of the safety management system in the organization, while root causes are defects of the safety culture in the organization. Finally, external causes that had an effect on the occurrence of the accident are analyzed, including external supervision factors, suppliers' products and service factors, natural factors, workers' family, inheritance, and growth environmental factors, as well as other factors such as politics, economy, culture, and law. The aim of accident analysis is to find out all causes that are responsible for the accident, i.e. all hazards (Fu et al. 2016a).

3.4 Application range

HFACS was originally designed for the investigation and analysis of military aviation accidents and then was gradually applied to analyze accidents in multiple domains, e.g., civil aviation (Li et al. 2008), marine traffic

Table 1 Comparison of causal factors of HFACS and the 24Model

Factors of the 24Model	The corresponding factors of HFACS
Unsafe acts	Errors; violations; crew resource management; personal readiness; inadequate supervision; planned inappropriate operations; failure to correct problem; supervisory violations
Unsafe conditions	Physical environment; technological environment; equipment/facilities in resource management
Lack of safety knowledge	Decision errors; technique failure in skill-based errors
Shortage of safety awareness	Routine violations; exceptional violations
Bad safety habits	Routine violations
Poor psychological status	Attention failure and memory failure in skill-based errors; adverse mental states; mental limitations; personal readiness
Poor physiological status	Perceptual errors; adverse physiological states; physical limitations; personal readiness
Defects of safety management system	Crew resource management; human resources and monetary assets in resource management; structures and policies in organizational climate; operations, procedures; and oversight in operational processes
Defects of safety culture	Culture in organizational climate
External causes	None

(Celik and Cebi 2009), railways (Reinach and Viale 2006), coal mines (Patterson and Shappell 2010), and medication and medical services (Mitchell et al. 2015). In the field of aviation, most of the literature directly applied HFACS to analyze human errors of the accidents. For example, Daramola (2014) used the HFACS as a conceptual framework to analyze 45 air transport accidents in Nigeria and found that skill-based errors, physical environment, and inadequate supervision were the most frequently occurring categories. Some researchers applied HFACS combined with other methods for accident analysis. For example, Wei et al. (2014) analyzed the human error in an aviation accident using HFACS combined with the expert subjective evaluation method and gray system theory. In other fields, researchers applied the modified HFACS for accident analysis. For example, Patterson and Shappell used a modified version of HFACS (HFACS-MI) to analyze 508 coal mine incidents and accidents to identify human factor trends and system deficiencies (Patterson and Shappell 2010). Soner et al. (2015) proposed a new approach, HFACS-FCM, combining fuzzy cognitive mapping and HFACS and demonstrated it on a fire-related deficiency sample database.

In contrast, the 24Model was first used for the investigation and analysis of coal mine accidents (Fu et al. 2013). It has been applied to analyze more than 800 coal mine accidents (Fu and Yang 2015; Gao and Fu 2015; Wang et al. 2016; Fu et al. 2017), including gas explosions, coal and gas outbursts, fires, floods, and roof accidents. The accident prevention training system was established based on the 24Model (Wang and Fu 2015). At present, this model has been applied to the chemical industry, public security, food, aviation, and other fields for accident investigation and analysis. For example, the 24Model was applied to analyze the Tianjin Port fire explosion accident, and the causes of the accident at all levels were identified (Fu et al. 2016b). Zhu et al. (2014) applied the 24Model to study 20 fire accidents on university campuses and found seven unsafe acts and corresponding habitual behavior, defects of the safety management system, and defects of the safety culture.

3.5 Accident prevention strategies

Through the practical application of HFACS, an accident analysis database can be formed. The forecast model can be obtained based on the database so as to formulate corresponding prevention measures before the accident occurs, whereas HFACS does not provide corresponding measures to predict and eliminate accident causes (Jiang 2015).

In the area of individual behavior control, the 24Model proposed a new method, namely "knowledge control,"

which takes case studying as a medium. Wang and Fu (2015) developed a case training system to control individual behavior with the aim of increasing organization members' safety knowledge by a case study to implement proactive accident prevention. In the aspect of organization behavior control, Fu et al. (2009) developed a safety culture analysis program, and Fu (2013) proposed a method of setting up a safety management structure and the operation method of safety management.

4 Comparison of the two models based on their applications

The HFACS and 24Model have been widely used in coal mine accident investigations and analyses. To illustrate the differences of the application effects between HFACS and the 24Model in accident analysis, a major coal and gas outburst accident (State Administration of Work Safety 2015) listed on the Web site of the State Administration of Coal Mine Safety in China was chosen as an example and analyzed by both models.

4.1 Description of a coal and gas outburst accident

The Xintian coal mine in Guizhou, China, is a coal and gas outburst mine adopting an inclined longwall comprehensive mechanized coal mining method. Before the accident, the No. 1404 tunneling face was driven into a complex geological tectonic belt without any improvement of the regional outburst prevention measures. The drilling coverage did not meet requirements, and the danger of a gas outburst was not eliminated. At 2:00 p.m. on October 05, 2014, the captain of the excavation team arranged for the team leader and gas inspector (who was unqualified and had no certificate) to lead eight workers to the No. 1404 tunneling face. At 5:17 p.m., the gas concentration at the tunneling face increased to 1.21%, and the duty monitor immediately reported this to the mine leadership. However, the mine leadership just required the on-duty dispatcher to report it to the chief engineer, and determination was found after receiving the report. They did not come to the dispatch office within 10 min to oversee an evacuation of the workers, which should have been done according to the regulations. The team leader determined that the gas concentration was caused by rib spalling, and the chief engineer reported it to the company's chief engineer. Prior to 6:04 p.m., the gas alarm situated in the working face rang three times in succession, and the on-duty dispatcher reported this according to regulations. However, the mine leadership accorded no importance to the event, and a coal and gas outburst incident occurred at 6:46 p.m. killing 10 workers.

4.2 Results of the accident analysis

The accident occurred in the Xintian coal mine, so the organization's internal factors were the factors within the scope of the Xintian coal mine, and other factors were the external causes. This paper applies both the HFACS and 24Model to analyze the coal and gas outburst accident and identify the causes at all levels. The results are shown in Table 2.

4.3 Comparative analysis of the application results

A comparison of the two models is made based on their applications to the accident analysis, including a comparison of the analysis process and the results of the coal and gas outburst accident. The following results are obtained.

The analysis process of the 24Model is more complex. It identifies the unsafe acts first, and then, it analyzes the other factors (including habitual behavior, defects of the safety management system, and defects of safety culture) that contribute to unsafe acts. Second, it analyzes the unsafe acts affecting the unsafe conditions and defects in the safety management system, and it continually analyzes the other factors leading to unsafe acts. It loops until all of the organization's internal factors are identified. Last, the external causes that have an effect on the accident should be analyzed. However, the HFACS determines other category factors according to the operators' unsafe acts, and it has a clear route of analysis. The comparative analysis illustrates that the specific application process for the accident analysis of the two models is different.

Moreover, some of the unsafe acts (e.g. "the head of the coal mine did not study gas outburst prevention measures on a monthly and quarterly basis") and external causes that are identified in the 24Model are not shown in HFACS. The contents of other factors are the same in both models, but their classifications differ. It is illustrated that the causes in the 24Model encompass the causes in HFACS, but the analysis results of the 24Model are more comprehensive.

In addition, psychological factors can be identified by interviewing relevant personnel when applying HFACS. In contrast, there is less interpretation and application of physiological and psychological factors in the 24Model. Since the information in accident investigation reports is limited, both models fail to identify the physiological and psychological factors.

Furthermore, HFACS's classification is based on analyzing the causes of aviation accidents. When the model is applied to coal and gas outburst accidents, the "crew resource management" category is not applicable. In contrast, the cause classification of the 24Model does not have

industry limitations. Thus, the cause classification of HFACS does not have versatility.

In summary, the applications of the two models in the coal and gas outburst accident analysis verify the results of the theoretical comparisons of the two models in terms of the accident analysis process, the comprehensiveness of accident causes, and the versatility of cause categories. What is more, the subjectivity and limitation of psychological factors analysis is a problem in both models.

5 Conclusions

This study conducted a theoretical and application comparison of the HFACS and 24Model and obtained the following conclusions.

(1) Both models have strong theoretical foundations. HFACS was proposed based on Reason's ideas and theory. The 24Model was proposed based on Heinrich's, Bird's, and Reason's accident causation models as well as organizational behavior theory. Thus, they are all scientific.

(2) The cause classification standards of the two models are different. The cause classification of HFACS is more practical than that of the 24Model. Both models contain individual factors as well as organization factors. Beyond that, the 24Model takes external factors into consideration, which makes the cause analysis of accidents more systematic and comprehensive.

(3) The accident analysis processes of both models are much the same. The accident analyst traces the causal factors from the direct causes back up to the other level factors. However, the analysis of unsafe acts using the 24Model is comparatively complex. Unsafe acts in the 24Model could be the result of unsafe conditions and defects of the safety management system, or it could be the cause of them.

(4) The promotion and application of the two models can be implemented in many areas. However, in most areas, with the exception of aviation, HFACS should be modified according to the actual conditions when used in accident analysis. Alternately, it is not necessary to modify the 24Model in the analysis of all area accidents. However, its versatility in areas, except for the coal mine category, still lacks the sufficient data to be supported.

(5) The two models have different ways of guiding accident prevention. HFACS explains the causal factors of the accident but does not give the corresponding implementing measures to predict and eliminate causes. On the other hand, the

Table 2 Results of accident causation analysis

24Model	Accident factors	HFACS
Unsafe acts	Tunneled with a tunneling machine after a warning of outburst	Exceptional violation
	Did not stop working or evacuate workers when gas concentration occurred	Supervisory violation
	Did not increase testing points in a complex geological tectonic belt when inspecting outburst prevention effect (only 5 testing points)	Failed to correct problem
	Acceptance of gas drainage drill holes was not strictly in accordance with regulations (part of the drill holes did not comply with the design)	Exceptional violation
	Did not stop tunneling and supplement or modify outburst prevention measures when the working face had folds and a change in coal seam thickness, etc	Failed to correct problem
	Violated construction process using hydraulic fracturing technology and implemented hydraulic fracturing job when tunneling	Planned inappropriate operations
	The head of the coal mine did not study gas outburst prevention measures on a monthly and quarterly basis	–*
	The gas inspector was not qualified and had no certificate	Exceptional violation
	Did not take effective measures to deal with unsafe conditions, with multiple occurrences of installing and checking methane sensors not in accordance with rules	Routine violation
	Management did not stop two workers from using each other's ID cards to register	Supervisory violation
Unsafe conditions	The working face had folds and changes in coal seam thickness, etc	Physical environment
	Rib spalling caused the gas concentration	Physical environment
Lack of safety knowledge	Unaware of the requirement to stop work when a warning for gas burst occurred	Exceptional violation
	The number of test points was not increased in a complex geological tectonic belt	Failed to correct problem
	The gas inspector was not aware of the need to evacuate workers when the gas concentration was higher than 1%	Exceptional violation
Shortage of safety awareness	The gas concentration was found to exceed the limit, but this was not dealt with	Supervisory violation
	Acceptance of gas drainage drill holes was against the rules	Exceptional violation
	Did not supplement outburst prevention measures after knowing about the change in geological conditions	Failed to correct problem
	Intentionally violated the construction process using hydraulic fracturing technology	Planned inappropriate operations
	Allowed unqualified personnel to work in an underground coal mine	Supervisory violation
Bad safety habits	Gas outburst prevention measures were not studied monthly and quarterly	
	Effective measures were not taken to deal with unsafe conditions; multiple occurrences of installing and checking methane sensors against the rules	Routine violation
Poor psychological status	–	–
Poor physiological status	–	–
Defects of safety management system	Grading provisions of the gas concentration exceeding limits was incorrect, and thus, there was no rule for evacuating workers immediately when the gas concentration exceeded limits	Organizational process
	A number of special operators worked without certificates	Resource management
	Regional outburst prevention measures were incomplete	Organizational process
Defects of safety culture	Not fully understanding the following safety concepts: safety importance; all accidents are preventable; safety depends on safety consciousness; role of safety regulations; leadership accountabilities; effect of management system	Organizational climate

Table 2 continued

24Model	Accident factors	HFACS
External causes	The company's leadership did not command that the coal mine evacuated workers immediately after receiving reports of the gas concentration exceeding limits	–
	Special approval for working face outburst prevention design, regional outburst prevention measures, and evaluation report was not compliant	–
	A special demonstration of the security and practicability of hydraulic fracturing technology was not conducted by the company	–

* Corresponding content does not exist

24Model presents the organization and individual behavior control methods, and the case training system of controlling individual behavior in accidents has been developed.

This study shows that the 24Model is scientific and has a certain practicability. The above conclusions are expected to be used by safety researchers and accident investigators in selecting the most appropriate model for scientific research and accident causation analysis. However, it is recommended that how the physiological and psychological factors affect direct causes and how the external factors affect an accident should be further studied, and the practicability and applicability of the 24Model should be proved by more practical applications of accident investigation and analysis.

Acknowledgements The authors gratefully acknowledge support from the State Key Program of the National Natural Science Foundation of China (No. 51534008).

References

Bird F. Management guide to loss control. Atlanta, GA: Institute Press; 1974.

Celik M, Cebi S. Analytical HFACS for investigating human errors in shipping accidents. Accid Anal Prev. 2009;41:66–75. doi:10.1016/j.aap.2008.09.004.

Daramola AY. An investigation of air accidents in Nigeria using the human factors analysis and classification system (HFACS) framework. J Air Transp Manag. 2014;35:39–50. doi:10.1016/j.jairtraman.2013.11.004.

Fu G. Safety management. Beijing: Science Press; 2013 (in Chinese).

Fu G. Studies on the structure of safety science. Melbourne: Safety Science Press; 2015 (in Chinese).

Fu, G. 2016.The modified 24Model and the definitions of unsafe acts and unsafe conditions. 29 March 2016. http://blog.sciencenet.cn/blog-603730-965647.html (in Chinese).

Fu G, Yang C. Comparative study of unsafe acts in case of water flooding accidents in domestic coal mines. J Saf Environ. 2015;15(4):166–71. doi:10.13637/j.issn.1009-6094.2015.04.036 (in Chinese).

Fu G, Zhang JS, Xu SR. Critical examination of safety science. Eng Sci. 2004;6(8):12–6 (in Chinese).

Fu G, Li XD, Li J. Common factors leading to accidents and behavior type research based on prevention. J Saf Environ. 2005a;5(1):80–3 (in Chinese).

Fu G, Lu B, Chen XZ. Behavior based model for organizational safety management. China Saf Sci J. 2005b;15(9):21–7. doi:10.16265/j.cnki.issn1003-3033.2005.09.005 (in Chinese).

Fu G, Li CX, Xing GJ, et al. Investigations into the impacts of enterprise safety culture and its quantitative measuring. China Saf Sci J. 2009;19(1):86–92 (in Chinese).

Fu G, Yin WT, Dong JY, et al. Behavior-based accident causation: the "2-4" model and its safety implications in coal mines. J China Coal Soc. 2013;38(7):1123–9 (in Chinese).

Fu G, Fan YX, Tong RP, et al. The universal methodology for the causation analysis of accidents (4th edition). J Accid Prev. 2016a;1:7–12 (in Chinese).

Fu G, Wang JH, Yan MW. Anatomy of Tianjin Port fire and explosion: Process and causes. Process Saf Prog. 2016b;35(3):216–20. doi:10.1002/prs.11837.

Fu G, Cao JL, Wang XM. Relationship analysis of causal factors in coal and gas outburst accidents based on the 24Model. Energy Procd. 2017;107:314–20. doi:10.1016/j.egypro.2016.12.160.

Gao P, Fu G. Analysis and prevention of unsafe acts in mine roof accidents. Ind Saf Environ Prot. 2015;41(7):67–70 (in Chinese).

Greenwood M, Woods HM. The incidence of industrial accidents upon individuals with special reference to multiple accidents. Industrial Fatigue Research Board, Medical Research Committee. Report No. 4. Her Majesty's Stationery Office, London. 1919.

Heinrich HW, Peterson D, Roos N. Industrial accident prevention. New York: McGraw-Hill Book Company; 1980.

Jiang H. Overview on HFACS and its application research. China Sci Technol Inf. 2015;5:13–4 (in Chinese).

Li WC, Harris D, Yu CS. Routes to failure: Analysis of 41 civil aviation accidents from the Republic of China using the human factors analysis and classification system. Accid Anal Prev. 2008;40:426–34. doi:10.1016/j.aap.2007.07.011.

Ma Y. Safety culture construction evaluation based on combination weighting and fuzzy topsis methods. Chem Eng Trans. 2016;51:715–20. doi:10.3303/CET1651120.

Mi FR, Yu JL, Li SX. Analysis and comparison of accident causation chains. China Public Secur Acad Ed. 2014;31(1):41–4 (in Chinese).

Mitchell RJ, Williamson A, Molesworth B. Use of a human factors classification framework to identify causal factors for medication and medical device-related adverse clinical incidents. Saf Sci. 2015;79:163–74. doi:10.1016/j.ssci.2015.06.002.

Patterson JM, Shappell SA. Operator error and system deficiencies: analysis of 508 mining incidents and accidents from Queensland, Australia using HFACS. Accid Anal Prev. 2010;42:1379–85. doi:10.1016/j.aap.2010.02.018.

Reason J. Human error. New York: Cambridge University Press; 1990.

Reinach S, Viale A. Application of a human error framework to conduct train accident/incident investigations. Accid Anal Prev. 2006;38:396–406. doi:10.1016/j.aap.2005.10.013.

Salmon PM, Cornelissen M, Trotter MJ. Systems-based accident analysis methods: A comparison of Accimap, HFACS and STAMP. Saf Sci. 2012;50:1158–70. doi:10.1016/j.ssci.2011.11. 009.

Shappell SA, Wiegmann DA. A human error approach to accident investigation: The taxonomy of unsafe operations. Int J Aviat Psychol. 1997;7:269–91. doi:10.1207/s15327108ijap0704_2.

Shappell SA, Wiegmann DA. The human factors analysis and classification system (HFACS). Report Number DOT/FAA/AM-00/7. Washington DC: Federal Aviation Administration. 2000.

Soner O, Umut A, Metin C. Use of HFACS–FCM in fire prevention modelling on board ships. Saf Sci. 2015;77:25–41. doi:10.1016/j. ssci.2015.03.007.

State Administration of Work Safety. China's work safety yearbook. Beijing: China Coal Industry Publishing Home; 2015 **(in Chinese)**.

Stewart JM. The turn around in safety at the Kenora pulp paper mill. Prof Saf. 2011;12:34–44.

Sui PC, Chen BZ, Sui X. Safety principles. Beijing: Chemical Industry Press; 2005 **(in Chinese)**.

Tang XS. Organizational behavior science. Beijing: China Railway Press; 2015 **(in Chinese)**.

Wang JH, Fu G. Individual behavior control technology and training system for coal mine accident prevention. Saf Coal Mines. 2015;46(S1):109–12. doi:10.13347/j.cnki.mkaq.2015.S1.027.

Wang JH, Zhang JS, Zhu K, et al. Anatomy of explosives spontaneous combustion accidents in the Chinese underground coal mine: Causes and prevention. Process Saf Prog. 2016;35(3):221–7. doi:10.1002/prs.11816.

Wei SX, Sui YC, Chen YC. Research into the human errors evaluation method of flight accidents based on HFACS. Aeronaut Comput Tech. 2014;44(2):50–3 **(in Chinese)**.

Wiegmann DA, Shappell SA. A human error approach to aviation accident analysis: The human factors analysis and classification system. Burlington, VT: Ashgate Publishing Ltd; 2003.

Zhu T, Fu G, Zhang S. Behavior reason analysis and prevention of campus fire accidents. Ind Saf Environ Prot. 2014;40(3):33–5 **(in Chinese)**.

19

Similarity measure of sedimentary successions and its application in inverse stratigraphic modeling

Taizhong Duan[1]

Abstract This paper presents a unique and formal method of quantifying the similarity or distance between sedimentary facies successions from measured sections in outcrop or drilled wells and demonstrates its first application in inverse stratigraphic modeling. A sedimentary facies succession is represented with a string of symbols, or facies codes in its natural vertical order, in which each symbol brings with it one attribute such as thickness for the facies. These strings are called attributed strings. A similarity measure is defined between the attributed strings based on a syntactic pattern-recognition technique. A dynamic programming algorithm is used to calculate the similarity. Inverse stratigraphic modeling aims to generate quantitative 3D facies models based on forward stratigraphic modeling that honors observed datasets. One of the key techniques in inverse stratigraphic modeling is how to quantify the similarity or distance between simulated and observed sedimentary facies successions at data locations in order for the forward model to condition the simulation results to the observed dataset such as measured sections or drilled wells. This quantification technique comparing sedimentary successions is demonstrated in the form of a cost function based on the defined distance in our inverse stratigraphic modeling implemented with forward modeling optimization.

Keywords Similarity quantification · Sedimentary succession · Inverse stratigraphic modeling · Global optimilization · Syntactic approach

1 Introduction

Quantitative study of sedimentary successions unavoidably involves the formal description of discrete or symbolic properties such as facies, rock texture, or structure, and until recent our ability has been still very limited on how to quantify such type properties, for instance, the difference or similarity between sedimentary facies successions from measured sections in outcrop, or drilled sections. More traditional ways to do such comparison are almost exclusively either qualitative or graphic such as a simple description "the two facies successions look very similar," or a plot representation usually used as shown in Fig. 1. On the other hand, economically efficient recovery of natural resources such as oil and gas demands better and formal quantification of geological models such as geocellular modeling in reservoir characterization.

In hydrocarbon reservoir modeling, geostatistical methods currently dominate, to large extent due to their data conditioning capacity, that is, the model can honor the observed dataset easily. In contrast, forward stratigraphic modeling has yet to be accepted as the major modeling technique in reservoir facies modeling as it seems it should have. Forward stratigraphic modeling is geological process based and is more relevant to petroleum reservoirs (Bosence and Waltham 1990; Granjeon and Joseph 1999; Griffiths et al. 2001), and this method has been developed since the 1960s (Harbaugh and Bonham-Carter 1970), compared to the geostatistics also developed since the 1960s (Matheron 1962, 1989). One of the main reasons

✉ Taizhong Duan
  duantz.syky@sinopec.com

[1] Petroleum Exploration and Production Research Institute, Sinopec, 31 Xueyuan Road, Haidian District, Beijing 100083, China

Edited by Jie Hao

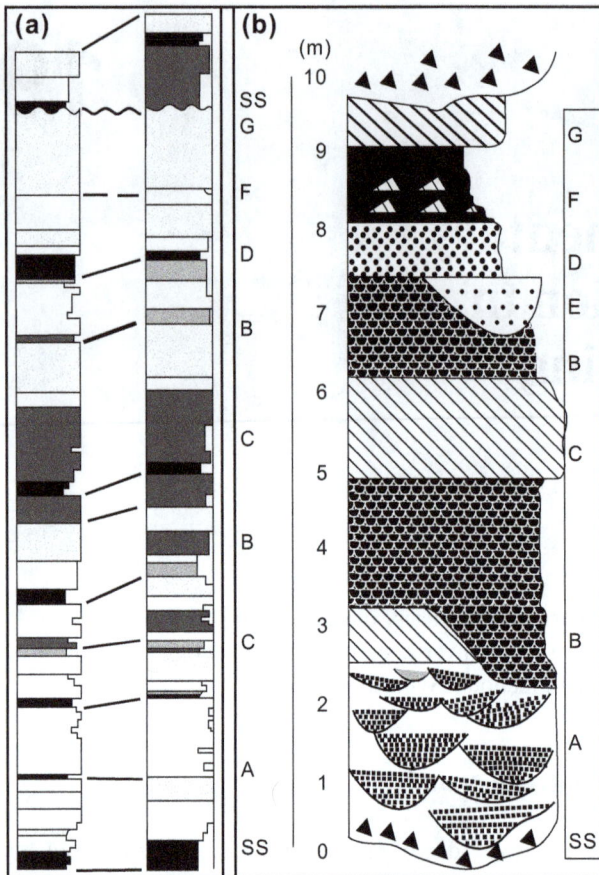

Fig. 1 Typical graphic representation of sedimentary facies succession. *Patterns* and *gray scale* represent different facies: **a** two "very similar" measured facies sections (modified from Kerans et al. 1994); **b** a plot channel model (modified from Cant and Walker 1976), with ordered facies codes to form a string presentation of the channel facies: SSABCBEDFG (*right*) or SSACBCBDFGSS (*left*)

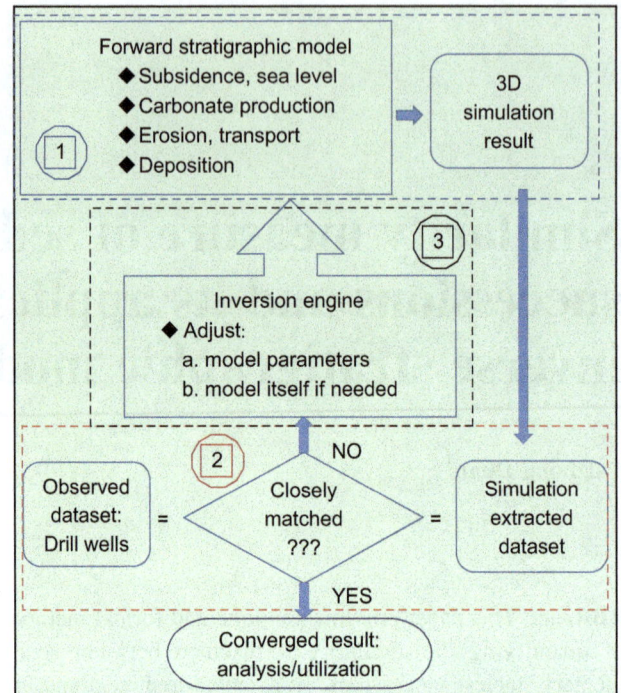

Fig. 2 ISM core techniques and workflow: *1* forward modeling generates 3D simulations; *2* comparing technique quantifies the matching between the simulated and observed datasets; *3* inversion engine updates new simulations for a better match

why this technique has been delayed to dominate in petroleum reservoir modeling is its inability to implement data conditioning. Since late 1990s, similar techniques but under different names were proposed to overcome the inability and initiated a new research front in computational stratigraphy and sedimentology, such as inverse stratigraphic modeling (ISM) (Griffiths et al. 1996; Lessenger and Cross 1996; Cross and Lessenger 1999; Duan et al. 2001a; Imhof and Sharma 2006; Charvin et al. 2009; Griffiths 2009; Charvin et al. 2011), adaptive modeling (Duan et al. 1998), modeling optimization (Bornholdt and Westphal 1998; Wijns et al. 2003, 2004), or model calibration (Falivene et al. 2014). However, the progress of these techniques, all of which will be called ISM afterward for simplicity, has been limited, and one of the major hurdles is still the data conditioning.

Therefore, in current version of ISM as shown in Fig. 2, among others, a critical technique needed to enhance the procedure is how to quantify the similarity or distance between simulated results and the observed dataset. That is,

a proper distance measure can speed up inversion and make it more robust and can lead to better practical application of geological process-based modeling in general. This paper presents a unique and formal method of quantifying the distance between sedimentary facies successions based on a discrete, or symbolic computing technique, combined with other numerical techniques. The formal definition of the distance measure between sedimentary facies successions will be presented first, and then, its application as the cost function in ISM will be demonstrated.

2 Definition of sedimentary facies successions distance

2.1 Formal representation of sedimentary facies successions

In order to quantify the distance or similarity between sedimentary facies successions, it is essential to define what a sedimentary facies succession is and what is the distance between the sedimentary facies successions or the sedimentary facies successions distance (SFSD) formally and quantitatively. A gene-typing technique was proposed for correlation of petrophysically derived numerical lithologies between boreholes (Bakke and Griffiths 1989; Griffiths and Bakke 1990). A syntactic methodology developed in pattern recognition (Fu 1982) was first proposed to formally describe the language of sedimentary rocks (Griffiths 1990), due to its ability in characterizing the naturally discrete or symbolic feature of sedimentary facies. The more detailed syntactic approach to the analysis of sedimentary successions for reservoir characterization can be found in Duan, Griffiths, and Johnsen (Duan et al. 1999, 2001a), among others. In the syntactic approach, a sedimentary facies succession (one-dimensional vertical stratigraphic section) is represented in a string of facies symbols, each of which contains one or more attributes, i.e., a string with attributes in symbolic computation language (Duan et al. 2001a). For instance, the bottom to top facies type of a typical channel sedimentary facies succession can be coded by symbols as shown in Fig. 1b, whereas the code of each facies can be associated with a number or an attribute representing the thickness of each facies in the parentheses as in the following attributed string of codes:

SS(0.5) A(3.0) C(0.5) B(2.5) C(1.5) B(1.3) E(0.5) D(0.4) F(1.5) G(0.3) SS(0.1)

Besides the normal facies, the erosional surfaces are treated as a special facies. SS(0.1) and SS(0.5) mean two erosional surfaces with 0.1 and 0.5 million years time gap, respectively, whereas A(3.0) and C(0.5) mean 3.0-m trough-cross-bedded sandstone facies and 0.5-m tabular-cross-bedded sandstone facies, respectively. Of course, each code may have more than one attribute, either numerical or symbolic, such as grain size, color, fossils, and mineral composition. It should be noticed that facies coding and attributing can be compensated with each other. For instance, if you already account for grain size in facies coding such as conglomerate, coarse sandstone, fine sandstone, siltstone, and mudstone, you may not need to repeat the same information by adding a grain size attribute for the above facies code.

The similar approach was first proposed for stratigraphic correlation between drilling wells or outcrop sections, which suffered from being incapable of handling the facies change problem between sections, and has been almost forgotten in the geological community. The so-called facies change dilemma in automatic strata correlation is that two sections of strata with the same facies can be defined equivalent, and two sections of strata with totally different facies also need to be defined equivalent if facies change occurs between, which is not mathematically sound. However, in stratigraphic inversion, similarity quantification between simulated and observed successions naturally avoids the facies change problem, two successions (simulated and observed) to be compared are fundamentally the same. As we understand, the application of this technique to the stratigraphic inverse modeling is like the similar technique's application to the comparison of

DNA or RNA in biology. Therefore, we believe this technique has great potential in improving stratigraphic inverse modeling.

2.2 Definition of distance between sedimentary facies successions

To define a distance or similarity measure between sedimentary facies successions, it is important to understand what characters are essential in distinguishing different sedimentary facies successions and what would be fundamental requirements for a distance definition mathematically.

From the point of view of sedimentology and stratigraphy, comparison of sedimentary facies successions should account for following aspects: (1) facies types and their division in sections, including special facies such as erosional surfaces; (2) the thickness of each identified and divided facies (maybe repeated) in sections; and (3) the vertical order or sequence of the coded facies in each sedimentary facies successions. When it is said that two sedimentary facies successions are equivalent, it means all three of the above-mentioned characters should be the same. For instance, following are two code strings X and Y from neighboring sedimentary successions:

X: SS(0.5) A(3.0) C(0.5) B(2.5) C(1.5) B(1.3) E(0.5) D(0.4) SS(0.15)
Y: SS(0.5) A(3.0) C(0.5) B(2.5) C(1.5) B(1.3) E(0.5) D(0.4) F(1.5) G(0.3) SS(0.1)

They are not equal, because firstly the string-X lacks "F(1.5) G(0.3)," implying that part of the channel top deposits may be eroded away. Secondly, its top SS attribute 0.15 is different from the string-Y's 0.1, implying that the time gap represented by the SS of the string-X is 0.05 million years longer than the string-Y's.

For the following two strings:

U: B(2.0) D(2.0) F(2.0) and
V: F(2.0) D(2.0) B(2.0)

They are not equal, because the vertical order of the facies code is different, though three facies types and their thickness are the same. Actually, string-U may represent upper channel deposits, while string-V may represent crevasse splay deposits.

A distance measure of sedimentary facies successions was first proposed in a syntactic approach (Duan et al. 2001a) which accounts for all three aspects of the above-mentioned characters in sedimentary succession comparison. It is based on a series of syntactic distance measures such as the Levenshtein distance (Levenshtein 1966), generalized Levenshtein distance (Fu 1986), and distance between attributed strings (Fu 1986). For details of the definition, refer to Duan et al. (2001b) from Definition 1 through to Definition 5 and related concepts. However, for continuation and readability, the main definition is reproduced as follows.

Definition 1 Let x and y be two attributed strings,

$$x = a_1 a_2 \ldots a_n$$
$$y = b_1 b_2 \ldots b_m$$

Corresponding attributes of x and y are denoted as:

$$x' = \left(a_1^1 a_1^2 \cdots a_1^k\right)\left(a_2^1 a_2^2 \cdots a_2^k\right) \cdots \left(a_n^1 a_n^2 \cdots a_n^k\right)$$

$$y' = \left(b_1^1 b_1^2 \cdots b_1^k\right)\left(b_2^1 b_2^2 \cdots b_2^k\right) \cdots \left(b_n^1 b_n^2 \cdots b_n^k\right)$$

It is assumed that each terminal symbol has k attributes. The distance between x and y is defined as:

$$d^{AS}(x, y) = \alpha d^{GL}(x, y) + \beta d^{A}(x, y)\left(\text{or } \beta d^{A'}(x, y)\right)$$

where α and β are two positive weights; $d^{GL}(x, y)$ is the generalized Levenshtein distance between x and y; and $d^{A}(x, y)$ is the attribute distance between x and y after optimal alignment with only insertion (when substitution is accepted, a variant $d^{A'}(x, y)$ is obtained) is carried out to make one string equal another syntactically. Let the transferred string be:

$$x_t = c_1' c_2' \cdots c_k' (y_t$$
$$= c_1'' c_2'' \cdots c_k'' \right) \{\max\{|x|, |y|\} \le k \le (|x| + |y|)\}$$

Then

$$d^{A}(x, y) = \sum \lambda_j d(A(c_j'), A(c_j'')) \quad j = 1, 2, \ldots, k$$

where $A(c_j')$ and $A(c_j'')$ are attribute vectors of c_j' and c_j'', respectively; $d(A1, A2)$ can be any p-norm distance ($p = 1$ used in our case); λ_j are weighting coefficients defined as:

1. $\lambda_j = \lambda_C(c_j', c_j'') = 1 c_j' = c_j''$ (continuation)
2. $\lambda_j = \lambda_I(c_j', c_j'') \ge 1 c_j' \ne c_j''$ (insertion)
3. $\lambda_j = \lambda_S(c_j', c_j'') \ge 1 c_j' \ne c_j''$ (substitution)

Note that when insertion occurs, the inserted symbol's attributes are assigned zero. If substitution is allowed and occurred, attributes remain the same in related symbols.

As Duan et al. (2001b) pointed out, the distance measure defined must satisfy following mathematical requirements (Kaufman and Rousseeuw 1990):

$D(1): d^{AS}(x, y) \ge 0$
$D(2): d^{AS}(x, x) = 0$
$D(3): d^{AS}(x, y) = d^{AS}(y, x)$
$D(4): d^{AS}(x, z) \le d^{AS}(x, y) + d^{AS}(y, z);$

For instance, the following calculation demonstrates the Rule $D(4)$ validation in our SFSD, whereas it is more obvious to validate the rule $D(1)$ to $D(3)$. Let:

x: $a(9)b(7)c(3)$

y: $a(2)b(7)c(3)$
z: $b(7)c(3)$
$\alpha = 1, \ \beta = 1$
$\omega(i, j) = 0 \ i = j$
$\omega(i, j) = 1 \ i \ne j$
$\omega(k, 0) = \omega(0, k) = 1; \ i, j, k = \{a, b, c\}$
$\lambda_j = \lambda_C(c_j', c_j'') = 1 \ c_j' = c_j''$ continuation
$\lambda_j = \lambda_I(c_j', c_j'') \ge 1 \ c_j' \ne c_j''$ insertion
$\lambda_j = \lambda_S(c_j', c_j'') \ge 1 \ c_j' \ne c_j''$ substitution

Then, the minimum-length alignment is

x_t [a b c] x'[9 7 3]
y_t [a b c] y'[2 7 3]
z_t [a b c] z'[0 7 3]

and distances are

$$d^{AS}(x, y) = \alpha d^{GL}(x, y) + \beta d^{A}(x, y)$$
$$d^{GL}(x, y) = 0; d^{A}(x, y) = |2 - 9| + |7 - 7| + |3 - 3| = 7;$$
$$d^{AS}(x, y) = \alpha \times 0 + \beta \times 7 = 7;$$
$$d^{GL}(x, z) = 1; d^{A}(x, z) = |0 - 9| + |7 - 7| + |3 - 3| = 9;$$
$$d^{AS}(x, z) = \alpha d^{GL}(x, z) + \beta d^{A}(x, z) = 10;$$
$$d^{GL}(y, z) = 1; d^{A}(y, z) = |0 - 2| + |7 - 7| + |3 - 3| = 2;$$
$$d^{AS}(y, z) = \alpha d^{GL}(y, z) + \beta d^{A}(y, z) = 3;$$
$$d^{AS}(x, y) + d^{AS}(y, z) = 7 + 3 \ge d^{AS}(x, z) = 10$$

where x, y, and z are strings; a, b, c are terminal symbols or facies codes; $\omega(i, j)$ weights for $d^{AS}(x, y)$, and $\lambda_C(i, j)$, $\lambda_I(i, j)$, $\lambda_S(i, j)$ are weights for $d^{A}(x, y)$ in symbol continuation, insertion, and substitution operations.

A dynamic programming algorithm (Fu 1986; Duan et al. 2001a) is modified to calculate the SFSD as defined in this section.

3 Application of sedimentary facies successions distance in inverse stratigraphic modeling

Distance measures between sedimentary facies successions can be of significant application in ISM (Duan et al. 1998, 2001a), among others (Duan et al. 2001a). The ISM can be especially beneficial to reservoir evaluation in the early stage of field development when the data quality or amount is not enough for geological reservoir modeling to take advantage of geostatistics-based geomodeling techniques.

Of great importance in ISM is quantifying mismatch between simulated results and the observations as shown in Fig. 2. Our SFSD defined in previous section provides a unique formal way to measure the mismatch properly: (1) it considers both syntactic and attribute distances between facies successions, i.e., facies type and their thickness; (2) it also naturally considers the vertical order of facies types in the succession, compares the succession as a whole, and does not need further within-succession time calibration between the simulated and observed successions; and (3) it can easily be adapted to accounting for time gaps associated with erosional surfaces as coded into the attribute.

To quantify the mismatch between simulated and observed successions, an absolute time framework needs to be established within the succession, so that facies/thickness formed in the same time interval can be compared each other. For instance, the strata formed between time-1 and time-2 in the simulated succession should be compared with the strata of the same time interval, time-1 and time-2 in the time-calibrated observed succession. The succession simulated by forward modeling very commonly contains time resolution, say, 5000 year (modeling time step), whereas the observed succession contains dated time resolution usually in million years, or at higher resolution hundreds of thousand years. The method published so far to calculate the mismatch of simulated and observed successions is to only compare the thickness of the smallest dated stratigraphic units such as a strata cycle (Cross and Lessenger 1999; Charvin et al. 2009), or the unit thickness maps (Falivene et al. 2014). Of course, in these methods, the higher resolution of time calibration the observed succession has, the more accurate the comparison of the simulated and observed successions is. However, it is almost impossible to time calibrate the observed succession with an order of modeling time step resolution. Therefore, the advantage of our method is very obvious, indicated by the properties (1) to (3) mentioned in the previous section. Moreover, the property (2) implies that the method does not need a higher-resolution chronostratigraphic framework within the observed succession to quantify the mismatch better.

The 3D carbonate forward stratigraphic model used in our ISM is energy and sediment flux based (Duan et al. 2000; Shafie and Madon 2008) and can simulate progradation, aggradation, and retrogradation of a carbonate platform simplified to account for the main factors controlling platform evolution such as basin subsidence, basement flexure, sea-level change, carbonate productivity, sediment transport, erosion, and deposition. The 2D model used is simplified from the 3D model as a tester simulating mainly the subsidence, sea-level change, carbonate productivity, and sedimentation.

3.1 Characterization of the cost function based on sedimentary facies successions distance

The sensitivity analysis of model parameters to the inversion process was run studying the landscape of the SFSD-based cost function in 3D ISM. Firstly, "an observed dataset" of 5 pseudowell facies sections was generated by extracting from a 3D facies model (reference model) which was simulated with known model parameters by the forward model. Then, a series of forward model runs was set up with all known parameters from the reference model fixed, but one selected parameter each time that did vary systematically across its known value in the reference model. Obviously, the change of the selected parameter to the known value used in the reference model will cause the simulation to be different from the reference model. Thirdly, the cost function values can be calculated against the varying selected parameter by quantifying the difference between the simulations and the reference model with the extracted 5 pseudowell sections based on the SFSD defined in previous section. Finally, the cross-plots between cost function values and the selected parameter were created as shown in Fig. 3.

In total, there were 29 model parameters sensitivity analyzed, with representatives shown in Fig. 3. Twenty-five out of 29 parameters are sensitive to the model inversion, with most showing the typical V-shaped curve (Fig. 3a, P17 curve as an example), and a few U-shaped (Fig. 3a, P18 curve), or L-shaped (Fig. 3b). The V- or U-shaped curves behave very similarly, both with a major minimum at the true parameter value of the reference model, and if multiminimums exist, the major one is much more significant than other smaller ones (also shown in Fig. 3c), which makes the convergence of model inversion much easier, and the inverted parameter values more accurate. The L-shaped curves, only a few of them, usually correspond to those model parameters, the value change of which beyond a specific limit no longer makes a contribution to the simulation results, and the true value of which is close to the limit, behaving just like a half-U curve.

The other 4 parameters seemingly insensitive to model inversion can be called as flat-shaped (Fig. 3d, P5 and P6 curves). But in fact in most cases, they are pseudoinsensitive, mainly caused by too large a sampling interval of parameter values in sensitivity analysis calculations. If high-resolution sensitivity is carried out, they would become sensitive to inversion. As shown in Fig. 3d, P5 and P6 curves will become more like P4 if their high-resolution curves are calculated.

The landscape of this cost function can be described as multimodel, stepped, probably noisy, and one-minimum dominated. This type of cost function is complex enough, but can be handled well in inversion with direct search algorithms of global optimization (Ingber and Rosen 1992; Storn and Price 1997). These features of the cost function based on SFSD have made our 3D model inversion possible with reasonably stable results.

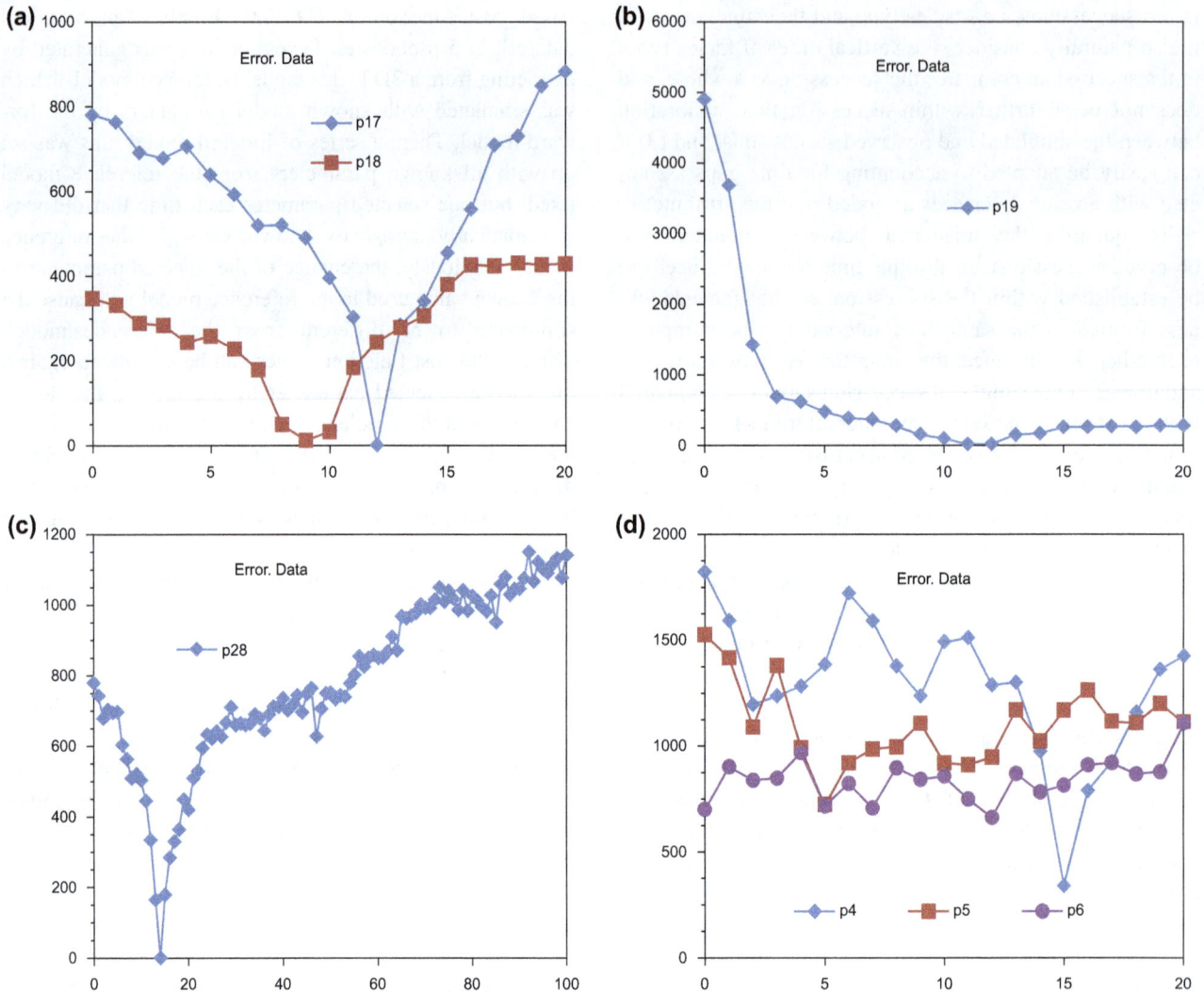

Fig. 3 Plots of cost function (*Y*-axis) to parameter value (*X*-axis) in sensitivity analysis. Function value was evaluated with systematic change of each parameter within assigned value range, while all other parameters fixed at their correct values of the reference model

3.2 Convergence behavior of stratigraphic inversion based on sedimentary facies successions distance

ISM is said converged if the cost function values become close to zero in acceptable ranges. A series of 2D and 3D ISM were run using a synthetic dataset, examining the convergence behavior of the inverse process by tackling the evaluations of the SFSD-based cost function during inversion. Both a genetic algorithm and simulated annealing were used as our inversion engine.

Our numerical experiments show that the stratigraphic inversion based on the SFSD cost function behaves very robustly for both 2D and 3D modeling as shown in Fig. 4. The behavior of the parameter inverting process can be grouped into four types. Figure 4a represents a most typical, straightforward parameter inversion in which the true value was inverted smoothly and the convergence was

relatively simple. The inversion process had correctly explored focusing around the true parameter value. Figure 4c represents the most complex convergence process. The inversion had worked on a local minimum for over half of its time and then focused on the other values a while before reaching the true value. Figure 4b represents an example between the two cases. The inversion had entered a local minimum briefly and then jumped out and started to focus on the true parameter value, though finally reached the true value after working on 5 other groups of values. Figure 4d represents a case in which even though the true value was outside of the assigned parameter range for searching, the inversion still can find the best value within the range.

The detailed analysis of convergence behavior, together with sensitivity analysis for each parameter, has helped guide the setup of model and parameters for inversion

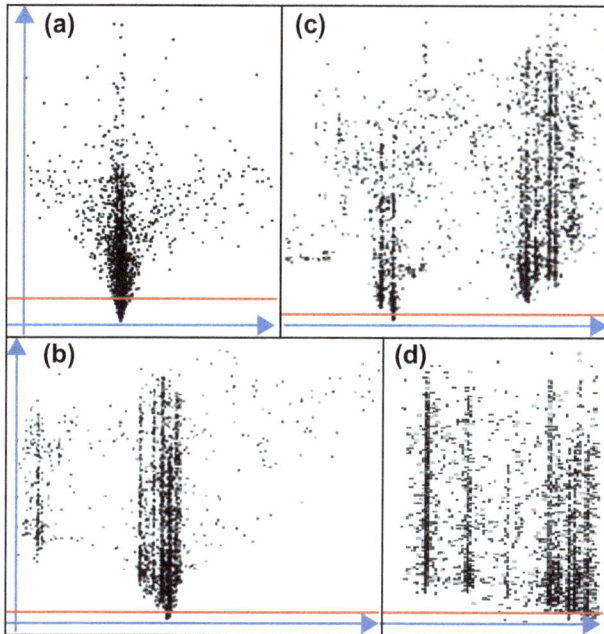

Fig. 4 Cost function value or error (*Y*-axis) to parameter value (*X*-axis) obtained in a converged inversion. *Each dot* represents one function evaluation during the inversion. *Horizontal lines* indicate the error threshold accepted for uncertainty analysis. Axis scale is normalized between 0 and 1

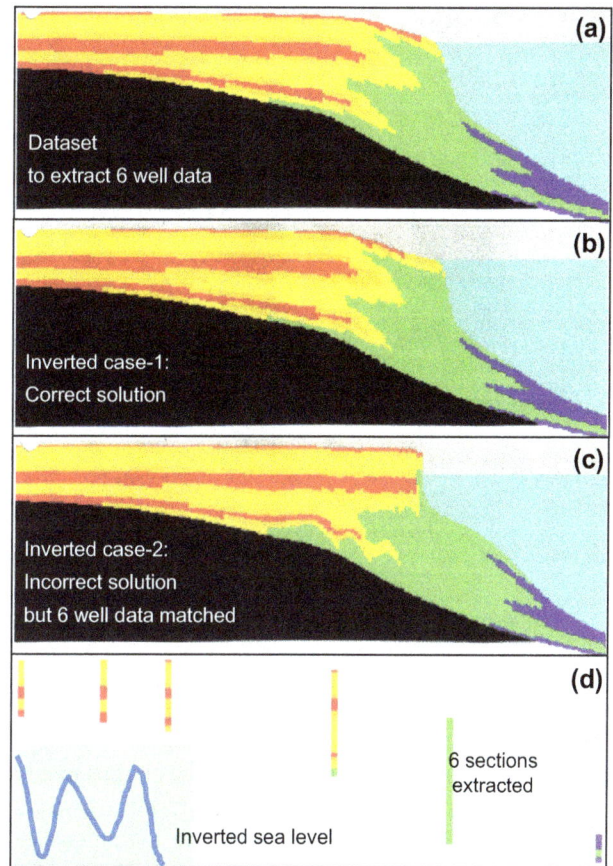

Fig. 5 Comparison of simulated results and observation (synthetic data). **a** Original carbonate dataset; **b**, **c** are two inversion results with inverted production rate shown in Fig. 6a, d, respectively; **d** extracted facies successions of 6 pseudowells as data

during our inverse modeling. For instance, those insensitive parameters in a specific time–space scale model will be excluded in inversion; more attention should be put on parameters with complex behavior in determining their searching ranges; and parameter ranges can be increased or decreased according to converging tendency from several short scoping modeling runs, so that the range is wide enough to include possible solutions, but narrow enough to speed up convergence.

3.3 Non-uniqueness of inverted results based on sedimentary facies successions distance

In a physical system such as a depositional system, non-uniqueness implies that more than one reason may cause the same consequence; for instance, an observed uncomformity may be caused either by a sea-level fall or by a tectonic lift. For the given observed sedimentary facies successions, can there be more than one set of model parameters found meeting the forward stratigraphic model? A series of runs of 2D and 3D ISM were also used to study the convergence behavior regarding the non-uniqueness issue of inversion, and multiple converged results were robustly achieved. In the 2D example, six vertical sections of the synthetic strata profile (Fig. 5a) were taken as our observed dataset (Fig. 5d), and parameter values related to basin subsidence, sea-level change, and carbonate productivity were inverted. It can be shown that inversions can

recover model parameter values correctly or closely with cost function errors close to zero (errors 3.0 to 8.1; zero is the minimum). There is almost no difference between the inverted strata profile and the original (Fig. 5a, b), or a difference difficult to identify graphically in many cases. For the specific parameter values, the inverted and original ones are also very close such as carbonate productivity (Fig. 6a–c). For 3D model experiments, very similar results are achieved.

However, non-unique solutions of the inversion are found as indicated in Fig. 5c, where the inverted profile is significantly different from the original though the cost function error is close to zero (3.0). The unexpected solution is caused by an unexpected inverted parameter productivity shown in Fig. 6d, which corresponds to a much higher productive rate than the original at specific water depths, causing an unusual sedimentary profile although at 6 observed locations, simulated and original are very close (converged in inversion). In practical study, this type of the non-uniqueness can be recognized easily as an incorrect solution by using additional geological information, such as the gradient of carbonate platform slope from seismic

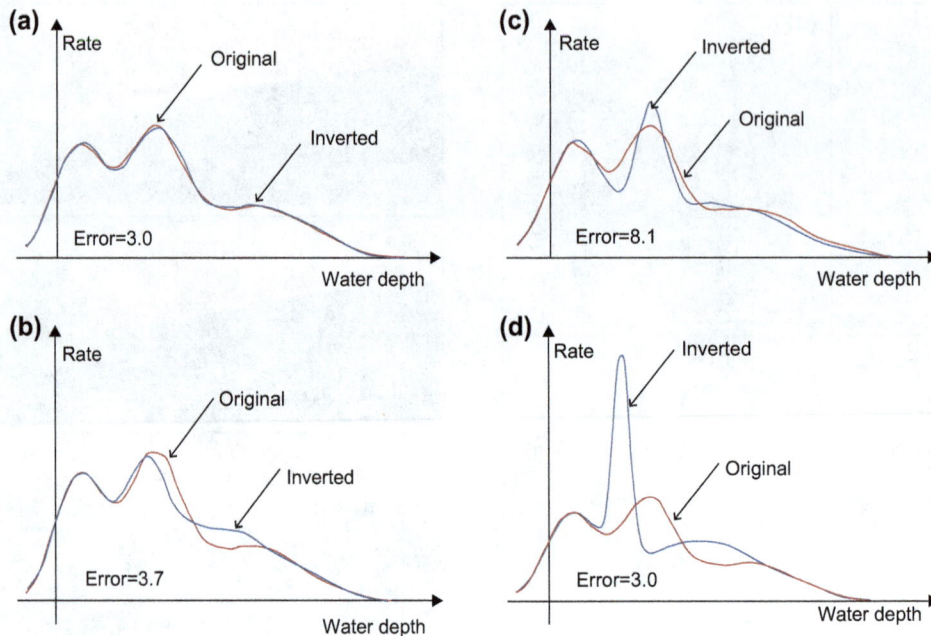

Fig. 6 Inverted carbonate production rate (*Y*-axis) to water depth (*X*-axis). **a–d** Curves corresponding to different error levels of inverted rates. *Note* **d** is associated with the inverted result shown in Fig. 5c. Axis scale is normalized between 0 and 1

data in our example, and it will not confuse our meaningful geological interpretation of the inversion results. However, this may imply that in simpler systems, the non-uniqueness may be a more common issue in inversion.

The non-uniqueness of a model solution means the same solution can be achieved by different sets of parameter values in modeling. Replacing of one set to another set of parameter values to generate the same modeling result is called equivalent-effect parameter substitution. No such similar, obvious non-uniqueness case as seen in our 2D inversions was found in over 10 inverted example realizations in our 3D ISM starting with different initial points of parameter space. The explanation may be that, in a system complex enough as in the 3D stratigraphic modeling, non-uniqueness caused by the equivalent-effect parameter substitution is less probable, compared to a simpler system as in the 2D model case. That is, the probability that equivalent-effect parameter substitution can generate non-uniqueness in a modeling system may be inversely proportional to the complexity of the system.

4 Conclusion

(1) A distance measure of sedimentary facies successions is formally defined based on syntactic presentation of attributed strings and symbolic computation; (2) application of the distance measure used as cost function in ISM is demonstrated with synthetic datasets to be robust in terms of convergence behavior of inversion in both 2D and 3D

modeling; and (3) therefore, the distance measure or other similar ones potentially would be very useful in facies or discrete feature-based inverse geological modeling.

Acknowledgements This research financially was supported by Colorado School of Mines, and are supported by the Science and Technology Ministry of China (2016ZX05033003), China Academy of Sciences (XDA14010204) and Sinopec (G5800-15-ZS-KJB016). The Petroleum Science editors and four anonymous reviewers are thanked for their constructive comments and suggestions. WB Zhang and PQ Lian are thanked for redrawing figures.

References

Bakke S, Griffiths CM. Interactive stratigraphic matching of petrophysically derived numerical lithologies based on gene-typing techniques. In: Collinson J, editor. Correlation in hydrocarbon exploration. London: Norwegian Petroleum Society, Graham & Trotman; 1989. p. 61–76. doi:10.1007/978-94-009-1149-9_7.
Bornholdt S, Westphal H. Automation of stratigraphic simulations: Quasi-backward modelling using genetic algorithms. In: Mascle A, et al. editor. Cenozoic foreland basins of western Europe: Geological Society (London) Special Publications. 1998;134(1):371–9. doi:10.1144/GSL.SP.1998.134.01.17.
Bosence D, Waltham D. Computer modeling the internal architecture of carbonate platforms. Geology. 1990;18(1):26–30. doi:10.1130/0091-7613(1990)018<0026:CMTIAO>2.3.CO;2.
Cant DJ, Walker RG. Development of a braided-fluvial facies model for the Devonian Battery Point Sandstone, Quebec. Can J Earth Sci. 1976;13(1):102–19. doi:10.1139/e76-010.
Charvin K, Gallagher K, Hampson GL, et al. A Bayesian approach to inverse modelling of stratigraphy, Part 1: method. Basin Res. 2009;21(1):5–25. doi:10.1111/j.1365-2117.2008.00369.x.
Charvin K, Hampson GL, Gallagher K, et al. Characterization of

controls on high-resolution stratigraphic architecture in wave-dominated shoreface-shelf parasequences using inverse numerical modeling. J Sediment Res. 2011;81(8):562–78. doi:10.2110/jsr.2011.48.

Cross TA, Lessenger MA. Construction and application of stratigraphic inverse model. In: Harbaugh JW, et al. editor. Numerical experiments in stratigraphy: recent advances in stratigraphic and sedimentologic computer simulations: SEPM. 1999;62(1):69–83.

Duan T, Cross TA, Lessenger MA. 3-D carbonate stratigraphic model based on energy and sedimentflux. In: AAPG annual convention, New Orleans, Louisiana, 2000.

Duan T, Cross TA, Lessenger MA. Reservoir- and exploration-scale stratigraphic prediction using a 3-D inverse carbonate model. In: AAPG annual convention, Denver, Colorado, 2001a.

Duan T, Griffiths CM, Cross TA, et al. Adaptive stratigraphic forward modeling: Making forward modeling adapt to conditional data. In: AAPG Annual convention and exhibition, Salt Lake City, Utah. 1998.

Duan T, Griffiths CM, Johnsen SO. Conditional simulation of 2-D parasequences in shallow marine depositional systems by using attributed controlled grammar. Comput Geosci. 1999;25(6):667–81. doi:10.1016/S0098-3004(98)00162-9.

Duan T, Griffiths CM, Johnsen SO. High-frequency sequence stratigraphy using syntactic methods and clustering applied to the upper Limestone Coal Group (Pendleian, E1) of the Kincardine basin, UK. Math Geol. 2001b;33(7):825–44. doi:10.1023/A:1010950814715.

Falivene O, Frascati A, Gesbert S, et al. Automatic calibration of stratigraphic forward models for predicting reservoir presence in exploration. AAPG Bullet. 2014;98(9):1811–35. doi:10.1306/02271413028.

Fu KS. Syntactic pattern recognition and applications. Englewood Cliffs: Prentice-Hall; 1982. p. 595–6.

Fu KS. A step towards unification of syntactic and statistical pattern recognition. IEEE Trans Pattern Anal Mach Intell v PAMI-8 1986; p. 398–404.

Granjeon D, Joseph P. Concepts and applications of a 3-D multiple lithology, diffusive model in stratigraphic modeling. In: Harbaugh JW, et al. editors. Numerical experiments in stratigraphy: recent advances in stratigraphic and sedimentologic computer simulations: SEPM Special Publication. 1999;62(1):197–210.

Griffiths CM. The language of rocks—an example of the use of syntactic analysis in the interpretation of sedimentary environments from wireline logs. In: Hurst A, Lovell MA, Morton AC, editors. Geological application of wireline logs: Geological Society of London, Special Publication. 1990;48(1):77–94. doi:10.1144/GSL.SP.1990.048.01.08.

Griffiths CM. What should an ideal objective function for 4D stratigraphic units look like? In: Presented at 18th Modsim,

IMACS, World Congress 09 international congress on modelling and simulation, Cairns, Australia. 13–17 July 2009.

Griffiths CM, Bakke S. Interwell matching using a combination of petrophysically derived numerical lithologies and gene-typing techniques. In: Hurst A, Lovell, MA, Morton AC, editors. Geological applications of wireline logs. Geological Society of London. 1990;48(1):133–51. doi: 10.1144/GSL.SP.1990.048.01.12.

Griffiths CM, Duan T, Mitchell A. How to know when you get it right: a solution to the section comparison problem in forward modelling. In: Proceedings numerical experiments in sedimentology. May 1996, University of Kansas.

Griffiths CM, Dyt C, Paraschivoiu E, et al. Sedsim in hydrocarbon exploration. In: Merriam DF, Davis JC, editors. Geologic modeling and simulation. New York, Kluwer Academic. 2001. p. 71–97. doi:10.1007/978-1-4615-1359-9_5.

Harbaugh J, Wand Bonham-Carter G. Computer simulation in geology. New York: John Wiley and Sons, 1970. p. 111–71.

Imhof M, Sharma AK. Quantitative seismostratigraphic inversion of a prograding delta from seismic data. Mar Pet Geol. 2006;23(7):735–44. doi:10.1016/j.marpetgeo.2006.04.004.

Ingber L, Rosen B. Genetic algorithms and very fast simulated reannealing: a comparison. Math Comput Model. 1992;16(11):87–100. doi:10.1016/0895-7177(92)90108-W.

Kaufman L, Rousseeuw PJ. Finding groups in data: an introduction to cluster analysis. New York: Wiley; 1990. p. 341–2.

Kerans C, Lucia FJ, Senger RK. Integrated characterization of carbonate ramp reservoirs using outcrop analogs. AAPG Bull. 1994;78(2):181–216.

Lessenger MA, Cross TA. An inverse stratigraphic simulation model—Is stratigraphic inversion possible? Energy Explor Exploit. 1996;14:627–37.

Levenshtein VI. Binary codes capable of correcting deletions, insertions and reversals. Cybern Control Theory. 1966;10(8):707–10.

Matheron G. Traité de géostatistique appliquée Tome 1, Editions Technip, Paris. 1962. p. 333–4.

Matheron G. Estimating and choosing. Berlin: Springer; 1989.

Shafie KRK, Madon M. A review of stratigraphic simulation techniques and their applications in sequence stratigraphy and basin analysis. Geol Soc Malays Bull. 2008;54(1):81–9.

Storn R, Price K. Differential evolution–a simple and efficient heuristic for global optimization over continuous spaces. J Global Optim. 1997;11(4):341–59. doi:10.1023/A:1008202821328.

Wijns C, Boschetti F, Moresi L. Inverse modelling in geology by interactive evolutionary computation. J Struct Geol. 2003;25(10):1615–21. doi:10.1016/S0191-8141(03)00010-5.

Wijns C, Poulet T, Boschetti F, et al. Interactive inverse methodology applied to stratigraphic forward modelling. Geol Soc Lond Spec Publ. 2004;239(1):147–56. doi:10.1144/GSL.SP.2004.239.01.10.

PERMISSIONS

LIST OF CONTRIBUTORS

Emre Artun and Kutay Köse
Petroleum and Natural Gas Engineering Program, Middle East Technical University, Northern Cyprus Campus, Kalkanli, Guzelyurt, 99738 Mersin 10, Turkey

Ali Aghazadeh Khoei
Petroleum and Natural Gas Engineering Program, Middle East Technical University, Northern Cyprus Campus, Kalkanli, Guzelyurt, 99738 Mersin 10, Turkey
University of Tulsa, Tulsa, Oklahoma, USA

Zhong Hong, Ming-Jun Su and Hua-Qing Liu
PetroChina Research Institute of Petroleum Exploration and Development (RIPED)-Northwest, Lanzhou 730020, Gansu, China

Gai Gao
Research Institute of Exploration and Development, PetroChina Changqing Oilfield Company, Xi'an 710018, Shaanxi, China

Zhao-Yang Kong, Xiu-Cheng Dong and Gui-Xian Liu
School of Business Administration, China University of Petroleum (Beijing), Beijing 102249, China

Qian Shao
School of Business, Tianjin University of Finance and Economics, Tianjin 300222, China

Xin Wan
Tangshan Iron and Steel Group Co., Ltd, Tangshan 261000, China

Da-Lin Tang
China Petroleum Enterprise Association, Beijing 100724, China

Hong Zhao
College of Mechanical and Transportation Engineering, China University of Petroleum, Beijing 102249, China

Yi-Xin Zhao
Department of Mining, China University of Mining and Technology, Beijing 100086, China

Zhi-Hui Ye
College of Petroleum Engineering, China University of Petroleum, Beijing 102249, China

Yu Yu, Liang-Biao Lin and Jian Gao
State Key Laboratory of Oil and Gas Reservoir Geology and Exploitation, Chengdu University of Technology, Chengdu 610059, Sichuan, China

Institute of Sedimentary Geology, Chengdu University of Technology, Chengdu 610059, Sichuan, China

Xia Zhang, Chun-Ming Lin, Ni Zhang and Jian Zhou
State Key Laboratory for Mineral Deposits Research, School of Earth Sciences and Engineering, Nanjing University, Nanjing 210023, Jiangsu, China

Yong Yin
School of Geographic and Oceanographic Sciences, Nanjing University, Nanjing 210023, Jiangsu, China

Yu-Rui Liu
Institute of Geological Sciences, Jiangsu Oilfield Branch Company, SINOPEC, Yangzhou 225009, Jiangsu, China

Chuang Li, Jian-Ping Huang and Zhen-Chun Li
School of Geosciences, China University of Petroleum, Qingdao 266580, Shandong, China

Rong-Rong Wang
Hisense (Shandong) Refrigerator Co. Ltd, Hisense, Qingdao 266580, Shandong, China

Liu-Yi Yin, Yu-Feng Hu and Hai-Yan Wang
State Key Laboratory of Heavy Oil Processing and High Pressure Fluid Phase Behavior and Property Research Laboratory, China University of Petroleum, Beijing 102249, China

Hossein Hamidi, Amin Sharifi Haddad, Roozbeh Rafati, Panteha Ghahri, Adi Putra Pradana, Bastian Andoni and Chingis Akhmetov
School of Engineering, King's College, University of Aberdeen, Aberdeen AB24 3UE, UK

Erfan Mohammadian
Faculty of Chemical Engineering, University Technology MARA, 40450 UiTM Shah Alam, Malaysia

Amin Azdarpour
Department of Petroleum Engineering, Marvdasht Branch, Islamic Azad University, Marvdasht, Iran

Shi-Yuan Li
School of Petroleum Engineering, China University of Petroleum, Beijing 102249, China

Janos L. Urai
Endogene Dynamik, Faculty of Geo-Resources and Materials Technology, RWTH Aachen University, Lochnerstrasse 4-20, 52056 Aachen, Germany

Xiao-Liang Bai and Shao-Nan Zhang
State Key Laboratory of Oil and Gas Reservoir Geology and Exploitation, Southwest Petroleum University, Chengdu 610500, Sichuan, China
School of Geoscience and Technology, Southwest Petroleum University, Chengdu 610500, Sichuan, China

Qing-Yu Huang
Research Institute of Petroleum Exploration & Development, PetroChina, Beijing 100083, China

Xiao-Qi Ding
College of Energy, Chengdu University of Technology, Chengdu 610059, Sichuan, China

Si-Yang Zhang
Department of Geology, University of Regina, Regina, SK S4S 0A2, Canada

Samyukta Koteeswaran, Josh D. Ramsey and Peter E. Clark
School of Chemical Engineering, Oklahoma State University, Stillwater, OK 74078, USA

Jack C. Pashin
Boone Pickens School of Geology, Oklahoma State University, Stillwater, OK 74078, USA

Valery Gulyayev and Natalya Shlyun
Department of Mathematics, National Transport University, Kiev, Ukraine

Fu-Yan Gao
Ningbo Institute of Technology, Zhejiang University, Ningbo 315100, China

Eric-J. Hu
School of Mechanical Engineering, University of Adelaide, Adelaide, SA 5005, Australia

Yong-Qiang Yang, Long-Wei Qiu, Zheng Shi and Kuan-Hong Yu
School of Geosciences, China University of Petroleum, Qingdao 266580, Shandong, China

Jay Gregg
Department of Geology, Oklahoma State University, Stillwater, OK 74075, USA

Zhi Li and Chun-Xi Lu
State Key Laboratory of Heavy Oil Processing, China University of Petroleum, Beijing 102249, People's Republic of China

Bao-Zhi Pan, Ming-Xin Yuan, Chun-Hui Fang, Wen-Bin Liu, Yu-Hang Guo and Li-Hua Zhang
Faculty of Geo-exploration Science and Technology, Jilin University, Changchun 130012, Jilin, China

Gui Fu, Jia-Lin Cao, Lin Zhou and Yuan-Chi Xiang
School of Resources and Safety Engineering, China University of Mining and Technology (Beijing), Beijing 100083, China

Taizhong Duan
Petroleum Exploration and Production Research Institute, Sinopec, 31 Xueyuan Road, Haidian District, Beijing 100083, China

Index

A

Accident Causation, 210-212, 215-217
Acoustic Wave Amplitude, 201-202, 205, 207
Acoustic Wave Velocity, 201-204, 207
Adaptive Singular Spectrum
Analysis, 92
Anionic And Cationic Polyacrylamides, 149
Artificial Neural Networks, 1, 6, 17

B

Blended Data, 92, 103-104
Blockage, 9, 49-53, 56-59, 61

C

Carbon Dioxide, 2, 112, 115, 118-119, 131
Carbonate Platform, 133-134, 137, 141, 144, 225
Chattanooga Shale, 149-151, 154-157
Co2 Sequestration, 112, 118
Coal And Gas Outburst Accident, 214-215
Compaction Recovery, 19, 21, 23-26, 28-31
Comparative Study, 210, 217
Critical States, 167, 172
Cyclic Pressure Pulsing, 1-3, 8, 17

D

Dainan Formation, 75-85, 87, 89-91
Directed Bore Hole, 160, 169
Dislocation Creep, 121-127, 130
Distribution Pattern, 75-76, 85, 87, 89
Dolomite, 30, 130, 133-137, 139-148, 151-152, 180-181, 183, 185-188, 190-194
Dolomitization, 133-134, 139, 141, 143, 145-148, 180, 183, 185, 191-194
Dolomitizing Fluids, 133, 139, 141, 143-144, 146-148, 185, 190
Drill String, 160, 162, 172

E

Eocene, 78, 180, 184-185, 188-189, 191-192, 194
Eroi, 32-34, 39-43, 45-48
Experimental Design, 1, 4, 17
Experimental Measurement, 111, 201
External Causes, 210, 212-213, 215, 217

F

Fcc, 195, 197, 200
Fluidization, 195, 200
Fracture Width, 5, 11-15, 201-208
Fractured Rock, 201
Friction Forces, 160-162, 164, 167-168, 172

G

Gaoyou Depression, 75-85, 87, 89-91
Glauconite, 63-64, 67-68, 72-73

H

Hammett Function, 106, 109-110
Hfacs, 210-218
High-frequency Waves, 112
Hydraulically-fractured Wells, 1

I

Imported Natural Gas, 32-33, 39, 43
Imported Oil, 32-33, 39, 41, 43, 47
Inverse Stratigraphic Modeling, 219, 222
Isolation Tool, 49-53, 55, 57, 59, 61

L

Lacustrine Carbonate, 180-181, 183, 185, 188-190, 193
Least-squares Migration, 92, 104
Lithological Compaction Unit, 19-20, 23, 25-27

M

Middle Ordovician, 133-134, 139, 141, 145-148
Modeling, 1, 3, 6, 16-17, 19, 50, 62, 93-94, 103, 119-122, 124-125, 130-132, 147, 161, 172, 179, 208-209, 219-227

N

Nitrogen Injection, 1, 5-6, 8-9, 11-12, 14
North Jiangsu Basin, 75-77, 91
Numerical Simulation, 49, 53, 124-128, 200, 202, 208

O

Oil And Gas Extraction, 32-33, 35, 37, 41, 198
Organic Salt, 106-111

P

Petroleum Coke, 174-176, 178-179
Petroleum Coke Water Slurry, 174

Power Law Creep, 120-121, 130
Pride Mountain Shale, 149, 152, 154-157
Pyrite, 63-64, 67-68, 72, 137, 140, 146, 151-152

Q
Quick Separation, 195-197, 199-200

R
Regularization, 92-94, 102, 104-105
Relative Volatility, 106, 110
Response Surface Methodology, 50, 62
Rheological Characteristics, 174, 176-177
Rock Salt Rheology, 120, 130

S
Salt Effect, 106-109, 111
Sedimentary Evolution, 75, 91
Sedimentary Facies, 64, 66, 69, 75-76, 85, 87, 89, 91, 135, 219-226

Sedimentary Succession, 84, 221
Sikou Sag, 180-181, 184-186, 190-192
Similarity Quantification, 219
Singular Perturbation, 160, 172
Slurry Rheology, 149, 179
Solubility, 112-119
Stability, 106, 121, 130, 150-151, 158-161, 167-170, 172-179, 200
Syntactic Approach, 219, 221

T
Transformation, 49-50, 59, 64, 90, 148, 162

U
Ultrasound, 112-119, 208
Unsafe Acts, 210-217

Z
Zeta Potential, 149-152, 154-159